Principles and Procedures for Rearing High Quality Insects

Edited by

JOHN C. SCHNEIDER
Department of Entomology & Plant Pathology
Mississippi State University
Mississippi State, MS 39762

MISSISSIPPI STATE
UNIVERSITY™

Mississippi State, MS

Cover illustration courtesy of Joe A. MacGown

MISSISSIPPI STATE UNIVERSITY
Mississippi State, MS 39762

LCCN 2009936047
ISBN 978-0-615-31190-6

Contents

Contributors

Current Contact Information

Michael A. Caprio, Department of Entomology & Plant Pathology, Mississippi State University, Box 9775, Mississippi State, MS 39762 (mac24@msstate.edu) (87)[1]

Muhammad F. Chaudhury, Screwworm Research Unit, USDA-ARS, U.S. Embassy-Panama, PSC 2 Box 4200, APO AA 34002 (Muhammad.Chaudhury@ARS.USDA.GOV) (121)

Frank M. Davis, Department of Entomology & Plant Pathology, Mississippi State University, Box 9775, Mississippi State, MS 39762 (FDavis@entomology.msstate.edu) (3, 307)

William R. Fisher, BASF Corporation, 26 Davis Drive, Research Triangle Park, NC 27709 (william.fisher@basf.com) (9, 43)

G. Douglas Inglis, Zoonotic Bacteriology, Agriculture and Agri-Food Canada, 5403-1st Avenue South, Lethbridge, AB T1J 4B1 Canada (Inglis@AGR.GC.CA) (149, 223)

Norman C. Leppla, Department of Entomology and Nematology, University of Florida, Institute of Food and Agricultural Sciences, P.O. Box 110620, Gainesville, FL 32611 (nclеppla@ifas.ufl.edu) (3, 289)

John P. Reinecke, 510 Colonial Circle, Starkville, MS 39759 (jpr20887@bellsouth.net) (71)

John C. Schneider, Department of Entomology & Plant Pathology, Mississippi State University, Box 9775, Mississippi State, MS 39762 (jcs1@msstate.edu) (3, 97)

Peter P. Sikorowski[2], Department of Entomology & Plant Pathology, Mississippi State University, Mississippi State, MS (149, 223)

[1]Page number on which author's contribution begins.
[2]Deceased.

The Insect Rearing Puzzle

Preface

This book provides an in-depth presentation of every major element comprising professional insect rearing programs. Its scope includes the entire range of insect rearing from small-scale, tabletop rearing to industrial-scale, mass rearing but focuses on medium-scale rearing, and within medium-scale rearing on the use of artificial rather than natural diets. By applying the principles and procedures discussed here, one can establish new rearing programs or improve existing ones. In addition, procedures are presented to either prevent or to diagnose and solve many of the problems that may arise in insect rearing.

The impetus for this book came from the five-day, insect rearing workshop, "Principles and Procedures for Rearing High Quality Insects" held at least annually since 2000 at the Department of Entomology and Plant Pathology, Mississippi State University, (http://www. irc.entomology.msstate.edu/). This workshop has provided the first comprehensive, university-level training course available in insect rearing. To date, 257 persons from a total of 22 countries have participated in the workshop (see Chap. 1 *Introduction* for characterization of the participants).

Regardless of size, any comprehensive insect rearing program should encompass all of the major areas illustrated as parts of the Rearing Puzzle on the Frontispiece. Adding an introduction and allowing the *Microbial Management* puzzle piece to represent the two chapters concerning diseases and diet contamination results in a total of 11 chapters.

The first four chapters are primarily concerned with the human element in insect rearing. Chap. 1 *Introduction* by N. Leppla, F. Davis, and J. Schneider describes the current status of insect rearing in society, characterizes professional insect rearing personnel based on participants in the insect rearing workshop, and predicts future applications for insect rearing. Chap. 2 *Insectary Design and Construction* by W. Fisher explains the design process for both medium- and large-scale insectaries, including layout, structural elements, environmental control systems, and equipment, as well as construction and "break in" of new facilities. Chap. 3 *The Insectary Manager* also by W. Fisher characterizes leadership styles, planning and organization, personnel issues, and crisis management. Chap. 4 *Health and Safety Issues in the Rearing of Arthropods* by J. Reinecke discusses hazards that are general to any laboratory or industrial setting as well as those hazards specific to insectaries. In addition, he discusses governmental health and safety regulations and their enforcement in the US.

Chap. 5 *Genetic Considerations and Strategies for Rearing High Quality Insects* by M. Caprio addresses changes in the nature of the insect itself during rearing. He considers heritable changes due to drift and selection and the measures available to ameliorate such changes.

The next two chapters are concerned primarily with the abiotic environment. Chap. 6 *Environmental Biology of Insect Rearing* by J. Schneider examines the effects of light, temperature, and humidity on taxes, circadian rhythms, diapause, developmental rates, and demographic variables. Chap. 7 *Insect Nutrition, Feeding, and Artificial Diets* by M. Chaudhury explains the mechanics and physiology of insect feeding and digestion as background considerations in the development of artificial diets.

Given that digestion is a microbe-dependent process, Chap. 7 is actually a transition between abiotic and biotic aspects of the rearing environment. The microbial biotic environment in insect rearing is treated comprehensively in two chapters by D. Inglis and P. Sikorowski. Chap. 8 *Microbial Contamination and Insect Rearing* characterizes the taxonomy of microorganisms contaminating artificial diets, consequences of contamination for the insect, sources of contamination, techniques for monitoring contamination levels, and management of contamination, including air filtration, sanitation, personnel education, and control within the diets themselves. Chap. 9 *Entomopathogens and Insect Rearing* presents an extensive catalogue of the taxonomy of disease agents of insects in culture, a discussion of disease development and manifestations, and a review of approaches to manage disease, including prevention and therapeutic measures.

The penultimate chapter is concerned with ensuring satisfaction of the end user. Chap. 10 *The Basics of Quality Control for Insect Rearing* by N. Leppla discusses colony establishment and maintenance, monitoring insect quality, colony evaluation, and total quality management.

Finally, Chap. 11 *Insect Rearing Production Systems, A Case Study: The Southwestern Corn Borer* by F. Davis, while focusing on the production component of a particular rearing program, functions as a "capstone" to the book by illustrating and synthesizing much of the information in the preceding 10 chapters.

The tabs and blanks of the pieces of the frontispiece puzzle represent the interdependence of the components of insect rearing programs. Whether directly or indirectly through a complex web of interactions, each component affects all of the others and must function properly for the whole to be successful. To emphasize this point, the puzzle piece corresponding to each chapter is reproduced on the first page of the corresponding chapter.

The components of insect rearing programs can be arranged hierarchically. *Production System* has been placed at the center of the puzzle in recognition of the fact that production of insects is the primary purpose of an insect rearing program. The next level in the hierarchy is composed of the four components that directly border *Production System*: *Facility Management*, *Nutrition & Diet*, *Environmental Biology*, and *Quality Control*. These components share the distinction of being absolutely required. Finally, the four components at the corners of the puzzle—*Health & Safety*, *Facility Design*, *Microbial Management*, and *Population Genetics*—each have some element of contingency, namely, the level of societal regulation, the stage of development of the rearing program, the occurrence of an agent of contamination or disease, and the purpose for which the insects are being produced, respectively. However, it bears repeating that the level of success of an insect rearing program depends absolutely on an integration of the contributions of all nine components. For example, in the case of artificial diets, microbial contamination is ubiquitous and in practice must be managed. Likewise, laboratory adaptation is ubiquitous and in practice must be monitored through quality control.

The person ultimately responsible for this book is Dr. Frank Davis, Adjunct Professor Emeritus of Entomology (Department of Entomology & Plant Pathology, Mississippi State University) and Research Entomologist (USDA/ARS Crop Science Research Laboratory, Starkville, MS; retired). Frank conceived the workshop and book, attracted the instructors and authors to contribute, and provided the coordination and enthusiasm to sustain the effort to completion. The generous support of Dr. Vance Watson (Emeritus Vice President, Mississippi State University) and of Dr. Clarence Collison (Head, Department of Entomology & Plant Pathology, Mississippi State University) are acknowledged. BASF Corporation made a generous financial contribution toward printing costs. Finally, Joe MacGown (Curatorial Assistant, Department of Entomology & Plant Pathology, Mississippi State University) prepared the rearing puzzle and puzzle pieces, Lois Connington (consultant) did the copy editing, and Robyn Hearn (Editor, Office of Agricultural Communications, Mississippi State University) did the page design.

John C. Schneider
Mississippi State, MS
17 June 2009

Chapter 1

Introduction

NORMAN C. LEPPLA
Professor and IPM Coordinator
Department of Entomology & Nematology
University of Florida, Institute of Food and Agricultural Sciences
Gainesville, FL

FRANK M. DAVIS & JOHN C. SCHNEIDER
Emeritus Adjunct Professor and Professor, respectively
Department of Entomology & Plant Pathology
Mississippi State University
Mississippi State, MS

I. History of Modern, Professional Insect Rearing

Professional insect rearing is a relatively recent subdiscipline in the field of entomology developed to satisfy the need for dependable supplies of high quality insects. Insect rearing was organized into a profession by connecting essentially isolated scientists and technicians who were rearing insects to support a variety of research and industrial activities (Dickerson and Leppla 1992). Examples of these activities included studying aquatic insects, conducting genetics research with *Drosophila* sp., maintaining housefly colonies to measure insecticide resistance, providing hosts for natural enemies, studying vectors of human diseases, producing silk and honey, and securing food for animals (Needham et al. 1937). The initial unification of the field was facilitated greatly by the United State Department of Agriculture, Agricultural Research Service (USDA/ARS), when it sponsored the 1958 national conference "Culture of Insects as a Requisite for Research on Insect Control" (unpublished report), and another in 1963, "Planning and Training Conference for Insect Nutrition and Rearing" (USDA/ARS 1963). A third conference held in 1974, "Genetics of Insect Behavior" (unpublished report), provided an opportunity to develop a national network of insect rearing specialists through the newsletter "FRASS."

The USDA/ARS provided further leadership in the development of insect rearing as a specific area of entomological specialization by organizing the 1980 national conference "Advances and Challenges in Insect Rearing" (USDA/ARS 1984). In his opening comments for the conference, E. F. Knipling, retired director of the Entomology Research Division of USDA/ARS, made the following statement: "Until appropriate insect colonies became available, important studies on insect physiology could not be undertaken, research on the genetics of economic pests was handicapped, and identification and synthesis of sex pheromones was delayed. The ability to rear insects facilitated research on plant resistance and made possible various field studies on the dispersal and behavior of insects. Finally, the field of biological control has been greatly advanced by having methods of rearing parasitoids and predators for introduction or augmentation and by the propagation of hosts for insect pathogens." Another special application of insect rearing to which E. F. Knipling provided substantial support was the rearing of enormous numbers of high quality insects to implement the sterile insect technique for managing invasive species, such as the screwworm, *Cochliomyia ho-*

minivorax (Diptera: Calliphoridae); boll weevil, *Anthonomus grandis* (Coleoptera: Curculionidae); pink bollworm, *Pectinophora gossypiella* (Lepidoptera: Gelechiidae); Mediterranean fruit fly, *Ceratitis capitata* (Diptera: Tephritidae); and several other tropical fruit flies. A new specialization of entomology, professional insect rearing, had to be created before the work it ultimately supported could be advanced.

II. Diversity of Reared Insects

The purposes for which insects are reared include education, personal collections, stationary exhibits, insect zoos, butterfly houses, pets, human and animal food, agricultural and medical research, pest management, and pollination (Leppla 2002). Worldwide, the number of insect species reared routinely under artificial conditions increased steadily from the 1950s to the 1980s but has since remained relatively constant at nearly 700 species in over 1,500 colonies (Edwards et al. 1987; Table 1). The six species with the greatest number of colonies comprise 10 percent of the total number of colonies: the housefly, *Musca domestica* (Diptera: Muscidae) (28); the American cockroach, *Periplaneta americana* (Blattodea: Blattidae) (26); the tobacco budworm, *Heliothis virescens* (Lepidoptera: Noctuidae) (25); the twospotted spider mite, *Tetranychus urticae* (Acari: Tetranychidae) (24); the confused flour beetle, *Tribolium confusum* (Coleoptera: Tenebrionidae) (21); and the yellow fever mosquito, *Aedes aegypti* (Diptera: Culicidae) (21).

Hunter (1997) listed more than 130 species of beneficial organisms available from 142 suppliers in North America: 95 located in the U.S., 33 in Mexico and 14 in Canada (Table 2). The most popular species were an egg parasite *Trichogramma pretiosum* (Hymenoptera: Trichogrammatidae) (77 suppliers); the green lacewing, *Chrysoperla carnea* (Neuroptera: Chrysopidae) (65); the convergent lady beetle, *Hippodamia convergens* (Coleoptera: Coccinellidae) (56); the brown lacewing, *C. rufilabris* (Neuroptera: Chrysopidae) (54); a whitefly parasite, *Encarsia formosa* (Hymenoptera: Aphelinidae)(54); a predaceous mite, *Phytoseiulus persimilis* (Acari: Phytoseiidae) (54); and the mealybug destroyer, *Cryptolaemus montrouzieri* (Coleoptera: Coccinellidae) (52). Many more species could be reared using the techniques developed for their close relatives. Leppla et al. (1982) projected that the then current techniques for rearing Lepidoptera could be used for at least 60 species in 16 families.

Table 1. Taxonomic distribution of species and colonies of reared insects (Edwards et al. 1987); Drosophila spp. excluded

Taxonomic Order	No. of Species	No. of Colonies
Lepidoptera	101	308
Coleoptera	78	266
Diptera	168	301
Hemiptera	90	206
Hymenoptera	119	169
Others	—	—
Acari	44	77
Anoplura	1	1
Mallophaga	3	3
Neuroptera	2	2
Orthoptera	64	203
Siphonaptera	2	7
Thysanoptera	4	11

Table 2. Number of species and suppliers of beneficial arthropods (Hunter 1997)

Type of Insect or Mite [Country (No. of Suppliers)]	Total No. of Species	Total No. of Suppliers
Predatory mites (US 59, Can 11)	17	70
Stored product pest including parasites and predators (P/P) (US 6)	9	6
Aphid P/P (US 57, Mex 16, Can 14)	17	87
Whitefly P/P (US 58, Mex 15, Can 13)	9	86
P/P for greenhouse pests (US 62, Can 12)	23	74
Scale and mealybug P/P (US 47, Can 9)	7	56
Insect egg parasites (US 55, Mex 27, Can 10)	12	92
Lepidoptera larval parasites (US 20, Mex 1)	6	21
Filth fly parasites (US 41, Can 7)	8	48
Other insect parasites (US 4, Mex 1, Can 1)	4	6
General predators (US 69, Mex 16, Can 14)	21	99
Weed controls (US 10)	36	10

III. Major Success Stories in Insect Rearing

Eradication of the screwworm from North America is one of the greatest historical successes in insect rearing (Dickerson and Leppla 1992). The flies were mass reared, sterilized with ionizing radiation, and released so that the sterile males could mate with wild-type females, thereby reducing reproduction below replacement rates. To achieve the initial production goal of 50–75 million sterile flies per week, more than 45,000 kg of raw meat and 7,500 liter of fresh beef blood were required each week to feed the larvae. The larval diet, composed of two parts red meat, one part citrated beef blood, and two parts water plus 0.24% formalin, was placed in electrically heated trays (Meadows 1985). Thus, rearing was labor intensive, expensive and extremely filthy due to the accumulation of waste and ammonia.

To reduce costs, the larval diet was shifted from beef and horse meat to whale, pig, and nutria, *Myocaster coypus*, meat (Taylor 1992). These alternatives soon became unacceptable, so insect rearing specialists developed a hydroponic diet composed of dried whole chicken egg, dried whole bovine blood, calf suckle (milk substitute), sucrose, dried cottage cheese, formalin, and water. This liquid diet was poured into vats to saturate cotton or acetate fiber mats that supported the developing larvae. The waste liquid was vacuumed off every four hours and replaced with fresh diet. The vats were maintained at about 35°C and 70% relative humidity and serviced by three shifts of workers per 24 hours, seven days per week, and 365 days per year.

Although this rearing technology successfully supported the screwworm eradication program, insect quality was highly variable. Production ranged from nine to less than three liters of pupae per vat, and periodically average pupal weight declined from more than 50 mg to less than 40 mg (Taylor 1992). The primary cause for this variability was the difficulty in replenishing the diet in the vats and the unpleasant and potentially unhealthy environment of the production facility. Consequently, insect rearing specialists developed a series of disposable gelled diets (Chap. 7 *Insect Nutrition, Feeding, and Artificial Diets*). The rearing system based on gelled diet produced higher quality insects (2% heavier pupae) at a direct, materials cost per insect that was 55% of the cost on the liquid diet. Moreover, the labor force for the facility was approximately halved, the work environment was vastly improved, and the waste was more easily disposed. The screwworm eradication program would not have succeeded without the many contributions of its pioneers in insect mass rearing.

One of the most important uses for insect rearing has been to screen plant genotypes for host plant resistance. This involves evaluating the results of interactions between insects and plants, either by measuring and evaluating the effects that the insects have on the plants or that the plants have on the insects. Before artificial rearing, natural populations were used for screening, especially for Lepidoptera. Unfortunately the supply of natural populations is too unreliable for substantial progress to be made. An adequate number of insects must be available at the same time as the susceptible plant growth stages to enable the desired interaction. Therefore, rearing programs were created to provide insects when they are needed for laboratory screening as well as field screening in sufficient numbers and of sufficient quality to provide the desired level of selection pressure.

In the 1950s and 1960s, the USDA/ARS established corn host plant resistance teams at Ankeny, Iowa, for the European corn borer, *Ostrinia nubialis* (Lepidoptera: Crambidae); Tifton, Georgia, for the corn earworm, *Helicoverpa zea* (Lepidoptera: Noctuidae), and fall armyworm, *Spodoptera frugiperda* (Lepidoptera: Noctuidae); and Mississippi State, Mississippi for the southwestern corn borer, *Diatraea grandiosella* (Lepidoptera: Crambidae) (Davis and Guthrie 1992). Each laboratory developed a rearing program based on the use of artificial diets and constantly exchanged information. By the early 1970s, these programs prospered to the point that commercial seed companies (e.g., Pioneer Hi-Bred International, Inc., DeKalb, and Funk Seeds International) could benefit by visiting their facilities. The seed companies made commitments to establish their own rearing capabilities, and the USDA/ARS laboratories transferred rearing technologies to them and provided assistance in building and maintaining their programs. This process has continued as federal rearing programs have advanced. In addition, these corn insect rearing programs have been adapted for cotton host plant resistance by expanding them to include the tobacco budworm and corn earworm (Jenkins et al. 1995).

Insect rearing has been essential for the identification of the sex pheromones of hundreds of species of pest insects. This research and its application to pest management could not have been accomplished without a reliable source of insects for extraction of the pheromone, use in bioassays to identify the active compounds, validation of the attractiveness of the synthetic attractant, and development of associated trapping technology. Two examples among many include the boll weevil and the southwestern corn borer, the research for both of which was done at USDA/ARS laboratories at Mississippi State, Mississippi. At the Gast Rearing Laboratory, thousands of boll weevils were reared in the 1960s to provide chemists with literally kilograms of frass from which to extract and identify the four major components of the pheromone grandlure (Tumlinson et al. 1969). Eventually the boll weevil was eradicated in large portions of the US cotton production area using grandlure in traps to detect the presence of adults at low densities. The substantial economic benefits to cotton farmers and, through reductions in environmental exposure to insecticides, the health benefits to society at large would have been delayed many years without the insect rearing support contributed by the Gast Rearing Laboratory (McKibben et al. 2001). At the R. W. Harned Research Laboratory, rearing of the southwestern corn borer was essential for identification of its sex pheromone (Hedin et al. 1986). Farmers, consultants, extension entomologists, and research scientists are using the pheromone and trap technology to monitor adult populations and study the northern distribution of the pest.

IV. Education in Insect Rearing

Continuous laboratory production of high quality insects depends on specialized knowledge, experience and talent, entailing both art and science, and business management skills (Singh and Ashby 1985, Singh and Clare 1992). Insect rearing professionals study the principles and procedures necessary to establish insect colonies, maintain their health, and optimize their production. Generally, they are required to design or renovate facilities; maintain and repair equipment; test alternative materials and supplies; store dietary ingredients without diminishing their nutritional quality; provide and monitor disease-free environments for insects and safe working conditions for employees; train and motivate employees; recycle or dispose of spent materials; and in the private sector, develop and improve marketing and public relations strategies. New insectary managers and the employees they supervise usually are not well trained in insect rearing and find it challenging to master the technical complexity. Consequently, they gain experience in insect rearing "on the job," build networks of colleagues with whom to share knowledge, pursue continuous education, and obtain periodic outside review for their insect rearing programs. To be successful, insectary managers must possess or quickly acquire mastery of all parts of the insect rearing puzzle (see Frontispiece).

The traditional approach to learning the art and science of insect rearing has been facilitated by a considerable amount of information published since 1950. However, formal educational opportunities generally have been lacking, whether at universities or in the private sector. In fall 2000, the five-day workshop, Principles and Procedures of Rearing High Quality Insects, was begun at Mississippi State University, Department of Entomology and Plant Pathology (http://www.irc.entomology.msstate.edu/) to provide the first comprehensive, university-level training course in insect rearing. Through fall 2008, 137 men and 120 women from a total of 19 countries have enrolled in the workshop. Of these, 62 have represented the following 22 countries: American Samoa, Argentina, Australia, Belgium, Brazil, Bulgaria, Canada, China, Cyprus, Germany, Honduras, Kenya, Mexico, Panama, Peru, Philippines, Singapore, South Africa, Spain, Switzerland, The Netherlands, and the United Kingdom. The 195 U.S. participants have come from a total of 34 states.

Information collected from workshop participants can be used to characterize professional insect rearing personnel on a worldwide basis. Many of the enrollees during the last four workshops have had an M.S. or Ph.D. in biology or a related field (Fig.1a). This level of educational background may reflect the level of competence required to systematically improve established colonies and solve the inevitable problems that arise. Career opportunities in insect rearing are quite diverse, with 58% of the workshop participants (N = 257) employed in the private sector and the balance employed at various educational and governmental levels in the public sector (including two individuals from international research centers). Half of the enrollees have been employed in insect rearing five or more years (Fig. 1b), and over two-thirds have been 33 years of age or older (Fig. 1c). Insect rearing professionals participating in this workshop tend to enjoy a high level of job satisfaction. The students have nearly all commented that insect rearing is never boring because insects are interesting and the rearing continually poses new challenges.

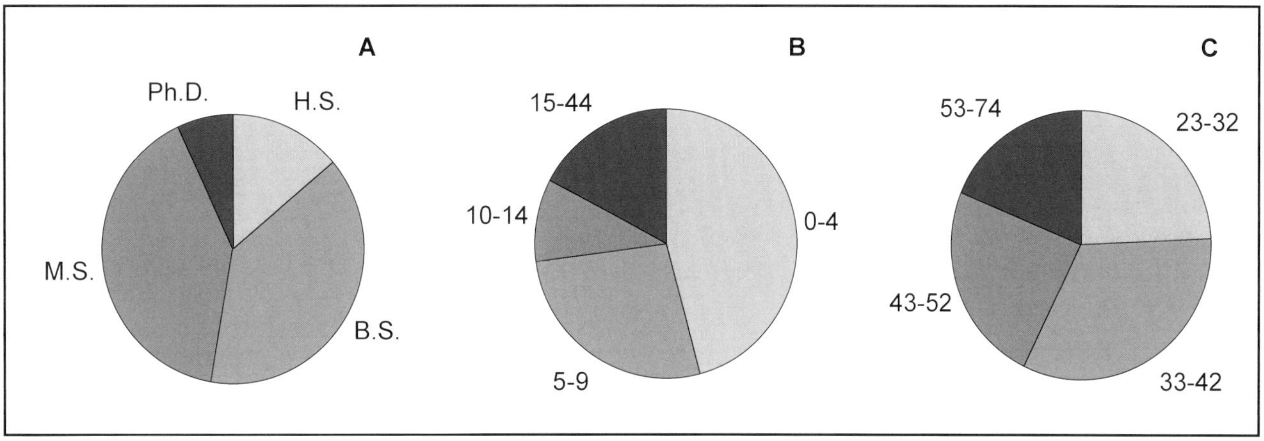

Fig. 1. Characterization of participants in the workshop Principles and Procedures for Rearing High Quality Insects held at Mississippi State University (2005–2008 workshops, N = 70; data not collected in earlier workshops): (A) Educational background, (B) Years of employment in insect rearing (range = 0–44), (C) Age (range = 23–74).

V. Future Opportunities and Challenges for Insect Rearing

The challenge of establishing professional insect rearing as a subdiscipline of entomology has been accomplished, but this is just the beginning. Demand for reared insects is increasing, particularly for research, pest management, education, and aesthetic purposes. Large numbers of insects are beginning to be used to develop bioengineered crops or to serve as convenient "biofactories" for producing useful proteins and other chemicals. Biological control agents are needed for use in integrated pest management to minimize the use of chemical pesticides. This includes insects as predators and parasites and as hosts to produce biopesticides and nematodes. Assuming that deleterious nontarget effects of genetically modified insects are not demonstrated, autocidal techniques employing genetically modified insects will provide advanced control of mosquitoes and other pests. The sterile insect technique has been proven effective for some pests but not implemented because conventional control measures are less expensive. However, with increased attention to the environmental costs of using insecticides, this could change. Butterflies and other insects are being used increasingly in ceremonies, especially weddings and funerals, in displays, and in educational programs.

The economic impact of the increasing demand for high quality, reared insects is multiplied by the proliferation of vendors marketing a wide array of insect rearing supplies and equipment, such as diet ingredients, complete diets, rearing containers, and diet preparation machinery. Thus, not only has professional insect rearing been established as an entomological subdiscipline but it is also of increasing economic importance.

References

Davis, F. M., and W. D. Guthrie. 1992. Rearing lepidoptera for plant resistance research, pp. 211–228. *In* T. E. Anderson and N. C. Leppla (eds.), Advances in insect rearing for research and pest management. Westview Press, Boulder, CO.

Dickerson, W. A., and N. C. Leppla. 1992. The insect rearing group and the development of insect rearing as a profession, pp. 3–7. *In* T. E. Anderson and N. C. Leppla (eds.), Advances in insect rearing for research and pest management. Westview Press, Boulder, CO.

Edwards, D. R., N. C. Leppla, and W. A. Dickerson. 1987. Arthropod species in culture. Entomol. Soc. Am., College Park, MD.

Hedin, P. A., F. M. Davis, J. C. Dickens, M. L. Burks, T. G. Bird, and A. E. Knutson. 1986. Identification of the sex attractant pheromone of the southwestern corn borer, *Diatraea grandiosella* Dyar. J. Chem. Ecol. 12: 2051–2063.

Hunter, C. D. 1997. Suppliers of beneficial organisms in North America. California Environmental Protection Agency, Department of Pesticide Regulation, Sacramento, CA (http:// www.cdpr.ca.gov/docs/pestmgt/ipminov/bensuppl.htm).

Jenkins, J. N., J. C. McCarty, Jr., and M. S. Moghal. 1995. Rearing tobacco budworm and bollworm for host plant resistance research. Technical Bulletin 208. Mississippi Agricultural & Forestry Experiment Station.

Leppla, N. C. 2002. Rearing of insects, pp. 975–979. *In* V. H. Resh and R. Carde (eds.), Encyclopedia of insects. Academic Press, San Diego, CA.

Leppla, N. C., W. R. Fisher, J. R. Rye, and C. W. Green. 1982. Lepidopteran mass rearing: an inside view, pp. l23–133. *In* Proc., FAO, IAEA Symposium, Sterile insect technique and radiation in insect control. Publication SM-255/14. International Atomic Energy Agency, Vienna, Austria.

McKibben, G. H., E. J. Villavaso, W. L. McGovern, and W. J. Grefenstette. 2001. United States Department of Agriculture—Research Support, Methods Development and Program Implementation, pp. 101–134. *In* W. A. Dickerson, A. L. Brashear, J. T. Brumley, F. L. Carter, W. J. Grefenstette, and F. A. Harris (eds.), Boll weevil eradication in the United States through 1999. The Cotton Foundation, Memphis, TN.

Meadows, M. E. 1985. Eradication program in the southeastern United States, pp. 8–11. *In* O. H. Graham (ed.), Symposium on eradication of the screwworm from the United States and Mexico. Entomological Society of America, College Park, MD.

Needham, J. G., P. S. Galtsoff, F. E. Lutz, and P. S. Welch. 1937. Culture methods for invertebrate animals. Comstock Publishing Company, Inc., Ithaca, NY.

Singh, P., and M. D. Ashby. 1985. Insect rearing management, pp. 185–215. *In* P. Singh and R. F. Moore (eds.), Handbook of insect rearing, vol. 1. Elsevier, New York, NY.

Singh, P., and G. K. Clare. 1992. Insect rearing management (IRM): an operating system for multi-species rearing laboratories, pp. 135–172. *In* T. E. Anderson and N. C. Leppla (eds.), Advances in insect rearing for research and pest management. Westview Press, Boulder, CO.

Taylor, D. B. 1992. Rearing systems for screwworm mass production, pp. 393–403. *In* T. E. Anderson and N. C. Leppla (eds.), Advances in insect rearing for research and pest management. Westview Press, Boulder, CO.

Tumlinson, J. H., D. D. Hardee, R. C. Gueldner, P. A. Hedin, and J. P. Minyard. 1969. Sex pheromones produced by male boll weevils: isolation, identification, and synthesis. Science 166: 1010–1012.

USDA/ARS. 1963. Proceedings of planning and training conference for insect nutrition and rearing. USDA, ARS, U.S. Government Printing Office, Washington, DC.

USDA/ARS. 1984. Advances and challenges in insect rearing. E. G. King and N. C. Leppla (eds.). USDA, ARS, U.S. Government Printing Office, Washington, DC.

Chapter 2

Insectary Design and Construction

WILLIAM R. FISHER
Principle Research Scientist
Team Leader, Insect & Plant Propagation
BASF Corporation
Research Triangle Park, NC

I. Introduction

To establish an insect colony in a rearing facility, founder populations are collected from the natural environment to which they are adapted and exposed to unusual conditions never before experienced by the countless generations that preceded them. To reduce costs, enable automation, and provide a more consistent food supply, the insects usually are fed on an "artificial" diet with exotic ingredients, including antimicrobial chemicals that may have deleterious side effects on the insects. To increase rearing efficiency, they are maintained at abnormally high densities, a situation that increases contact among individuals and the potential for disease outbreaks. To help prevent the spread of pathogens, eggs may be treated with formaldehyde or sodium hypochlorite, reducing hatchability. To facilitate ease of handling, eggs may be oviposited on substrates such as gauze or wax paper.

These conditions may push the insects to the limit of their adaptive potential, threatening their very survival. Individuals without adequate adaptive flexibility do not survive these pressures past a few generations. Others—albeit changed from the wild population—are maintained continuously as long as appropriate conditions are met. If these conditions are not met, a healthy culture can change into one at risk. To reduce this risk, the insect-production facility should be designed to consistently provide conditions supporting success of the insect cultures it contains.

An insectary is a highly specialized building, and designing one can be a daunting task. Considerable thought must be given to all details of the project. Otherwise, there is the likelihood that design deficiencies will become "permanently" incorporated into the completed facility, resulting in a continuous and long-term impact on the rearing program. These shortcomings may be resolved after construction with the expenditure of additional resources, which may be difficult to obtain. Even if these resources are available, correcting deficiencies may not be possible within the physical constraints of a newly completed facility. It is always easier to eliminate deficiencies from a Computer Aided Design (CAD) file with a few strokes on a keyboard than it is from the building with a saw and hammer.

Each rearing facility is unique, and there is no standardized insectary plan that can be applied to all. However, there are design features that should be incorporated into all insectaries to help ensure a successful production program. These features, their interrelationships, and their interactions with insectary operation will be described in this chapter.

Two types of rearing facilities will be highlighted here. First is the small to midsize, multispecies insectary frequently used for research in government, academia, and industry. A typical midsize facility is 100–300 m^2, produces thousands to hundreds of thousands of insects of several species each week, and employs about 3–8 full-time workers. The second type of facility is for mass production in which hundreds of millions to billions of insects of a single species are produced each week. This "biofactory" may have an area as great as 20,000 m^2 and employ hundreds of individuals.

II. The Design Process

An insectary performs a variety of functions that support the insects, rearing processes, and insectary employees. A thorough understanding of these functions is essential for developing an appropriate design, estimating costs, scheduling activities, managing the construction project, and minimizing the need for costly change orders.

A. Functional Goals

The functional goals of all insectary facilities will include the following:

- Contain the insects within designated areas.
- Protect the insects from pathogens, parasitoids, predators, and competitors (Chap. 9 *Entomopathogens and Insect Rearing*).
- Protect the insects' diet from microbial contamination (Chap. 8 *Microbial Contamination and Insect Rearing*).
- Provide optimal environmental conditions for insect growth and development (Chap. 6 *Environmental Biology of Insect Rearing*).
- Provide sufficient capacity to produce the number of insects required.
- Provide proper environmental conditions for worker safety and comfort (Chap. 4 *Health and Safety Issues in the Rearing of Arthropods*).
- Provide space, utilities, and proper environmental conditions for equipment.
- Provide a facility layout that maximizes production efficiency.
- Provide flexibility to accommodate future program changes.

B. Owner Responsibilities

To achieve these functional goals, the requirements for successful insect rearing must be identified, quantified, and incorporated into design specifications. Much of this information is unique to insect rearing and unfamiliar to most architectural/engineering (A/E) firms. Thus, the owner of the new facility and/or his representative(s) who are experienced in insect rearing must develop the goals and requirements and assist the A/E in understanding and implementing them into the design. The following items should be addressed (Sec. IV and Sec. V):

- Maximum rates of production for each species to be reared and the amount of "housing" space required per insect produced for each growth stage.
- Environmental conditions within insect-holding areas including the required range and permissible level of variability of temperature, relative humidity, photoperiod, air flow pattern and velocity, filtration, and air-exchange rates.
- Amount of food required per insect produced and the amount of production and storage space required per unit of food used.
- A schedule of insectary operation that organizes employee traffic, distributes demands for equipment, and reduces contamination potential.

- Requirements for isolation of species from one another, from diet/food plant production areas, and from the environment external to the insectary.
- Requirements for additional spaces to support the species being reared (e.g., maintenance shop, packaging area, and quarantine area for insects received from the field or other insectaries).
- The potential of human health and safety hazards associated with the insects and rearing procedures (e.g., wing scales and urticating hairs, mold spores, dusts from dry diet ingredients, chemicals, and steam).
- Kinds and quantities of waste to be disposed.
- The need for facilities outside the insectary proper (e.g., shipping and receiving dock, waste disposal pick-up areas, and greenhouses).
- Plans to expand or modify the insectary in the future.

For additional help in developing functional goals, the owner may find the following suggestions useful:

- Visit other insectaries for ideas about facility design, equipment, procedures, and traffic flow plans. Identify the advantages and disadvantages within each program, and use this information in the design process.
- Prepare a scale drawing of the proposed facility for positioning laboratory furniture and equipment within rooms, determining adequacy of room sizes, visualizing procedures and flow plans, locating utilities, identifying potential sources of contamination, and evaluating needs for isolation.
- Summarize specifications for each room by listing room type and function, procedures, environmental conditions, utilities, isolation requirements, and equipment needed. For example, Table 1 lists specifications for an oviposition room containing adult Lepidoptera, and Table 2 lists those for a clean workroom where artificial diet is prepared.

As discussions between the owner and the A/E progress, the A/E will provide design-concept sketches that evolve into a final design, which is used in the bid process to secure contractors. Drawings should include demolition plans, if needed; room dimensions; permanently installed equipment and casework; finish schedule including construction materials for floors, walls, ceilings, and their coatings; door schedule, listing types, composition, surface treatment, and sizes, along with details of the frame and jambs; framing plan; electrical system including receptacles, circuits with loads and amperage, backup power; lighting schedule with specifications for fixtures, bulbs, voltage, and fixture supplier; air-handling equipment with schedules for humidifiers, air distribution, exhaust fans, air-pressure differentials, air velocity, exhaust-hood connections, temperature and humidity specifications, and redundancy or backup equipment (or system) to maintain environmental conditions if the primary system fails; plumbing with location of vents, drains, waste lines, supply lines, sinks, hose bibs, gas, and steam; and a list of equipment such as hoods, humidifiers, autoclave, and diet dispenser requiring utilities.

The owner has additional responsibilities in a design project. He or she assists in developing a realistic schedule for project milestones, such as completion of the first drawings, start and end of construction, receipt of the certificate of occupancy, and resolution of all postconstruction issues. The owner inspects the site, evaluates progress, documents the construction process, and attends meetings held by the construction manager when status of the work is discussed with the various subcontractors.

At some point during the construction phase, a change in the approved design may be required. The owner evaluates each change and its potential impact on the construction schedule, budget, and facility function. Changes can be costly and disruptive to project momentum, particularly if they require significant time for additional design work and approval. In this case, insectary construction may be stopped with contractors transferring employees to other jobs, creating uncertainty as to when work on the insectary will resume. Disruptions like this can be minimized by identifying facility functions, preparing technical requirements, and assisting A/E in understanding and implementing them early in the design process.

When construction is completed, the owner walks through the facility with the construction manager and produces a punchlist. The list includes aspects of construction that need to be completed or fixed, their location, a schedule for completion, and clear assignment of responsibility for resolution, which may reside with the contractor, sub-contractor, supplier, or owner of the building. All items on

Table 1. Example of specifications for an oviposition room (3.0W x 3.7L x 2.4H m) for adult Lepidoptera

Environment	Utilities	Equipment
Temperature: 26°C ± 2°C (21–29°C possible) with ≤ 1°C stratification 0.3–1.8 m off floor	Electrical: 120-v service with emergency backup circuit for scale-collection equipment, lights, and photoperiod timer; duplex receptacles 1.5 m off floor in middle of each wall except 1.8 m off floor on back wall (for black light trap)	Scale collectors: laminar-flow isolators (1.8 x 0.9 m) for holding cages
Humidity: 70% ± 5% RH (40–80% possible) from steam generator in HVAC duct	Water: hot and cold at tub sink and cold at hose bib by door with covered floor drain in center of floor	Disposal cabinet: container (1.2 x 0.6 m) for disposing of used egg sheets and nonreusable cages
Thermostats, humidistats, and monitoring sensors: located in or near insect-holding zone but away from doors and external walls		Storage cabinet: small with hinged, glass-panel doors on wall over sink
Light: 14:10 LD (24–h timer located outside door) with 1,076 lux at 0.9 m off floor from cool white fluorescent bulbs in moisture and insect tight, flush mounted ceiling fixtures		
Air: HEPA filtered (95% efficiency at 0.3 µm) at a rate of 15 changes/h (including 10% fresh air) and a pressure negative to adjacent work rooms/hallways		

Table 2. Example of specifications for a clean workroom (3.7W x 11.0L x 2.5H m) for preparing artificial diet

Environment	Utilities	Equipment
Temperature: 22°C (20–24°C acceptable)	Electrical: 120 v service except 240 v for steam kettle	Steam kettle: 120-liter with canopy hood to exhaust steam and diet ingredient dusts
Humidity: 40% RH (20–60% acceptable)	Water: hot and cold at sink and floor drain under kettle	Diet dispenser: automatic
Light: 1,076 lux at 0.9 m off floor from cool white fluorescent bulbs in moisture and insect tight, flush mounted ceiling fixtures	Steam: 1.1 kg/cm^2 for steam kettle	Workbench: general use (0.8 x 2.1 m) with two sinks (0.4 x 0.6 m, one with disposal) and base cabinets
Air: HEPA-filtered (99.9% at 3 µm) at a rate of 15 changes/h (including 10% fresh air and no air from rest of facility) and a pressure positive to adjacent work rooms/hallways	Compressed air: 2.8 kg/cm^2 for diet dispenser	Workbench: diet handling (0.8 x 6.1m) with base cabinets and wall-mounted cabinets (0.3 m deep) with hinged, glass-panel doors
		Chemical fume hood: small (1.2 m) with 120 v electrical supply but no water
		Canopy hood over kettle to control steam and dusts from diet mixing

the list must be completed before the contractor leaves the site, unless there is a written agreement to delay it past that point.

C. Test and Balance

As discussed later in this chapter, proper environmental conditions produced by the HVAC (heating, ventilating, and air conditioning) system are essential to a successful rearing program. This system is the most complex part of the facility and is subject to deficiencies during design, equipment selection, and installation. Thus, the owner of the new building must be assured that the system is installed properly and operates according to design specifications. To that end, an independent testing and balancing agency should be used for evaluation. The Associated Air Balance Council (AABC) is a non-profit association of independent, certified test and balance agencies that provide these services (http://www.aabchq.com/). They will inspect and test piping and filtration systems; ductwork; and the chiller, compressor, boiler, pumps, fans, and controls. They will also check for air leaks; balance air pressures; and measure temperature, stratification, relative humidity, air exchange rates, and pressure gradients. The owner of the new insectary must interact closely with individuals of the commissioning agency so the latter understand what the system requirements are and why they are important to successful rearing.

III. Preparing the Facility for Initial Use

After the certificate of occupancy has been received and before insects are moved in, address the following items:

- Clean up dust generated during construction. Cutting and sanding sheetrock creates a fine powder that is hazardous to employees and insects. Watch for a significant buildup on horizontal structures such as the top of doorframes.
- Check the floor for cracks. Inspect the junction between the floor and the wall—it should be tight with no gaps. Inspect for complete seals around lighting fixtures, and look for broken fixtures and air diffusers.
- Check walls and ceilings for tightness and complete coverage of paint or other type of surface coating. If epoxy paint is used, test it by lightly running a moist paper towel over the surface. If the pigment comes off, it was improperly mixed and must be reapplied. Repaint before the insects and/or plants are placed in the room, and vent toxic fumes to the outside.
- Clean and top-dress the floors.
- Test the operation of all electrical circuits, ground fault interrupters, and back-up circuits. Label circuits, and clearly identify which ones are supported by emergency power.
- Set time clocks that control light cycles and monitor them for several days to determine if they are operating properly. For mechanical timers, tighten the on/off tabs securely so they do not loosen and fall off. In electronic models, install a battery for back-up power before programming the unit. Make sure each timer is controlling the appropriate lights or equipment. Label each timer with what it controls, the electrical circuit it is on, and the desired time period for each on/off cycle.
- If not done by a commissioning agency, check calibration of thermostats and humidistats against desired set points for temperature and humidity. Monitor conditions in each insect-holding room until conditions stabilize. Check air-pressure differentials between rooms using a hot-wire anemometer or nontoxic smoke. Evaluate temperature stratification within insect-holding zones.
- Monitor the temperature in refrigerators and freezers with a data logger or maximum/minimum thermometer to determine if the units are operating properly. Check operation of hoods (air velocity) and autoclaves (temperature, pressure, and treatment time). Test the diet preparation and dispensing equipment.
- Initiate preventative maintenance programs for equipment such as autoclaves, balances, biosafety cabinets, laminar-flow hoods, and other units using high-efficiency filtration.
- Check the plumbing for leaks and proper water pressure and temperature. Make sure floor drains are clear and can handle the volume of water anticipated. Clean the aerators in faucets to remove grit and other construction debris (this will be required regularly for several months after the facility becomes operational). Check sink hangers to ensure they are properly positioned and securely tightened.

- Determine if all doors operate smoothly and if they seal completely when closed. Make sure the locks work.
- Test safety equipment, including eyewashes and safety showers.
- Supply each room with the materials required for completion of all activities that will be conducted within. Avoid sharing materials among rooms to reduce the potential for contamination.

IV. Insectary Components

An insectary is made up of a variety of components such as walls, floors, rooms, electrical and plumbing systems, utilities, and air conditioning equipment. The key to a successful facility is in selecting the appropriate components and effectively incorporating them into the design. In this section, the major components of an insectary will be identified and described.

A. Structural Elements

Structural elements consist of materials that are used in the construction of a building. This section will focus on internal elements such as surfaces, doors, windows, built-in casework, and utilities.

1. Walls and Ceilings. Walls and ceilings form barriers that separate colonies, contain escapees, isolate procedures, and protect colonies from harmful contaminants. Thus, wall corners and wall/ceiling junctions should be tight fitting and properly sealed, surfaces should be smooth and uncluttered to prevent the accumulation of dirt and debris, and utility pipes should be run inside the walls with all penetrations sealed.

Ceiling-access panels should not be located in holding rooms or clean rooms. Suspended ceilings should not be used in insect-holding rooms because they cannot be sealed to exclude contaminants or water.

Appropriate construction materials include concrete blocks finished with flush joints and painted smooth or gypsum wallboard over wood or metal studs; metal studs are preferred in the construction of high-humidity rooms. Several types of waterproof coatings or coverings will withstand cleaning with bleach, detergent, or high-pressure water sometimes required in insectary operations. These include vinyl sheeting and rigid plastic panels, where each seam is sealed to form a smooth, continuous, impenetrable surface. Epoxy paint is also a good choice because it can be formulated for crack bridging, impact resistance, low odor, high gloss, and antimicrobial activity. Water-based latex paint is acceptable in areas where surfaces are not regularly washed down. It is low odor, tough, simple to clean up, and easy to use for quick touch ups. Installing bumper rails on walls will prevent damage from carts or racks.

2. Floors. Floor coverings should be nonporous, resistant to chemicals such as detergents and sodium hypochlorite, easily cleaned, nonslip, monolithic, hard enough to prevent damage from chewing insects, and capable of supporting point loads from carts and racks. Catalyzed epoxy resin makes a durable floor. It can be sprayed, rolled, or trowled on in a monolithic layer. The surface should be treated so it is not slippery, yet smooth enough that it is not difficult to mop. Vinyl sheeting also is a good selection if the seams between the sheets are carefully welded to form a continuous, impenetrable surface. If concrete floors are used, they should be painted or otherwise sealed to improve cleanliness and minimize the formation of dust.

Drains are especially useful in insectary rooms when walls and floors are washed down. Drains that are not being used can be covered to prevent sewer gases from getting into the building. A trap can be installed in rooms where soil and other debris are frequently washed into the drain. The decision to install floor drains should be made early in the design phase because they are difficult, expensive, and messy to install after the flooring has been put down. Floors should slope smoothly and gently toward the drain and should be free of depressions where water may accumulate. Too great a slope will cause carts and racks to roll toward the drain. When the building is operational, floor drains should be cleaned regularly to prevent the build up of organic matter that supports microbial and arthropod contaminants.

Cove molding is not recommended to cover the floor/wall junction. It may not fit tightly against the floor and/or wall, creating an ideal place for insects and dirt to accumulate. In one insectary, over the years, last-instar tobacco budworm, *Heliothis virescens* (Lepidoptera: Noctuidae), crawled behind the molding, under the linoleum, and chewed cavities in the subflooring to make pupation cells, leaving the floor soft and spongy. Ants, beetles, mites, and spiders joined the budworm population, creating a biological community that made it impossible to keep the holding room clean.

Instead of cove molding, the floor covering can be extended about 10 cm up the wall to form a smooth cove that makes cleaning easier. A beveled edge where the floor covering and wall meet will minimize dirt build up. With an epoxy floor, the top of the cove can be feathered against the wall. Then, the wall paint can overlap the feathered area by 2 cm or so making for an impenetrable floor/wall junction.

A raised door threshold leading to insectary rooms makes cart transport difficult and provides a great place for the accumulation of dirt, insects, and microorganisms. Specifying a smooth and level floor at the doorsill instead of a threshold will allow the use of an automatic-sealing unit that is activated when the door is closed. This unit consists of a spring-loaded rubber strip that lifts up when the door is opened and seals against the floor when it is closed (e.g., McMaster-Carr 8403A56).

Wax or other top dressing should be applied to the floor before the insects and equipment are moved into the room. Some of these floor treatments produce toxic fumes that can affect insects and plants. Treating the floor when the room is empty eliminates this problem and results in a more thorough application.

3. Doors. Doors should operate smoothly and safely. Door systems represent a potential route for the spread of contaminants within the facility, so doors should seal properly when closed. Hollow-core, steel doors are often used in insectaries because they are sturdy and not affected by humidity like wooden doors that shrink and swell. Door width is determined by the activities occurring in the room, the size of the equipment being moved in or out, and the amount and type of traffic that will be passing through the opening. A 0.9-m door to holding rooms is adequate for normal uses, including moving small carts (e.g., 0.5 x 0.9 m) in and out. To accommodate larger equipment, a wider single door will provide an opening of up to 1.2 m while a set of double doors has an opening of 2.4 m or more. In large facilities, roll-up doors may be required between rooms. The main entrance should be at least 1.2 m wide.

Doors that open automatically are useful when materials are frequently carried by employees or transported on carts. These doors are activated when electronic sensors detect a person in the area or when a wall-mounted plate is pushed. Automatic doors are more convenient and safer than hinged doors, but are considerably more expensive.

Automatic doors should not be installed at a main entrance, particularly if it opens into a hallway with considerable foot traffic. Otherwise the doors will open more frequently than needed, exposing the insectary to increased contamination potential. Automatic doors should have a "hold-open" feature, a built-in doorstop, or a slow closing rate to keep the door open while carts are being moved in or out. A bumper railing on each door will protect it from carts.

Magnetic gaskets similar to those on refrigerators are often used around the perimeter of steel doors to seal small gaps and reduce the movement of contaminants between adjacent rooms. Such tight closures also help control temperature stratification by restricting the flow of colder air into a holding room from an adjacent room or hallway. Automatic door seals help prevent the movement of insects and other contaminants under the doors.

A pocket door is a good choice when there is inadequate space for a hinged door to operate. These sliding doors require different sealing mechanisms than conventional doors. For example, a brush-type door seal prevents the movement of contaminants in and out of the rooms (e.g., McMaster-Carr 1130A54). Steel pocket doors are heavy and the top section of the frame must be constructed to handle the weight.

Unless there are compelling safety issues, doors with windows are not recommended for insect-holding rooms, particularly when adjacent areas are lighted at night. Light coming into the rooms during scotophase may interfere with normal light cycles and alter the behavior of the insects (Chap. 6 *Environmental Biology of Insect Rearing*). In rooms where insects are not present, windows should be installed in doors to help prevent accidents when the doors are opened.

Frequently used external doorways should not be located on a wall facing the prevailing winds to prevent leaves, dirt, insects, and other debris from blowing directly into the facility. Setting the door back from the building face will help prevent a strong wind from making the door difficult to use and keep precipitation from getting into the facility. Vegetation growing just outside external doors should be limited, particularly those plants frequented by aphids, thrips, mites, predators, parasitoids, diseased insects, or other organisms that can affect the colonies inside.

Security must be considered wherever a door is installed. Access to the main insectary should be restricted to minimize the introduction of contaminants into the facility. This can be accomplished with a regular lock and key, a magnetic reader that unlocks the door with a key card or small button, or a keypad-activated lock. Regardless of the locking mechanism used, the door must be simple to operate from the inside during an emergency. With a secure entrance and limited access to the insectary, doors to internal rooms may not have to be locked unless there are special circumstances that require additional restrictions. Signs should be posted on doors indicating restricted access. They should include information for contacting individuals within the insectary and those responsible if there is an emergency. Installing a telephone at or near the main insectary entrance will support access restrictions.

4. Lighting. Moisture- and insect-tight light fixtures help maintain cleanliness. Flush-mounted units are easiest to keep clean. Fiberglass fluorescent fixtures with full gasketing and captive latches (e.g., McMaster-Carr 1616K92) have been used successfully in holding rooms with high levels of humidity. These units are not flush-mounted but can be caulked against the ceiling to eliminate hiding places for insects. Bulbs in any fixture should be simple to change. Some 4-bulb fixtures have 12 screws holding a very heavy cover that must be removed. This type of fixture can discourage maintenance and is awkward and heavy enough to be a safety hazard.

5. Windows. Windows provide natural light and allow employees a refreshing break from repetitive procedures. All windows should be insulated to prevent condensation and the growth of mold. Nonopening windows should be used to prevent entry of contaminants. South-facing windows often require external louvers, awnings, and/or reflective shades to help control solar radiation entering the facility. Windows in insect-holding rooms are discouraged because they make it difficult to maintain temperature, airflow, and a consistent photoperiod. In addition, they may cause undesirable insect behavior as in the case of the boll weevil, where windows caused overcrowding in the brighter parts of the oviposition cages (Griffin and Lindig 1978).

6. Cabinets. Built-in, floor-mounted cabinets provide storage space below and a work surface on top. They come in various widths and can be configured with cupboards and/or drawers. Worktops should be nonporous, easily cleaned, hard, scratch resistant, and water tight. Light-colored surfaces are recommended so dirt and insects are easier to see. They can be made of stainless steel, plastic laminate, painted wood, epoxy, or composite. Most can be ordered with backsplashes, coved corners, and lips in the front to help control spills. Procedures that will be performed on the worktop will determine the cabinet height. For operations that require standing, a cabinet height of 0.9 m is appropriate. If the employee will be sitting on a tall (e.g., 0.7-m) laboratory stool, a 0.9 m high cabinet with a knee space can be used. Lower cabinets are usually 0.8 m high and are used while sitting at knee spaces.

Wall-mounted cabinets are about 0.3 m deep and usually installed so that cabinet bottoms are about 0.5 m above the top of floor-mounted cabinets. Cabinet doors help keep items clean. Hinged doors provide more space inside the cabinet than sliding doors, and they do not have a track where dirt can accumulate.

7. Sinks. Sinks are used for a wide variety of activities throughout an insectary. They can be wall mounted, floor mounted, or dropped into an opening in a countertop. Sinks are manufactured in stainless steel, epoxy, and plastic and can be constructed as one piece with the countertop, thereby eliminating a seam that encourages the buildup of contaminants. Rounded internal corners make sinks easier to clean. The type of procedures performed in the sink will determine the size and number of bowls. Sinks that are too small for the job are inefficient, messy, and frustrating to use. A scullery sink is a large, single-, double-, or triple-bowled unit. These are useful in workrooms where cages and other large pieces of equipment are washed. A mop-service sink (e.g., McMaster-Carr 2180K1) can be installed in closets where cleaning supplies are kept. They are installed on the floor and are only a few inches deep, so minimal lifting is required to empty a mop bucket.

A drainboard and pegboard, mounted on the wall behind a sink, are useful for draining glassware, cages, etc. after washing. A small area near the sink should be provided for supplies, such as hand soap, moisturizing lotion, and sponges.

Mixing faucets are installed where control over water temperature is important (e.g., surface sterilization of eggs). Unless a hose needs to be attached, an aerator operates better in an insectary than a serrated nozzle because it provides greater flow at lower pressures. Some codes require that faucets with serrated nozzles be supplied with a vacuum breaker to prevent contaminated water from backing up into the water supply. Faucets can be installed on the horizontal deck surface, on the back splash, or the wall behind the sink. The latter location makes it easier to keep the area underneath clean. A high-pressure wash-down sprayer is appropriate for large sinks and multibowled sinks. Such sprayers can be moved freely to wash the largest items (e.g., McMaster-Carr 2686K13). A standard faucet can be added to the wash-down unit for use when the sprayer is not needed.

8. Insulation. Walls and ceilings of insect-holding rooms should be insulated to maintain temperature for several hours when the air-handling system is inoperable or power is lost. Greater insulation may be required in holding rooms with south-facing walls and in buildings in climates where weather extremes regularly occur. Floor insulation prevents condensation during winter in rooms maintained at a warm temperature and high humidity. It also reduces temperature stratification exacerbated by a cold floor (Griffin 1984).

9. Utilities and Other Systems. Consider which of the following utilities are required and where they should be located within the facility.

a. Alarms and other safety equipment. Fire and emergency-evacuation alarms, emergency back-up lights, fire extinguishers, eyewashes, safety showers, and fume hoods should be installed as needed or directed by local codes (Chap. 4 *Insectary Health and Safety*). The last three items are of particular importance if formaldehyde or other dangerous chemicals will be used.

b. Gases. Compressed air may be required for transferring insects, cleaning cages, or running automated equipment such as a diet dispenser. Carbon dioxide is often used for anaesthetizing insects to make handling easier. Natural gas may be needed for use with gas stoves and burners.

c. Communications. Computer connection to a local network and the Internet via wired or wireless technology can be used to communicate with customers, suppliers, and to obtain information. Intercoms are used to communicate internally within the insectary. They improve efficiency and reduce the potential for contamination. Phones should be installed in hallways and workrooms and not in insect-holding rooms.

d. Electrical. In the United States, 120 volts are required for normal equipment and appliances while greater voltages are needed for equipment such as diet-mixing kettles or autoclaves with integral steam generators. There should be at least two separate electrical circuits in each room so that if one becomes inoperable, another is still available. Label each receptacle with the corresponding circuit number.

Ground-fault circuit interrupter (GFCI) devices shut off the flow of electricity within milliseconds when current leakage is detected. They are essential safety devices when electrical receptacles are located near sources of water. All circuits in some insectaries are protected with GFCI. Watertight covers prevent short-circuiting of electrical outlets in rooms whose walls and floors are washed down.

Emergency power should be supplied for critical uses such as refrigerators, freezers, scale-collection devices, growth chambers, and lights in insect-holding rooms. Several receptacles throughout the facility should also be on emergency power so that fans, heaters, and portable lights can be used during power outages. Labeling these receptacles will make them easier to locate during an emergency. Lights that turn on automatically when power is lost will enable employees to move safely within the facility and can provide illumination in areas where work must continue even with the power off.

e. Garbage disposal. Disposals are useful when bulky organic debris is discarded down the drain, such as spent diet removed from reusable containers. These units, including commercial stainless steel models, can corrode and seize up over time if sodium hypochlorite is regularly poured down the drain.

f. Steam. Steam is used for autoclaves, heating-jacketed mixing kettles, some humidifiers, steam cleaners, and reheating air entering insectary rooms through heat-exchange coils. Steam sometimes contains chemicals to control the growth of fungi and bacteria in condensate return lines. Determine if these chemicals are harmful to plants or insects located where steam is used.

g. Water. Water of consistent quality is critical for the preparation of insect diets (Reinecke 1984, Cohen 2004). Processes using water should be identified along with the type (e.g., hot, cold), quality (e.g., deionized, distilled, tap), and quantity required for each process. The rate of water delivery must be appropriate for the process being performed, particularly when filling large containers or using a wash-down system.

h. Vacuum. Vacuum at very low pressure is valuable when harvesting or transferring small flying insects such as leafhoppers, whiteflies, and parasites.

B. Environmental Control Systems

The air-handling system, also called the heating, ventilating, and air conditioning (HVAC) system, maintains the temperature and relative humidity, as well as the velocity, flow pattern, cleanliness, exchange rate, and pressure gradients of the air.

Selecting the appropriate HVAC system is a challenging task made even more difficult because of differing opinions among engineers with considerable experience in rearing systems. For example, Owens (1984) recommends individual air-conditioning units for each insect-holding room while Griffin (1984) states that central heating and cooling are preferable.

To obtain the system best suited for a rearing program, the selection process must be a collaborative effort between the design engineer and the building owner. The owner must provide the design engineer with pertinent functional goals and information on financial constraints, and the design engineer must provide the owner with information on the advantages and disadvantages of each HVAC option. (ASHRAE Handbook 2000). This is particularly true in buildings that house biological organisms where conditions generated by the HVAC arguably have the greatest impact of any variable on program success.

1. Environmental Monitoring System. This system tracks temperature and humidity throughout the facility including critical equipment such as incubators, freezers, and refrigerators. It also monitors fan operation, pressure drop across filters, and other HVAC variables. It can be programmed to trigger an alarm when limits are exceeded and notify appropriate individuals (e.g., insectary manager and/or the maintenance department). The most desirable system is user friendly and comes with software that tracks and stores the data, allows some data manipulation, presents summary information in a variety of report formats, and supports the transfer of data to a program such as Microsoft Excel®.

2. Temperature. Insects are poikilotherms, and their body temperature approximates and varies with the temperature of their surroundings. As a result, an insect's developmental rate increases as the temperature rises. A precise amount of heat is required for an insect to achieve developmental milestones, for example, the point when an egg hatches. The warmer the environment the faster this point is reached (Chap. 6 *Environmental Biology of Insect Rearing*).

There are four temperature characteristics that should be considered when developing specifications for an insectary. They are the desired temperature set point, temperature variation around that set point, temperature stratification, and the range of possible desired temperatures.

The temperature set point is the average temperature to which insects are exposed. This is usually the setting on the thermostat within an insect-holding room. This temperature produces a "comfortable" rate of development, one that is not so slow that it adversely impacts program efficiency, or so fast that it imparts metabolic stress on the insects. In many insectaries, the desired temperature set point is between 25 and 27°C. To maintain the desired set point, each holding room should have an independent thermostat. High- and low-temperature cutouts should be installed in the HVAC system to shut it down and prevent damage to the colonies if the temperature exceeds a specified upper or lower threshold. Temperature variation is the amount that the temperature fluctuates around the desired set point. For most rearing operations, ± 2°C is adequate and technologically achievable.

Temperature stratification occurs when warmer, lighter air rises and accumulates at the top of a room, while colder, heavier air settles to the bottom. Stratification results in temperature differences between the top of an insect-holding rack and the bottom, with insects developing faster the higher they are on the rack. Two to three °C can create a difference in developmental rates that can adversely impact production quality. A gradient of 0.6°C is acceptable. Stratification is exacerbated by cool air entering the holding room from adjacent rooms through a poorly-sealed door or from a cool, uninsulated floor (Griffin 1984).

An additional consideration is the range of possible desired temperatures that may be required in a holding room at some point in the future. Knowing this range will help the engineer design a system with adequate capacity to support changes in the rearing program. A common range for a multispecies facility where species are replaced from time to time is 20 to

30°C. It will be less in a facility dedicated to rearing one species.

Finally, a portable heater can be used to maintain temperatures in smaller holding rooms when the air-handling system is inoperable. The heater can be plugged into an external thermostat (e.g., McMaster-Carr, 1940K9) that is more accurate than the one provided with the heater.

3. Humidity. Relative humidity is the percentage of moisture in the air relative to the total amount it can hold at a particular temperature. It can be measured by a hygrometer or psychrometer. Several factors determine the relative humidity within the facility, including the season of the year, amount of outside air brought into the facility, temperature of the conditioned air before it reaches the holding rooms, temperature set points, and the amount of moisture added by humidification equipment (Chap. 6 *Environmental Biology of Insect Rearing*).

Humidity affects insect colonies in a variety of ways (Table 3), including the rate of water evaporation from larval foods. Artificial diet should dry out slowly over the larval development period, keeping it palatable but not so wet that it supports mold growth. By the time pupae are harvested, the remaining diet, as well as the frass, should be dry enough that it does not stick to the collection equipment.

As with temperature, the proper humidity must be maintained with minimal variation to achieve developmental consistency. The desired humidity set point will vary with species and different developmental stages of the same species. For example, Lepidoptera adults require 60 to 70% RH for maximum egg production and fertility, while larvae feeding on artificial diet require 40 to 50%. Maintaining appropriate humidity in the winter is challenging. Humidity can drop to single digits when dry, cold outside air is heated before it enters the holding rooms. Each holding room should have a humidistat that will control the humidity about the desired set point independent of other rooms.

Humidity variation is the amount of fluctuation around the desired humidity set point. For most rearing operations ± 5% is adequate. The range of possible desired humidities includes the highest and lowest humidity that may be required at some point in the future. It will dictate the type and capacity of humidification/dehumidification equipment to be installed. A common range is 40 to 80%.

While there are a number of humidification systems on the market, a steam system is most reliable and effective at maintaining the desired conditions. The steam can be introduced directly into the room or through the HVAC duct system. Centrifugal atomizers are not recommended because their limited capacity is appropriate only for the smallest rooms. In addition, they are difficult to keep clean, especially in an adult Lepidoptera holding room where wing scales mix with the water to form a gel-like material that prevents the unit from working properly.

Table 3. Some consequences of abnormally high and low humidity in insect rearing

Species/location	High humidity	Low humidity
Corn earworm, *Helicoverpa zea*/larval holding	Mold on diet resulting in reduced feedingand increased mortality of larvae and pupae	Dessication of diet resulting in developmental delay and small adults
Tobacco budworm, *Heliothis virescens*/oviposition	Delamination of cardboard cages resulting in escape of adults	Dessication of pupae, adults, and eggs resulting in increased mortality, wing deformities, and reduced egg production and hatch
Corn rootworm, *Diabrotica virgifera virgifera*/larval holding	Mold on corn seedlings resulting in reduced feeding, low yields of adults, and establishment of mite pests (e.g., *Tyrophagous* spp.)	Dessication of corn seedlings resulting in reduced feeding and low yields of adults
Corn rootworm, *D. virgifera virgifera*/adult holding	Mold on diet and on eggs resulting in reduced egg production and hatch	Dessication of soil in oviposition dishes resulting in reduced oviposition and hatch

Table 4. Lighting specifications for three locations in an insect rearing facility

Location	Light intensity (lux)	Working distance from light source (m)	Photoperiod (L:D) (h)
Insect-holding room	807–1,076	1.2 from ceiling-mounted fixture	14:10
Workroom	1,706	1.2 from ceiling-mounted fixture	Lights on when employees working
Light cart	4,306–5,382	0.3 from cart-mounted fixture	14:10

Under some conditions, e.g. lower temperatures, a dehumidifier may be needed to eliminate excessive humidity from holding rooms. A fan pulls air across cold coils where moisture condenses and drips into a pan for manual disposal or drains directly into the sewer system via a hose. For manual operation, when the pan fills up, the unit shuts off until it is emptied. This can be problematic in humid conditions because the dehumidifier will operate unattended only for short periods of time.

4. Light. The lighting system controls the quality, intensity, and periodicity of the electromagnetic environment. Light influences insect development and behavior, and sources of light should be installed in each holding room (Chap. 6 *Environmental Biology of Insect Rearing*).

Light intensity is a measure of the brightness at a specific distance from the source and can be measured by an inexpensive, portable light meter. Table 4 shows the intensities of cool white fluorescent bulbs at three locations within a multispecies rearing facility. In rooms with high ceilings or where greater light intensity is required, banks of lights may be mounted vertically on the walls. In addition, carts in which each shelf is independently lighted can be used for improved illumination (Fig. 1), or light fixtures can be hung by an adjustable system that allows them to be moved up or down as needed (Hampson 1978, Etzel 1978).

Photoperiod is the number of hours of light and the number of hours of dark during a 24-h period (Table 4). Each insect-holding room should be equipped with a timer to control the photoperiod of the ceiling- and/or wall-mounted lights. Lights and timers should be backed up with emergency power, particularly if the facility is prone to electrical outages that last more than a few hours.

Additional lighting may be required for specific applications. For example, 100-watt incandescent bulbs with reflectors are placed directly outside oviposition cages for species such as the imported cabbageworm, *Pieris rapae* (Lepidoptera: Pieridae), which requires intense light and heat to stimulate feeding, mating, and oviposition behavior (W. R. Fisher, unpublished). Some rearing facilities use night-lights to stimulate oviposition by adult Lepidoptera (Debolt and Peterson 1978). A household timer plugged into a receptacle can be used to control the on/off cycle for these additional lights.

5. Air Handling. Air handling involves the movement of air within rooms and throughout the facility, incorporation of fresh air, differential air pressures, and filtration to reduce airborne contaminants.

a. Airflow Patterns. Airflow pattern is the route that conditioned air takes as it passes through a room. Two primary types of patterns occur in insectaries. In a laminar-flow system, air moves in one direction at a uniform low velocity. Air enters the room through large banks of filters that cover the entire ceiling or one of the walls and exits through the floor or the opposite wall (Figs. 2a and 2b). Perforated panels located on the downwind side of air plenums have also been used to create laminar-flow rooms. These systems are effective at maintaining clean conditions and minimizing temperature variation. However, in a horizontal system, cleanliness can be compromised if a dirty operation is performed upwind from a cleaner operation. Laminar flow is considerably more expensive to install and maintain compared to nonlaminar designs.

Nonlaminar, turbulent flow occurs when the air flows in several directions and at different velocities. Nonlaminar flow is more typical of insect-holding rooms because it is simpler and less costly when compared to laminar-flow designs. Ideally, air enters the room from a vent(s) in the ceiling and exits through return ducts placed near the floor on either side of the room (Bell et al. 1981; Fig. 2c). The air sweeps across the room to create relatively uniform conditions without requiring high velocity to mix the air.

When the supply and return vents are in the same plane (e.g., both on the ceiling), the air short-circuits and does not deliver conditioned air to the entire room (Fig. 2d). This pattern results in stratification, creates "dead" air space where minimal air movement creates conditions that encourage the growth of mold, and does little to clean the air as it passes through the room. This situation can produce extreme temperature gradients.

Ceiling fans create a gentle downward flow of air within holding rooms to minimize temperature stratification. They can be used with smooth, uncluttered ceilings that are high enough so the blades do not create a safety hazard. Portable fans also improve air circulation for more uniform temperature and humidity within holding rooms.

Some insect-holding rooms incorporate both laminar-flow and turbulent-flow characteristics. In one case, air supplied from a plenum enters a 3.7 x 4.9-m room through a perforated ceiling, flows down towards the floor, and exits the room via 11 return registers located on opposite walls near the floor. It then enters return ducts behind false walls and is filtered and reconditioned before being recirculated back into the room (Klassen and Gentz 1971).

Harrel et al. (1979) constructed two rooms, one with air entering and exiting through perforated walls on opposite sides of the room and the other with air entering from a perforated ceiling and exiting through vents at varying heights on opposite walls. Airflow velocities were the same throughout both rooms and the temperature variation was ≤ 1.1°C.

Bell et al. (1981) constructed a 3.7 x 8.8-m insect-holding room incorporating a laminar-flow system. Air was supplied through nine supply vents, each fitted with a 95% efficient HEPA filter. Vents were located at the end of the room in three rows of three vents each. The air exited through six registers on the opposite wall. The airflow rate was ~1.4 m^3/min, and temperature variation was ± 1.0°C throughout the room.

b. Ventilation. Ventilation is the process of supplying air to dilute and remove contaminated air, eliminate odors, control humidity and condensation, pressurize rooms, control airflow within the facility, and remove air heated by equipment such as steam kettles and autoclaves. Ventilation is measured by the number of air changes per hour.

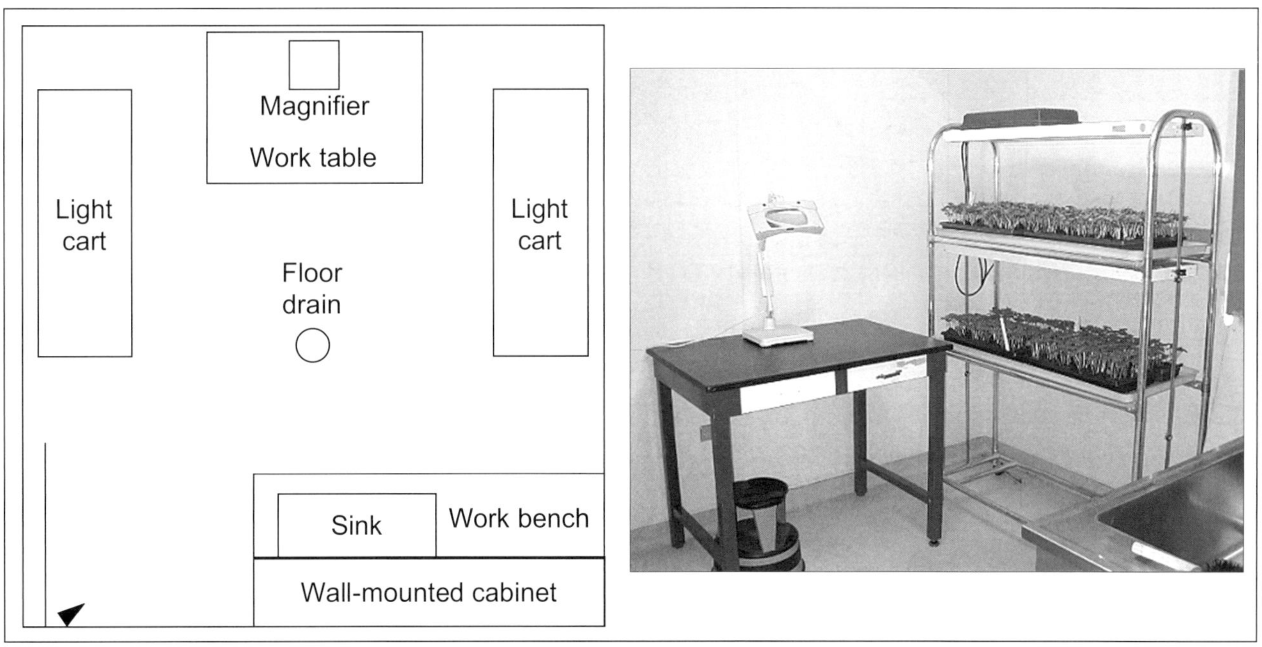

Fig. 1. Floor plan and photo of a 3.0 x 3.0-m room for maintaining a single aphid colony reared on cotton seedlings. Two light carts provide enough space for holding 6 flats of 24 pots each with an average of 5 seedlings per pot. The worktable with lighted magnifier is used to scan plants for undesirable pests such as thrips and spider mites that could threaten the colony. The solid black arrowhead indicates the direction from which the photo was taken.

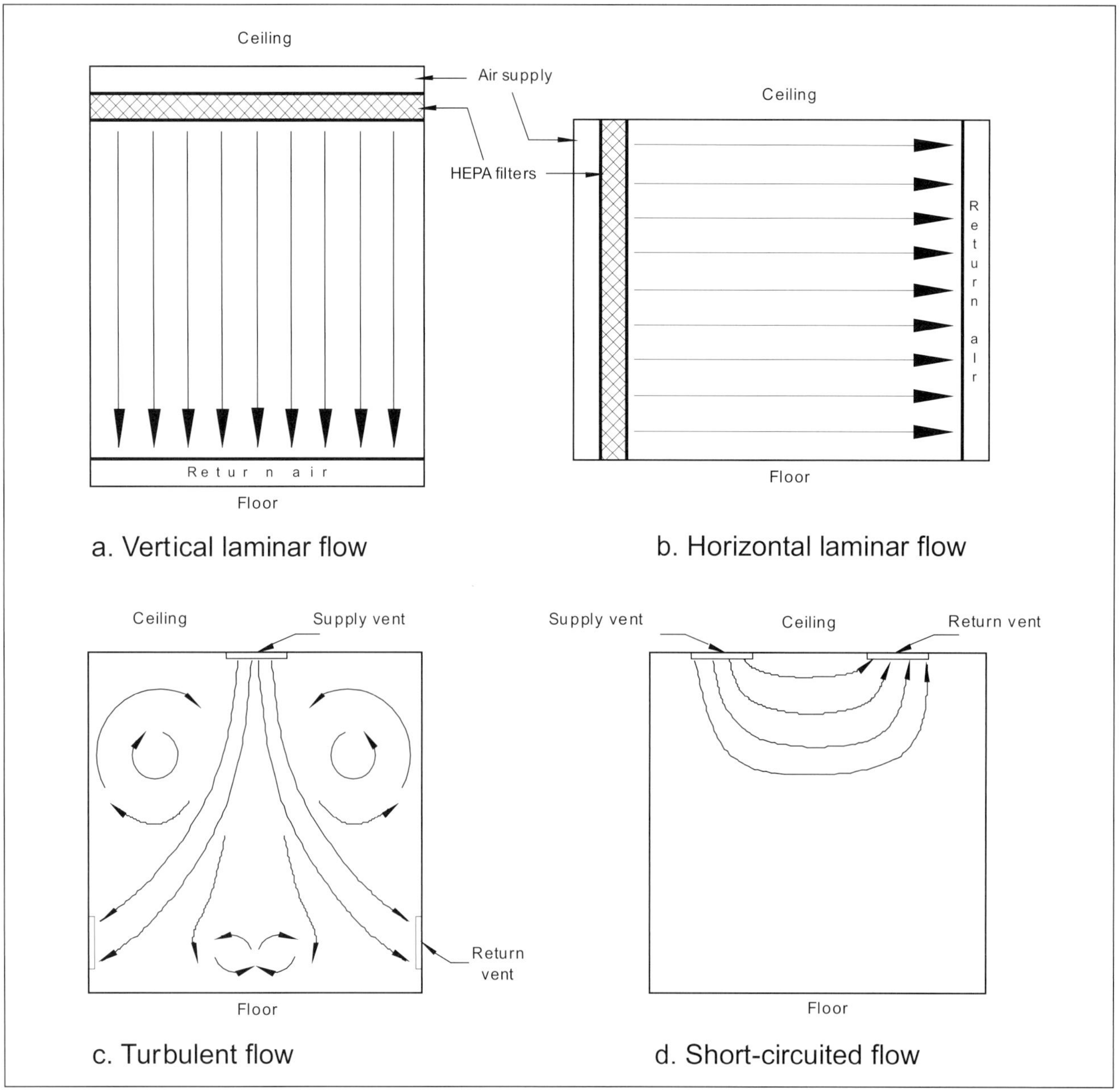

Fig. 2. Airflow patterns in insectary rooms. The two laminar-flow patterns (a, b) provide more uniform temperature throughout a room than nonlaminar flow; however, they are more expensive to install and maintain. In a turbulent flow pattern (c), air enters the room from ceiling height and exits near the floor to create a sweep of air across the room. Air entering and exiting a room at the same height (d) can result in short circuiting that is ineffective in conditioning the entire room.

In a single-pass ventilation system, after conditioned air moves into and through a room, it is exhausted to the outside, and none of it is recirculated. A recirculating system, on the other hand, reuses some of the previously conditioned air by retaining it within the HVAC system. Recirculated air may include contaminants that were picked up during previous passes inside the building. Thus, it represents a greater threat to colonies than air from a single-pass system. Recirculated air is not allowed in animal facilities (NIH 1999).

Ten to thirty complete changes of air per hour are recommended for achieving desired environmental conditions, providing fresh air, and helping remove airborne contaminants in animal-holding rooms (ASHRAE Handbook 1987) and in insect-holding rooms (Owens 1984). The greater the change rate, the less time required to purge airborne contaminants from a room. For instance, 28 min are required to remove 90% of airborne contaminants with five air changes per hour compared to 14 min with 10 changes and five min with 30 changes (CDC 1994). With insects susceptible to airborne contaminants, Owens (1984) suggests using at least 30 changes per hour to help rid the room of these contaminants.

On the other hand, excessive air-exchange rates can increase the level of airborne dust and debris, result in faster desiccation of artificial diets, create a noisy environment, and increase costs. Owens (1984) states that 20 air changes per hour are more than necessary to maintain the required temperature and humidity. To save on installation and operating costs of a high-exchange system, he designed a method to allow up to 75% of the air to bypass the conditioning equipment.

Seventeen air changes per hour works well in a holding room for *H. virescens* larvae (W.R. Fisher, unpublished). The room is 2.2 x 3.3 x 2.7 m high. Air enters the room through the supply vent at 3.4 m^3/min and exits through the exhaust vent at 5.5 m^3/min creating negative pressure relative to the adjacent hallway. Because of this gradient, 2.1 m^3 per min enters the holding room through spaces around the door. Larvae are held in 32-well trays with 19 ml of artificial diet and one to three larvae in each well. Trays are covered with a Mylar film punched with 0.03-mm to 0.05-mm diameter holes at a density of 67 per 3.9 cm^2. Trays are separated vertically by 2.5 cm to allow for proper air and moisture exchange. This combination of variables results in ideal conditions for the insects. The diet dries out slowly over time but is still a good nutrition source through the 14 days to pupation. Diet and frass continue to dry out until pupae are harvested, about 26 days after set up. The result is a soft pliable medium that is neither so dry it entombs the pupae nor so wet it sticks to forceps to slow the collection process. This rate of air exchange also effectively eliminates odors and the buildup of stale air.

c. Pressure Differentials. Air entering a room from an air-handling system creates pressure. If the pressure in one room is greater than that in an adjacent hallway, air will flow from the room into the hallway. If the pressure is less in the room, air will flow in the opposite direction. If the pressures are the same there will be

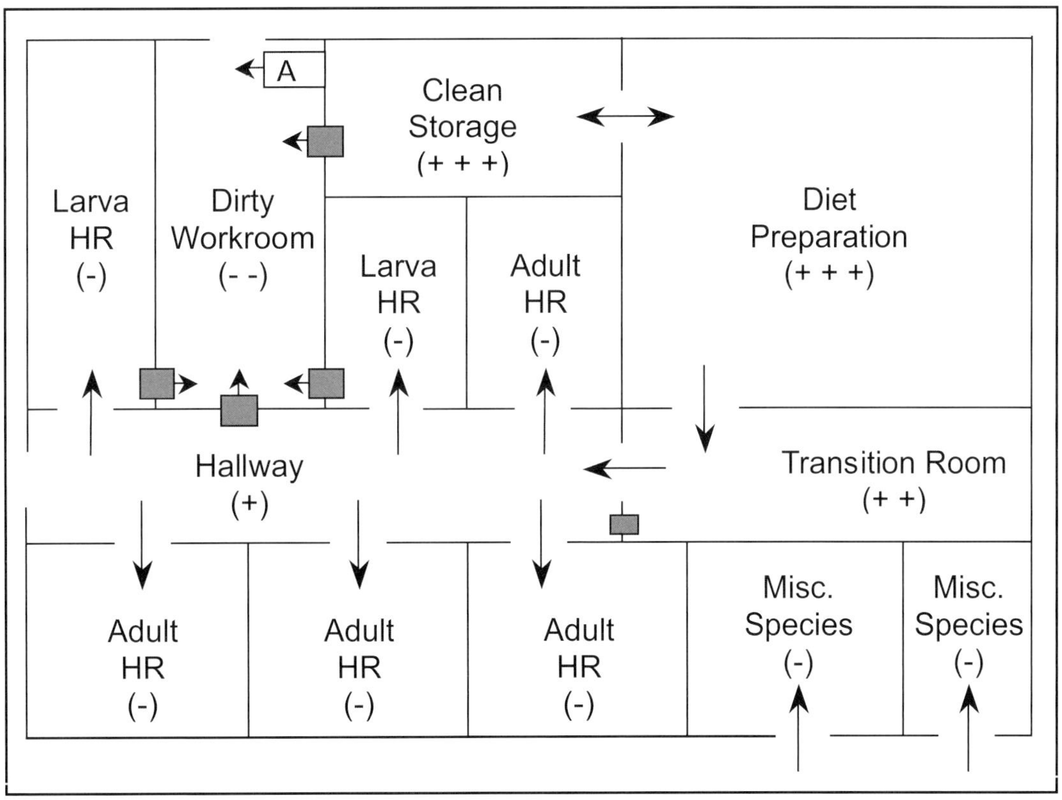

Fig. 3. A multispecies rearing facility where air-pressure differentials were used to limit the movement of airborne contaminants between rooms (Fisher 1984). Cleanest rooms have the greatest pressure (+++) while the dirtiest room has the lowest (--). Arrows indicate the direction of airflow from higher to lower pressure. Air moving from the hallway into the holding rooms (HR) helps contain contaminants within the holding rooms. This airflow could be reversed to prevent hallway contaminants from entering the holding rooms. Autoclave (A); passthroughs (■).

no net air movement. Pressure differentials can be used to inhibit the spread of airborne microorganisms. Fig. 3 illustrates an airflow plan where the greatest pressure occurs in clean rooms (+++) to inhibit airborne contaminants from entering and the least pressure is in the dirty workroom (--) to contain contaminants.

A pressure difference of 0.025 mm of water is enough to create air movement between two rooms, 1.30 mm is recommended between clean and dirty rooms, and 0.51 to 0.76 mm is required between clean rooms and rooms of intermediate cleanliness to control the spread of airborne pathogens (CDC 1994). Wang (2001) states that the minimum air velocity between rooms of different cleanliness should be between 4.6 and 15.2 m/min with the door open. Achieving effective pressure differentials may be difficult in older buildings where rooms are not well sealed.

d. Air Cleanliness. Microorganisms and particulates generated within an insectary can have adverse effects on insects and employees. Contaminants such as bacteria and fungi can compete with insects for the food provided them (Chap. 8 *Microbial Contamination and Insect Rearing*). Pathogens such as bacteria, fungi, protozoans, and viruses can reduce insect quality, production efficiency, and under the right conditions, destroy cultures (Chap. 9 *Entomopathogens and Insect Rearing*). Particulates including wing scales, dusts from artificial diets, soil, vermiculite, and fungal spores are unhealthy for employees (Chap. 4 *Health and Safety Issues in the Rearing of Arthropods*). Microorganisms and particles are small and become airborne easily, allowing them to be transferred through the facility by normal air movement. Equipment to control these contaminants should be located as close to the process that creates them as possible. The use of hoods and point-source air filtration systems are discussed in Chap. 8 *Microbial Contamination and Insect Rearing*.

Whole facility air filtration is an additional means to keep the insectary air clean. Filter efficiency is determined by the percentage of particles of a specific size that are retained by a filter. Thus, a 99.9% rating at 0.3 μm means that the filter captures essentially all particles such as bacteria, mold spores, protozoans, and most viruses that are 0.3 μm or larger.

Owens (1984) recommends using a 99.9% high-efficiency particulate air (HEPA) filter with at least 30 air exchanges per hour in a facility where the insects are highly susceptible to airborne contaminants. HEPA filters with an efficiency of 95% have also been used effectively in insectaries. Other filters that don't have the disadvantages of HEPAs (high initial cost, high replacement cost, and a high pressure drop) have also been used successfully (Davis and Jenkins 1995).

Lower efficiency filters are placed upstream to remove most of the larger particulates that will quickly clog high-efficiency filters and reduce their effective life. A simple disposable filter can increase the life of a HEPA filter by 25%. However, if a 90% extended-surface filter follows that disposable filter, the life of the HEPA can be extended almost 900% (CDC 1994). This can represent a significant savings, as high-efficiency filters are expensive. HEPA filters must be tested regularly to ensure they are operating properly. Any leaks in these filters can greatly diminish their effectiveness. Austin (1994) lists 19 types of leaks that can occur including cockroaches chewing holes in the filter material.

HEPA filters should be placed as close as possible to the room they service. If air is recirculated within an insectary, it should be passed through HEPA filtration to prevent contaminants from being transported back into the rooms. Lower efficiency filters may be needed in return vents to prevent escaped insects from being transported in the exhaust stream to other parts of the facility or outside the building.

Filtration within HVAC ducts is not the only way to clean the air. The action of any laminar-flow hood provides supplemental filtration that helps clean the environment in which it is located. For example, a 1.8 x 1.2-m hood with an airflow rate of 30 m/min will clean about 4,000 m^3 of air per hour.

C. Miscellaneous Equipment

Equipment aids employees in the effective and safe performance of rearing procedures. A list should be prepared of all the major equipment to be installed in the new facility including a description (model number, inventory number, and current location, if not new), intended use, vendor (if new), dimensions, and requirements for utilities, backup power, and ventilation. The list should also contain the intended location for each piece of equipment so that the most effective flow plan for materials and products is established. A scale drawing of the facility will simplify the process of placing the equipment. The following is a list of equipment commonly found in rearing facilities. Pieces of equipment for more specific applications are found in Griffin (1984), Harrell and Gantt (1984), Harrell and Griffin (1981), Bell et al. (1981), Singh and Moore (1985), Sikorowski et al. (1984), Nordlund and Greenberg (1994), Nakamori et al. (1992), and Fisher and Finney (1964). Goodenough (1984) discusses equipment needs as they relate to materials handling.

1. Autoclaves. A steam autoclave kills organisms with high temperature and pressure. It is used to sterilize reusable cages, containers, and other items; aid in the preparation of microbiological media; kill insects prior to disposal; heat water for a variety of purposes; and preheat diet-dispenser equipment prior to use. A pass-through, double-door autoclave is valuable in helping maintain cleanliness (Sec. IV. D. 2. b). An ethylene oxide unit is used to sterilize plastic and other materials that would be destroyed in a steam autoclave.

2. Dishwashers. Standard-sized household dishwashers are useful for small containers, cages, glassware, and equipment. However, for larger items these units are not efficient. Larger washers, like those used in restaurants, are often more appropriate for insectary applications.

3. Freezers. A freezer can be used to kill insects in cages or bags prior to disposal, hold alcohol specimens, and store chemicals and diet ingredients. Standard upright models with about 0.6 m^3 volume are ideal for smaller items (e.g., 0.03-m^3 screen cages) while walk-in units are necessary for larger items.

4. Hot-water Heaters, Demand. If hot-water tanks are unavailable or inadequate, a wall-mounted unit provides hot water quickly, where it's needed, and with no large tank. There are several models available with various electrical requirements, volumes, and maximum temperatures.

5. Light Carts. These carts have several adjustable shelves and a bank of fluorescent lights to provide for better plant growth. Cart dimensions are approximately 1.4 m wide x 0.6 m deep x 1.8 m high. They are available in nursery supply catalogs.

6. Light Traps. Black lights are attractive to many insects, making them ideal for use with insect traps. Two types are useful to help control contaminants in an insectary. The electrocution type (e.g., McMaster-Carr 6995T14) uses an electrified grid in front of the lights to kill insects that come in contact with it. The dead insects drop off the grid and are collected in a pan at the bottom of the unit. The second type captures insects on cardboard coated with sticky glue (e.g., McMaster-Carr 98195T71). The cardboard is easily changed when it's full. This style is quieter and cleaner and captures smaller insects than the electrocution type. Fisher and Finney (1964) describe a black light trap built into the wall between an anteroom and an insect-holding area of a quarantine facility.

7. Mixing Kettles. Kettles are used for mixing artificial diets. In a jacketed kettle, steam enters an envelope surrounding the kettle bowl and produces boiling water very quickly. The steam source comes from a centralized system or from an integral generator. In kettles without a steam jacket, water is boiled elsewhere and then added along with the dry ingredients.

8. Passthroughs. These are essentially boxes mounted through the common wall of adjacent rooms

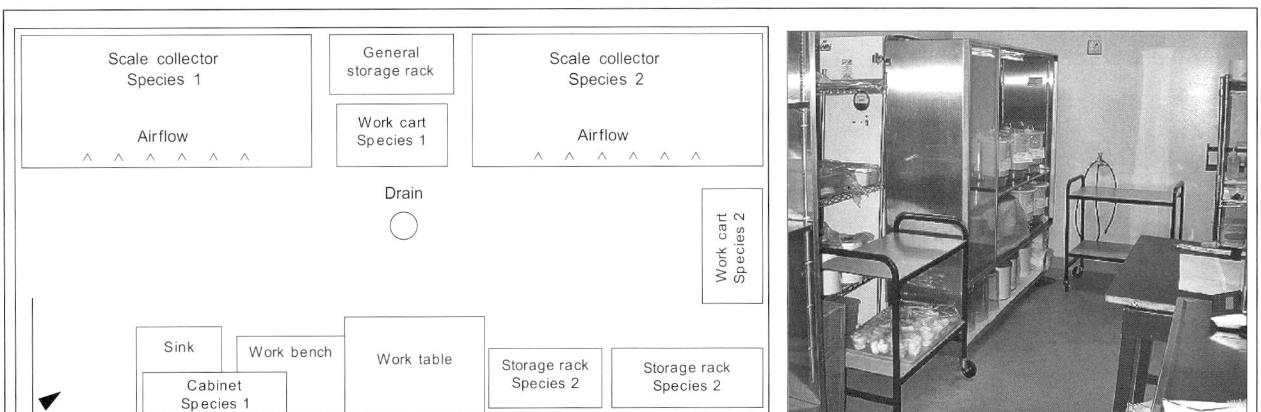

Fig. 4. Floor plan and photo of a 3.0 x 6.1-m room for holding adult Lepidoptera in 3.8-liter cardboard cages for one species and 5.7-liter polycarbonate cages for another. Each population is housed in a separate laminar-flow isolator (inflow) that captures wing scales produced by the moths. Each species has its own work cart and rack for storing supplies. The solid black arrowhead indicates the direction from which the photo was taken.

with a door on each side of the box. They are used to transfer insects, cages, and containers from one room to another with minimal contamination spread. Passthroughs are often equipped with interlocks that prevent both doors from being opened at the same time and ultraviolet (UV) lights that sterilize interior surfaces after a transfer is made. They may be built in when the wall is constructed or can be purchased premade and inserted into the wall (e.g., Clean Air Products, CAP 18 W).

9. Refrigerators. Refrigerators are required for storing items such as raw diet ingredients, prepared diets, and microbiological media plates. They also can be used for slowing insect development and chilling insects to make handling easier. Refrigerated storage is also required for host material such as melons, fruits, and potatoes in the production of hosts for beneficial organisms.

10. Scale Collectors. Insect scales and other minute debris must be removed from the air in the insectary because they are potential allergens for susceptible employees; and they harbor microorganisms, including pathogens, on their surfaces that can be harmful to the insects.

Animal containment enclosures (e.g., NuAire, NU-605-500; Fig. 4) remove scales and other airborne debris as air is drawn from the room and across the cages, into a bank of prefilters, through a HEPA filter, and back into the room. These units are mobile and can be moved for cleaning and changing filters. They should be certified annually to ensure they are operating according to specification. Scale collectors should be placed on an electrical circuit that is backed up by emergency power. Other types of collectors are described in Davis and Jenkins (1995), Owens (1984), Wirtz (1980), Wolf (1984), Fisher and Finney (1964), Stewart (1984), and Chap. 11 *Insect Rearing Production Systems, A Case Study: The Southwestern Corn Borer.*

D. Compartmentalization

An insectary is a highly specialized facility requiring unique rooms and areas for maintaining and processing insect cultures. The most important rooms, and those common to essentially all insectaries, are insect-holding rooms, workrooms, and storage rooms. They will be discussed in detail. Other rooms often found in insectaries are discussed briefly at the end of this section.

1. Insect-Holding Rooms. An insect-holding room should be considered a refuge for the cultures, a place where the insects can remain relatively undisturbed under optimal environmental conditions. Each room should provide a barrier that prevents the intrusion of harmful contaminants from other areas of the building, while it contains escapees from cultures within its own walls. In general, insect-holding rooms should not be used for performing specific rearing procedures because of the risk of spreading contaminants. Workrooms with the proper contaminant-containment equipment and a more employee-friendly environment are better suited for these procedures.

a. Number of Holding Rooms. For multispecies rearing, more small rooms are better than fewer large ones. This offers greater flexibility in isolating populations and providing more varied environments. There should be enough insect-holding rooms to separate at least the following groups:

- Populations susceptible to the same diseases, such as closely related species or different strains of conspecifics (e.g., an insecticide-resistant strain and a susceptible strain of the same species).
- Species difficult to distinguish. For example, tobacco budworm and corn earworm, *Helicoverpa zea* (Lepidoptera: Noctuidae), eggs and larvae are difficult to differentiate. Mixing them up is possible if they are located in the same room.
- Adults and larvae of the same species that require different environmental conditions. Some adult Lepidoptera need 60–70% RH to maximize their reproductive potential, while their larvae require 45–50%. Putting adults and larvae in separate rooms also reduces the risk of losing the entire colony if there is an equipment failure in one of the rooms.
- Parasites, predators, or pathogens and the host colonies' reproductive population. Different buildings may be required for adequate separation in this situation.
- Colonies of phytophagous insects that feed on the same plants as other species in culture. For example, the two-spotted spider mite, *Tetranychus urticae* Koch (Acari: Tetranychidae), is an aggressive colonizer and will displace less competitive species such as the cotton aphid, *Aphis gossypii* (Homoptera: Aphididae). In an insectary with these species, not only should each colony have a separate room, but there should also be as much distance between the rooms as possible. In addition, employees should be instructed to work with the mite colony the last thing each day to prevent cross-contamination.

Backup holding rooms should be included in the design. They can be used to establish a clean secondary colony when the primary colony becomes contaminated, temporarily hold colonies when maintenance is performed on one of the permanent holding rooms, establish a small colony for research purposes so the main population is not jeopardized by the experiments, handle the overflow population when a colony is increased past normal production levels, and provide flexibility to the rearing program when changes occur in the future.

b. Size of Holding Rooms. The size of a holding room is determined by the amount of space required for the equipment that will hold the insects in addition to the amount of room required to access and move it. Consider, for example, a 1.5-m-long x 0.8-m-wide x 1.5-m-high rack unit that holds 288, 32-well larval rearing trays, which are separated vertically by 2.5 cm to allow adequate air circulation (Fig. 5). With an average of 2 insects per well, the monthly production is over 18,000 pupae. This rack unit is accessed only from the front so the back is placed against a wall. Additional space is required to "use" the rack. This includes 0.3 m on either end to move it for cleaning and another 1.2 m in front to operate carts and to insert and remove trays from the shelves. The total minimum space required for 18,000 pupae is 4.2 m^2 or a room 2.0 m wide x 2.1 m long. In this case, the insect-holding space is 28.6% of the available floor space.

The area required to produce any number of pupae per month can be calculated using this basic rack unit. For example, a yield of 100,000 pupae would require 6 larval rack units. The total space would be 14.3 m^2 of which 7.2 m^2 is insect-holding space. The room could be 2.8 x 5.1 m with three units located on each of the long walls. Here the insect-holding area represents 50.4% of the room space. Space efficiency increases because the additional area in front and on the sides of each unit is shared with the unit on the opposite wall and those adjacent to it.

If insect-holding units are placed in three rows of two units each, the room size would be 4.8 x 5.1 m or a total of 24.5 m^2. Insect-holding space drops to about 35% because two, 1.2-m work spaces are required in front of the units instead of just the one as in the longer room.

These are minimum room sizes and do not account for additional space needed for a bench, table, or sink. The area surrounding complex, stationary, or difficult to move rearing units should be increased to simplify the cleaning process and allow access by personnel and maintenance equipment. A 20–30% in-

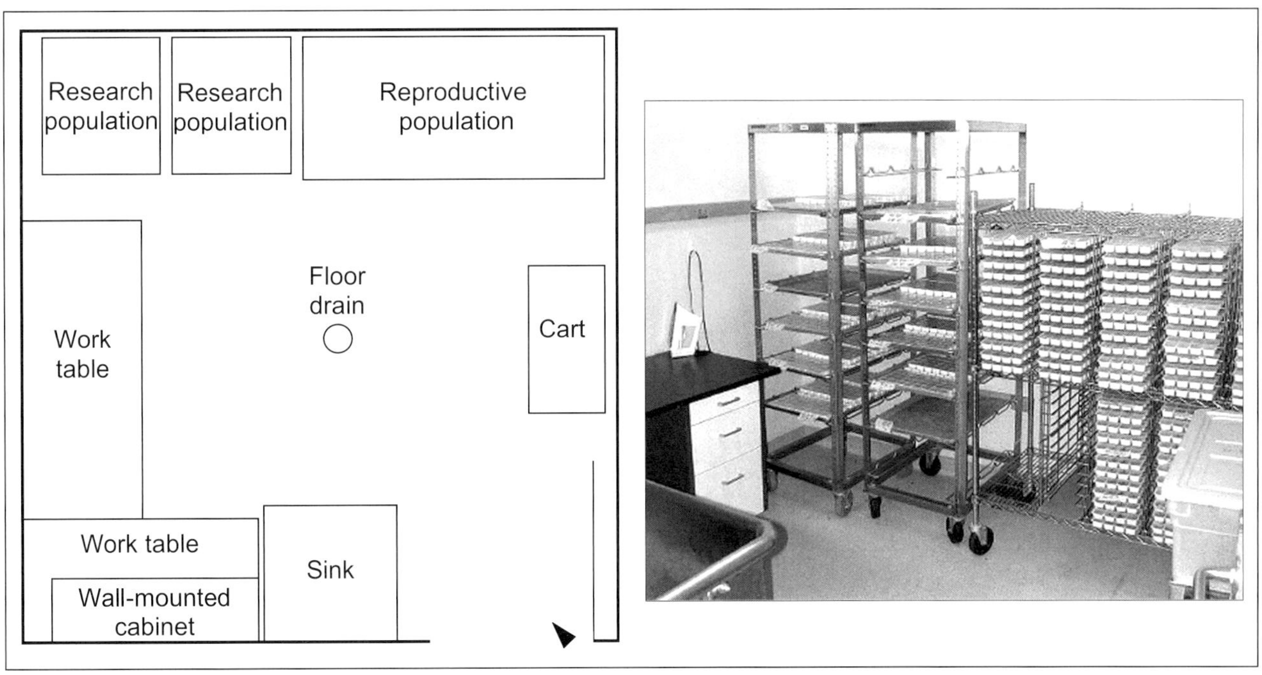

Fig. 5. Floor plan and photo of a 3.0 x 3.0-m room for holding Lepidoptera larvae fed artificial diet in 32-well trays. Trays are vertically separated by 2.5 cm to allow controlled moisture loss from the rearing wells and prevent the growth of mold. The solid black arrowhead indicates the direction from which the photo was taken.

crease in room size over immediate needs will improve flexibility. The additional space can be used to accommodate an increase in the insect population, implement improvements that may require more space (e.g., arrange racks and cages to optimize airflow patterns), and provide more flexibility if there is a change in use for the room.

Ceiling height should be 2.1 m to facilitate retrieval of insects in open-room cultures and to reduce the volume of conditioned air required to meet specifications (Fisher and Finney 1964). Ceilings at 2.4 m are adequate for other facilities; however a higher ceiling may be required to accommodate specialized equipment, including insect-holding racks and ceiling fans. Alschuler (1963) states that a minimum of three feet clearance is required between the bottom of the ceiling diffusers and the top of cages for proper air distribution in animal facilities.

2. Workrooms. These rooms are designed and equipped for performing specific rearing procedures in isolation. Access to a workroom should be restricted to the individuals who work inside. Environmental conditions best suited to employee comfort are maintained with a temperature of about 24°C and no control over humidity. Workrooms are classified as "clean" or "dirty" based on the type of procedures that is performed within.

a. Clean Workrooms. A clean workroom supports procedures that cannot be exposed to contaminants generated from other parts of the insectary or from outside the insectary. A room where agar-based artificial diet is prepared, dispensed, and allowed to cool is an example. Protecting the diet from exposure to contaminants during this period is an effective and necessary means of reducing undesirable microorganisms from the rearing system.

A clean room is isolated from the rest of the facility with access only by passing through at least two doors. Supply air is filtered through a HEPA system at 95–99% efficiency, and a positive air pressure is maintained relative to adjacent dirtier rooms or hallways. Wall and floor coverings should be durable and easy to clean. Prepared diet can be transferred to other parts of the insectary via a passthrough that protects the clean room from the intrusion of contaminants.

The diet-preparation room in Fig. 6 is an example of a clean workroom. From 100 to 280 liters of high-quality diet were routinely produced each week in this room (473 liters per week capacity) to support a multispecies rearing program. The room is 3.7 m wide x 11.0 m long. It has a 150-liter steam-jacketed kettle for mixing diet. There is a hood over the kettle to remove steam, dusts, and chemical fumes generated during mixing. A semiautomated diet dispenser is located next to the kettle at one end of a 6.1-m-long workbench. Diet is pumped directly from the kettle into rearing trays, which are stacked and moved to the opposite end of the table where they cool for several hours. Smaller batches of diet are prepared either in an 18.9-liter steam kettle or a 3.8-liter Waring® blender. Diet premix and prepared diet are stored in a 1.4-m^3 refrigerator. Other fixtures and equipment in

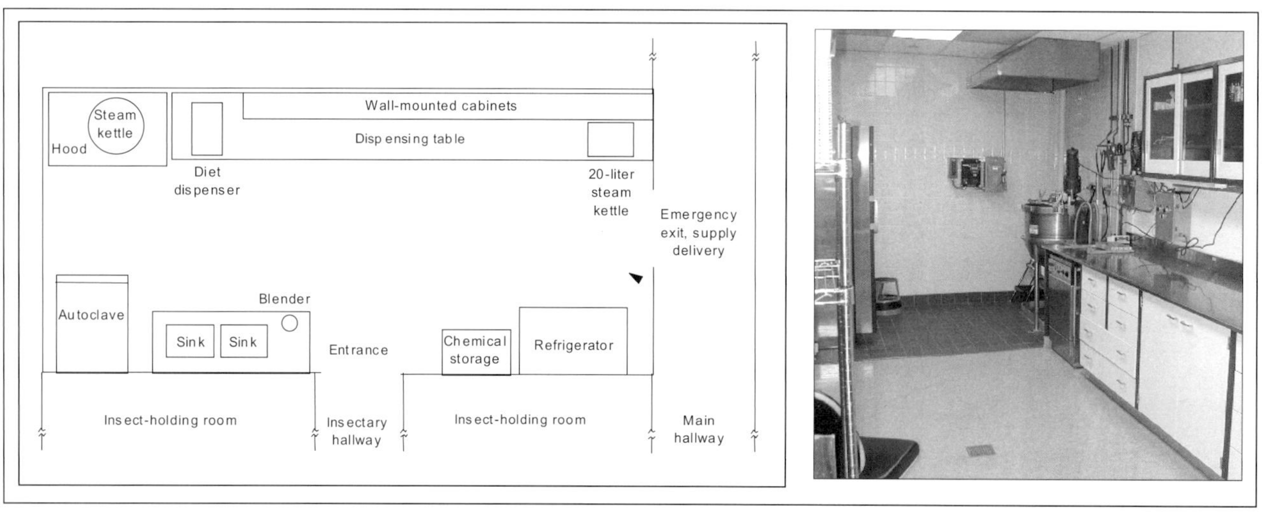

Fig. 6. Floor plan and photo of a "clean" workroom for preparing and dispensing artificial diet. The solid black arrowhead indicates the direction from which the photo was taken.

the room include a steel cabinet for storing corrosive dietary ingredients, a 2-bowl stainless steel sink with disposal, and an autoclave. There are two doors into this room: one leading to the hallway of the main facility and the other leading to the insectary hallway.

The room was monitored for microbial contamination for more than 6 months. Media plates exposed to room air during diet preparation (about 2.5 h) averaged 1.4 bacterial or fungal colonies per plate with as few as 0 and as many as 25 colonies. There is some risk of greater contamination when one of the doors is opened or the autoclave is being used. The door leading to the main insectary has magnetic seals that prevent insects and airborne organisms from entering when the door is closed. Positive pressure in the clean room forces air into the adjacent hallway when the door is opened. The door leading to the main corridor is used only for emergencies and delivering supplies. The clean-room autoclave is the only one in the insectary; and all trays, cages, and other equipment must be taken into the diet room for steam sterilization. This was a mistake in the insectary design, so protocols were developed to minimize the impact. They directed that reusable items be thoroughly washed outside the clean room and autoclaved on days when diet was not being prepared.

b. Dirty Workrooms. A dirty workroom is the site of procedures requiring handling of potentially significant sources of contaminants and is designed to prevent these contaminants from moving to other parts of the facility. These sources include oviposition cages and larval rearing containers, in which insects lived, fed, excreted, and perhaps died. The rearing containers also include used diet possibly contaminated with microorganisms. Harvesting pupae and washing and sterilizing reusable containers occur within this room.

Fig. 7 illustrates a dirty workroom from a 3- to 4-species insectary (148.6 m^2) in California (Fisher 1984). Pupae are transferred from adjacent insect-holding rooms directly into this 2.4 x 5.2-m room via passthroughs. Both passthroughs are equipped with an ultraviolet lamp that is illuminated for 15 min after each transfer to kill microbial contaminants inside. Pupae are separated from used diet in a fume hood that further isolates contaminants and protects workers from airborne debris. Reusable containers are washed in a large sink, sterilized in a double-door autoclave, if heat stable, and stored in a clean room adjacent to the diet-preparation room. Heat-labile reusables are washed in the sink, sterilized in a dunk tank containing sodium hypochlorite solution, rinsed, placed in the passthrough next to the autoclave, and stored in the clean room. Entrance to the dirty room is isolated from the rest of the facility, and access is restricted to only those individuals who work inside. Air pressure is neg-

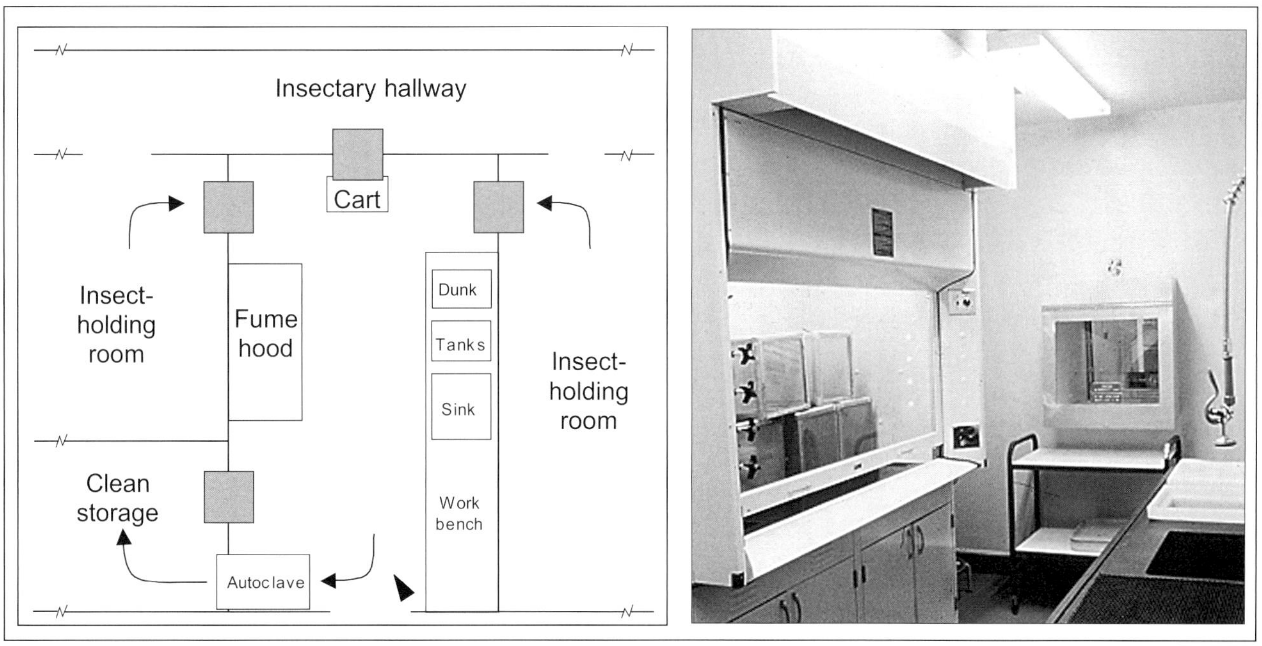

Fig. 7. Floor plan and photo of a "dirty" workroom designed to contain contaminants and prevent them from spreading to other parts of the insectary (Fisher 1984). See Sec. IV. D. 2 for a description of the traffic flow through this room. Passthroughs (▩). The solid black arrowhead indicates the direction from which the photo was taken.

ative relative to the rest of the facility, and the air leaving this room is discharged directly outside and not recirculated.

Separate clean and dirty workrooms are ideal, particularly for a facility that houses insects that feed on artificial diet and are susceptible to serious diseases. However, budgetary and space constraints may preclude having separate rooms. In this case, alternatives are available. For example, 1) a clean room can be used for diet preparation while a fume hood, located in another part of the facility, is used for dirty work; 2) a laminar-flow hood (outflow) can be used for clean procedures while a fume hood isolates dirty activities; and 3) a clean area can be established at one end of a room with an area for dirty work at the opposite end. With any isolation plan, activities should be scheduled so that work in the clean area is completed before beginning activities in the dirty area.

3. Storage Rooms. Storage rooms provide the space, appropriate environmental conditions, and security for holding materials used in the insectary. Diet ingredients, prepared diet, operating supplies, and equipment require the greatest amount of storage space.

Raw ingredients and prepared diet are usually stored in a freezer or refrigerator set between 1.7 and 4.4°C. Approximately 4.0 m^3 is required to store premixed raw ingredients and prepared diet for the weekly production of 70–80 liters. A two-month supply of premix, which is received from an outside source, can be stored in 1 m^3 of refrigerated space. If the diet were mixed on site, additional space would be required for storing containers of individual ingredients. A 1.4-m^3 double-door refrigerator can hold over 100 diet-filled, 32-well rearing trays (17.8 cm x 35.6 cm x 2.5-cm high) commonly used today.

A separate building for storing and processing diet ingredients may be required in larger operations to reduce contamination within the main insectary (Griffin 1984, Goodenough 1984). Such a building should be designed to expedite the movement of materials; keep the items dry and protected from rodents, insects, and birds; avoid temperature extremes; supply adequate ventilation for controlling dusts; and have a refrigerated area for perishable items.

Storage space for operating supplies and equipment is often underestimated. A 2- to 3-month reserve of supplies gives an adequate buffer for unexpected delays in the next shipment. Much of the equipment used in an insectary is unique or made of expensive materials and should be stored when not in use. These include "homemade" acrylic cages, stainless steel equipment, racks, and cages modified for specific purposes.

4. Other Rooms/Areas. Requirements for other insectary rooms are determined by the purpose of the program, its size, number of employees, quarantine or isolation requirements, and the insectary's autonomy from other buildings. In general, larger, more independent insectaries require additional rooms for support activities and employee comfort and services. The following are examples of rooms and areas that have been used in insectaries.

- *Work Stations*—For computers, printers, scanner, etc.
- *Visitor Display Area*—A space near the entrance to the facility where photographs of the rearing operations, insect exhibits, an outline of the facility, and other illustrations that summarize the purpose and operations of the program can be displayed.
- *Environmental Monitoring Center*—A central site where data is recorded for temperature, humidity, and other conditions within the facility.
- *Environmental Chamber Room*—A place to house multiple chambers to isolate them, along with their noise and heat, from the rest of the facility.
- *File Storage Room/Library*—For storage of current reports, data sheets, historical records, and literature. A desk or worktable should be included in this room.
- *Greenhouses/Head House*—Given that some phytophagous insects must be reared on plants, an in-house supply of the appropriate plant material may be a necessity. There should be a specific route for transporting plant material from the greenhouse to the insectary. This may require a covered walkway or corridor, especially in rainy or snowy locations. If the greenhouse and insectary are in close proximity, a barrier should be installed between them to reduce the movement of pests from one to the other.
- *Maintenance Shop*—An area for fixing equipment, making prototypes, etc. is important in some insectaries and essential in others. The need for this space, its size, and equipment required to operate it are determined by the type of rearing

program as well as its independence from other facilities on site that may already have these resources.

- *Packaging Area*—Staging area for preparation of insects to be shipped.
- *Quarantine Facility*—Very specific rules and regulations govern the construction and operation of a quarantine facility to handle exotic insect species. Early in the design phase, check with the U.S. Department of Agriculture, Animal and Plant Health Inspection Service (APHIS) or similar agencies in other countries for requirements.
- *Insect Receiving Area*—Insects received from field collections or other insectaries represent a potential for the introduction of disease or other possible hazards to established colonies and should be isolated from them. The receiving area should be equipped to temporarily rear the insects, independently of the primary rearing rooms, until the risk they pose to other colonies has been determined.
- *Vestibules*—Doors at either end of a vestibule permit an individual to enter through one door and close it before opening the other door into a room, thereby reducing the probability of contaminants entering or leaving the room.

E. Alternatives to Conventional Facilities

Prefabricated environmental growth chambers are often used to hold insect colonies (Leppla and Ashley 1978). Each unit is self-contained with the equipment needed to control temperature, humidity, and light. Walk-in units with a laminar airflow design for minimal temperature stratification and reduced airflow velocity are well suited for insect rearing. Reach-in chambers are used for maintaining small cultures or holding specific developmental stages of larger colonies. With a footprint about the size of a standard refrigerator, they can be located in most rooms or hallways in an insectary.

Lawson and Headstrom (1978) describe an inexpensive insectary for the production of nuclear polyhedrosis virus, a potent insect pathogen. Each room is an independent "building" that is physically separated from the others to help prevent the spread of virus. Prefabricated buildings (mobile homes) can be used for rearing (Brewer et al. 1978, Bell et al. 1981, Stewart 1984). However, Griffin (1984) states that these buildings are not suitable for large, permanent facilities.

F. Layout and Traffic Flow

After determining room requirements, the next step is to decide where each room should be located in the overall design. Rooms should be positioned so that the movement of materials, products, and employees follows a specific route that reduces the spread of contaminants (Fisher 1984, Fisher and Leppla 1985). The effectiveness of this traffic flow plan is enhanced by the use of sanitary transfer points, such as passthroughs and double-door autoclaves, where materials and products can be moved safely between rooms. A specific traffic flow plan is especially important for species that are susceptible to disease, produce considerable debris such as wing scales, and are reared in reusable containers. A plan can be developed around numerous rooms or can be created within one room.

1. Multiple Workrooms. A California insectary illustrates a traffic flow plan that was fundamental in the elimination of disease and dietary contamination from Lepidoptera cultures (Fisher 1984; Fig. 8). The 148.6-m^2 facility was divided into three major areas: clean rooms, insect-holding rooms, and a dirty workroom. The flow begins when sterile larval rearing containers are taken from the clean storage room to the diet-preparation room, filled with diet, infested with surface-sterilized eggs, and covered. These rearing units are then transferred to the main hallway via a passthrough and taken to the appropriate holding room where they remain until pupation.

After pupation, rearing units are put in one of the passthroughs leading to the dirty workroom where pupae are harvested (Sec. IV. D. 2. b). Insect-holding rooms 1 and 2 have direct passthrough access so that insects from these rooms can be transferred without exposing the rest of the facility to contaminants. Materials from other holding rooms enter the dirty workroom through the passthrough in the main hallway. In the workroom, larval rearing units are taken from the passthrough to the fume hood for pupa harvest. Reusable containers are washed, sterilized, and transferred into the clean storage room via the double-door autoclave or passthrough.

Pupae are placed in clean cups with tight-fitting lids and passed into the egg and pupa treatment room where they are washed in sodium hypochlorite to destroy surface contaminants. They then are placed in sterile cages, passed to the main hallway and taken to the appropriate adult holding room for eclosion and

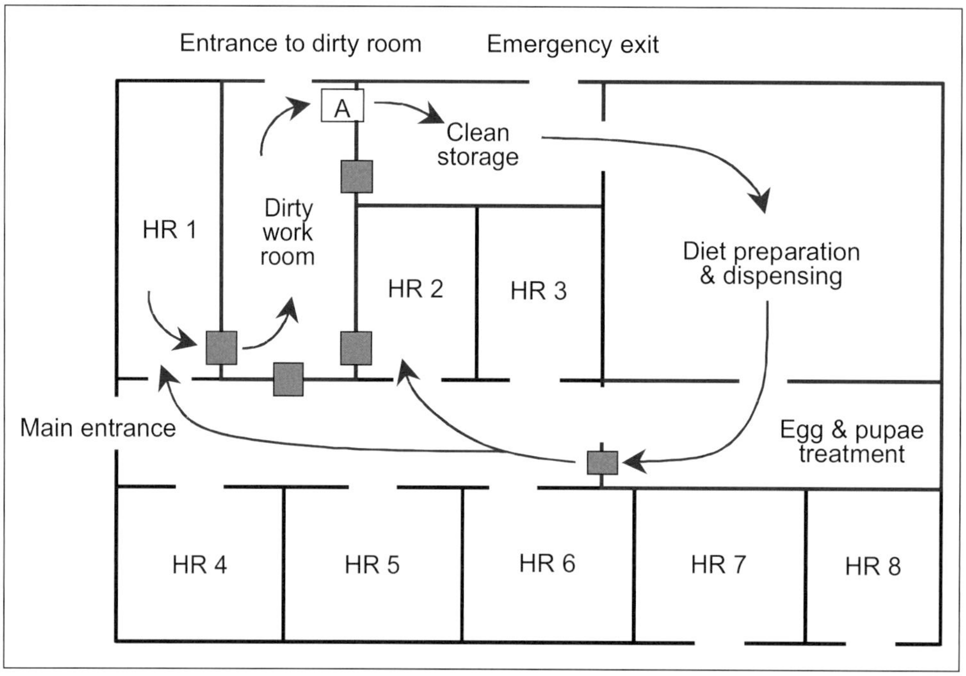

Fig. 8. Traffic flow plan for materials and products in an insectary with separate "dirty" and "clean" workrooms (Fisher 1984). Arrows indicate movement of insects, cages, and containers. Double-door autoclave (A); insect-holding room (HR); passthroughs (■).

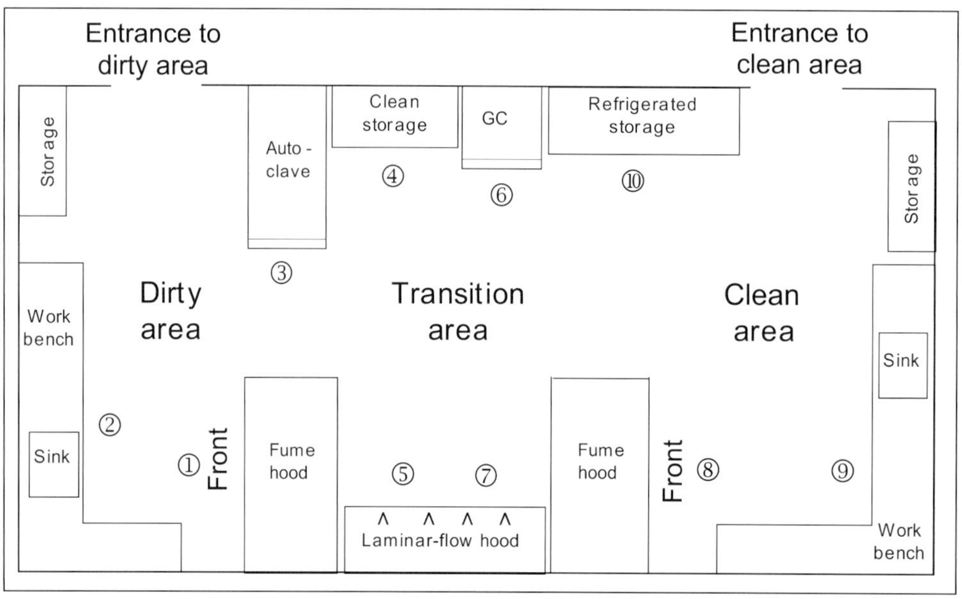

Fig. 9. Flow plan for the movement of insects in a single insectary workroom, in which both "clean" and "dirty" work areas have been incorporated. Circled numbers indicate equipment where various tasks are performed. All work in the clean area is completed before work begins elsewhere in the room. Growth chamber (GC).

oviposition. Egg sheets are collected from these cages, placed in clean plastic bags, and passed to the egg and pupa treatment room, disinfected in sodium hypochlorite, and set up on diet in larval rearing units.

2. Single Workroom. Not every program has the space or budget for a multiworkroom flow plan, but every program needs separation between different levels of contamination. A flow plan can be designed to fit into a single workroom, where both clean and dirty procedures are performed (Fig. 9). The room illustrated is 4.9 by 9.1 m with two entrances, one for the dirty area at one end and the other for the clean area at the opposite end. A "transition" area is located in between. The flow of materials is as follows (circled numbers correspond to locations within the room).

Eggs are collected from oviposition cages outside the workroom, placed in clean plastic bags, and taken to the fume hood ① in the dirty area where they are surface sterilized. Eggs are taken from the fume hood to the transition area where they are dried in the laminar-flow hood (outflow) ⑤. When dry, the eggs are put into cups and placed in the growth chamber ⑥ until hatch. Neonates are transferred to containers with fresh diet in the laminar-flow hood (outflow) ⑦ and taken to a larval holding area outside this room. Pupae are collected and surface sterilized in the fume hood ①, set up in clean cages from storage ④, and taken to an oviposition area outside this workroom.

Used rearing units are cleaned in the fume hood ①, washed at the sink ②, autoclaved ③, and stored in a cabinet ④ in the transition area. The same process is used when preparing adult oviposition cages for reuse.

Diet is prepared inside a fume hood in the clean area ⑧ and dispensed into containers on the adjacent workbench ⑨. Raw diet ingredients and dispensed diet are stored in the refrigerator ⑩.

The fume hood in the dirty area is used to surface sterilize eggs, harvest pupae, and clean reusable materials. Because it is used for more than one procedure, the hood must be thoroughly cleaned after each use. There are no physical barriers to restrict movement within this design, so an employee could inadvertently carry dirty larval rearing trays into the clean entrance and expose freshly prepared diet to contaminants. Therefore, protocols governing the movement of materials, insects, and employees must be developed and implemented.

Regardless of the degree of isolation or the complexity of the flow plan, incorporating these features into the facility design is most practical and effective at the earliest phase of the project. Success can only be achieved with a thorough knowledge and understanding of what operations and activities will be performed in each area.

G. Phytophagous Cultures

Phytophagous insects in culture may include aphids, mites, whiteflies, thrips, and other species maintained on living host material. Because arthropod pests can parasitize, prey upon, or compete with cultured, phytophagous species, contaminant control and isolation of colonies are of paramount importance.

Parasites and predators are a common concern in phytophagous colonies. For example, in one insectary, a predatory fly, *Aphidoletes aphidimyza* (Diptera: Agromyzidae), entered the main building through an outside door, flew into a poorly isolated holding room where green peach aphids, *Myzus persicae* (Homoptera: Aphididae), were maintained, and destroyed 30% of the aphid population. To save the colony, a parallel population was established in a backup room physically separated from the contaminated colony. Each plant taken into this room was inspected under a dissection microscope for predators. This process took several months and considerable additional effort.

In a facility where multiple phytophagous species are reared, individuals from one culture can pose a significant contaminant potential to another culture, particularly with colonies that thrive on the same plant host (e.g., thrips invading a whitefly culture; or twospotted spider mites infesting a peach aphid colony). These "pests" can adversely affect a pure culture by competing for food and space, particularly if they are aggressive colonizers with a life cycle shorter than that of the species being invaded. Physical isolation, vigilance, and time-consuming salvage procedures are often required to prevent the loss of colonies.

Arthropods that live and breed in soil also can be important contaminants in phytophagous cultures. In one program, fungus gnats (Diptera: Mycetophilidae), inadvertently brought into the insectary with plants from the greenhouse, invaded corn rootworm, *Diabrotica spp.* (Coleoptera: Chrysomelidae), pupation containers filled with rich, moist soil. The gnat population exploded in this medium, and flies quickly spread throughout the multispecies facility. They were found in larval rearing trays, adult cages, oviposition dishes, air-conditioning ducts, and inside a fire alarm box. It took months of intense trapping to finally rid the insectary of this pest.

Pathogens are usually of little concern in phytophagous cultures, so a sophisticated traffic flow plan is not required. However, colonies must be adequately isolated. Fig. 10 illustrates isolation techniques for such cultures. While phytophagous colonies are featured as examples, these techniques can be used for any species.

Closed, properly sealed doors in insect-holding rooms are the primary barriers. Additional means of isolation can be added. For example, a vestibule doubles the number of doors leading into Room 1, significantly increasing the degree of isolation. A sticky or black-light trap in the vestibule will capture arthropods and indicate how effective isolation is. Whiteflies might be reared in Room 1.

Reach-in chambers in Room 2 provide additional isolation for colonies held within them. Thrips or other small flying insects that take up little room can be kept in chambers like these. A door that divides the main hallway into two areas and doors at either end of the hallway provide barriers to the movement of contaminants from within the facility or from out-of-doors. For example, an aphid colony in Room 3 is better protected from parasites behind two or three doors than behind just one if Hallway A leads directly to an outside exit.

Distance is also an effective means to isolate colonies. It is risky to hold, for example, twospotted spider mites in Room 5 and an aphid colony feeding on the same host in Room 4. The close proximity may allow mites to reach the aphid plants and disrupt or destroy the less competitive homopterans. Greater isolation could be achieved by placing the aphids at least as far away from the mites as Room 3. Room 4 can then contain insect colonies such as cockroaches that would be unaffected by the intrusion of phytophagous mites. Mites can be further isolated by incorporating a second door in Room 5 that leads into Hallway B instead of Hallway A. The original door can be temporarily taped to prevent contaminants from escaping (or entering) Room 5. This affords the opportunity to employ operational isolation in which the primary mite caretaker maintains the colony without ever entering Hallway A. Two doors in a room such as 4 also provide the flexibility to accommodate program changes in the future.

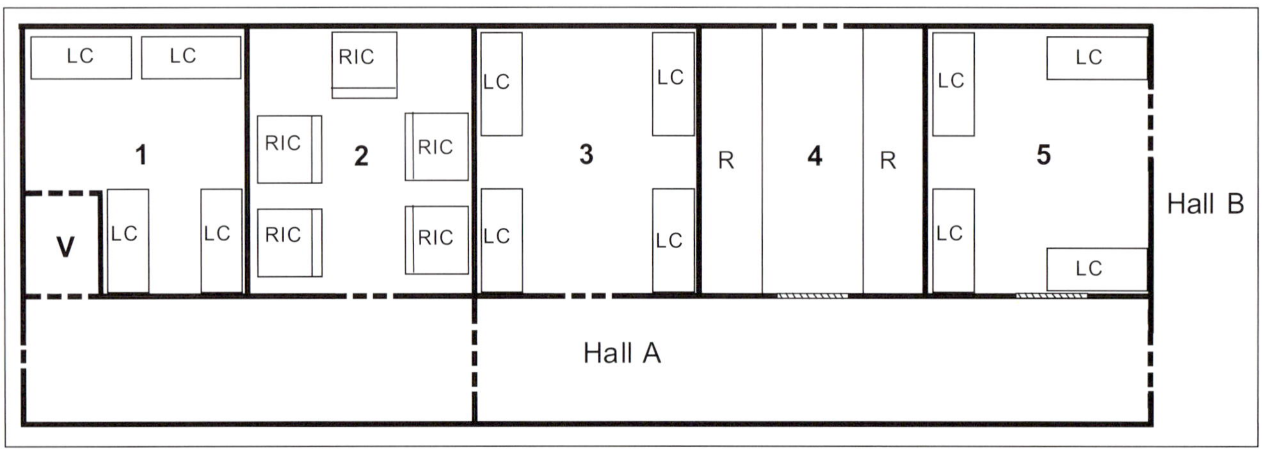

Fig. 10. Group of hypothetical insect-holding rooms for illustrating various techniques to isolate colonies. Techniques include a vestibule (V) that isolates Room 1 from Hallway A, reach-in chambers (RIC) in Room 2 that provide additional isolation for colonies held within them, a door that divides Hallway A into two areas, and installation of a second door in rooms 4 and 5. LC, light cart; R, rack for holding insects; dashed lines, primary door; rectangle with stripes, alternate door.

V. Large-Scale Insectaries: Unique Design Considerations

Mass-rearing programs have unique requirements that cannot adequately be addressed simply by scaling up a smaller operation. The facilities are large and expensive to build. They are usually highly automated, use artificial diet, require considerable resources, operate 24 hours a day, and have sophisticated methods development and quality control sections. Further, large insectaries have an impact on the environment, resources, and the people living around them. All of these considerations influence the design and operation of such facilities.

Many large-scale facilities are in operation today. There are facilities producing billions of egg parasites, *Trichogramma spp.* (Hymenoptera: Trichogrammatidae), per month in Russia and China (Greenberg et al. 1996). Hendrichs and Robinson (2003) list 14 insect species, including fruit flies, tsetse flies, moths, and the screwworm, reared at 22 biofactories throughout the world in support of projects involving the release of sterilized males. The weekly production from all these insectaries ranges between 2 and 3 billion males. Other facilities are listed on the website of the International Atomic Energy Agency (IAEA 2008).

A. Size

Mass-rearing programs require large facilities with the capacity to continuously produce the required number of insects. The main rearing facility for the screwworm in Chiapa de Corzo, near Tuxtla, Gutierrez, Mexico, is larger than 18,580 m^2, approximately the size of 2.5 football fields. Much of this building consists of a "secure" section that protects the screwworm colony from contaminants and prevents fertile flies from escaping. The secure section contains rooms for the various developmental stages of screwworm. For example, the primary larval rearing room is 3,530 m^2, the pupa initiation room is 576 m^2, and the adult holding room is 2,230 m^2 with a capacity of 600 cages, holding 36 million flies. A second adult holding room of similar size is located outside the main rearing facility for use at peak production times.

Facilities for large rearing programs include all support activities that midsize insectaries usually share with the parent organization. Thus, large programs must include a dependable source of water, wastewater treatment, utilities, solid-waste disposal, shipping and receiving, administration, motor pool, backup power, maintenance, security, and means to meet employee needs such as a cafeteria(s), restrooms, showers, and medical services.

Space for these support activities can be considerable. At the Tuxtla facility, for example, metal, woodworking, and machining shops take up 669 m^2, and the main warehouse is 5,388 m^2 or roughly the size of a football field. The methods development and quality control sections alone are about 1,300 m^2. Two 2.2-ha lagoons treat wastewater generated by the plant.

B. Minimizing Cost per Insect Produced

High production at low cost is critical for large-scale programs, particularly for parasites and predators that must be priced competitively with existing pest-control methods. To achieve this type of program, the focus is on keeping costs down, preventing insect failures in the factory and the field, and continuously incorporating new technologies to improve efficiency and lower costs further. Thus, automation, inexpensive artificial diet, methods development, and quality control are critical aspects for any large-scale program; and the design of the facility should reflect their importance.

1. Automation. Replacing or supplementing personnel with machines can reduce operating costs, increase throughput, improve consistency, and reduce human exposure to contaminants. Automation has been achieved in various programs for producing rearing containers, packaging, and diet and for egg and larva distribution, egg and pupa collection, and field release of insects (Nordlund 1998). The boll weevil production facility (Mississippi State, Mississippi) was designed to provide 10 million beetles per week and was replete with automated equipment for all aspects of the program (Harrell and Griffin 1981, Sikorowski et al. 1984). Nakamori et al. (1992) describes a 4,274-m^2 automated facility built to eradicate the melon fly, *Dacus cucurbitae* (Diptera: Tephritidae), from Okinawa. The plant can produce 200 million flies a week. Greenberg et al. (1996) discuss automation and rearing processes for *Trichogramma* spp. wasps in China, the former Soviet Union, western Europe, and the United States.

2. Diet. Feeding billions of insects is the most costly part of a mass-rearing program. Having a relatively simple and economical artificial diet whose use can be automated is critical to reducing rearing costs. Waage et al. (1985) states that for entomophage cultures an artificial diet helps reduce insect disease and problems of asynchrony that often occur when plants, hosts, and entomophagous insects must be reared simultaneously. In addition, such a diet permits a rapid increase of insect populations without the time lag associated with a similar increase of hosts.

In the mid 1980s when the screwworm facility in Mexico was operating at peak production of about 500 million pupae per week, 113,400 kg of dry ingredients were used to make over 821,000 liters of larval diet per week (Marroquin 1985). While the six-component diet was easy to make, the sheer volume dictated an automated system to handle, transfer, mix, and blend the ingredients and dispense the prepared diet.

Feeding costs can be reduced significantly either by replacing expensive ingredients or lowering their concentrations, incorporating efficiencies based on nutritional studies, or by implementing new technologies. The following example illustrates the impact that a single diet modification had on the cost and operations of the screwworm facility.

In 1986, a gelling agent was incorporated into the diet for mass production. It created a semisolid diet that had distinct advantages over the liquid diet used previously. Using the new diet eliminated acetate mats, required less diet to produce the same number of larvae, reduced labor (liquid diet was replaced every four hours), reduced the volume of liquid waste that needed to be treated, and eliminated the need to remove old diet before refeeding. The gelled diet resulted in 2% heavier pupae, 32% higher yields, 54% less diet, an 88% reduction in labor (50% reduction in work force from 231 to 108 workers per day and a reduction in shifts from 3 full shifts per day to 1 full and 2 partials), and savings of $1.45 million per year. In addition, spent diet created income potential as cattle feed and fertilizer (Taylor 1992).

The cost for conversion from liquid to gelled diet was estimated to be between $500,000 and $750,000, including changes in rearing pans, racks to hold the pans, larval collection procedures, transport of diet ingredients and prepared diet, and feeding stations. New mixers and dispensers were also required. However, savings from the new diet in the first year were enough to cover the conversion costs (Taylor 1992).

3. Methods Development. Methods development applies research results and new technologies to increase operational efficiency, reduce costs, and improve efficacy of released insects. This section of a mass-rearing facility is a scaled-down version of the primary rearing facility, which allows for testing of separate, smaller colonies under nearly identical conditions as in the primary facility without jeopardizing the primary colony. Also in this section, the population size of a new strain that will be mass produced is increased before being transferred into the main rearing area.

4. Quality Control. The quality control function is important in any size insect rearing facility, but it is particularly critical to large programs because 1) insect-control programs depend on regular releases of high-quality insects, 2) facilities of this size must operate by the most efficient and economical means with minimal insect failures, and 3) there is a considerable investment of resources that needs to be protected.

In the Mexican facility, the quality-control group consists of 20 employees who monitor the program throughout the rearing process and field release. They work in a basic wet lab with a large, walk-in environmental chamber for holding insects. The group takes 500 pupae from each cohort and evaluates longevity with and without food, percent adult emergence, sex ratio, sterility, and ability to fly. They also monitor dietary ingredients, such as the blood component, which can be contaminated. Pure bovine blood is specified from vendors, but blood from pigs, chickens, or horses occurs in some batches at a level that adversely affects fly production. In addition, ingredients are tested for moisture content, pH, solubility, particle size, and bacterial content.

The quality-control group also monitors environmental conditions (temperature, humidity, and lighting), microbial contamination within the facility, quality of water being discharged to the environment, and the effects of changes made in rearing procedures, diet, automation, and replacement of strains (Chap. 10 *The Basics of Quality Control for Insect Rearing*).

C. Site Selection

When selecting the site to construct a mass-rearing facility, consideration must be given to the impact it will have on the surrounding communities and environment. Meyer (1987) addresses many of the issues listed below.

1. Location Relative to Customers. Reliable overnight delivery companies make it possible to inexpensively ship small numbers of insects anywhere in the country as well as many foreign locations. In this case, location of the facility is not critical. For larger-scale production, close proximity to the release site(s) is desirable to reduce shipping costs and minimize damage to the insects during transit.

2. Skilled and Unskilled Labor. This issue is of particular concern to large production programs that employ hundreds or thousands of individuals. For example, it was difficult to find qualified technicians or individuals that could be trained as technicians within the Tuxtla vicinity to maintain the more sophisticated machinery. This may have slowed the mechanization of the operations. On the other hand, it was not difficult to find unskilled labor for the plant, which at one point employed 1,200 individuals, many of whom had a third or fourth grade education (Marroquin 1985).

3. Terrain. A level site is less costly to prepare for construction than a hilly site. However, Fisher and Finney (1964) suggested building an insectary underground or partially underground to reduce heating and cooling costs. Smaller insectaries can be located in the basement of larger buildings to realize similar cost savings.

4. Water Supply. Water is an important resource used for a variety of purposes in all rearing programs. The greatest amount of water is often used to prepare artificial diets. The water must be of consistent quality to minimize batch-to-batch variability (Reinecke 1984, Cohen 2004). At peak production in the Mexican screwworm facility, more than 2.6 million liters of water were used each day for employees (drinking and showering), rearing (preparing diets, cleaning, automated collection procedures, and laboratory use), firefighting, and grounds maintenance (Marroquin 1985).

5. Sewage Disposal. The considerable volume of water used in large-scale programs must be treated before it is released back to the environment. For example, the Mexican facility has two large lagoons to aerate and filter water before it is returned to the river (Meyer 1987).

6. Electrical Power. All insectaries require an uninterrupted supply of electricity with backup power supplied to important equipment, rooms, and chambers. At the Tuxtla site, a diesel generator produces 2.2 MW of energy when commercial electricity is unavailable. This is enough electricity to power more than 2,000 homes.

7. Highways. All insectaries must be accessible for employees and other individuals responsible for smooth insectary operations. Snow, ice, or other deterrents cannot prevent workers from tending the insects on a daily basis. In larger facilities, the roads must provide a safe and reliable means for trucks loaded with perishable insectary products to leave the factory at regularly scheduled times for the airport and other dispersal sites. Likewise, trucks with critical rearing supplies must be able to get to the facility several times each week.

8. Public Transportation. In Tuxtla, few employees live close to the facility and many rely on public transportation to get to and from work. Some facilities own buses or other vehicles to transport their employees.

9. Urbanized Areas. Odors from rearing processes, composted diet, and water-treatment lagoons may cause problems with local residents. The public may perceive as a hazard those insects that have escaped from the factory, particularly if they were "irradiated" or are "predators" or "wasps." Constructing the facility away from developed areas will minimize these issues.

10. Agricultural Areas. Fisher and Finney (1964) state that an insectary should be located at least 400 m from agricultural land because the buffer will reduce the potential for insecticide drift that could harm colonies, lower the risk of undesirable arthropods getting into the facility, and reduce the potential for escaped host species damaging crops.

11. Climate. Marroquin (1985) lists climate as an important criteria for site selection. The greater the disparity between ambient and desired conditions, the greater the energy use. In Tuxtla, the volume of air required to meet specific temperature and humidity needs for the 10 primary rearing rooms and the two hot rooms is over 54,000 m^3. The average temperature for the rearing rooms is 31.4°C while that for the two hot rooms is 76.7°C. If this insectary had been constructed in a four-season climate, energy costs may have been too costly.

12. Environmental Impact. Large rearing programs can have a significant impact on the environment through water use and treatment, solid waste disposal in landfills, incineration, composting, and attendant odors. An analysis may be required to determine what impact, if any, the proposed building would have on the local environment. Such an analysis was performed for the proposed construction of a new screwworm rearing facility in Pacora, Panama, that will provide a barrier of sterile flies at the Darien Gap along the border with Colombia (USDA/MRP/APHIS 2001). Examples of conclusions from that analysis include:

- Water taken from the Pacora River and used in the plant will be treated to meet US drinking water standards.
- The treated effluent will not endanger shrimp rearing downstream from the site.
- Sludge from settling lagoons will be removed, dried, and incinerated.
- The incinerator will be equipped to remove particulates and undesirable gases.
- Incineration ash will be disposed of in a landfill lined with a high-density barrier and concrete.
- Irradiation equipment will be regularly checked to comply with Nuclear Regulatory Commission regulations.
- Workers will be monitored for radiation exposure and will have regular physicals.
- Minority and low-income populations will not be affected disproportionately by the construction of this building.
- There will be no negative impact on endangered or threatened species.

D. Screwworm Insectary, Tuxtla, Mexico: An Example

The oldest large-scale rearing facility is located near Tuxtla, Gutierrez in southern Mexico. It has been in operation since 1976 to support the sterile-release program for the screwworm. At peak capacity in the mid 1980s, more than 500 million flies were produced each week. Production has fallen to approximately 100 million flies a week as eradication continues in Central America and some of the Caribbean islands. The history of the Tuxtla plant is long and filled with many examples of improvements in insect quality and production efficiency. It will be used to illustrate many of the unique requirements of large-scale rearing.

The main production facility is divided into two sections—the secure section and the nonsecure section (Fig. 11). These will be considered in turn.

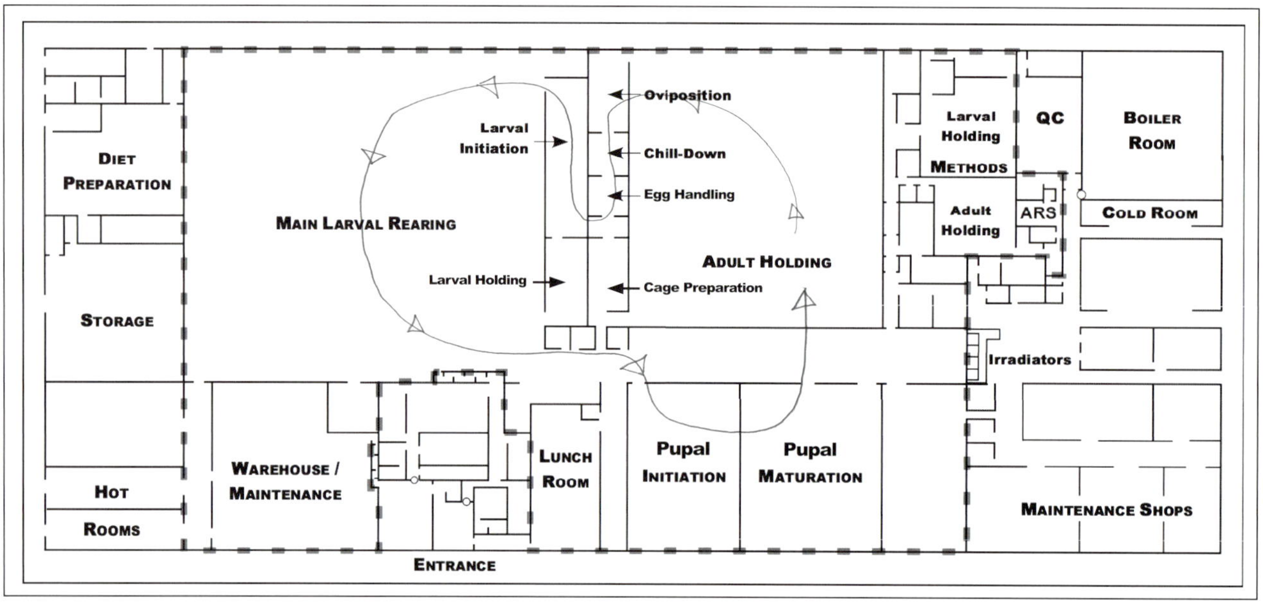

Fig. 11. Floor plan of the screwworm rearing facility near Tuxtla, Gutierrez, Mexico. The dashed line indicates the perimeter of the secure area, which prevents all nonirradiated insects from escaping to the outside. The general flow of insects occurs successively through the following rooms within the secure area: adult holding, oviposition, chill down, cage preparation, larval initiation, main larval rearing, pupal initiation, pupal maturation, and back to adult holding (for the brood colony). Pupae for field release are transferred via a passthrough into the nonsecure area where they are irradiated, taken to the cold room, and packaged for shipment. The secure section also contains methods development, which is separated into larval and adult holding areas. The nonsecure section includes maintenance shops, a boiler room, two "hot" rooms, storage, quality control, diet preparation area, and quality control room (QC).

1. Secure Section. This section of the Tuxtla facility consists of all rooms in which any stage of nonirradiated screwworm may be present. It is isolated from other parts of the facility and from outdoors to prevent fertile screwworms from escaping and attacking local animals. This isolation also keeps outside contaminants from getting into the rearing area and jeopardizing the health of the screwworms or reducing the efficiency of the operations. For example, shortly after initiation of production, houseflies, *Musca domestica* (Diptera: Muscidae), contaminated the screwworm culture and thrived on the diet. Significant resources were wasted inadvertently rearing houseflies until they were eliminated (Meyer 1987).

The main entrance is the heart of the security effort. It consists of a suite of rooms designed to prevent the movement of contaminants while allowing passage of personnel. Employees and visitors entering the area sign in at a guard station and leave any personal items in a locker. The guard then unlocks the first door leading either to the male or female locker room where clothing is removed and stored in a cabinet. After donning clean jumpsuits and boots provided by the facility, the guard unlocks another door to allow entrance into the secure area.

When returning from the secure section, individuals wash off their footwear in water and high-pressure air and pass through an air shower. They again remove all clothing and shower to remove any contaminants that may be present. All clothing worn in the facility and the towels used after showering are collected and laundered within the secure area. Finally, the guard unlocks the door leading back to his station where personal items are retrieved.

In the adult holding room, the temperature is maintained at 26°C, while humidity is 55%. Ceiling fans minimize stratification. Light fixtures are mounted upside-down to provide indirect uniform light. This reduces light gradients that can alter the behavior of the flies. The concrete floor is sealed and painted with orange lines to indicate aisles where cages are placed. This organizes the room and keeps all cages together from the same set up date.

Adult emergence begins about 24 h after pupae are put in the room. Mating begins in another 24 h and ends around 5 days later. Oviposition begins when females are about 7.5 days old . At that time, cages are rolled from the adult holding room into the 111.5-m^2 oviposition room. An oviposition device is placed inside each cage. It consists of a rectangular aluminum pan equipped with heating elements in the bottom and sides. An absorbent pad, soaked in spent larval diet, is placed in the bottom of the pan. Wooden boards, also treated with old diet, are placed on top of the pan's edges. Two 12-V incandescent lights are mounted in a hood, which is placed over the pan to provide dim lighting. When all cages are outfitted with these devices, the room lights are turned off. The dim lights, heat from the aluminum pan (34°C), and odor of the spent larval diet attract females that lay masses of eggs near the upper edges of the wooden boards. Oviposition occurs for 4 h.

Two-tube, electrocution-type black-light traps hang from the ceiling of the oviposition room to control escapees. A series of duplex receptacles along the walls several feet off the floor provide electricity for the oviposition units.

At the end of 4 h, the cages are rolled into the adult chill-down room, which is maintained between 2 and 4°C. Flies quickly become immobilized and the egging device is removed from the cage. Caged flies remain in the cold room for 24 h and are killed when the cages are washed.

In the egg preparation room, eggs are scraped from the wooden boards with a spatula, divided into 1 g groups (~ 20,000 eggs), and the groups placed on pieces of paper toweling. One such group is then put in each rearing tray on the surface of 3 liters of diet and held on a rack in the larval initiation room. There are 22 trays per rack. This room is maintained at 38°C and 90% relative humidity. The eggs hatch within 12 h. After 32 h, racks are moved to the diet dispensing area (in the main larval rearing room next to the wall of the diet preparation room) where each tray receives another 7 liters of fresh diet. The racks are then taken into the larval holding room, which is maintained at 37°C and 60% relative humidity. Larvae feed for about 20 h; and when they reach a weight between 5.5 and 7.5 mg each, six more liters of fresh diet are dispensed into the rearing trays. The racks are then wheeled into the main larval rearing room.

The main larval rearing room is the largest room within the secure area, accounting for about 3,530 m^2. It is divided into more than 20 lanes defined by central "humps" that act to properly align the larval rearing racks between U-shaped concrete trenches that run the entire length of each lane on both sides, similar to the gutters in a bowling alley.

After about 16 h, mature larvae begin crawling out of the rearing trays and drop down into the trenches to the sides of the rack. Larvae are scooped up six times a day, placed in a utility truck, and wheeled to the pupal initiation room. Racks remain in the main rearing area for another 2 to 3 days until most larvae have crawled off. Racks are then taken to the washing area where a semi-automated steam process removes spent diet, remaining larvae, and other debris from the trays and racks.

In the 576-m^2 pupal initiation room, larvae are scooped up from the utility truck in 3-liter units (~30,000 larvae) and placed in trays with sawdust for pupation. After 24 h, pupae are mechanically separated from the sawdust and transferred through a large pipe opening into the pupal maturation room. Three liters of young pupae are put into screen-bottomed trays that are placed in racks hanging from a ceiling-mounted monorail. After 5 days, about 5% of the pupae produced are returned to the adult holding room for colony maintenance.

Pupae to be irradiated remain in the maturation room slightly longer until they are about 12 h from adult eclosion. Forty thousand pupae are then placed in a canister and transferred through the wall of the secure area directly into one of four irradiators where they are treated with 5,000 rads of cesium-137 for about 2.5 min.

Other areas within the secure section include methods development, a maintenance shop, storage, several offices, lunch room, and laboratories for the Agricultural Research Service personnel who collect, maintain, cross, and develop screwworm strains that are best suited for the release program.

2. Nonsecure Section. The remaining space in the main insectary consists of the nonsecure section where there are no limitations on the movement of personnel. Diet is prepared in a 501.6-m^2 room adjacent to the main larval rearing room within the secure area. This proximity allows for diet to be pumped through the wall for dispensing directly into rearing trays. This room contains equipment that automates the preparation process. An adjustable steel platform is used for raising bags of ingredients to the proper working height. Equipment then picks up a bag and empties it into a conveyer leading to a conical mixer. Two homogenizers then mix all ingredients before the completed diet is pumped into the secure area.

Other rooms within the nonsecure area include cold storage for diet ingredients and shops for metalworking, woodworking, and machining. These three shops total about 668.9 m^2 and can make essentially anything that is required for the program. Also included are a small warehouse, the pupa irradiation site with four Husman irradiators, a cold room for storing treated pupae prior to packaging, quality-control laboratories, an equipment room with boilers, and several offices.

Two "hot rooms" provide a sanitary means to move materials in and out of the secure section. Each is 167.2 m^2 and held at 76.7°C. A tow truck can enter the rooms through roll-up doors at either end. These rooms are used as "ovens" to kill any organisms coming into or out of the secure area. For example, used materials such as spent sawdust being transported out of the secure area remain in one of these rooms for at least 24 h to kill screwworms that may be present. In addition, large water tanks located near the ceiling of these rooms provide hot water for showers.

Other on-site features include a second adult emergence room similar in size to that in the main insectary; an electric generator for backup power; an equipment room housing additional boilers; a warehouse for storing equipment, diet ingredients, and other supplies; a motor pool; a two-story administration building; a medical office; a waste water treatment plant; a compost facility; and a security station at the entrance to the site.

VI. Conclusions

Designing a complex building such as an insectary can be a daunting task. The process begins by identifying the functions that the facility will be expected to perform. Technical requirements necessary to achieve the functions are then established by the owner and presented to the design team. The design and construction phases proceed with fewer disruptions if all people involved in the project have a clear understanding of these requirements.

One key to a successful facility lies in selecting the appropriate facility components and effectively incorporating them into the design. The air-handling system maintains the environmental conditions critical to successful rearing. There is a wide selection of air-handling equipment and systems available, and the owner and members of the design firm must communicate effectively to develop the most appropriate one. Insectary rooms isolate insects and activities, and their location within the design establishes a traffic flow pattern that allows the movement of employees, insects, and supplies throughout the building while preventing the spread of contaminants.

The needs of large-scale rearing facilities cannot be met simply by "gearing up" a smaller operation. Factors that make them different include operational costs, building size, automation, simplified diet, and the need for strong methods development and quality control sections. The process of selecting the site for a large-scale facility must include consideration of local labor availability, terrain, water supply, sewage disposal, electrical supply, and climate.

Mention of distributors and their products is made only for the purpose of providing examples and does not constitute an endorsement or a recommendation by the author.

Acknowledgements

I thank members of the Mexican American Commission for the Eradication of Screwworm for discussions about the eradication program and rearing facilities: John Welch, John Spencer and his staff, and Henri Charpentier at the Tuxtla, Gutierrez facility in Mexico and Steve Skoda in Panama. I would also like to thank Hassan Oloumi, Nigel Armes, and Norm Leppla for their helpful suggestions and Frank Davis for the opportunity to participate in his vision of an insect-rearing workshop.

References

Alschuler, J. H. 1963. Air treatment in research animal housing. Lab. Animal Care 13: 321–331.

ASHRAE Handbook. 1987. Heating, ventilating, and air conditioning systems and applications. American Society of Heating, Refrigerating and Air Conditioning Engineers, Atlanta, GA.

ASHRAE Handbook. 2000. Systems and equipment. American Society of Heating, Refrigerating and Air Conditioning Engineers, Atlanta, GA.

Austin, P. R. 1994. Design and operation of pharmaceutical bio-cleanrooms and aseptic areas. Contamination Control Seminars, Livonia, MI.

Bell, R. A., C. D. Owens, M. Shapiro, and J. R. Tardif. 1981. Mass rearing and virus production, pp. 599–655. *In* C. C. Doane and M. L. McManus (eds.), The gypsy moth: research toward integrated pest management. Forest Service, SEA, APHIS Tech. Bull. 1584.

Brewer, F. D., C. W. Gantt, and D. F. Martin. 1978. Media-preparation and brood-colony facility, pp. 72–75. *In* N. C. Leppla and T. R. Ashley (eds.), Facilities for insect research and production. USDA, SEA Tech. Bull. 1576.

(CDC) Centers for Disease Control. 1994. Guidelines for preventing the transmission of *Mycobacterium tuberculosis* in health-care facilities, 43 (RR13).

Cohen, A. C. 2004. Insect diets—science and technology. CRC Press, New York, NY.

Davis, F. M., and J. N. Jenkins. 1995. Management of scales and other insect debris: occupational health hazard in a lepidopterous rearing facility. J. Econ. Entomol. 88: 185–191.

Debolt, J. W., and M. A. Peterson. 1978. Rearing facility for vegetable and sugarbeet insects, pp. 64–66. *In* N. C. Leppla and T. R. Ashley (eds.), Facilities for insect research and production. USDA, SEA Tech. Bull. 1576.

Etzel, L. K. 1978. University of California Quarantine Facility, Albany, pp. 45–46. *In* N. C. Leppla and T. R. Ashley (eds.), Facilities for insect research and production. USDA, SEA Tech. Bull. 1576.

Fisher, T. W. and G. L. Finney. 1964. Insectary facilities and equipment, pp. 381-401. *In* P. DeBach (ed.), Biological control of insect pests and weeds. Reinhold, New York, NY.

Fisher, W. R. 1984. Production of insects for industry: the Dow Chemical rearing program, pp. 234-239. *In* E.G. King and N.C. Leppla (eds.), Advances and challenges in insect rearing. USDA, ARS, U.S. Government Printing Office, Washington, DC.

Fisher, W. R., and N. C. Leppla. 1985. Insectary design and operation, pp. 167–183. *In* P. Singh and R.F. Moore (eds.), Handbook of insect rearing, vol. 1. Elsevier: New York, NY.

Goodenough, J. L. 1984. Materials handling in insect rearing, pp. 77–86. *In* E. G. King and N. C. Leppla (eds.), Advances and challenges in insect rearing. USDA, ARS, U.S. Government Printing Office, Washington, DC.

Greenberg, S. M., D. A. Nordlund, and E. G. King. 1996. Mass production of *Trichogramma* spp.: experiences in the former Soviet Union, China, the United States and western Europe. Biocontrol News and Info. 17: 51N–60N.

Griffin, J. G. 1984. Facility and production equipment, pp. 11–52. *In* P. P. Sikorowski, J. G. Griffin, J. Robeson, and O. H. Lindig (eds.), Boll weevil mass rearing technology. University Press of Mississippi, Jackson, MS.

Griffin, J. G., and O. H. Lindig. 1978. Pilot facility for mass rearing of boll weevils, pp. 75–78. *In* N. C. Leppla and T. R. Ashley (eds.), Facilities for insect research and production. USDA, SEA Tech. Bull. 1576.

Hampson, M. C. 1978. Controlled-environment room, pp. 30–32. *In* N. C. Leppla and T. R. Ashley (eds.), Facilities for insect research and production. USDA, SEA Tech. Bull. 1576.

Harrell, E. A., and C. W. Gantt. 1984. Automation in insect rearing, pp. 74–76. *In* E. G. King and N. C. Leppla (eds.), Advances and challenges in insect rearing. USDA, ARS, U.S. Government Printing Office, Washington, DC.

Harrell, E. A. and J. G. Griffin. 1981. Facility for mass rearing of boll weevils: Engineering aspects. USDA. SEA. ATT. AAT-S-19.

Harrell, E. A., W. D. Perkins, and B. G. Mullinix, Jr. 1979. Environmental rooms for insect rearing. Trans. Am. Soc. Agric. Engineers 22: 922–925.

Hendrichs, J., and A. Robinson. 2003. Sterile insect technique, pp. 1074–1079. *In* V. H. Resh and R. T. Carde (eds.), Encyclopedia of insects. Academic Press, San Diego, CA.

(IAEA) International Atomic Energy Agency. 2008. URL: http://www-naweb.iaea.org/nafa/ipc/index.html

Klassen, W., and G. Gentz. 1971. Temperature-constant and temperature gradient-free insectary: design and operation. J. Econ. Entomol. 64: 1334–1336.

Lawson, F. R., and R. L. Headstrom. 1978. Small plant for production of *Trichoplusia* NPV, pp. 37–39. *In* N. C. Leppla and T. R. Ashley (eds.), Facilities for insect research and production. USDA, SEA Tech. Bull. 1576.

Leppla, N. C., and T. R. Ashley [eds.]. 1978. Facilities for insect research and production. USDA, SEA Tech. Bull. 1576.

Marroquin, R. 1985. Mass production of screwworms in Mexico, pp. 31–36. *In* Symposium: eradication of the screwworm. Misc. Publication # 62. Entomological Society of America, College Park, MD.

Meyer, N. L. 1987. History of the Mexico-United States Screwworm Eradication Program. Vantage Press, New York, NY.

Nakamori, H., H. Kakinohana, and M. Yamagishi. 1992. Automated mass production system for fruit flies based on the melon fly, *Dacus cucurbitae* Coquillett (Diptera: Tehpritidae), pp. 441-454. *In* T. E. Anderson and N. C. Leppla (eds.), Advances in insect rearing for research and pest management. Westview Press, Oxford.

(NIH) National Institutes of Health. 1999. Vivarium design policy and guidelines. Division of Engineering Services. URL: http://des.od.nih.gov

Nordlund, D. A. 1998. Capacity and quality: keys to success in the mass rearing of biological control agents. Natural Enemies of Insects 20: 169–179.

Nordlund, D. A., and S. M. Greenberg. 1994. Facilities and automation for the mass production of arthropod predators and parasitoids. Biocontrol News and Info. 4: 45N–50N.

Owens, C. D. 1984. Controlled environments for insects and personnel in insect-rearing facilities, pp. 58-63. *In* E.G. King and N.C. Leppla (eds.), Advances and challenges in insect rearing. USDA, ARS, U.S. Government Printing Office, Washington, DC.

Reinecke, J. P. 1984. Nutrition: artificial diets, pp. 391–419. *In* G. A. Kerbut and L. L. Gilbert (eds.), Comprehensive insect physiology, biochemistry, and pharmacology. Pergamon, Oxford.

Sikorowski, P. P., J. G. Griffin, J. Roberson, and O. H. Lindig. 1984. Boll weevil mass rearing technology. University Press of Mississippi, Jackson, MS.

Singh, P. and R. F. Moore [eds.]. 1985. Handbook of insect rearing, vols. 1 & 2. Elsevier, NewYork, NY.

Stewart, F. D. 1984. Mass rearing the pink bollworm, *Pectinophora gossypiella*, pp. 176–187. *In* E.G. King and N.C. Leppla (eds.), Advances and challenges in insect rearing. USDA, ARS, U.S. Government Printing Office, Washington, DC.

Taylor, D. B. 1992. Rearing systems for screwworm mass production, pp. 393–403. *In* T. E. Anderson and N. C. Leppla (eds.), Advances in insect rearing for research & pest management. Westview Press Boulder, CO.

(USDA/MRP)/APHIS) United States Department of Agriculture/Marketing and Regulatory Programs/Animal and Plant Health Inspection Service. 2001. Panama Cooperative Screwworm Eradication Program: environmental analysis, July 2001. URL: http://www.aphis.usda.gov/ppd/es/pdf%20files/screwea.pdf

van Lenteren, J. C., M. M. Roskam, and R. Timmer. 1997. Commercial mass production and pricing of organisms for biological control of pests in Europe. Bio. Control 10: 143–149.

Waage, J. K., K. P. Carl, N. J. Mills, and D. J. Greathead. 1985. Rearing entomophagous insects, pp. 45–66. *In* P. Singh and R. F. Moore (eds.), Handbook of insect rearing, vol. 1. Elsevier, New York, NY.

Wang, S. K. 2001. Handbook of air conditioning and refrigeration, 2nd ed. McGraw-Hill, New York, NY.

Wirtz, R. A. 1980. Occupational allergies to arthropods—documentation and prevention. Bull. Entomol. Soc. Am. 26: 356–360.

Wolf, W. W. 1984. Controlling respiratory hazards in insectaries, pp. 64–69. *In* E. G. King and N. C. Leppla (eds.), Advances and challenges in insect rearing. USDA, ARS, U.S. Government Printing Office, Washington, DC.

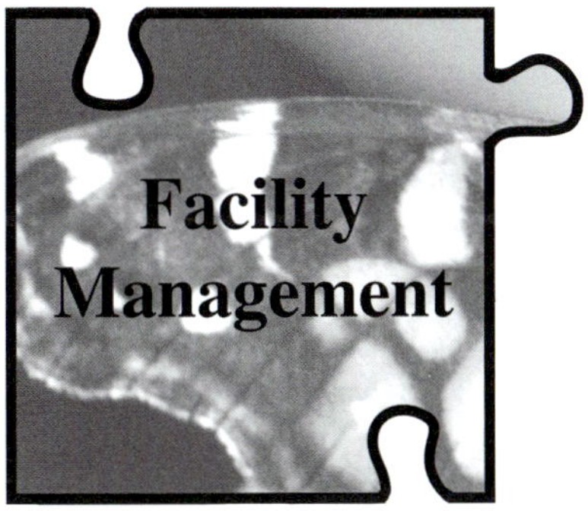

Chapter 3

The Insectary Manager

WILLIAM R. FISHER
Principle Research Scientist
Team Leader, Insect & Plant Propagation
BASF Corporation
Research Triangle Park, NC

I. Introduction

On-the-job training for new insectary managers is valuable experience, but with limited guidance available, this method has resulted in the early end of many potentially successful careers, either through frustration or poor performance associated with uncertainty about the job and how it is to be performed. This chapter will fill that void by presenting practical guidelines for insectary management.

Leppla (1984) and Singh and Ashby (1985) describe the various components of insectary management systems and how these components are interrelated. Leppla (1984) outlines a systems-management approach for large-scale insect rearing operations in support of insect population suppression programs. Such a program includes the functions of identification, collection, and colonization of the appropriate insects from the field, continuous rearing, treatment (e.g., irradiation), transport to the use site, and deployment. It also includes quality control involving the monitoring and evaluation of these functions in order to provide feedback about the program and suggest improvements. Finally, it includes elements such as proper management style, training, and communications to achieve the desired program over time. The systems approach to management of insect rearing on any scale is based on integrating all aspects of the rearing program rather than considering the operation as an aggregation of smaller, more or less independent units. This approach has many advantages, including improved communication and pooling the talents of program personnel.

Singh and Ashby (1985) introduced the concept of insect rearing management (IRM), which consists of the following seven major elements: objectives, colony establishment, design of the rearing laboratory, research and the development of techniques, resources, quality control, and production. All elements are linked together and modified as needed through management. The authors give an example of IRM using their codling moth program with a section on an evaluation of the process from receipt of an order for insects through supply to the customer.

The insectary manager's responsibilities include linking technology, production, and user groups, and managing available resources to achieve program objectives (Schwalbe and Forrester 1984). The authors discuss management of time, space, personnel, materials, and funds; the development of operating plans; monitoring insect quality; and identifying areas for improvement. The authors conclude that if the manager does not fully appreciate the broad scope of his or her responsibilities, day-to-day problem solving will be required at the expense of longer-term planning and smoother operations.

Fisher (1984) states that the manager's role includes training employees, scheduling daily activities to minimize equipment conflicts and contamination potential, evaluating insect quality, and taking a proactive approach to identify and solve problems. The manager should view the program as a dynamic system that evolves as new technology and techniques are developed.

The manager takes a more expansive view of the program than other insectary personnel who focus their efforts on specific rearing activities for which they employ technical skills and task-related management. Their view of the program is usually short term and more narrowly focused. The manager is involved with the general, nontechnical activities that keep the program operating smoothly and aligned with customer goals. Basic to this effort, the manager maintains a longer-term vision of what the program should be and provides the leadership to guide it toward fulfilling that vision. Thus, success for an insectary manager is not measured by how well he or she performs specific technical tasks but by how effectively the team accomplishes their responsibilities in support of the organization's goals.

The manager's broader participation incorporates additional responsibilities such as procuring and organizing resources, solving administrative and technical problems, making informed decisions, planning and scheduling activities, and providing ideas to improve insect quality and program efficiency. He also represents the rearing group to management, customers, and peers; adapts to changing circumstances in ways that support program goals; builds and maintains the insectary team; supports employees; and develops a network of individuals who assist in program success.

Working with and through other people to accomplish the objectives of both the organization and its members is fundamental to the management process (Montana and Charnov 2000). Thus, successful managers require effective interpersonal skills, including communication, leadership, delegating, coaching, team building, work management, and customer focus skills. Other useful competencies are adaptability, analytical thinking, creativity, initiative, quality orientation, and strategic thinking. Conceptual skills are used for planning, anticipating problems, and identifying obscure resources. The need for technical skills varies with the size of the program. Generally, the larger the rearing operation, the more the manager's responsibilities will differ from the technical tasks of other personnel.

II. Leadership

A leader inspires and motivates members of the insectary team and guides them in a direction that will accomplish the goals of the rearing program. Leadership requires a vision of where the program is headed and what resources are required to move it in that direction. This vision is formulated through an understanding of customer goals and the role the insectary plays in helping the customer successfully achieve those goals. It is a blueprint for the future and is used in the planning process to create goals, objectives, and procedures that drive the program and keep it aligned with customer requirements. This process is discussed in Sec. IV.

For implementation, the vision must be effectively communicated, through both words and actions, to other members of the insectary network. It is reinforced regularly as the manager guides the team, organizes resources, solves problems, makes decisions, facilitates rearing operations, and helps prioritize activities.

A. Leadership Qualities

1. Adaptability. A manager must be adaptable and perform effectively under changing or uncertain circumstances. Upset is inevitable and some, if not all, employees will require guidance to ensure their job gets done properly during these times. The manager's response should be consistent with program goals and provide stability, reassurance, and confidence for the staff.

However, change, if managed properly, can bring improvements in insect quality and program efficiency. Constructive changes are deliberate and calculated modifications to improve the program. They are thoroughly and systematically evaluated before implementation to minimize negative effects on the program. Evaluation should include adequate experimentation using a parallel colony or test population, rather than the entire population, to avoid endangering the culture. When planning for change, the manager should ask questions such as: What are the benefits and the risks if the change is made? Is the change worth making if the potential benefits are minimal? What long-term consequences may be expected? Are resources, space, and time available to make the change? To maintain it? What impact will the change have on other tasks?

2. Delegation. Delegation is the process of assigning a specific task to another insectary employee. For the manager, the ability to delegate allows more time to accomplish his or her primary responsibilities. For the assignee, it increases knowledge, experience, and confidence. The process includes defining the job in unambiguous terms, selecting the individual best suited for the job, training the employee, and making the necessary resources available. The manager gives the assignee the authority to perform the job, holds him or her accountable for the results, shows trust and confidence in the individual, and follows up to ensure that the process proceeds as intended. Handling a crisis, employee performance reviews, personnel issues, or setting program priorities are not jobs that should be delegated. On the other hand, delegated jobs should be more than trivial activities so they pique interest and make the experience rewarding for the employee.

3. Networking. The ability to build a good relationship with insectary employees is an essential responsibility of the manager. Establishing beneficial relationships with individuals outside the insectary is just as important. These individuals include customers, vendors, and depending on whether the rearing program is part of a larger organization, persons in human resources, purchasing, shipping and receiving, and maintenance. A tour of the facility and discussions about rearing in general will increase understanding of what is involved in insectary operations. For example, representatives from human resources will better appreciate the competencies required for new employees and will be more effective at screening applicants. Purchasing agents will become more familiar with insectary equipment and operating supplies, enabling them to more intelligently discuss rearing needs with suppliers. Maintenance personnel will gain a better understanding of the importance of environmental conditions and their relationship to successful rearing. The network of outside contacts should include individuals with experience in management, insect pathology, nutrition, and other relevant areas.

B. Leadership Styles

The manager should choose a leadership style that is most appropriate for the circumstances. The autocratic or dictatorial style is used when employee experience is limited, as with employees new to insect rearing or those learning a different job within the insectary. This style is also useful during a crisis when quick action and coordinated effort are required. However, the dictatorial style is not appropriate for normal, day-to-day management and can create resentment among employees because it implies they are unable to accomplish their tasks successfully without managerial control.

At the other end of the spectrum is a "hands-off" style that essentially lets the employees manage operations. For this approach to work, employees must be well trained and have the confidence and experience to make decisions. This style is appropriate when, for example, the manager has little experience or prefers to work in his office and remain isolated from operations. However, this is a dangerous long-term style because an aloof manager may miss signs of impending problems. Managing the production of biological organisms requires an engaged individual who regularly observes the colonies, facility, employees, and direction of the program.

Between these two extremes is a style in which the manager and employees team up to solve problems and guide the program together. A team is a group of employees who work together to achieve common goals, share the workload, exchange information, support one another, and recognize the value of teammates. A team in a small or mid-size program consists of the entire rearing staff, while a group organized around a specific function, problem, or project may constitute a team in larger operations.

The team approach brings additional experience and knowledge, as well as different perspectives, to bear on issues. Usually more ideas are generated by a team than by the same individuals working separately because each member benefits from the knowledge, work, and support of the others. Getting everyone involved motivates them and gives them a greater personal stake in making the operations successful. Employees also learn more about the program and feel more comfortable and confident in making decisions. This management style builds stronger bonds between the manager and employees.

III. Problem Solving

In insect rearing, a problem is any abnormal situation with the potential to adversely impact the ability to supply the desired number of quality insects in a timely manner. The cause(s) of a problem may be rooted in one or more parts of the program, including processes, equipment, employees, facility, and the insects themselves. When a problem has multiple causes, simple solutions are usually precluded, and resolution requires following up to determine if the proposed solution works and does not result in unintended effects.

A common approach to problem solving consists of several stages. They include defining the problem, searching for causes, identifying possible solutions, selecting the most appropriate one, evaluating its potential effect on the program, implementing the solution, and following up (Davis et al. 1992, Montana and Charnov 2000, Anonymous 2002). Finding the causes of the problem is frequently the most challenging step, particularly if the root of the problem lies embedded within multiple facets of the program. Thus, I will focus here on how to gather information to find the source(s) of the problem.

Begin by systematically investigating the extent of the problem. Discuss the situation with employees that work with the insects. Develop a clear, precise description of the symptoms and how they are different from normal. Ask probing questions to extract information the employee may have forgotten or considers unrelated. Read the literature and consult with others who may be familiar with the problem.

Focus attention on the smallest unit impacted. To illustrate, say the percentage of malformed pupae has increased above normal. Do symptomatic insects occur in a single container from just one cohort? If so, focus on variables that can affect an individual rearing unit such as insect density, cleanliness of the container, placement of the lid, container location within the holding room, and air exchange (e.g., were holes in the lid blocked?). Did affected insects occur in multiple containers within one cohort or in multiple containers in multiple cohorts? Were the high numbers of deformed pupae from cohorts initiated on one day, more than one day, or every day? Did abnormalities occur in insects that fed on a single batch of artificial diet or more than one batch of diet? Did all symptomatic insects come from a single insect-holding room or from other rooms as well? Was there a difference in symptoms between rooms? Did symptoms begin and end abruptly? Were the affected insects set up when the normal caretaker was on vacation? In a multiple species program, were other colonies affected in the same way; and, if so, what do they have in common that may have caused the problem? Did developmental rates, weights, eclosion success, egg production, wing deformity, or other insect characteristics change when the percentage of malformed pupae increased?

Look for temporal patterns and relationships. Is there a tendency for the symptoms to occur under certain circumstances but not under others? Do they occur randomly? Do they occur each day of the week, just on Mondays, or only during October or November? Do the symptoms get progressively worse, better, or stay about the same over time?

Check rearing labels and records for abnormal events (e.g., power outages, temperature spikes). Review quality control data to determine at what point the symptoms began. Did symptoms occur after a significant temperature spike or implementation of a change in procedures? Did they occur after introduction of a new species into culture, a new employee began working, or a significant change in the weather?

Determine through historical records whether a similar problem has occurred in the past and what was done to resolve it.

Investigate the possibility of a disease causing the abnormal insects (Cantwell 1974, Goodwin 1984, Soares 1992, Lacey 1997; Chap. 9 *Entomopathogens and Insect Rearing*). Send a sample of the affected insects to a pathologist (e.g., the insect pathology services available at Mississippi State University). Preliminary results are usually available within 24–48 hours.

Determine if equipment malfunction may have contributed to the problem. Are the autoclave, laminar-flow hood, and photoperiod timers working properly? Ask maintenance personnel about changes in the air-handler, utilities, and other systems that service the insectary. For example, were air or water filters recently replaced? Was a new water deionization system installed? Has the quality of water used to prepare diet changed?

Determine the role artificial diet may have played in the reduction of usable pupae. What was the highest temperature during the preparation process for the batch used to feed the cohort with lower yields? Was there a difference from normal in the gel strength or stratification of ingredients? Was there a significant change in temperature within the diet storage area? When was the diet purchased? When was it first opened? Was there a change in the diet ingredients or suppliers of those ingredients? Have other insectaries that use the same artificial diet (if it is purchased from a common supplier) experienced similar problems? Was an abnormal color or odor noticed during diet preparation or after insects began feeding on it? Did the diet surface have shiny or sticky spots (bacteria), or "wooly-looking" green, black, or white areas (fungi)?

Flow charts, schedules, and other visualization aids can facilitate problem solving. A diagnostic or troubleshooting chart provides a systematic way to help identify problems (Leppla and Fisher 1989). Quality-control graphs reveal slowly developing trends and relationships that may be too subtle to detect otherwise (Leppla and Ashley 1989). The graph of an insect characteristic can be overlaid on a temperature or humidity chart for the same time period to indicate possible causal relationships.

If a problem suddenly threatens the rearing program, decisions about its solution should be based on the best information available, even though it may be incomplete. The problem must be followed closely to determine if the actions taken were appropriate. The solution can be changed if future information sheds additional light on the problem or additional ideas come to mind.

IV. Planning and Organization

Planning is a process in which customer goals and resources within the rearing program are integrated into a set of operational plans for producing insects that satisfy customer requirements. Regular planning keeps the program responsive to new opportunities, modifications of customer goals, and changing resources.

A. Planning Process

The process begins by reviewing customer goals, formulating a mission statement for the rearing program that will meet those goals, and developing a set of rearing goals that are aligned with the mission statement. Rearing goals are general, long-term targets that guide insectary activities in the desired direction. For example, goals may be to: 1) meet customer specifications for the number and quality of insects, and the timeliness of their delivery; 2) maintain flexibility to respond to varying customer and marketplace needs; and 3) support change for the improvement of product quality and program efficiency.

Rearing goals are broken down into objectives that describe how the goals will be completed. Objectives are assigned specific actions or tasks to achieve the objective. For example, an objective may state that "insect quality for each shipment to the customer will meet or exceed standards, as determined by a series of tests agreed to by supplier and customer." One of these tests may involve regularly evaluating the eclosion success of adult flies. Specific actions for this test may consist of "selecting 500 puparia, placing them in the appropriate container, holding the insects at precise temperature and humidity for 24–48 h, monitoring eclosion, and determining the percent successful eclosion." The combination of all tasks required to successfully rear an insect constitutes a standard operating procedure (SOP) for that species (Sec. IV. B.).

The final steps in the planning process include determining the frequency with which the procedures will be performed, assigning procedures and/or individual tasks to employees (Sec. IV. F.), evaluating and coordinating available resources (Sec. IV. G.), training employees so they can successfully complete their responsibilities (Sec. V. B.), and setting performance objectives for individual employees (Sec. V. C.).

B. Standard Operating Procedures

A standard operating procedure (SOP) is a set of detailed work instructions for the completion of routine and/or repetitive activities (EPA 2001). Following an SOP reduces variability and facilitates consistency in quality and integrity of insectary products. In addition, an SOP is used as a framework for training new employees (Sec. V. B.), a database for scheduling employee activities (Sec. IV. C.), a check list for specific activities, as well as a reference for quality-control standards, development of employee objectives, performance reviews, and documentation of the evolution of rearing procedures.

An SOP includes procedures and the specific activities or tasks required for successful completion of the procedures. Tasks are listed in the order they are to be performed (Table 1). Entries should be concise, easy to read, and written by individual(s) who have performed the work described. SOPs should also include the frequency with which the activities are performed, standards and tolerances used to monitor quality, equipment and supplies, environmental conditions, hazards, and vendors. Changes in an SOP should be documented along with the date and the reason for the change.

C. Scheduling Procedures

A schedule lists the day and time procedures will be performed. It is used to assign specific tasks to employees, manage the use of equipment, promote consistency when performing procedures, conduct time audits, and assist in operational planning for the future. Spreadsheet programs such as Microsoft Excel™ are useful for making schedules.

An SOP is the source of much of the information used in scheduling. For example, in Table 2, procedures for lepidopteran species were copied from the sample SOP in Table 1 and pasted into column B along with procedures for other species/activities. Greater detail can be displayed by listing the specific tasks associated with each procedure in a separate column; however, inclusion of these details results in a much longer document so they are often excluded. Additional columns in this schedule include species/activity (col. A), an index of the order in which the tasks are to be completed (col. C), the employee responsible for performing a given task (col. D), and the days of the week it is performed (cols. E–I).

A species/activity schedule shows procedures for each species or activity in the order they should be performed for each day of the week (Table 2). It can be used by new employees for taking notes during training and following the flow of activities for each culture. This schedule can be made by sorting by column A, then column C.

An employee schedule (Table 3) lists procedures by employee for each day of the week. It can be used to quickly reassign activities when an employee is unexpectedly absent or when preparing for planned absences. It requires sorting by column D, column A, and column C, in that order.

A day-of-the-week list (Table 4) displays the responsibilities for each employee on a specific day and in the order in which they should be performed. It is valuable when preparing schedules for holidays or when equalizing daily responsibilities among employees. A day-of-the-week list is prepared by sorting the data by the day of interest (e.g., column E, Monday) and sorting that subset by column D, column A, then column C, in that order.

A time-audit list is used to determine how long various procedures take to complete (Table 5). This list can be used to equalize employee workloads, justify new employees, evaluate efficiencies made in the program, and help in scheduling when employees are absent. It is made using data from the species/activity schedule (Table 2) except the last seven columns for days of the week are deleted and three new columns are added: time required to complete procedure (col. K), time required for each employee to complete their responsibilities within a species/activity (col. L), and total time required to complete procedures for each species (col. M).

A priority list is made by assigning a priority number to each procedure for every species (Table 6). It identifies those procedures that must be completed on a specific day and those that can be put off until a later

Table 1. Partial standard operating procedure for rearing a lepidopteran.

Procedure (schedule)	Tasks	Additional information
Harvest pupae **(M–F)**	• Use featherweight forceps to remove pupae from wells. • Regularly clean forceps in bleach solution (25 ml 5.25% NaOCl in 75 ml water).	• Perform in a fume hood to contain contaminants.
Evaluate quality of pupae **(M–F)**	• Count the number of unusable insects from all trays, including: larvae (dead, alive), intermediates, pupae (dead, deformed, discolored, soft, small, no movement, or other). • Count the number of wells with fungal or bacterial contaminants. • Weigh and calculate the average weight (g) of 2 randomly selected, usable pupae per tray. • Record this quality-control information on a monthly data sheet.	• NOTIFY manager if the number of unusable insects is > 2.5 per tray. If black "V"-shaped larvae are present, immediately re-cover tray and NOTIFY manager. • NOTIFY manager if the average weight per pupa is < 0.300 g.
Surface sterilize pupae **(M–F)**	• Pour between 743 & 758 ml of lukewarm (24–27°C) tap water into a 41-cm-dia plastic tub. • Add between 148 & 152 ml of household bleach (5.25% NaOCl). • Gently add pupae to the bleach solution. • Keep pupae in the bleach solution for 3 min while swirling gently with gloved fingers. • Pour solution and pupae through a 4.8-mm sieve in the sink. • Rinse the pupae in running lukewarm tap water for 1 min. • Line a second 41-cm-dia plastic tub with paper toweling. • Pour the pupae onto the paper towel and allow to dry. • Weigh all pupae and enter the total weight on the data sheet.	• Use CAUTION when handling bleach; wear safety glasses, gloves, and a lab coat. • Use a gentle flow of water to prevent damage to pupae.
Set up oviposition cages **(M–F)**	• Use 2 cages for each date of pupae collected. • Cut a 12-cm hole in the lid for each cage. • Label each cage with the dates for set up of the larval rearing trays and the adult cage. • Place between 34 & 36 g of pupae in each cage. • Cover top of cage with an oviposition sheet. • Secure sheet with rubber band. • Place plastic lid on cage.	• Use container # 50 TB and lid # 50 TBL from Pieris Paper Products. • Lids can be reused after soaking in germicidal detergent for 1 h. • Use 20 x 20 cm cotton gauze (#2020 fine, manufactured by Cotton Products).

Table 2. Schedule of weekly procedures sorted first by species/activity and then by order. Letters at the top of the figure are for column identification. Abbreviations: coleopteran (col), homopteran (hom), lepidopteran (lep), and preparation of artificial diet (diet). This schedule is used when assigning procedures and training new employees.

A	B	C	D	E	F	G	H	I
				Schedule				
Species/ Activity	Procedure	Order	Worker	M	Tu	W	Th	F
col	Transfer adults to oviposition cages (morning)	5	Pete	x	x	x	x	x
col	Collect third instars for set up of pupation boxes	10	Pete	x	x	x	x	x
col	Transfer larval set-ups to fresh corn	15	Pete	x		x		x
col	Remove old corn from previous transfers	20	Jeff	x		x		x
col	Wash out tubs generated from transfers	25	Jeff	x		x		x
col	Set up corn tubs	30	Jeff	x	x	x	x	x
col	Set up newly hatched larvae	35	Pete	x	x	x	x	x
col	Transfer egg-hatch dishes	40	Pete	x	x	x	x	x
col	Set up pupation boxes	45	Pete		x		x	
col	Prepare egg dishes	50	Gail		x		x	
col	Remove egg dishes from oviposition cages	55	Pete	x		x		x
col	Separate eggs from soil	60	Pete	x		x		x
col	Service oviposition cages	65	Pete	x	x	x	x	x
col	Transfer adults to oviposition cages (afternoon)	70	Pete	x	x	x	x	x
diet	Measure ingredients	5	Andy	x			x	
diet	Add required volume of water and bring to boil	10	Andy	x			x	
diet	Add dry ingredients to water	15	Andy	x			x	
diet	Blend for 2 min	20	Andy	x			x	
diet	Add additional ingredients	25	Andy	x			x	
diet	Blend for 2 additional min	30	Andy	x			x	
diet	Dispense diet into rearing containers	35	Kacie	x			x	
hom	Prepare flats for planting	5	Jeff		x		x	
hom	Plant seeds	10	Jeff		x		x	
hom	Place flats in appropriate position in greenhouse	15	Jeff		x		x	
hom	Water established plants as needed	20	Jeff	x	x	x	x	x
hom	Monitor all plants for undesirable pests	25	Donna	x		x		x
hom	Select plants to use for aphid propagation	30	Donna	x		x		x
hom	Place cut plants on top of new plants	35	Donna	x		x		x
hom	Water new plants as needed	40	Donna	x		x		x
hom	Supply researchers with aphids as needed	45	Donna	x	x	x	x	x
lep	Harvest pupae from 8, 32-well trays	5	Chris	x	x	x	x	x
lep	Evaluate quality of pupae	10	Chris	x	x	x	x	x
lep	Surface sterilize pupae	15	Chris	x	x	x	x	x
lep	Set up oviposition cages	20	Chris	x	x	x	x	x

Table 3. Schedule of weekly procedures sorted first by worker, then by species/activity, and then by order. Letters at the top of the figure are for column identification. Abbreviations: coleopteran (col), homopteran (hom), lepidopteran (lep), and preparation of artificial diet (diet). This list is useful for quickly reassigning activities when an employee is unexpectedly absent or when preparing for planned absences.

A	B	C	D	E	F	G	H	I
				Schedule				
Species/ Activity	Procedure	Order	Worker	M	Tu	W	Th	F
diet	Measure ingredients	5	Andy	x			x	
diet	Add required volume of water and bring to boil	10	Andy	x			x	
diet	Add dry ingredients to water	15	Andy	x			x	
diet	Blend for 2 min	20	Andy	x			x	
diet	Add additional ingredients	25	Andy	x			x	
diet	Blend for 2 additional min	30	Andy	x			x	
lep	Harvest pupae from 8, 32-well trays	5	Chris	x	x	x	x	x
lep	Evaluate quality of	10	Chris	x	x	x	x	x
lep	Surface sterilize pupae	15	Chris	x	x	x	x	x
lep	Set up oviposition cages	20	Chris	x	x	x	x	x
hom	Monitor all plants for undesirable pests	25	Donna	x		x		x
hom	Select plants to use for aphid propagation	30	Donna	x		x		x
hom	Place cut plants on top of new plants	35	Donna	x		x		x
hom	Water new plants as needed	40	Donna	x		x		x
hom	Supply researchers with aphids as needed	45	Donna	x	x	x	x	x
col	Prepare egg dishes	50	Gail		x		x	
col	Remove old corn from previous transfers	20	Jeff	x		x		x
col	Wash out tubs generated from transfers	25	Jeff	x		x		x
col	Set up corn tubs	30	Jeff	x	x	x	x	x
hom	Prepare flats for planting	5	Jeff		x		x	
hom	Plant seeds	10	Jeff		x		x	
hom	Place flats in appropriate position in greenhouse	15	Jeff		x		x	
hom	Water established plants as needed	20	Jeff	x	x	x	x	x
diet	Dispense diet into rearing containers	35	Kacie	x			x	
col	Transfer adults to oviposition cages (morning)	5	Pete	x	x	x	x	x
col	Collect third instars for set up of pupation boxes	10	Pete	x	x	x	x	x
col	Transfer larval set-ups to fresh corn	15	Pete	x		x		x
col	Set up newly hatched larvae	35	Pete	x	x	x	x	x
col	Transfer egg-hatch dishes	40	Pete	x	x	x	x	x
col	Set up pupation boxes	45	Pete		x		x	
col	Remove egg dishes from oviposition cages	55	Pete	x		x		x
col	Separate eggs from soil	60	Pete	x		x		x
col	Service oviposition cages	65	Pete	x	x	x	x	x
col	Transfer adults to oviposition cages (afternoon)	70	Pete	x	x	x	x	x

Table 4. Schedule of Monday procedures sorted first by worker, then by species/activity, and then by order. Letters at the top of the figure are for column identification. Abbreviations: coleopteran (col), homopteran (hom), lepidopteran (lep), and preparation of artificial diet (diet). This list is valuable when preparing schedules for holidays or equalizing daily responsibilities among employees.

A	B	C	D	E	F	G	H	I
Species/ Activity	Procedure	Order	Worker	Schedule				
				M	Tu	W	Th	F
diet	Measure ingredients	5	Andy	x			x	
diet	Add required volume of water and bring to boil	10	Andy	x			x	
diet	Add dry ingredients to water	15	Andy	x			x	
diet	Blend for 2 min	20	Andy	x			x	
diet	Add additional ingredients	25	Andy	x			x	
diet	Blend for 2 additional min	30	Andy	x			x	
lep	Harvest pupae from 8, 32-well trays	5	Chris	x	x	x	x	x
lep	Evaluate quality of pupae	10	Chris	x	x	x	x	x
lep	Surface sterilize pupae	15	Chris	x	x	x	x	x
lep	Set up oviposition cages	20	Chris	x	x	x	x	x
hom	Monitor all plants for undesirable pests	25	Donna	x		x		x
hom	Select plants to use for aphid propagation	30	Donna	x		x		x
hom	Place cut plants on top of new plants	35	Donna	x		x		x
hom	Water new plants as needed	40	Donna	x		x		x
hom	Supply researchers with aphids as needed	45	Donna	x	x	x	x	x
col	Remove old corn from previous transfers	20	Jeff	x		x		x
col	Wash out tubs generated from transfers	25	Jeff	x		x		x
col	Set up corn tubs	30	Jeff	x	x	x	x	x
hom	Water established plants as needed	20	Jeff	x	x	x	x	x
diet	Dispense diet into rearing containers	35	Kacie	x			x	
col	Transfer adults to oviposition cages (morning)	5	Pete	x	x	x	x	x
col	Collect third instars for set up of pupation boxes	10	Pete	x	x	x	x	x
col	Transfer larval set-ups to fresh corn	15	Pete	x		x		x
col	Set up newly hatched larvae	35	Pete	x	x	x	x	x
col	Transfer egg-hatch dishes	40	Pete	x	x	x	x	x
col	Remove egg dishes from oviposition cages	55	Pete	x		x		x
col	Separate eggs from soil	60	Pete	x		x		x
col	Service oviposition cages	65	Pete	x	x	x	x	x
col	Transfer adults to oviposition cages (afternoon)	70	Pete	x	x	x	x	x
col	Prepare egg dishes	50	Gail		x		x	
hom	Prepare flats for planting	5	Jeff		x		x	
hom	Plant seeds	10	Jeff		x		x	
hom	Place flats in appropriate position in greenhouse	15	Jeff		x		x	
col	Set up pupation boxes	45	Pete		x		x	

Table 5. Time audit for procedures listed in Table 2 sorted first by species/activity, then by worker, and then by order. Letters at the top of the figure are for column identification. Abbreviations: coleopteran (col), homopteran (hom), lepidopteran (lep), and preparation of artificial diet (diet). This information can be used to equalize employee workloads, justify new employees, evaluate program modifications, and schedule activities when employees are absent.

A	B	C	D	J	K	L
Species/ Activity (S/A)	Procedure	Order	Worker	Time per procedure (min)	Time per worker per S/A (h)	Total time per S/A (h)
col	Prepare egg dishes	50	Gail	35	0.58	3.53
col	Remove old corn from previous transfers	20	Jeff	5	0.42	
col	Wash out tubs generated from transfers	25	Jeff	10		
col	Set up corn tubs	30	Jeff	10		
col	Transfer adults to oviposition cages (morning)	5	Pete	10	2.53	
col	Collect third instars for set up of pupation boxes	10	Pete	25		
col	Transfer larval set-ups to fresh corn	15	Pete	15		
col	Set up newly hatched larvae	35	Pete	5		
col	Transfer egg-hatch dishes	40	Pete	2		
col	Set up pupation boxes	45	Pete	15		
col	Remove egg dishes from oviposition cages	55	Pete	10		
col	Separate eggs from soil	60	Pete	45		
col	Service oviposition cages	65	Pete	15		
col	Transfer adults to oviposition cages (afternoon)	70	Pete	10		
diet	Measure ingredients	5	Andy	20	0.72	1.30
diet	Add required volume of water and bring to boil	10	Andy	15		
diet	Add dry ingredients to water	15	Andy	2		
diet	Blend for 2 min	20	Andy	2		
diet	Add additional ingredients	25	Andy	2		
diet	Blend for 2 additional min	30	Andy	2		
diet	Dispense diet into rearing containers	35	Kacie	35	0.58	
hom	Monitor all plants for undesirable pests	25	Donna	5	0.45	1.37
hom	Select plants to use for aphid propagation	30	Donna	2		
hom	Place cut plants on top of new plants	35	Donna	10		
hom	Water new plants as needed	40	Donna	5		
hom	Supply researchers with aphids as needed	45	Donna	5		
hom	Prepare flats for planting	5	Jeff	20	0.92	
hom	Plant seeds	10	Jeff	20		
hom	Place flats in appropriate position in greenhouse	15	Jeff	5		
hom	Water established plants as needed	20	Jeff	10		
lep	Harvest pupae from 8, 32-well trays	5	Chris	35	1.10	1.10
lep	Evaluate quality of pupae	10	Chris	5		
lep	Surface sterilize pupae	15	Chris	10		
lep	Set up oviposition cages	20	Chris	15		

time. This list is useful when time is limited during the day (e.g., when an ice storm delays employee arrival at work for several hours) or when there is a shortage of employees (e.g., when a sizeable number of employees are out sick). Prepare a priority list before it is needed, when there is an opportunity to calmly decide what should be done under potentially adverse conditions.

Table 6 can be made using the same format as the time audit (Table 5), except columns L and M are deleted and three columns are added: priority (col. N) and total estimated time by priority level (cols. O, P). Procedures that must be completed on a given day are given a priority of 1. For example, eggs must be collected before they hatch, plants must be watered, moths must be fed, mites must be transferred to new plants or they will exhaust resources.

Procedures that can be put off until the end of the day, or even the next day, are given a priority of 2. For example, harvesting pupae can be put off if they are normally collected 2–3 days before eclosion. Procedures that can wait until the next day or longer are given a priority of 3. For example, diet preparation would be a priority-3 procedure if there is enough diet remaining from the previous batch to allow for the delay. This is a good example of why a procedure such as diet preparation should not be put off until the last minute. A leeway of several days provides the flexibility to respond to abnormal circumstances with minimal disruption to the program. Finally, the time required to complete all procedures in each of the priority levels is calculated and used in managing the available work force.

D. Planning for Contingencies

A contingency plan helps prevent or reduce disruptions to the rearing program caused by unfavorable circumstances that may occur in the future. The proactive approach of identifying potential problems allows time to either eliminate the conditions that cause the problem or at least to minimize the impact they have on the program.

Preparing for an unexpected employee absence, for example, can save time and reduce anxiety in team members who must cover the absence. Weekly or monthly schedules that show critical dates for rearing processes can be posted near the holding room for that species. When the primary caretaker is absent, other employees can tell at a glance what needs to be done. The monthly schedule in Table 7 is based on the larval development rate for a species of Lepidoptera at the temperature required by the SOP. Output from the computer program that generated this schedule indicates which procedures are to be performed on a given day. For example, on Mon 10 April several activities will be performed: eggs will be set up on fresh diet to initiate a new cohort, larvae from eggs that were set up on Sat 1 April will be transferred to rearing cups, larvae from eggs that were set up on Sat 25 Mar will be transferred to pupation boxes, and pupae from eggs that were set up on Mon 13 March will be harvested and placed in oviposition cages.

To update this list, a date is entered into the first date cell, and the dates for the procedures are computed automatically based on the numbers entered in the "Period (d) egg harvest to task" entries. These represent the development time in days required before the indicated procedure can be performed. For example, it takes nine days for the first larval transfer, another seven days before last instars are transferred to pupation boxes, and then 12 more days before pupae are harvested. This type of schedule is simple to prepare in Microsoft Excel™ or other spreadsheet programs and a month or more of data can be calculated in seconds.

There are other ways to prepare for employee absences. A week's worth of operating supplies can be stored at or near the point of use so employees do not have to search when time is of the essence. Employees should be cross-trained so that anyone on the team can cover for an absent employee. If there is a choice of doing something today and putting it off until tomorrow, do it today. Today's circumstances are known, and tomorrow's cannot be predicted, half of the staff may be absent with the flu.

Consider "absence" scenarios to help focus thinking on how absences will be handled. Use the employee schedule (Table 3) to list responsibilities for each employee. How will their responsibilities be divided when they are absent? Are the replacements trained, and do they have the information needed to complete the procedure? If the absent person is responsible for shipping insects to outside users, for example, who will select the insects, pack them, prepare the paperwork, ensure that the insects get to the shipping department on time, track the shipment, and follow up with the shipper and recipient?

Individuals outside the insectary can be identified, trained and called in to help when schedules are

Table 6. Priority list for Monday procedures (Table 4) sorted first by priority, then by species/activity, and then by order and a time audit by priority level. Abbreviations: coleopteran (col), homopteran (hom), lepidopteran (lep), and preparation of artificial diet (diet). Highest priority procedures that must be completed today (1), intermediate priority procedures that should be completed today if time permits (2), and lower priority procedures that can be completed tomorrow or later (3). This list is useful when several employees are out sick and there is inadequate help to complete the normal schedule of activities.

A	B	C	D	J	N	O	P
Species/ activity (S/A)	Procedure	Order	Worker	Time per procedure (min)	Priority	Total time per priority level (h)	
col	Transfer adults to oviposition cages (morning)	5	Pete	10	1	level 1	2.67
col	Collect third instars for set up of pupation boxes	10	Pete	25	1	level 2	2.32
col	Transfer larval set-ups to fresh corn	15	Pete	15	1	level 3	2.30
col	Remove old corn from previous transfers	20	Jeff	5	1		
col	Set up corn tubs	30	Jeff	10	1		
col	Set up newly hatched larvae	35	Pete	5	1		
col	Set up pupation boxes	45	Pete	15	1		
col	Prepare egg dishes	50	Gail	35	1		
col	Service oviposition cages	65	Pete	15	1		
col	Transfer adults to oviposition cages (afternoon)	70	Pete	10	1		
hom	Water established plants as needed	20	Jeff	10	1		
hom	Water new plants as needed	40	Donna	5	1		
col	Transfer egg-hatch dishes	40	Pete	2	2		
col	Remove egg dishes from oviposition cages	55	Pete	10	2		
hom	Prepare flats for planting	5	Jeff	20	2		
hom	Plant seeds	10	Jeff	20	2		
hom	Place flats in appropriate position in greenhouse	15	Jeff	5	2		
hom	Monitor all plants for undesirable pests	25	Donna	5	2		
hom	Select plants to use for aphid propagation	30	Donna	2	2		
hom	Place cut plants on top of new plants	35	Donna	10	2		
lep	Harvest pupae from 8, 32-well trays	5	Chris	35	2		
lep	Evaluate quality of pupae	10	Chris	5	2		
lep	Surface sterilize pupae	15	Chris	10	2		
lep	Set up oviposition cages	20	Chris	15	2		
col	Wash out tubs generated from transfers	25	Jeff	10	3		
col	Separate eggs from soil	60	Pete	45	3		
diet	Measure ingredients	5	Andy	20	3		
diet	Add required volume of water and bring to boil	10	Andy	15	3		
diet	Add dry ingredients to water	15	Andy	2	3		
diet	Blend for 2 min	20	Andy	2	3		
diet	Add additional ingredients	25	Andy	2	3		
diet	Blend for 2 additional min	30	Andy	2	3		
diet	Dispense diet into rearing containers	35	Kacie	35	3		
hom	Supply researchers with aphids as needed	45	Donna	5	3		

Table 7. Six-week schedule for rearing a lepidopteran species. This schedule lists those procedures to be performed on each cohort by day. It is useful when the primary caretaker is absent and another team member covers for him or her.

		Period (d) egg harvest to task at 26.7°C & 45% RH		
		9	16	28
		Tasks for cohorts with indicated egg harvest dates		
Day Date	Harvest eggs[a]	Larvae to cups	Larvae to pupation boxes	Harvest pupae
Mo 3 Apr	Y	Sa 25 Mar	Sa 18 Mar	Mo 6 Mar
Tu 4 Apr	Y	Su 26 Mar	Su 19 Mar	Tu 7 Mar
We 5 Apr	Y	Mo 27 Mar	Mo 20 Mar	We 8 Mar
Th 6 Apr	N	Tu 28 Mar	Tu 21 Mar	Sa 9 Mar
Fr 7 Apr	N	We 29 Mar	We 22 Mar	Su 10 Mar
Sa 8 Apr	Y	NA[b]	NA	NA
Su 9 Apr	Y	NA	NA	NA
Mo 10 Apr	Y	Sa 1 Apr	Sa 25 Mar	Mo 13 Mar
Tu 11 Apr	Y	Su 2 Apr	Su 26 Mar	Tu 14 Mar
We 12 Apr	Y	Mo 3 Apr	Mo 27 Mar	We 15 Mar
Th 13 Apr	N	Tu 4 Apr	Tu 28 Mar	Sa 18 Mar
Fr 14 Apr	N	We 5 Apr	We 29 Mar	Su 19 Mar
Sa 15 Apr	Y	NA	NA	NA
Su 16 Apr	Y	NA	NA	NA
Mo 17 Apr	Y	Sa 8 Apr	Sa 1 Apr	Mo 20 Mar
Tu 18 Apr	Y	Su 9 Apr	Su 2 Apr	Tu 21 Mar
We 19 Apr	Y	Mo 10 Apr	Mo 3 Apr	We 22 Mar
Th 20 Apr	N	Tu 11 Apr	Tu 4 Apr	Sa 25 Mar
Fr 21 Apr	N	We 12 Apr	We 5 Apr	Su 26 Mar
Sa 22 Apr	Y	NA	NA	NA
Su 23 Apr	Y	NA	NA	NA
Mo 24 Apr	Y	Sa 15 Apr	Sa 8 Apr	Mo 27 Mar
Tu 25 Apr	Y	Su 16 Apr	Su 9 Apr	Tu 28 Mar
We 26 Apr	Y	Mo 17 Apr	Mo 10 Apr	We 29 Mar
Th 27 Apr	N	Tu 18 Apr	Tu 11 Apr	Sa 1 Apr
Fr 28 Apr	N	We 19 Apr	We 12 Apr	Su 2 Apr
Sa 29 Apr	Y	NA	NA	NA
Su 30 Apr	Y	NA	NA	NA

[a]Y = "Yes," N = "No"

[b]NA = "Not applicable"

tight. In one program, two individuals from another team were trained in all rearing activities, and when an insectary employee was absent, an hour or two of their help in the morning stabilized daily activities. In another program, one person was on call and worked when an employee was sick or on vacation. The individual had considerable experience in arthropod rearing and was interested in "getting out of the house to make a few bucks" now and then. This was an excellent arrangement for her, and it provided stability and continuity to the rearing program.

E. Organization

In a large-scale production program, operations are divided into units that are more manageable, focused in their responsibilities, and responsive to problems within the unit. For example, procedures may be grouped by process such as production, quality control, methods development, procurement of supplies, diet preparation, and insect treatment (e.g., sterilization via radiation treatment). A manager or supervisor heads each unit.

In a midsize insectary, a single manager is accountable for the entire operation. Tasks are grouped by employee, with each primary caretaker responsible for production, quality control, and methods development for the species assigned to them. Procurement should be placed in the hands of a single employee who controls all aspects of ordering, making the process more efficient and easier to track. Likewise, one employee should be accountable for preparing artificial diet for all colonies to reduce variability among batches and to minimize the potential for contamination.

In some smaller programs, the role of manager and other members of the rearing team are not clearly defined because an exclusive management position is not warranted. In this case, the manager may be responsible for performing technical tasks on a regular basis in addition to managerial activities. However, he or she must continue to take a broader view of the program, ensure directional stability, and adequately separate these two roles so that one does not overshadow the other.

F. Assigning Responsibilities

Assigning responsibilities requires knowledge of each employee's experience, performance, aspirations, and skills. To gain this knowledge, an employee profile is formulated by answering a series of questions about each worker. For example: How many species have they reared? What problems have they encountered and solved? How have their colonies fared under their supervision? Do they have supervisory or managerial experience, or do they have an interest in this area? What skills do they have? Which skills need to be strengthened? Is more training required? If so, what type of training? Are they a leader or a follower? Are they a team player? Do they show initiative, or do they prefer to be told what to do? Do they like change, or do they fear it? Is there a mismatch between responsibilities and their personality or interests? The employee profile indicates where within the program each employee should do the best job and be of most benefit.

The following example from reorganization of a four-employee rearing program will illustrate the process of assigning responsibilities. Reorganization was initiated because production consistency and product quality were slowly dropping over a period of several months. In addition, employee enthusiasm for their work was declining. The first employee had considerable experience working in an insectary and was rarely absent. Ironically, she was not very good at keeping insects alive. She lacked intuition, initiative, and conscientiousness and had a shallow understanding of basic rearing concepts. Problems with her insects continued despite coaching and other attempts to help her. Her confidence dropped, and she was easily distracted.

After program reorganization, her responsibilities no longer included difficult or complex rearing processes; and her authority to make changes in the program was reduced. Her responsibilities focused on support activities such as ordering and tracking supplies, ensuring availability of supplies to other team members, monitoring environmental conditions, recording abnormal circumstances, and autoclaving reusable materials. She also prepared artificial diet with assistance, represented the insectary during tours and outside presentations, and was caretaker of three insect colonies that required minimal rearing skills. Specific projects focused on improving observation and quality-orientation skills.

The second employee had ten years of experience rearing a wide variety of arthropods, was resistant to change, did not deviate from procedures or from a daily schedule, and organized work areas logically and efficiently. He also had a good "feel" for each colony, made detailed observations, and provided accurate data. He loved the job but was overwhelmed when asked to maintain species with complex scheduling requirements. After program reorganization, his new responsibilities included activities where consistency and conscientiousness were important, such as in surface sterilizing Lepidoptera eggs and pupae, maintaining specialized workstations, monitoring changes in colonies, and generating data for quality-control purposes. He also sent insects to outside users and was responsible for four insect colonies that required a patient and meticulous caretaker. Resistance to change shown by this employee was generally a benefit to the program, although it made improvements more difficult to implement. Thus, this individual was encouraged to be more flexible and adaptable.

The third employee had only one year of rearing experience but had supervised a biological laboratory before she began raising a family. She reentered the work force as a settled, secure, reliable, motivated, and hard-working individual. Her responsibilities focused on helping the manager evaluate ideas and potential changes through experimentation and moving proven changes into daily practice. She helped solve problems, scheduled work for weekends and holidays, and handled all insects that were received from outside sources. She also maintained four of the more complex colonies because of her understanding of rearing principles and her exceptional organizational skills. After reorganization, increasing reliance on this employee within a supervisory role required training in interpersonal relationship skills.

The fourth employee had 30 years of rearing experience, including management of programs in eight different facilities involving many different species and rearing systems. His responsibilities were to manage the technical and personnel aspects of the program, generate ideas to improve product quality and production efficiency, anticipate and solve problems, design facilities and equipment, and develop and implement quality control measures and SOPs. He was also the caretaker of two minor colonies that ensured active participation in the program on a regular basis. Program reorganization required no changes in his duties.

After the reorganization, employees felt more comfortable; and colonies recovered because the workers were no longer trying to perform above their anxiety thresholds.

G. Coordinating Resources

Resources are the building blocks used to construct, maintain, and improve a rearing program. The most important resources include people, facilities, equipment, space, time, budget, and operating supplies. When coordinating resources, the manager evaluates and optimizes the use of available resources to create and maintain a successful program. This is an ongoing process because rearing programs are dynamic. Resource needs will change as objectives are modified, as new resources and technologies become available, and as experience increases. This is how the program evolves and improves over time.

The manager evaluates available resources by answering questions such as: How well do resources support the tasks required to achieve program goals? Which resources are strong, which are weak? Can one resource be used to compensate for a weaker one? How can the weaker resources be strengthened? How can goals be achieved while waiting to improve resources? What alternative resources exist? Are resources available that have never been considered? How can the resources available at this time best be coordinated to provide product of acceptable quality to the customer?

To illustrate compensation for a weaker resource by a stronger one, consider a midsize rearing program in which disease outbreaks occur sporadically because the facility lacks the proper physical barriers to prevent the movement of pathogens. Until the facility can be modified to provide proper barriers, the manager should strengthen other resource(s) to offset the deficiency. He should focus first on employees because they can be trained to take a variety of protective actions and they can observe and report results. Employees should be educated in basic insect pathology with a focus on symptomology of diseased insects in culture, impact of disease on the rearing program, characteristics of the most destructive disease agents, and the manner in which pathogens can be transmitted within the facility. Employees should be trained to detect abnormal insect behavior; morphological changes such as color, size, texture, and shape of the body or wings; malodors from larval or adult cages; and changes in insect variables such as fecundity, fertility, pupal yields, and adult eclosion. The manager should require the employees to record unusual events such as the incidence of extreme environmental conditions, changes in procedures, employee absences, date of receipt, and use of new supplies. Employees could also begin monitoring the facility for the presence of airborne contaminants such as fungi and bacteria and attempt to determine their sources. Additional measures that can be used to compensate for resource deficiencies, particularly as they relate to the control of insect disease, are discussed in Chap. 2 *Insectary Design and Construction* and Chap. 9 *Entomopathogens and Insect Rearing*.

The insectary manager should regularly review program objectives, profile available resources, identify additional assets that may be needed, and prepare justifications to procure them. In the meantime, he or she should manipulate and adjust those resources that are presently available to shape the day-to-day program.

V. Managing Personnel

Managing people is recognized as a difficult responsibility for any manager, and insectary management is no different. The process can be time consuming and frustrating, and there is little practical information on how best to do it, particularly within the unique structure of a rearing program. However, managing people is one of the most important responsibilities an insectary manager has because success ultimately boils down to one common denominator—the people who do the work. They regularly handle the insects, perform procedures, make direct observations about the colony, and represent the first line of defense against developing problems.

A. Hiring

At a minimum, the process of hiring a full-time employee should involve preparing a job description, advertising the position, receiving and reviewing résumés, selecting candidates for interviews, scheduling and conducting interviews, checking references, selecting the most appropriate candidate, making an offer, and negotiating compensation. In larger organizations, the process may also involve filling out request forms, getting the appropriate signatures, discussing candidates among interviewers, and administering a physical and drug test. Also, the new employee may be required to attend an orientation to the new work environment. Hiring can be a lengthy process taking months to complete, but the completion date can be moved earlier by beginning as soon as possible and regularly following the status of the process, giving it a push as needed.

A job description includes the principal responsibilities for the position, education level, amount and type of experience, and competencies required. It should reflect current and future needs and include enough detail so the candidate knows not only what his or her responsibilities will be but also which skills will be used to complete them. For example, take the following list of responsibilities: monitor (observational skills) quality of insect colonies, analyze key data for insects and rearing processes (analytical skills), and regularly report (communication skills) to the manager in a format (organizational skills) that indicates subtle changes or developing trends (quality assessment) and provide feedback for improving insect quality or program efficiency (follow up).

Available positions can be announced through personnel agencies, job notices at local colleges or universities, newspaper ads, headhunters, and websites such as entsoc.org (Entomological Society of America), Monster.com, and CareerBuilders.com. In addition to such formal sources, several additional, informal sources of candidates may be available. These include referrals, children of employees (especially attractive for part-time or temporary needs in the summer when school is out), retirees, and contacts at other insectaries.

1. Qualifications. Qualifications that should be considered for individuals seeking an insectary job include education, experience, and competencies.

a. Education. Education exposes individuals to a formal, structured environment with responsibilities and opportunities that are not available in other settings. As the amount of education increases, so should the individual's ability to work independently, perform analyses, make decisions, and solve problems. Advanced education focuses attention and provides additional skills, personal interactions, and other experiences that can be of benefit in certain positions within the rearing program.

b. Experience. Experience comprises the knowledge, skills, and understanding that an individual has accumulated over time. It is important to find an individual with the right experience, but it is just as important to find a person who will apply that experience in the new position to benefit the rearing program. Ideally, a candidate for an insect rearing position possesses experience in biology, entomology, and rearing, with some knowledge of insect pathology, nutrition, sanitation, facility design and operation, quality control, operation of equipment, environmental control and monitoring, and insectary management. However, there are few people with any rearing experience—much less the ideal—so requirements should be realistic. If set too high, there is the possibility of overlooking a candidate with excellent potential.

c. Competencies. The functional and personal competencies that a candidate possesses, such as those discussed below, are often good indicators of future employee success. Evaluate candidates for these qualities as part of the interview process.

Employees should be adaptable and continue to work successfully when confronted with change or other disruptions to the normal work routine. Such employees understand the workload, appreciate priorities, know the time required to complete activities, and assist other team members during uncertain times. They also anticipate change and prepare for it.

Communication is the transmission and receipt of information between individuals. Effective and trustworthy communication among all members of the insectary team is fundamental to a successful program. It consists of effective organization of thoughts, concise and clear exchange of information, use of a vocabulary that is understood by both parties, and active listening. Active listening includes paying attention to the speaker, allowing him or her to finish without interrupting, avoiding prejudgment, asking questions for clarification, and summarizing what was said to determine if the message was received properly.

Trainability is the capability and desire to learn about insectary principles, procedures, goals, and teamwork and to incorporate that knowledge into daily activities and decision-making.

Individuals should at least demonstrate a desire to learn about computers, be willing to take on computer projects, and understand that the computer is not an end in itself but rather a means to more efficiently and effectively accomplish their responsibilities.

Employees should focus on their customers and be patient and flexible with them. They should also understand customer requirements and constraints, work with them to improve insectary products, and help them improve their prospects for success.

Initiative is taking action prior to being told to do so or before someone else takes the lead. People with initiative have the desire to do more, make things better, think of new ideas, and keep busy with productive activities. They are also interested in taking steps to learn more and to improve themselves. This often results in perspective and an understanding of the general needs of the insects, facility, and other employees.

Employees must possess integrity and be trustworthy, reliable, and adhere to policy. They must admit when they have made mistakes and be willing to learn from them. Integrity also includes conscientiousness (accuracy and completeness of data), confidentiality, professionalism when dealing with others, and taking responsibility for personal actions. Each employee is a window for the manager to view basic insectary operations, and the employees must provide reliable, relevant data for him or her to properly evaluate the program.

Safety is freedom from injury and health hazards. Because an insectary presents some unique human hazards (Chap. 4 *Health and Safety Issues in the Rearing of Arthropods*), employees should demonstrate an awareness of these issues, perform their job according to policy and written protocols, recognize unsafe conditions and either fix them or report them to the relevant individuals, and encourage others to operate safely.

Teamwork is the joint action of employees by which the interests and ideas of each individual are focused to successfully achieve group goals. It is characterized by agreement on common goals and working together to achieve them, accepting personality differences within the group, helping others, sharing information and knowledge, and making good use of free time. Effective team players understand the entire program and what other people need to get their jobs done. They also perform quality work when substituting for an absent team member.

2. Interviewing Candidates. If possible, more than one person should interview each candidate, including someone who is good at quickly and accurately assessing the honesty, integrity, and seriousness of people. Each interviewer should be provided with the job description, résumé, application, and an evaluation form. This form ensures that each candidate is evaluated using the same criteria.

Interviewers should be enthusiastic and excited about the position, the rearing program, and the hiring organization. This will relax the interviewee and increase interest in the job. Getting to know the candidate and how they handle themselves will be easier with a challenging interview. Leading questions or those that can be answered with a yes or no should be avoided. The interviewee should speak about 80% of the time. Spending the whole interview telling the candidate what they will be doing will provide little information for selecting the right candidate.

Understanding the candidate's experience and how they performed in the past is the best indicator of how they will perform in the future. Previous experience should be analyzed, keeping in mind that just because someone has worked in an insectary, does not necessarily mean they know how to rear insects. Specific questions about their experience will emphasize what they did and how they did it. Responses will show the level of understanding, approach to rearing and problem solving, interpersonal skills, and interest and enthusiasm for the work. Appendix 1 lists sample questions that can be asked during an interview.

Questions to evaluate observational skills and entomophobia can be handled at the same time. Using a large, colorful insect like a late-instar tobacco hornworm, *Manduca sexta* (Lepidoptera: Sphingidae), ask if they would mind holding it. Pay attention to their reaction: Are they afraid? Squeamish? Do they drop it, or fling it across the room in disgust? Are they interested? Excited? Do they look it over thoroughly? How accurately can they describe the color, shape, size, and behavior of the insect?

3. Choosing Among Interviewees. After all the interviews have been completed, interviewers should discuss their impressions of each interviewee and seek a consensus on whom to hire. Discussions should center on the following questions: Do the interviewees meet the most important needs in the program? Are they interested and enthusiastic about the job? Are they personable? Will they make good team players? Do they have the initiative required to get the job done? Are they stable, reliable? Do they bring something different, and perhaps refreshing, to the program? Will they supplement the skills and experience of others in the group? How do their personal references think they will fit into the program and perform?

The desired educational requirements, experience, and personal competencies all may not come together in one person. Prioritizing the requirements and determining the degree of flexibility for each will assist in identifying an appropriate candidate even though that person may not fully meet every requirement.

B. Training Employees

Proper training is fundamental to the success of employees and the rearing program. Take the time to train properly because there will be a considerable payoff in the future. Be specific when training and let the person know exactly what is required of them. There are several categories of training.

Initial training is general preparation of new employees and should include health and safety as well as rearing principles. Discussions include specific hazards in the insectary (e.g., wing scales, proper use of formaldehyde, and the dangers of steam) and the use of personal protective equipment. Procedures for chemical spills, ergonomics, emergency evacuations, waste disposal, and reporting procedures are also covered. Initial training continues with a discussion of general rearing principles including techniques for handling insects, the team concept, sanitation, organization of work areas, contaminants and their impact on the program, product quality, efficiency of operations, and variability and its control.

Specific training in the SOPs for a species or a process begins shortly after initial training. A hard copy of an SOP should be given to the trainee to follow the progression of procedures and take notes for future reference. Training includes demonstrating how the procedure is performed and giving reasons why things are done in a particular way. Specific activities should be highlighted if they are health or safety hazards, areas of special difficulty, or critical monitoring points for quality control. The new employee then performs the procedure. Supervision continues until the trainee is comfortable with the procedures. However, following up in the future will ensure that procedures are still being completed properly.

Cross-training of employees outside their routine areas of responsibility is highly desirable. Cross-training provides the employee with a better understanding of the overall program and creates greater confidence. It also allows work to be shifted to different employees when needed with less stress.

Ad hoc training is required to handle specific issues. For example, one insectary had an unusual infestation of whiteflies, mites, thrips, and aphids on greenhouse plants used for maintaining phytophagous cultures. Some employees were unfamiliar with these pests and the impact they had on the colonies, so a special training session was prepared. It included identification, importance, type and location of damage, feeding behavior, reproductive capacities, length of life cycle, and capacity of the pest to compete with the cultured species. This training enabled employees to spot the presence of a pest and discard the plants before they were taken to the insect-holding room. It was an effective means of control until the greenhouse infestation was resolved.

Insect disease is another example for specific-issue training. Topics could include description of pathogens and their potential impact on insect quality and operational efficiency, transmission of disease organisms, and specific procedures to prevent the establishment and spread of disease.

Advanced training provides the employee with greater focus, knowledge, and experience in specific rearing-related topics. For example, the weeklong insect rearing workshop held annually at Mississippi State University offers lectures on rearing systems, diets, genetics, quality control, insectary design, insectary management, and health hazards, as well as lectures and laboratories on microbial contaminants, insect diseases, and environmental biology. In addition, workshop participants can use the Pritam Singh Rearing Library of over 7,000 entries on insect rearing, discuss specific problems or issues with instructors, and visit several local insectaries for an in-depth look at equipment, insectary design, and rearing techniques.

C. Facilitating Performance

The manager and employee should agree on performance objectives that will guide their work during the year. This agreement establishes what employees will do and what resources the manager will provide to support their efforts. Performance objectives list employee's responsibilities and assist the manager in organizing and distributing the workload, assigning timelines, evaluating performance, and motivating employees. Objectives should include a specific description of the responsibilities to be completed, a standard by which success can be measured, and a timeline for completion. The amount of detail in each objective depends on the employee.

For experienced employees, objectives are written in a simple, straightforward and somewhat general format to allow flexibility in their completion. For others, objectives are expanded to include specific details so there is no confusion about what is to be done. An example of an expanded objective is presented in Appendix 2.

D. Evaluating Performance

Employee performance is evaluated against the performance objectives that outline what the individual is to accomplish within the workplace during a specified period of time.

1. Documentation. Documentation of employee performance can be one of the most important things a manager does. It is used to supplement performance reviews, justify awards for good performance, initiate action such as probation or termination, and help visualize patterns of questionable performance. Documentation may even be introduced into evidence in a legal proceeding if a personnel issue gets to that point, a not necessarily remote possibility these days.

Employee records should contain entries for good and poor performance, such as achievements, awards, work above and beyond what is expected, behavioral problems, excessive absenteeism, and problems interacting with team members. Documentation must be clear and easily understood. Documentation of a specific event consists of the date and time of day of the occurrence, the circumstances surrounding it, and the impact it had on the program and on the insectary team.

2. Reviewing. Formal reviews are usually given at the end of the year or near the anniversary of the employee's start date. However, an informal review can be given any time. The objective of the review is to acknowledge good and poor performance and to generate a plan to help the employee improve.

Information generated during the year from direct observations, discussions with customers and teammates, relevant reports and memos, as well as other documentation is used in the review. Pose questions about each objective. For example, if an objective was to maintain a specific insect colony, ask: Was complete responsibility taken for the colony? Were adequate numbers produced to meet the needs of the reproductive colony and the customers? Was the manager notified when deviations from SOPs occurred? Were standard procedures violated without approval of the manager? Were population levels approximately constant, or were there large swings in numbers? Were substitute caretakers updated and properly trained to handle the colony when necessary? Was the colony jeopardized by neglect? How were problems handled? Were appropriate data kept to help identify and solve problems? Were time-related aspects of the objective completed successfully? Did the employee do more, or less, than was expected?

Identify those areas where expectations were exceeded and those where improvement is needed. Indicate the impact their performance had on the team and on program goals. Assign a "grade" for each objective.

Continue the review process by asking questions about competencies. For example, under the customer focus competency: Does the employee understand the needs of the customer? Was the employee frustrated with customer demands? How was any frustration handled? How effective were his or her interactions with customers? Did the individual meet customer needs in a timely manner? Is he or she interested in improving insectary products to better meet customer needs? Did the employee seek feedback from the customer and use it to improve service? Would the customer consider the insect supplier reliable and responsive to his or her needs? How did the employee respond when a customer changed his or her order?

At the end of the review, the manager assists the employee in developing a plan for improving his or her performance and contribution to the program. This requires the development of an action plan and commitment by the manager to provide the resources needed to complete it.

a. Good Performers. A rearing program is often considered a stepping stone to other opportunities outside the insectary, but good performers who want to remain in the insectary should be encouraged to do so. Make the insectary an end in itself where employees can move up the "rearing ladder" to fulfill their career objectives. Employee achievements associated with different job levels from entry level to manager are listed in Appendix 3. It can be used as a guideline in developing skills and preparing employees to take greater responsibilities within the insectary management structure.

b. Poor Performers. Poor performers create issues that can adversely impact the program. Dealing with these issues may be the most difficult part of a manager's job. Because of the emotional nature of this issue, the potential legal implications, and the fact that most managers have little training in this area, it is strongly recommended that working with poor performers be done with the full knowledge and assistance from the insectary manager's superior (if applicable) and other appropriate individuals. Further, be certain about the need to follow through with this process. If a poor performer is terminated, there is no guarantee that the next person hired will be any better.

With documentation showing repeated issues of poor performance, the manager should make every effort to assist the employee to correct the problem. These attempts usually include discussions with the employee in which the manager points to a specific issue and indicates why it is of concern. The two agree on what acceptable performance is, and the employee promises to improve under the watchful eye of the manager. If this scenario occurs several times without improvement, the manager should proceed to the next stage of dealing with the problem.

The employee is objectively evaluated using documentation to ensure proper recall of the events and their impact on the program. Answer questions such as: Is this a temporary issue that may not last very long? Are there circumstances that can explain the poor performance (e.g., a death in the family, a divorce)? What are the strengths and weaknesses of this person? Do the strengths outweigh the weaknesses? Is the issue a true performance problem or is it a reflection of bias? Is there a capacity and desire to improve? How long has this been an issue? Is it increasing in frequency? How many discussions have there been with the employee? How seriously does the issue influence the program? The team?

In a face-to-face meeting with the employee, the manager expresses concerns about performance, referencing supporting documentation. The manager should make it clear to the employee that poor performance caused this situation and that it is his or her responsibility to resolve it. At the same time, the manager should indicate a willingness to help the employee improve performance. The manager lists possible consequences of failure to improve, including a poor review, reduced salary increase, probation, or termination.

The manager then indicates that the employee's performance will be monitored for signs of improvement and selects a date to meet again, say in two weeks, to review progress and to initiate further action, if required. If the employee shows a genuine interest in improving, continue to follow his or her progress until acceptable performance appears to be ongoing. If performance does not show improvement, discuss options with the appropriate individual(s) in the organization (if applicable).

VI. Managing in a Crisis

A crisis threatens the capability of the insectary to supply customers with products of satisfactory quality. It can occur suddenly or develop over a period of months. The duration of a crisis can be short term or long term, and its effect may involve one or more species in culture.

A short-term crisis is an acute situation that may result in poor quality product for a brief period of time and usually does not threaten the colony with total loss. A short-term crisis may occur when an ingredient is left out of a single batch of artificial diet or important equipment breaks down and cannot be replaced for a week. In each case, the crisis is resolved when the abnormal conditions are rectified. The impact, however, may be felt an insect generation or two into the future. A short-term crisis can become long term if the cause is not identified and resolved in a timely manner.

A long-term crisis is a chronic situation that can result in unavailability of product, reduced quality, and increased costs for an indefinite period of time. A long-term crisis can result in total loss of colonies. Disease, especially those caused by viruses, can affect colonies for generations and is probably the most common cause of long-term crises. Chronic stress is another cause. The physiology of unstressed insects is focused on normal growth and development. Under stressful conditions, the focus is redirected toward protection from the stressor. The resulting physiological changes can adversely affect insect quality and the way the organisms perform in their intended role. There are a variety of stressors including inadequate nutrition, overcrowding in rearing containers, and chronic environmental problems (Steinhaus 1963). See Chap. 10 *The Basics of Quality Control for Insect Rearing*, Chap. 6 *Environmental Biology of Insect Rearing* Environmental, and Chap. 7 *Insect Nutrition, Feeding, and Artificial Diets* for additional information.

The most effective way to handle crises is to prevent them from occurring. Regularly review the program and its resources to anticipate what problems are most likely to occur and how they may be avoided. Prepare graphs or other visual representations of the data to help detect developing problems. Keep records of abnormalities and/or changes in the program. Train employees to detect changes in insect characteristics, conditions within the facility, and equipment function. Educate employees about contaminants, the symptoms they cause, and the damage they can do to insect colonies. Perform regular preventative maintenance of critical equipment such as laminar-flow hoods, autoclave, diet dispenser, and automation equipment.

Keep track of products that are "new and improved." Manufacturers may make subtle modifications that could have a negative impact on rearing procedures. For instance, for decades household bleach contained 5.25% sodium hypochlorite. Now there are products with higher concentrations of sodium hypochlorite plus fragrances, surfactants, and other components. Regularly check product labels and record any changes that have been made, along with the date the product was used for the first time. If necessary, call the manufacturer to get more details.

To prevent a crisis or minimize its effect, maintain an insect population larger than needed under normal conditions. The size of a colony is determined by the number of individuals required for customer requests, those needed as reproductives to perpetuate the colony, and an additional number to compensate for inherent fluctuations in the population. Generally, the more uncontrolled variables that occur in a rearing program, the greater the risks to the insects, and the larger the buffer population should be. Also, larger buffers should be provided for more valuable cultures that would be difficult to reestablish if lost. The buffer may consist of an additional number of individuals in each cohort and/or an increase in the frequency that the cohorts are initiated. Considering a colony with a 30-day life cycle, it is riskier to set up only one cohort per generation than to set up 4 cohorts, one each week. However, as the risk drops, the cost increases; and each manager must determine the practice that maximizes the benefit/cost ratio for them.

Establish a parallel colony in a separate area if the insects are unique and the risk of crisis is high. If possible, hold the second colony in a room whose conditions are maintained by a different heating, venting, and air conditioning (HVAC) system than that of the first colony.

When a crisis occurs, do not panic. Do get intimately involved by observing procedures, colonies, insect behavior, and abnormal trends in key insect characteristics. Profile available resources and determine which will best solve the problem and be the least disruptive to the program. Are there adequate supplies, space, and time to increase the size of the colony? Can another colony be established in a separate location? Is there sufficient time and personnel to handle the problem? Are insects available from another source so that customer needs can still be met?

In the case of disease, the decision to save or replace an affected colony depends on its value, uniqueness, importance to customers, availability of acceptable replacement colonies, facility design, and the likelihood of successfully eliminating the pathogen. While the idea of colony replacement sounds relatively simple, there is no guarantee of success.

Many pathogens can remain infective in the rearing facility for years. All surfaces must be carefully and correctly cleaned and disinfected (Chap. 9 *Entomopathogens and Insect Rearing*). Even then the disease can reemerge. Consider that the replacement insects may themselves be diseased. If disease reoccurs, review program resources, paying particular attention to the building design and procedures that may be inadequate to break the life cycle of disease organisms. This exercise is important whether the colony will be replaced or an attempt will be made to save it.

Take action based on current information. Time is important when dealing with a crisis, so make decisions based on the best information available at the time. However, do not be afraid to reevaluate the situation when more details become available.

Declare an emergency if warranted. This opens up options that are not normally considered. For example, temporarily delegate responsibilities to others, request additional resources, restrict access to the insectary, cancel vacations, change procedures, and reevaluate the risk/benefit ratio associated with the program. Greater risks may be required during emergencies.

Notify customers of the crisis. If it affects them, tell them, otherwise a credibility issue may damage the relationship. If the customers are knowledgeable about the rearing process, this will be easier to do because they will know that insect production is not immune to problems. They may not like it, but they should be able to understand it. Be confident, and do not blame people or events. Do not make promises that may not be kept. Characterize the problem for the customers: What it is? How does it impact rearing operations? How does it impact them? How long might the problem last? What alternatives are available?

After a crisis has been resolved, learn from the experience. Could anything have been done to prevent the incident? What were the first signs that indicated a crisis was imminent? How were the customers affected? How was the situation handled? Are there things that could have been done better? Are there potential crises within the program that can be defused by the experience gained?

VII. Conclusions

The primary responsibilities of the insectary manager are to lead, solve problems, plan, organize, manage personnel, and manage crises. Success comes with a vision of what the program should be and how it will benefit the customer. The vision establishes a direction for the program and provides the basis for the development of goals, objectives, and specific tasks. The manager uses an appropriate style to move the program in the desired direction. Solving problems requires information that the manager acquires from a variety of sources including observations, rearing records, historical information, insectary employees, experimentation, and visualization aids. Planning and organizing includes the preparation of standard operating procedures and various schedules derived from them. It also incorporates contingency planning, structuring the program and organizing resources for an efficient operation, and assigning responsibilities.

Managing personnel resources involves hiring qualified individuals and training them, as well as facilitating, documenting, and evaluating their performance. It also includes rewarding good performers and dealing with poor performers in a way that is least disruptive to the program. To prevent a crisis, the manager should think and act proactively by anticipating adverse situations and making preparations to prevent them from occurring. When a crisis does occur, there are many questions to answer, but there are also many sources of information that will aid in a solution to the problem.

Insectary management does not stop when a successful program is achieved. Rather, it demands a continuous effort to keep pace with the complex, dynamic nature of raising biological organisms.

References

Anonymous. 2002. Every manager's desk reference. Alpha Books, Indianapolis, IN.

Cantwell, G. E. [ed.]. 1974. Insect diseases, vol. 1. Marcel Dekker, New York, NY.

Davis, B. L., C. J. Skube, L. W. Hellervik, S. H. Gebelein, and J. L. Sheard. 1992. Successful manager's handbook. Personnel Decisions, Inc., Minneapolis, MN.

(EPA) Environmental Protection Agency 2001. Guidance for preparing standard operating procedures (SOPs). EPA QA/G-6.

Fisher, W. R. 1984. The insectary manager, pp. 295–299. *In* E. G. King and N. C. Leppla (eds.), Advances and challenges in insect rearing. USDA, ARS, Government Printing Office, Washington, DC.

Goodwin, R. H. 1984. Recognition and diagnosis of diseases in insectaries and the effects of disease agents on insect biology, pp. 96–129. *In* E. G. King and N. C. Leppla (eds.), Advances and challenges in insect rearing. USDA, ARS, Government Printing Office, Washington, DC.

Lacey, L. [ed.]. 1997. Manual of techniques in insect pathology. Academic Press, New York, NY.

Leppla, N. C. 1984. Systems management of insect-population suppression programs based on mass-production of biological-control organisms, pp. 292–294. *In* E. G. King and N. C. Leppla (eds.), Advances and challenges in insect rearing. USDA, ARS, Government Printing Office, Washington, DC.

Leppla, N. C., and T. R. Ashley. 1989. Quality control in insect mass production: a review and model. Bull. Entomol. Soc. Am. (Winter): 33–44.

Leppla, N. C., and W. R. Fisher. 1989. Total quality control in insect mass production for insect pest management. J. Appl. Entomol. 108: 452–461.

Montana, P. J., and B. H. Charnov. 2000. Management. Barron's Educational Series, Inc., New York, NY.

Schwalbe, C. P., and O. T. Forrester. 1984. Management of insect production, pp. 300–303. *In* E. G. King and N. C. Leppla (eds.), Advances and challenges in insect rearing. USDA, ARS, Government Printing Office, Washington, DC.

Singh, P., and M. D. Ashby. 1985. Insect rearing management, pp.185–215. *In* P. Singh and R. F. Moore (eds.), Handbook of insect rearing, vol. 1. Elsevier, New York, NY.

Soares, G. G. 1992. Problems with entomopathogens in insect rearing, pp. 289–322. *In* T. E. Anderson and N. C. Leppla (eds.), Advances in insect rearing for research & pest management. Westview Press, Boulder, CO.

Steinhaus, E. A. [ed.]. 1963. Insect pathology, an advanced treatise, vol. 1. Academic Press, New York, NY.

Appendix 1. A list of questions that can be asked to help evaluate the suitability of a candidate for a position within an insectary.

- What courses did you like best at school? Which were the most difficult?
- Describe a problem that you solved recently. How did you approach the problem and what were the results?
- What has been the biggest achievement in your career so far? What steps did you take to achieve it?
- Give an example of an idea you had to improve the job you were doing. What did you do with that idea?
- Describe a time when you feel you went beyond your job requirements. What did you do and what happened?
- Have you dealt directly with customers in previous jobs? What did you like most about dealing with them? Least?
- Describe a situation where you had to deal with a difficult customer. What did you do and how did it work out?
- How do you control mistakes in your work? What do you do when you make a mistake—if no one else notices? If it's obvious you made it?
- Have you been involved in a job where you repeat the same procedures day after day? What do you do to keep your concentration when performing routine procedures?
- Why would you think that keeping focused on the job is important when you're responsible for keeping organisms alive?
- Do you like to schedule your activities each day, or do you prefer to have a schedule prepared for you?
- Describe a situation when you ran out of time before you completed your job for the day. What did you do? How did it work out?
- What is teamwork? Have you worked as part of a team before? What did you like most about the team approach? Least?
- What are some of the important contributions you made to a team?
- Describe your current job responsibilities. What do you like most about this job? Least?
- What are your strengths, and how will they benefit the rearing program?
- What have you done at previous jobs to improve your performance?
- When are you most satisfied with a job? Least satisfied? Why?
- Why do you want this job?
- What do you picture yourself doing in five years?
- Describe your understanding of the job that you're applying for. Do you have any questions about it?

Appendix 2. Example of an expanded objective designed to ensure employee understanding of their responsibilities: process requests for insects from other insectaries.

- Work with customers to determine their needs.
- Keep a list of all requests with the date, species, number of individuals and stage of development required, the expected receipt date, and actual receipt date.
- Search for suppliers using established sources, the internet, and other contacts.
- Maintain an updated list of current suppliers.
- Notify customers about the status of their request within two days of receiving it.
- Apply for permits from the USDA, APHIS to ship the insects. Maintain a file for permits granted and one for those rejected.
- Order the insects through standard purchasing channels.
- Make arrangements to have the insects delivered as soon as they arrive.
- Notify the insectary manager and customer immediately if there is a problem receiving the shipment.
- Follow up with the shipper and/or supplier to resolve problems. Keep the customer up to date with any problems.
- Prepare the insects for use as required by the customer.
- Monitor insect quality and immediately notify the insectary manager and customer if a problem is detected.
- Handle invoices for signature and payment. Keep copies in a file.
- Ensure that suppliers get paid within 2–4 weeks after receipt of the invoice.
- Summarize insect requests received in a report due five days after the last day of the each month.

Appendix 3. Employee competencies associated with increasing levels of responsibility.					
	Entry Level	**Junior**	**Intermediate**	**Senior**	**Manager**
Adaptability	Copes with routine change.	Copes with non-routine change.	Helps others cope with non-routine change.	Coordinates the response to non-routine change.	Anticipates and prepares for change; displays effective leadership and stabilizes the program during nonroutine change.
Customer focus	Meets customers' needs for his primary species; communicates regularly with customers.	Ensures that customers' needs are met for all insectary products requested.	Understands the insectary's role in maintaining insect quality and meeting customer needs.	Manipulates insectary variables to meet customer needs.	Anticipates customers' needs; sets standards for customer focus and interactions.
Independence	Requires some direct supervision while work is performed, plus regular checks to ensure work is completed correctly; notifies supervisor of any uncertainties or problems that may develop.	Requires little, if any, direct supervision; infrequent checks on completed work may be required; finds things to do for the insectary or the team when their assigned work is completed.	May require some supervision at the beginning of new projects; has some supervisory experience.	Can supervise others; works independently on specific projects; works with manager to implement program objectives and helps develop long-range projects; makes some daily operations decisions; knows when to involve manager.	Operates program independently and within the framework established by customer objectives, safety guidelines, ethical and moral codes, and policy of the organization.
Initiative	Completes responsibilities properly and in a timely fashion; asks questions when necessary.	Learns more about the colonies and insectary operations; accomplishes more than assigned responsibilities; communicates needs to supervisor.	Generates ideas to improve product quality and production efficiency; seeks opportunities for self-improvement.	Requests help to pursue areas of interest; continuously improves experience and knowledge; implements ideas to improve program.	Works continuously to improve all resources available to the program.

Appendix 3 (cont.). Employee competencies associated with increasing levels of responsibility.

	Entry Level	Junior	Intermediate	Senior	Manager
Problem solving	Recognizes a problem when it occurs; generates reliable data to help the supervisor identify and solve problems.	Recognizes problems; collects information to help solve the problem; presents information to supervisor; assists supervisor in solving the problem.	Understands the causes of problems; differentiates problems based on type and frequency of occurrence; recognizes problems with all colonies as well as with facilities.	Spots problems quickly and assists in solving them; begins to anticipate problems based on previous experience; differentiates major from minor problems.	Spots problems quickly and directs the effort to solve them; identifies a situation as a major or minor problem, or not a problem at all and reacts accordingly; anticipates problems and takes measures to prevent them from occurring; leads the problem-solving process.
Quality of work	Follows established procedures for routine responsibilities; understands what makes a quality product; completes all details; checks work for accuracy and completeness.	Broadens view to entire program—all procedures and all rearing principles; has good observational skills; generates quality control (QC) data.	Assists manager in the establishment of quality parameters; generates QC data.	Generates QC data; maintains databases for QC information; assists manager in evaluating data; reports to customers about quality issues.	Establishes standards for QC and production efficiency; establishes a means to regularly monitor these standards; develops correlations between quality and program variables; trains employees in QC; reports to management about quality issues.
Teamwork	Follows directives to work with others.	Extends assistance to others; shares information about all colonies and other insectary resources.	Anticipates team and employee needs; appreciates the sense of team achievement.	Directs the team effort over the short term in absence of manager.	Uses divergent personalities and interests of insectary employees to strengthen the team through synergy.
Understanding of rearing principles	Follows established protocols whether rearing principles are understood or not.	Uses basic rearing principles to guide the performance of daily activities.	Teaches team members rearing principles; desires to learn more advanced rearing principles.	Applies advanced rearing principles including of relevant information from the literature; can teach principles to others.	Applies principles at all levels to coordinate all aspects of the program whether in routine situations or in crisis; educates management and customers about principles.

Chapter 4

Health and Safety Issues in the Rearing of Arthropods

JOHN P. REINECKE, Retired
Research Entomologist/Safety and Occupational Health Officer
USDA-ARS
Starkville, MS

I. Introduction

The need for insectaries is increasing and with that comes the potential for greater health and safety problems. My purpose in this chapter is to offer a variety of means to eliminate or reduce work-related health and safety problems in the insectary.

Along with other industrial facilities, insectaries are quite capable of generating safety, health, and environmental problems. Although most hazards in the insect-rearing workplace are common to those in other work situations, there are aspects of rearing insects that pose special risks, principally allergens and pathogens. Although other risks will be treated, the focus here will be on risks peculiar to insectaries.

I wish to set the tone for this chapter. "Safety" is not a term to be demeaned or belittled. In the past, I have heard the paraphrased question, "Should I work, or should I do this safety stuff?" Thankfully, times have changed, and one can do both, as well as see that others do too. In fulfilling my objectives, I intend to preach a little along with presenting facts and dogma from others that I believe to be important. One should consider the current status of safety in America as a long and hard-won series of incremental victories for the worker with a few outstanding breakthroughs, including mine safety, child labor laws, Workers Compensation, and the Occupational Safety and Health Administration (OSHA). As great a benefit as our current Safety, Occupational Health, and Environmental Protection (SOHEP) laws are, I believe there is an advantage in keeping a sense of humor about safety and related aspects such as your health and the environment's well-being. This is humor derived from serious matters. Thus, use cheerfulness, optimism, and enthusiasm, but avoid flippancy, frivolity, or making sport. Even though effective SOHEP costs time, effort, and money, consider it a wonderful gift that most workers in the past have not had and, in many countries, still lack. So, when one is working on safety with others, follow the rules and regulations with astuteness but also with some humor to make the process more palatable.

There are other nations that may surpass SOHEP in the United States (US) in some regards (Corn 1992). However, since the US is a reasonably representative example of a nation with a well-developed SOHEP program and because of my familiarity with it, I will dwell on the US in this chapter.

It is not my intent to consider all of the safety and health factors in the workplace. There is a vast amount of information concerning the numerous SOHEP laws and regulations in force in the US. Such information, happily, is readily available from a variety of sources on the Internet and in print form. The current SOHEP procedures published by OSHA and the North Atlantic Free Trade Agreement (NAFTA) are actually works in progress and undergo change as new research information and other forces dictate. The information on the Internet tends to keep pace with such changes.

A particularly important problem, the development of allergies to insects among insectary workers, is a phenomenon documented in numerous publications. Bellas (1999) cited over 150 publications on allergy-inducing insects not including stinging species, but this phenomenon is often not addressed as a preventative safety and health issue. Rather, it is left to physicians to deal with the ailing worker. Measures to reduce or avoid the development of allergies in insectary workers will be a major component of this chapter.

A. Safety, Occupational Health, and Environmental Protection (SOHEP)

During the Industrial Revolution, management did not consciously design the work place to harm employees, but they did not equate spending time and effort on safety as a way to increase profits. The accidental loss of life, limb, and health was accepted as part of the risk the common laborer must take to make a living. Little recourse was available to the worker or his family to gain any restitution from the employer. This had been going on, one assumes, since there were industries in which to lose life, limb, and health. The US lagged behind many European countries in workers' safety for an embarrassing number of years, in part because the US courts were conservative and favored management. For example, the loss of health, job loss, and slow death due to miner's black lung disease became so prevalent in England in the 1800s that the public outcry forced the British Parliament to act in 1870 (Guarnieri 1992). England's action is cited as the prod or starting point of safety awareness in the US. However, only after World War II did some states in the US begin to adopt laws regulating coal mine dust, and it was not until 1969 that federal protections were adopted. Unfortunately, in many countries even today, deplorable working conditions still prevail.

Vigorous efforts to avoid work-related accidents followed the institution of workers' compensation, an idea the US borrowed from Europe that guarantees payment of medical bills derived from work-related injuries. In the US, workers' compensation was adopted state by state and was not universal until 1949. Workers' compensation, with its set fees, avoided far more expensive lawsuits, and employers found that a safe working environment resulted in significant savings. Strong safety and health programs in the workplace are now mandatory in the US with the exception of family-owned-and-operated businesses (those with no unrelated employees).

Insurance companies and punitive litigation have worked in concert with federal and state agencies to bring about improved worker protections. Insurance companies demand that their industrial clients develop, and actively maintain, a good safety program to protect their workers and all who come in contact with their products or services. These requirements, plus paying workers' compensation, increase the cost of doing business in the US significantly. This fact should be no surprise considering the scope of safety measures in areas such as food, transportation, waste management, construction, water, and health care, that protect US citizens. It should also be no surprise that management in the US sometimes seeks to derive relief by moving production facilities to carefully selected countries where SOHEP laws and regulations are less strict.

B. Regulation and Enforcement

The US has created a number of federal agencies and congressional acts that provide information and means to enforce SOHEP of which the most important is OSHA. States must set their SOHEP standards so that they are "at least as effective as" federal standards. Most states adopt standards effectively identical to the federal ones. These federal regulations are published in several volumes available from the Government Printing Office (telephone: 888-293-6498). Fortunately, OSHA regulations are available without charge at http://www.osha.gov/index.html.

The National Institute for Occupational Safety and Health (NIOSH) is responsible for conducting research and making recommendations for the prevention of work-related injury and illness. It is available at http://www.cdc.gov/niosh/ and is a source of information on chemical safety, safety equipment such as respirators, and common safety hazards in the workplace (e.g., latex rubber and formaldehyde).

Federal and state enforcement of regulations can range from official reprimands to fines, business closure, and even jail terms for criminal negligence. Punishments are more often issued by governmental levels below the federal. Thus, there is a strong incentive to be aware of state and local requirements, particularly in waste management. Currently, the severity of penalties levied is not uniform across the 50 states. Some states are easily provoked to action while others seem to be blind to all but the most egregious infractions. Hopefully, the general trend for increased sensitivity to occupational health problems and the deteriorating environment will increase enforcement even in the least enthusiastic states.

What follows is an effort to present as much information as possible for solving safety-related problems in the insectary. The equipment and techniques discussed in this chapter will be more or less useful depending on one's current objectives, budget, and physical plant.

II. General Occupational Health Hazards and Their Prevention

In large part, all workplaces have similar, potential hazards. We become so familiar with our work environment that we tend to be oblivious to the presence of many hazards. These are separated into electrical, mechanical, thermal, acoustical, and chemical hazards in this chapter. The following brief discussions of these five hazard categories are intended to stimulate interest and awareness concerning accident prevention.

A. Electrical

Electrical hazards can be effectively eliminated by the use of ground-fault circuit-interrupter (GFCI) devices. A GFCI device can be integrated into a wall outlet, a circuit breaker, or an electrical cord (Fig. 1A–C). GFCI devices function very well in wet or dry environments and act to break the circuit before injury or fire can occur. Electrical equipment still needs to be maintained in good working operation, but the shock hazard can be eliminated. Having a GFCI device integrated into the circuit breaker takes the worry out of having to remember to use independent plug-ins or to find a GFCI wall outlet.

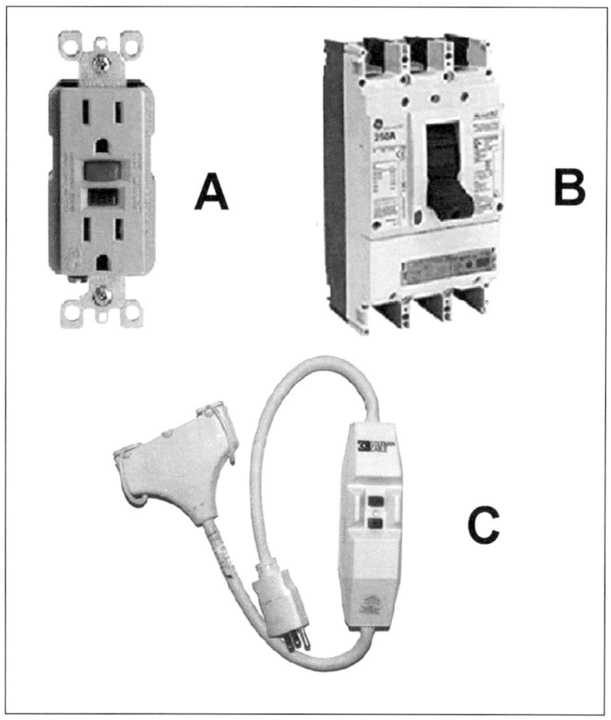

Fig. 1. Ground-fault circuit-interrupter (GFCI) devices used to eliminate shock hazards by breaking the circuit at the first instant of a short circuit. The circuit is reestablished by pressing the reset button. (A) Standard wall receptacle. (B) Circuit breaker capable of monitoring and closing down a number of electrical receptacles. (C) Part of an extension cord.

B. Mechanical

Most workplace injuries are due to mechanical hazards—those things that bump, squash, slice, abrade, puncture, and trip people. Mechanical hazards are very common and difficult to identify because they make up part of one's familiar surroundings and generally have not caused a problem in the past. One tends to automatically compensate for such hazards rather than to explicitly identify and neutralize them. Mechanical hazards commonly include unshielded power drives (turning shafts or belt drives); improperly stacked supplies; presses without safety features (Fig. 2); unsafe handling or disposal of sharp items such as needles, blades, and broken glass (Fig. 3); catapulted objects causing eye or other bodily injury; unsecured pressure hoses, cords, or other trip hazards; equipment or supplies that are too heavy for safe lifting or are inappropriately lifted.

The great majority of injuries resulting from mechanical hazards are due to haste, fatigue, inattentiveness, inadequate training, and lack of appropriate safety gear (e.g., helmets, shoes, goggles). They can be neutralized by removal, alteration, shielding, signage, training, and safety gear. Removal is the best but often least acceptable solution. What cannot be removed may be altered or shielded to restrict or control contact. Flagging hazards with appropriate warning signs is an easy solution. Although warning signs tend to decrease in effectiveness over time, they may be rejuvenated in a timely fashion to draw fresh attention to the hazard.

The toppling or rapid discharge of compressed gas cylinders is such a serious mechanical hazard that it deserves specific consideration. OSHA has devoted a section to this subject [29 CFR 1910.101(a)(b)(c)] stating, in part, "Each employer shall determine that com-

Fig. 2. Tray-forming machine with cautionary signs indicating a mechanical hazard but with no physical restraints designed to prevent a worker's hands from being crushed by the activated machine. The level of safety is frequently a judgment call. If the machine works slowly and hands are commonly not placed between the plates, the warning signs may be suitable.

Fig. 3. Disposable waste bin for broken glass. Broken glass, needles, and razor blades should never be placed in standard trash receptacles.

pressed gas cylinders under his control are in a safe condition to the extent that this can be determined by visual inspection." Standard, compressed gas cylinders, which contain a variety of gases at $6x10^6$ to $14x10^6$ Pa (900 to 2,200 psi), are 145 cm tall, 23 cm in diameter, and weigh about 50 kg when filled. A falling cylinder is capable of crushing bone upon impact. If the valve is knocked off when a cylinder falls, the cylinder may become a jet-propelled missile capable of going through a concrete wall. Consequently, cylinders must be securely fastened to a suitable surface (Fig. 4) when in use, and a valve protection cap must be in place when a cylinder is moved. Gases can be combustible, explosive, corrosive, poisonous, inert, or a combination of the above. A rapid discharge may be able to displace the air in a room before a worker can exit. Regulators or valves in good working order as well as appropriate tubing must be used to control gas flow. Pressurized gas cylinders demand respect and good training, and equipment is required for their proper handling.

Fig. 4. Improperly secured compressed gas cylinders. The two cylinders on the left are not secured to the wall.

C. Thermal

Thermal hazards in an insectary are commonly encountered with cooking, cleaning, and sterilizing equipment and with flammable chemicals. Hazardous equipment includes hot plates, drying ovens, and autoclaves. Flammable chemicals include alcohol and ether. Steam under pressure represents both a thermal and a mechanical hazard. Safe and appropriate equipment, barriers, warning signs, protective apparel, and training are common means of avoiding thermal hazards.

Smoke detectors, either commercial or consumer grade, should be appropriately mounted throughout the building and routinely tested. Emergency equipment for extinguishing combustible materials should be readily available and properly maintained. This includes fire extinguishers of proper size and type (Fig. 5) mounted on the wall to meet OSHA compliance standards. Training in the use of such emergency equipment should be provided to all staff members, and emergency evacuation drills should be developed and practiced routinely. Fire departments may provide on-site training in the proper use of fire extinguishers, cardiopulmonary resuscitation (CPR), and first aid. Appropriate and timely use of a fire extinguisher can be essential for job security—no building, no job. However, one must know when using a fire extinguisher is useless and personal safety is at risk.

Fig. 5. Improperly located, class ABC, dry chemical fire extinguisher. This fire extinguisher is out of compliance because access is blocked on the left. Filled with monoammonium phosphate, class ABC extinguishers are suitable for most fires and thus require less thought under stress; but they are hazardous to computers and similar sensitive equipment. The tag attached to the top of the extinguisher records the dates the extinguisher was checked and when it was recharged.

D. Acoustical

Acoustical hazards are not commonly considered life threatening. However, the deleterious effects of noise can range from stress that decreases worker productivity to permanent auditory nerve damage. The defense against noise hazards is to remove the noise source, relocate personnel from the noisy area, or reduce the decibel level. If none of these measures is possible, hearing protectors, from disposable to industrial grade, should be made available, and signs urging their use should be present (Fig. 6). The permissible decibel limit is somewhat complex, and an industrial safety engineer may be needed to make accurate determinations concerning this hazard. For instance, 90 dBA (decibels weighted for sensitivity of the human ear) is permissible for 8 hours of exposure and 105 dBA is permissible for 1 hour exposure. Exposure to impulsive or impact noise should not exceed 140 dBA. A full explanation of acceptable noise levels in the workplace is available in OSHA 29 CFR 1910.95.

E. Chemical

Commonly, insectaries possess a rather conservative chemical inventory. Nevertheless, workers handling chemicals should be trained to do so properly for their own safety as well as that of others. This training may be formal or specific, on-the-job training. Workers should be fully aware of the potential hazards of the chemicals with which they are working, and reference information on each chemical should be close at hand. Cleanliness is important, and spilled chemicals should be cleaned up as soon as possible. To better avoid mishaps, consider establishing a policy of treating all chemicals, particularly spilled chemicals, as hazardous. Proper protective apparel should be available and used. Emergency showers and eyewash stations, if called for, should be present and routinely tested (Fig. 7). Appropriate spill kits should be strategically located in chemical use areas (Fig. 8A, B). The stockpiling of chemicals should be avoided, particularly in work areas; and appropriate disposal methods should be employed that are compliant with federal and local regulations.

The Chemical Abstracts Service (CAS) number, a number unique for each chemical, should be present on chemical containers, and the chemical's Materials Safety Data Sheet (MSDS) should be easily accessi-

Fig. 6. Sign warning people to wear hearing protectors. It is common to have disposable earplugs available in the vicinity of acoustic hazards for use by people who do not have reusable hearing protectors at hand.

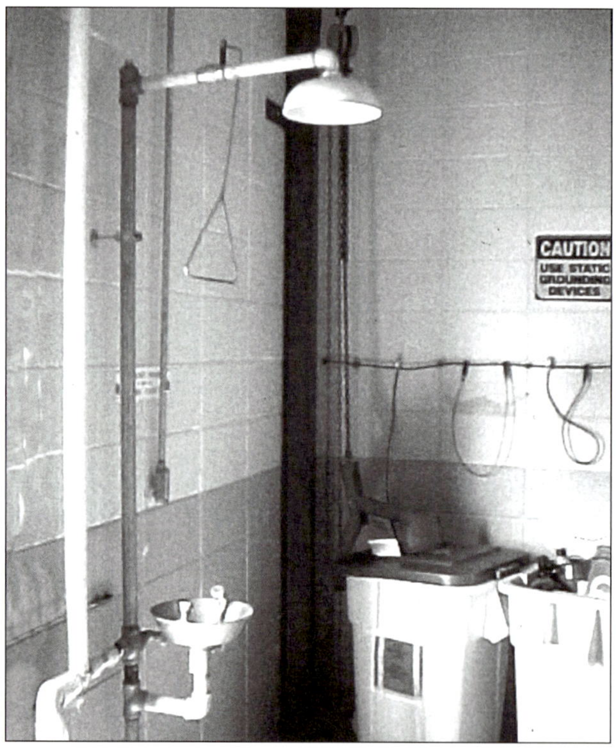

Fig. 7. Combined emergency shower and eyewash station. These safety items must be tested on a regularly scheduled basis. All safety equipment must be fully operational; and if found to be defective, immediately repaired or removed.

ble. This even includes certain items purchased at a grocery store, e.g., sodium hypochlorite (Clorox™) and ammonia.

The MSDS includes the minimum required information about the chemical, including synonyms of the chemical name; personal protection equipment (PPE) required in handling, such as gloves, gowns, and eye protection; fire and explosive hazards; common routes of entry into the body; acute effects; carcinogenicity; teratogenicity; clean-up procedures; and methods of disposal. The MSDS data for the more hazardous chemicals should be augmented with information from other sources.

The MSDS is routinely shipped by the supplier to the user along with the chemical. However, if there is a breakdown in this procedure or in the case of chemicals purchased locally, the MSDS can be obtained through the internet. Some sites charge for their service while others offer the information free of charge.

Another source of chemical hazard can occur if certain insecticides are used in the insectary. Sometimes, infestations of opportunistic insects such as cockroaches, grain beetles, and ants must be eliminated; and extreme caution is needed to avoid adverse effects on personnel (probably unrealized at the time) as well as effects on the insect being reared (expressed as behavioral changes and lowered reproductive rate, if not premature death). Other means of pest control such as baits and traps may be suitable. Of course, the best control is prevention: a clean insectary offers less incentive for pest invasion.

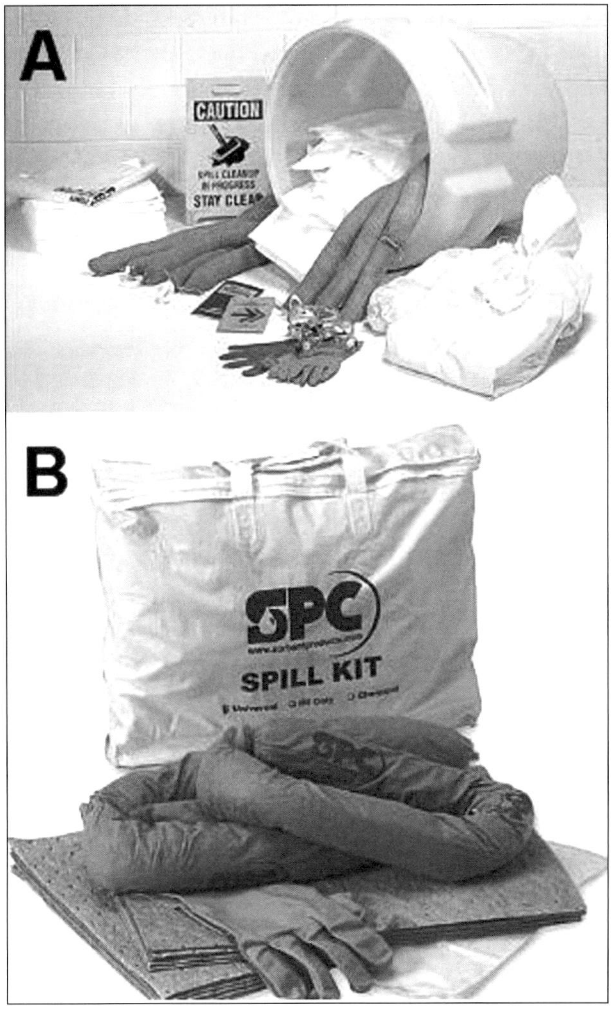

Fig. 8. Spill kits. (A) Designed to deal with a large quantity of material. (B) Designed to deal with a small quantity of material.

III. Occupational Health Hazards Peculiar to Insectaries

There are two occupational health hazards peculiar to insectaries: 1) allergic sensitivity reactions to insect-produced allergens and 2) diseases caused by microbes associated with insect diet and frass. Bellas (1990) estimated that 30% of the staff in facilities rearing insects and other arthropods has allergic sensitivity to the organisms being reared. Given this high incidence, the emphasis in this section will be on allergic sensitivities.

A. Allergens

Insect-induced allergic reactions can be life threatening. An estimated 12.4–16.8% of the population of the United States have suffered from anaphylactic shock, primarily due to insect stings; and of these it is estimated that 0.002% have died (Stark and Sullivan 1986). The most common anaphylactic reaction is sudden cardiovascular collapse, which usually leads to death in a matter of minutes if left untreated (Neugut et al. 2001). Of 115 insectary personnel with insect-induced allergies, Wirtz (1980) reported that 3 had suffered anaphylactic shock. Insectary staff should be able to recognize the symptoms of anaphylactic shock and be prepared to call 911 immediately and administer emergency treatment, which may include CPR and intramuscular injection of epinephrine (Brown 1995, Amerecho 2005).

A large percentage of the US population suffers from non–life threatening allergies to one or more products in their environment. Allergies including asthma have increased in the last decade, and scientists are studying hygiene and other factors to explain the rise (Renz et al. 2006). In the US, allergic sensitivity to pollen affects approximately 13% of the population according to the National Institute of Allergy and Infectious Diseases (NIAID). Six percent suffer from asthma, which is often provoked by airborne antigens of arthropod origin.

Anywhere arthropods are present in large numbers (e.g., dust mites, grasshoppers, cockroaches), humans can develop allergic sensitivities to them. This can complicate the process of finding the source of allergic sensitivity problems in insectary personnel. Did an employee develop her insect allergy in the insectary or outside of it? Fortunately, interest in allergic reactions to products found in insectaries seems to have increased in the last 15 or so years (Platts-Mills et al. 2005); and allergists have developed the means to determine whether a worker's allergy is associated with the insect being reared. A skin prick or patch test for allergies is considered accurate, reasonably quantitative (e.g., graded from 0 to 4+ or – to +++) and comparatively inexpensive (Ten et al. 1995). The radioallergosorbent (RAST) test measures allergen-specific immunoglobulin E (IgE) antibodies in the blood. There are newer tests measuring IgE, but all are more costly. It would be very useful to have an allergy test history for all insectary workers. A financially feasible group rate might be arranged with a medical unit that would run an annual allergy test against potential allergens supplied by the insectary.

In insectaries, as well as in rearing facilities for other arthropods such as spiders, mites, and shrimp (Bellas 1990; Kagen et al. 2004), scales, spines, other body parts, and waste products such as *frass* (excrement), and *meconium* (excreta from newly emerged adults) are present and act as allergens. This particulate detritus can exist as airborne particles at concentrations that depend on the work being done, humidity (static electricity charge suspension), and the protective equipment and procedures in use. As airborne particles, insect allergens easily invade the respiratory system and eyes. On surfaces, allergens may need simply to come into contact with the skin. Additional routes of exposure include bites, stings, and punctures by sharp spines. In a survey of 136 institutions involved in arthropod rearing or with employees potentially exposed to arthropod rearing (e.g., university entomology departments), Wirtz (1980) reported that 50 of the 84 respondents reported a total of 115 individuals, each with at least one allergic sensitivity. Of these 115 affected personnel, airborne (75.6%) and contact (63.5%) mechanisms were the most common means of conveying allergens. Bellas (1999) cited 470 publications in which allergy or hypersensitivity to a total of 150 insect species had been reported in 820 papers published over a 93-year period. These data suggest that some insects are inherently more allergenic than others.

In 1980, the National Institute of Occupational Safety and Health (NIOSH) conducted an evaluation of health hazards among insectary workers of the Agricultural Research Service. A questionnaire was designed and used to assess the prevalence of symptoms possibly related to allergenic airborne particles associated with raising colonies of insects in laboratory settings. At 61% (52 of 85) of the responding facilities, at least one worker reported current or past symptoms of work-related allergy. Five of the facilities reported 10 or more employees with such symptoms. Of employees working directly with insects, 33% (168 of 507) reported symptoms of work-related allergies compared with 9% (22 of 246) of workers who had little or no direct contact with insects. The most prevalent symptoms included sneezing or runny nose (73%), eye irritation (68%), skin irritation or skin rash (41%), coughing (38%), wheezing (26%), and shortness of breath (24%). Forty-four percent consulted physicians because of symptoms, and treatment was prescribed for 83% (69 of 83). Twenty-two percent of those suffering from work-related allergies either discontinued working with the insect or transferred to another work area or job. The insects most often involved were Lepidoptera (66% of 282 multiple responses) (NIOSH 1983).

A case was published in NIOSH (1983) that dramatically illustrates the tendency of some insectary personnel to develop multiple allergic sensitivities when appropriate protective measures are not taken:

> A worker experienced nasal irritation and congestion, cough, and episodes of shortness of breath with chest tightness within one-half hour after exposure to scales and "frass" (debris or excrement) from several insects. After working for 2 years with the *Anthonomus grandis* (boll weevil), the worker, because of these symptoms, was trans-

ferred to a job that involved working with *Heliothis* species. The symptoms recurred within 3 years of working with moths, and resolved after transfer to a job involving work with *Musca domestica* (housefly). The symptoms occurred again within 2 years' work with this insect. Symptoms recurred within 5 months after another job change to work with *Cochliomyia homonivorax* (screwworm) species. The worker's serum immunoglobulin E level was markedly elevated to 2,060 (normal less than 780), and eosinophil counts were elevated. Chest x-ray and pulmonary-function tests, including methacholine challenge, were normal. Allergy skin testing was positive for housefly and moth extracts and for extracts from the adult and larval stages of the screwworm fly. When the employee transferred to a job that did not involve insect-related work, the symptoms finally disappeared.

B. Pathogens

Microbes are a diverse and ancient part of the Earth's biota, and mankind generally thrives in their presence. Some insectaries, however, have experienced massive microbial contamination of the diet and frass. Such events can represent a health hazard to the insectary staff, particularly those workers whose jobs include cleanup. Our immunological defenses may be overpowered by massive exposure to an otherwise nonthreatening microorganism. Appropriate precautions must be taken to protect the staff during clean-up procedures.

A common contaminant of moist insect diets is the fungus *Aspergillus niger*, whose black fruiting bodies can produce clouds of spores if not properly contained. This fungus, of which there are 11 closely related variants, is ubiquitous but generally harmless. However, high airborne densities of *A. niger* can pose a health risk to insectary staff. Large amounts of inhaled *A. niger* spores can cause *aspergillosis*, a dangerous lung disease. *A. niger* is also a common cause of *otomycosis*, fungal ear infection, which causes pain and hearing loss that may be permanent in severe cases. These maladies and other complications may be encountered by people cleaning up after massive outbreaks of *A. niger* and other microorganisms in insectaries (Redlich et al. 1997, Hodgson 2000).

Another microorganism that may pose a human health risk in insectaries is *Serratia marcescens*. This bacterium produces a characteristic red pigment that is readily apparent *en masse*. It can cause septicemia as well as infections of the eye, respiratory system, and urinary tract. In the insectary setting, it is a danger primarily to immunocompromised individuals.

Other commonly benign microorganisms may become problematic when they are encountered *en masse*. Since species identification is rare, it is appropriate to treat a massive microbial breakout with the care and respect due a possible pathogen. In short, use appropriate personal protection equipment (PPE) when cleaning up any insects or rearing materials contaminated with microorganisms.

IV. Prevention of Hazards Peculiar to Insectaries

Since the mid 1960s, I have worked in several insectaries that gave little to no consideration to protecting workers from insect allergens and pathogens. Although in the intervening years these problems have received increasingly greater attention and progress has been made (e.g., Davis and Jenkins 1995), sanitation can be an elusive goal. When repeated attempts to achieve asepsis fail, such efforts can become burdensome, and an insectary can remain deficient in this regard.

Preventing exposure of insectary personnel to potential allergens and pathogens can be accomplished by sanitation of the insectary environment or by use of personal protection equipment (PPE). The focus in this section will be on PPE.

A. Sanitation

Various aspects of insectary sanitation have been covered elsewhere, including insectary design to control personnel traffic and air flow patterns (Chap. 2 *Insectary Design and Construction*), surface sanitation (Chap. 8 *Microbial Contamination and Insect Rearing*), and air filtration (Chap. 9 *Entomopathogens and Insect Rearing*, Chap. 10 *Insect Rearing Production Systems, A Case Study: The Southwestern Corn Borer*).

B. Personal Protection Equipment (PPE)

OSHA has seven sections of regulations devoted to PPE (29 CFR 1910.132–138). The PPE categories covered here include gloves, goggles/face masks, and respirators. Because PPE continually undergoes modification and improvement, one must take the information provided here as a starting point only.

1. Gloves. As OSHA states in 29 CFR 1910.138, "Employers shall select, and require employees to use, appropriate hand protection when employees' hands are exposed to hazards such as those from skin absorption of harmful substances; severe cuts or lacerations; severe abrasions; punctures; chemical burns; thermal burns; and harmful temperature extremes."

I wish to focus on the examination glove, which is the glove most often in use in the insectary. Handling spiny insects such as cockroaches or grasshoppers can result in punctures of the skin that lead to the development of allergic sensitivities. Allergies can also develop due to contact of the skin with the waterproofing hydrocarbons or other materials produced by the insect. These gloves, used by those who need sensitive touch, are often made of latex rubber. Similar, but latex-free, gloves are made of nitrile (acrylonitrile butadiene rubber) or vinyl (polyvinyl chloride).

There are an increasing number of people who have developed allergic sensitivity to latex, including, very notably, latex gloves with powder used as a lubricant (Beezhold and Beck 1992, Tomazic et al. 1994). The powder becomes contaminated with latex, or a chemical used in its production, and the contaminated powder is easily dispersed in air and inhaled. A worker's sensitivity to latex can be severe, costing some their jobs, profession (e.g., nursing), and more. The American College of Allergy, Asthma, and Immunology (ACAAI) considered that 6% of the US population has evidence of latex sensitivity as indicated by an increase of anti-latex IgE antibodies in their blood (ACAAI 1995). The federal government is being pressured to have the presence of latex noted on all products containing natural rubber. It is highly advisable to use nonlatex gloves in the workplace. Exam gloves made of nitrile or vinyl have become quite competitive in quality and cost with latex. At the very least, latex gloves that contain a lubricant powder should be avoided. Replacement of other latex products such as rubber tubing should be considered as well.

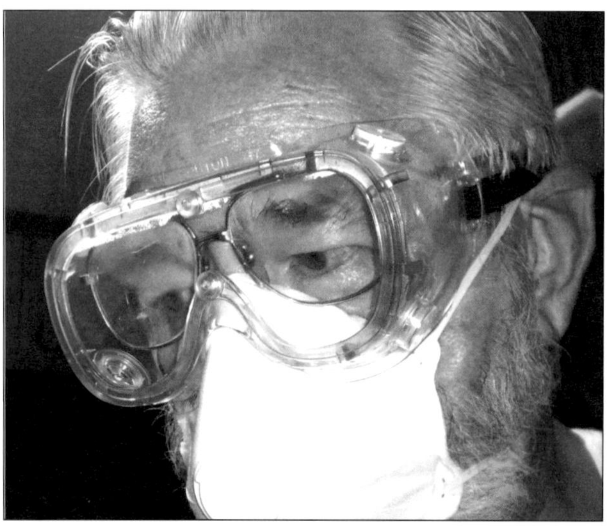

Fig. 9. Combination of single-use mask with soft-sided goggles and prescription spectacles. The wearer has a beard, which reduces the effectiveness of the mask. Further, although the goggles allow for passive ventilation, the lenses of the spectacles were not treated with detergent, allowing some condensation to form.

2. Goggles/Face Masks. OSHA states in 29 CFR 1910.133(a)(1), "The employer shall ensure that each affected employee uses appropriate eye or face protection when exposed to eye or face hazards from flying particles, molten metal, liquid chemicals, acids or caustic liquids, chemical gases or vapors, or potentially injurious light radiation."

Goggles and face masks are commonly worn to protect people from mechanical damage to the eyes. In an insectary, airtight or air-filtered goggles may be considered appropriate to protect the eyes from airborne allergens (Fig. 9). A worker currently without allergies may be annoyed at this level of protection. Fogged eye protectors can be a major annoyance and cause noncompliance. Spreading a minute and invisible coating of concentrated dishwashing detergent on the inside surface of the lenses prevents the beading of any moisture that may be present. Workers who already have an insect allergy should find relief from eye irritation using airtight or air-filtered goggles.

3. Respirators. OSHA states the following in explanation of its position that respirators are not the solution of first choice in dealing with airborne hazards (OSHA URL):

> Engineering and work practice controls are generally regarded as the most effective methods to control exposures to airborne hazardous substances. OSHA considers the use of respirators to be the least satisfactory approach to exposure control because:
>
> - respirators provide adequate protection only if employers ensure, on a constant basis, that they are properly fitted and worn.

- respirators protect only the employees who are wearing them from a hazard, rather than reducing or eliminating the hazard from the workplace as a whole (which is what engineering and work practice controls do).
- respirators are uncomfortable to wear, cumbersome to use, and interfere with communication in the workplace, which can often be critical to maintaining safety and health.
- the costs of operating a functional respiratory protection program are substantial—including regular medical examinations, fit testing, training, and the purchasing of equipment.

Considering this statement, it seems appropriate to manipulate the physical plant to maintain clean air. This is best done via original building design but can be managed through retrofitting. This subject is dealt with in Chap. 2 *Insectary Design and Construction*.

The proper use of respirators for personnel working in dirty areas of an insectary must be taken seriously. Bellas (1981) stated, "The reduction of exposure to insect detritus is the chief means by which sensitization can be avoided or the effects of allergies can be minimized." The use of respirators is one way this reduction can be accomplished. For our purpose here, there are three basic types of respirators: 1) disposable, negative pressure face masks; 2) replaceable-filter, negative pressure respirators; and 3) positive pressure respirators.

Single-use, negative pressure face masks should be NIOSH approved and perform as well as reusable masks in the fit test. Consequently, only higher quality single-use masks with a one-way valve to prevent unseating of the mask during exhalation and to reduce moisture buildup are recommended (Fig. 10). Primarily designed to remove particulate matter, they can also be obtained with a fabric treatment such as activated charcoal to reduce some noxious odors as well. All negative pressure masks and respirators are designed for people who do not have compromising facial hair.

Negative pressure respirators fitted with one or two disposable filters selected for specific purposes, such as some hazardous vapors and airborne particulate material, are reusable (Fig. 11A, B). These respirators should be routinely cleaned and stored as directed by the manufacturer and the filters replaced in a timely manner. Full-face, negative pressure masks

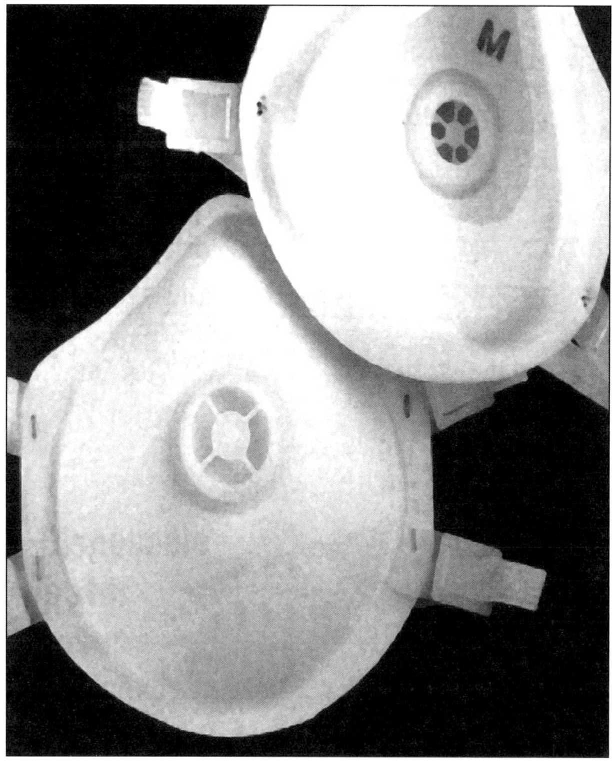

Fig. 10. Single-use, negative-pressure, particulate mask.

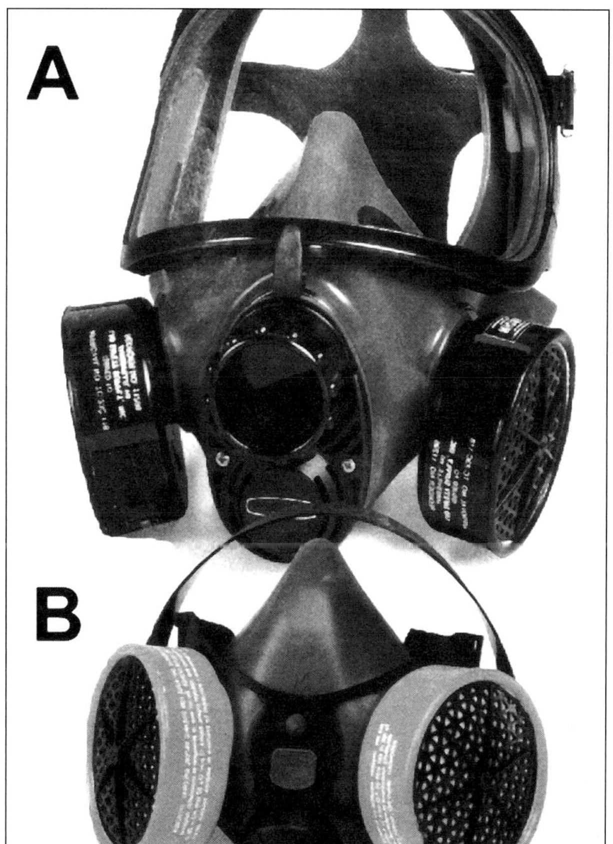

Fig. 11. Reusable, negative-pressure respirators with disposable filters. (A) Full-face respirator. (B) Half-face respirator.

should be considered for their added ability to protect the eyes of the worker. Some full-face masks allow for prescription lenses. Half-face, reusable respirators and single-use masks can allow one to wear separate eye protectors at the same time (Fig. 9).

Positive pressure respirators are for use by people who are physically unable to wear a negative pressure respirator (doctor's orders) or who have a preference for this type in work situations that favor it (Fig. 12). Filtered air may be supplied by flexible tubing from an external, stationary, blower-filter source or by a portable, battery-operated, blower-filter apparatus worn by the person. Although these masks may look ideal, the wearer may find them exhausting to wear for long periods.

A procedure has been established in OSHA 29 CFR 1910.134 to have each worker who wears a negative pressure respirator to be tested at least once a year. In brief, the process involves a knowledgeable physician's signature approving a worker's physical ability to wear one or another type of negative pressure respirator, instruction on the proper care and use of the respirator, and a testing procedure to prove that the specific respirator worn by the worker is functioning properly. An atomizer is used to disperse a bitter tasting airborne particulate (Bitrex™) into a small tent surrounding the mask wearer (Fig. 13). The face mask wearer must be previously checked for sensitivity to the bitter-tasting chemical being injected into the test

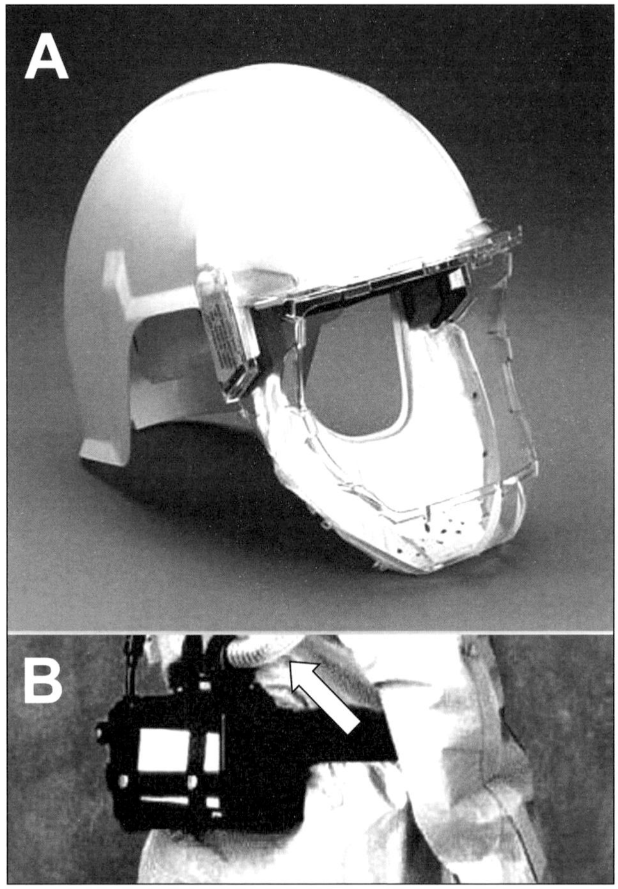

Fig. 12. Positive-pressure respirator. (A) Helmet/air distribution mask. (B) Battery-powered blower, which supplies filtered air to the helmet/air distribution mask via a flexible hose (see arrow). Filtered air may also be supplied by a satellite blower system through a hose extension from another location.

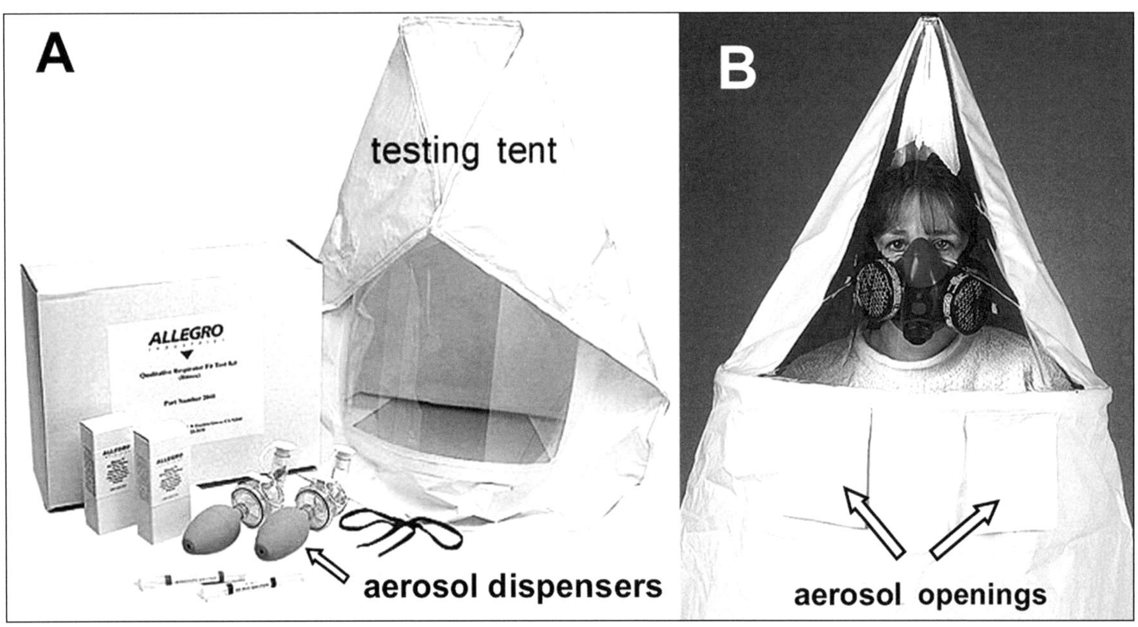

Fig. 13. Respirator leak testing equipment. (A) Testing tent. (B) Test tent in use.

tent. If the wearer can detect the bitter taste during the test, adjustments are made and the test repeated until none is detectable. Documentation is required to be OSHA compliant. With the possible exception of a physician's examination, all of OSHA's requirements for users of respirators may be accomplished by trained in-house staff. Alternatively, there are private businesses that specialize in certifying OSHA compliance.

V. Administration of Insectary Safety

The National Safety Council (NSC) has estimated that the total average cost (direct and indirect) of a lost time injury is around \$39,000, and of a work-related fatality \$1,240,000 (NSC URL). The costs of accidents include lost expertise, slowed production, and payment of workers' compensation. Having a well-designed and run safety, occupational health, and environmental protection (SOHEP) program can reduce these costs.

However, the rate of increase in safety benefit declines with increased investment. For example, "absolutely" clean air may be prohibitively expensive while "almost" clean air is financially feasible and provides a level of safety not significantly less than the former. Insectary managers must decide how to strike the balance between the costs and benefits of their facility's SOHEP program—within the constraint of meeting minimum mandatory standards. It should now be obvious that there are usually several approaches to deal with any given safety issue. Also, even within a given approach, the cost can fall into quite a wide range. For instance, site-built equipment may be as functional as manufactured but much less costly. The expertise and imagination that management uses to make SOHEP decisions will determine the cost-benefit ratio of the resulting program.

A. Best Management Practices for Safety

Several management practices that will facilitate insectary safety are listed below. Most of these practices are discussed in Chap. 3 *The Insectary Manager*. Two will be discussed below in more detail.

1. Show personnel that managers and supervisors vigorously support the safety program by word and deed, including funding.

2. Encourage a positive attitude throughout the insectary that supports good safety behavior.

3. Keep rules and procedures for personnel reasonably uncomplicated and current, and have a hard copy of their SOPs (Standard Operating Procedures) in each worker's possession.

4. Assign a safety officer, and arrange for his/her proper training, which is shared with the rest of the staff.

5. Document the safety program and any occurrence concerning safety and health, and maintain a good file system.

6. Conduct reasonably frequent safety training for all personnel, and have the safety files accurately reflect the safety training completed by each person.

7. Emphasize to supervisors their responsibility in developing and maintaining the safety program (the employer is ultimately responsible for SOHEP).

B. Safety Officer/Committee

If the insectary is part of a larger organization, a Safety Committee should already exist. If this is the case, a part time safety officer from the insectary should be a member of that committee and be specialized in assessing the particular safety hazards of insectaries. An independent insectary with several personnel should consider training one or more staff members to serve as safety officers. Smaller, independent insectaries should have at least one person assigned to safety duty. Family-owned and run insectaries should work together to minimize safety and health hazards to which family members are exposed, even though, technically, such businesses do not fall under the federal and state safety regulatory umbrella.

Having safety equipment and supplies distributed throughout the workplace does not guarantee that the products will work when needed. Safety equipment such as emergency eye-wash stations, respirators, and emergency showers must be inspected on a routine basis. This inspection and maintenance should be carried out by the safety officer or his designee. Fire extinguishers and fume hoods pose special problems in that they should be inspected and serviced by properly equipped and trained technicians at specified intervals.

Job-related mishaps in an insectary that result in injury or sickness should be reviewed in depth by the safety officer or, preferably, by a safety committee. The resulting report should evaluate the causes of the problem and suggest preventative measures on the basis of which management will take action. If workers' compensation is requested, a detailed report to the appropriate governmental authorities will be required.

C. Safety Training

Insectary managers should ensure that the safety officer receives appropriate training and that the safety officer passes this information on to the rest of the personnel in safety training sessions. The training sessions should be held annually for all personnel and on an individual basis for each new employee.

The standard operating procedure (SOP) is an instrument for management not only to maintain consistency in production but also to promote personnel safety. Every SOP document should include a simple statement of any safety hazards inherent in the procedure and the appropriate safety procedures and PPE necessary to avoid injury. Every worker should receive and have access to copies of relevant SOPs so that the statement, "You didn't tell me that!" carries no credence if uttered by an injured worker. The safety officer should reevaluate the safety aspects of all SOPs annually or whenever an SOP is modified.

VI. Conclusions

The effectiveness of an insectary's SOHEP program depends on both the management and staff. A worker who does not currently have an allergy and does not believe he or she will develop one will find wearing PPE somewhat irksome. The extra effort and expense associated with the elimination or significant reduction of allergy problems associated with detritus and other factors that induce sensitivity (e.g., latex rubber, airborne antibiotics, microbials) is no small matter. Management needs to believe in the worthiness of these health objectives and be fortunate enough to have personnel and advisers that can optimize safety and health while minimizing cost. The issue of employees rejecting part of the PPE program and perhaps even signing a document to that effect should be considered with serious reservations. For instance, what effect will an inconsistent safety program have on the rest of the staff?

Building and air handling system design may reduce investment in PPE, time and effort necessary for proper PPE application, and waste in improper PPE application (e.g., leaky face masks or respirators). Before hardware and procedures have been developed and put to use, there should be a plan for determining their effectiveness that does not depend on the appearance of allergies among the workers. This might be done with some form of quantitative airborne detritus test (Chap. 8 *Microbial Contamination and Insect Rearing*).

It is understandable that effective SOHEP-type programs are more difficult to support in some countries or situations. In such instances, it is hoped the material presented in this chapter will encourage the development of a program that can be supported by administrators and staff. Much can still be done with information, training, and a positive attitude to make the insectary safer.

Differences in the insect species being reared, insectary design, and machinery and techniques used for rearing produce wide variation in safety and health problems. In short, each insectary has its own set of SOHEP problems to solve. The challenge addressed in this chapter is to provide information useful in the selection and adaptation of procedures, equipment, and supplies useful in obtaining a higher level of safety. Share your findings on what does work with the insect rearing community and keep a sense of humor about what does not.

References

Amerecho, R. 2005. Anaphylactic shock. Internet J. Health 4(2).

ACAAI (American College of Allergy, Asthma and Immunology). 1995. Latex allergy: an emerging health-care problem. Ann. Allergy, Asthma and Immunology 75: 19–21.

Beezhold, D., and W. Beck. 1992. Surgical glove powders bind latex antigens. Archives of Surgery 127: 1354–1357.

Bellas, T. E. 1981. Insects as a cause of inhalation allergies: A bibliography. Commonwealth Scientific and Industrial Research Organization, Australia, Division of Entomology Report No. 25: 1–64.

Bellas, T. E. 1990. Occupational inhalant allergy to arthropods. Clin. Rev. Allergy 8: 15–29.

Bellas, T. E. 1999. Insects as a cause of inhalant allergies: a bibliography 1900–1993. CSIRO Division of Entomology Report No. 25. Commonwealth Scientific and Industrial Research Organization, Canberra, Australia.

Brown, A. F. 1995. Anaphylactic shock: mechanisms and treatment. J. Accid. Emerg. Med. 12: 89–100.

Corn, J. K. 1992. Response to occupational health hazards: a historical perspective. Van Nostrand Reinhold, New York, NY.

Davis, F. M., and J. N. Jenkins. 1995. Management of scales and other insect debris: occupational health hazards in a lepidopterous rearing facility. J. Econ. Entomol. 88: 185–191.

Guarnieri, M. 1992. Landmarks in the history of safety. Journal of Safety Research. 23: 151-158.

Hodgson, M. 2000. Sick building syndrome. Occup. Med. 15: 571–585.

Kagen, S. L., M. D. Chapman, and R. Muthiah. 2004. Insects, arachnids and human allergic disease, pp. 205–242. *In* M. I. Levine and R. F. Lockey (eds.), Monograph on insect allergy, 4th ed. American Academy of Allergy, Asthma and Immunology, Milwaukee, WI.

(NSC) National Safety Council. (www.nsc.org/resources/issues/estcost.aspx).

Neugut, A., A. Ghatak, and R. Miller. 2001. Anaphylaxis in the United States: an investigation into its epidemiology. Arch. Intern. Med. 161: 5–21.

(NIOSH) National Institute of Occupational Safety and Health. 1983. Health Hazard Evaluation Report no. GHETA 81-121. National Institute for Occupational Safety and Health, Morgantown, WV.

(OSHA) Occupational Safety and Health Administration. (www.osha.gov/SLTC/etools/respiratory/exposure_control/exposure_control.html).

Platts-Mills, T. A. E., E. Erwin, P. Heymann, and J. Woodfolk. 2005. Is the hygiene hypothesis still a viable explanation for the increased prevalence of asthma? Allergy 60 (suppl. 79): 25–31.

Redlich, C. A., J. Sparer, and M. R. Cullen. 1997. Sick-building syndrome. Lancet 349: 1013–1016.

Renz, H., N. Blümer, S. Virna, S. Sel, and H. Garn. 2006. The immunological basis of the hygiene hypothesis. Chem. Immunol. Allergy 91: 30–48.

Stark, B. J, and T. J. Sullivan. 1986. Biphasic and protracted anaphylaxis. J. Allergy Clin. Immunol. 78: 76–83.

Ten, R. M., J. S. Klein, E. Frigas. 1995. Allergy skin testing. Mayo Clin. Proc. 70(8): 783–784.

Tomazic, V., E. Shampaine, A. Lamanna, and T. Withrow. 1994. Cornstarch powder on latex products is an allergen carrier. J. Allergy Clin. Immunol. 93: 751–758.

Wirtz, R. A. 1980. Occupational allergies to arthropods—documentation and prevention. Bull. Entomol. Soc. Am. 26: 356–360.

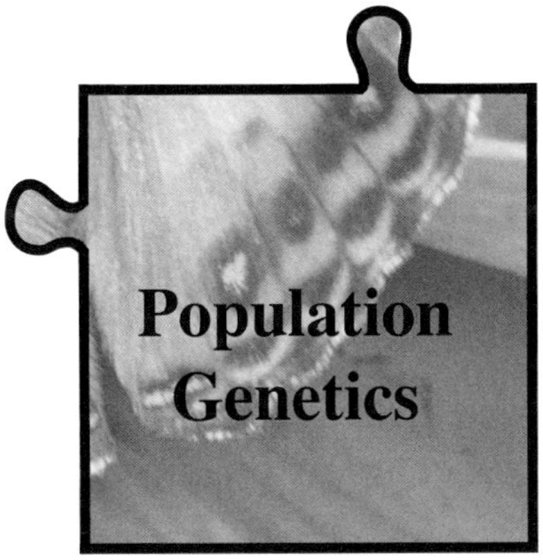

Chapter 5

Genetic Considerations and Strategies for Rearing High Quality Insects

MICHAEL A. CAPRIO
Professor of Entomology
Department of Entomology and Plant Pathology
Mississippi State University
Mississippi State, MS

I. Introduction

At the most fundamental level, the principles of insect rearing are dependent on genetic and evironmental factors. Consequently, applied population genetics is one of the two primary pillars on which insect rearing is based.

The same population genetic principles apply to caged populations as to wild or feral populations. The primary changes that occur during domestication are a result of restricted population size and exposure of the population to an entirely novel environment. All rearing managers must accept the fact that genetic changes will occur in their populations; for all practical purposes, such changes are inevitable (Mangan 1992, Mason et al. 1987). As a rearing specialist, one's job will be to manage those changes; and the techniques one adopts to manage them will depend on how the colony is to be utilized. There is no single optimal way to manage a colony. For example, if one intends to release individuals into wild populations and expects them to mate with wild insects, then the colony should be maintained as genetically close to the wild population as possible. At the other extreme, a geneticist may desire to develop multiple inbred strains, strains that contain no genetic variation whatsoever. Neither colony is intrinsically "better" than the other. They are each extremely important for their particular use, and the rearing conditions for each would be radically different.

II. Processes of Genetic Change

Our interest in this chapter is in heritable changes, changes in genetics that are inherited from generation to generation. More specifically we are interested in the phenotypic expression of such changes, changes that result in differences in some measurable property of the colony. One must remember that changes can occur in populations that do not reflect underlying genetic changes but are rather the result of environmental variation. Changes in temperature or diet can affect survivorship, pupation rates, and in some cases even sex ratio. Such phenotypic changes can be easily corrected because no underlying change has occurred in the genetic constitution of the colony. These environmentally induced phenotypic changes may be similar to changes resulting from genetic changes (e.g., behavior, reproductive capacity, longevity), but only genetic changes will be passed on through subsequent generations. One should be aware that there may also be genotype by environment interactions in which genetic differences are only manifested under specific environmental conditions. These types of changes may be of interest when the reared insects are to be used in environments other than the laboratory, for the effects of genetic changes that are not obvious in the laboratory environment could assume major importance in the new environment (most often the field).

To understand why colonies change genetically, one must understand what forces or mechanisms are acting on the populations. In the domestication of insects, two primary mechanisms exist that will lead to changes in allele frequencies (proportion of copies of a given variant of a gene in the population). The first is *genetic drift*, or the random variation in allele frequencies due to sampling between generations. The second mechanism is *selection*, which occurs as a colony adapts to its new environment. The direction of change in allele frequencies with genetic drift is random, and changes in duplicate colonies are unlikely to be identical. In contrast, changes due to selection are directional and likely to be repeated in all duplicate colonies. As an aside, there are other processes such as *assortative mating* (individuals of like genotype choosing to mate with each other more often than one would expect by chance alone, e.g., humans mate assortatively by height) that change the frequency with which alleles for different genes are associated with one another. We will not deal with such processes here as they do not in themselves result in changes in allele frequencies.

A. Genetic Drift

Genetic drift is a result of random sampling of alleles between generations. As many of us know from statistics, increasing population size (analogous to sample size) will decrease the effects of random sampling (Fig. 1a, b, Fig. 2 a, b). The effects of drift are most

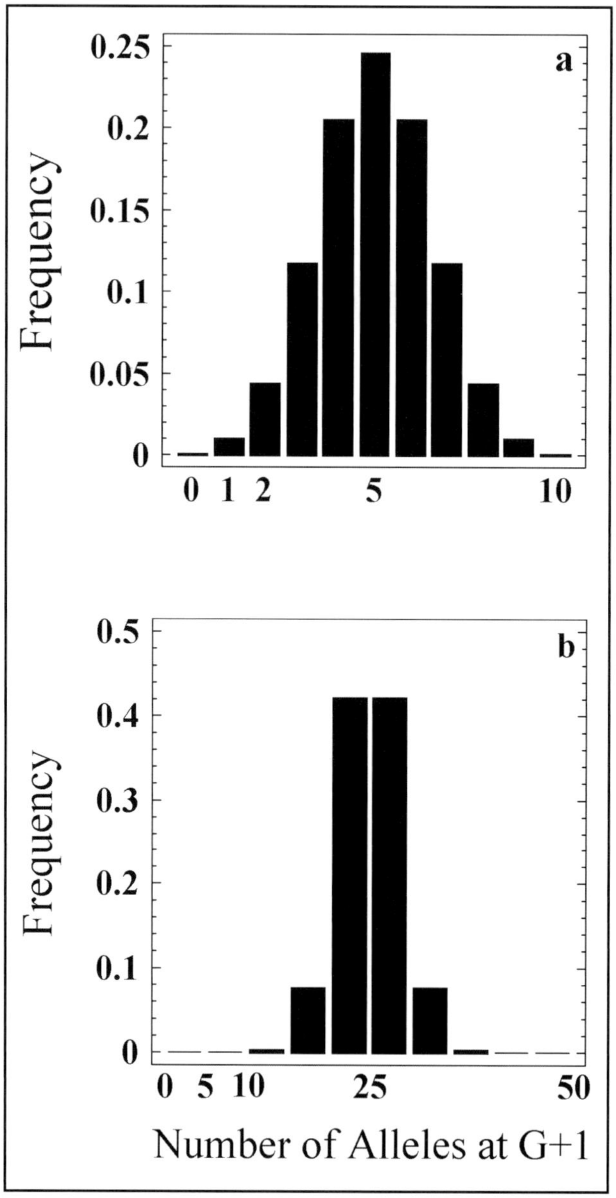

Fig. 1. The sampling probabilities for an allele at frequency 0.5 in generation G with diploid population sizes: a) 5 and b) 25.

important in small populations (proportional to $\frac{1}{N^2}$, where N is the number of diploid individuals in the population). An important difference between selection and drift is that genetic drift affects the entire genome equally, while selection only acts at specific loci. This has practical implications, as the effects of drift can be quantified by examining a sufficient number of randomly selected loci. The effects of selection can only be determined for specific loci, and it is difficult if not impossible to be certain that changes at other loci have not occurred.

As previously noted, changes due to drift are random; and the changes in allele frequencies between replicate colonies will be unpredictable. In the absence of mutation, population genetics theory predicts that a colony will eventually become fixed for one allele at each genetic locus. The probability that an allele will become fixed in any given colony is directly proportional to its initial frequency. Thus, if an allele is at a frequency of 30% in a source population, one would expect it to become fixed in 30% of the colonies established from this source population. The implication is that if equal numbers of insects are combined from many replicate colonies to establish a new composite colony, the expected allele frequencies in this composite colony will be identical to the original source population (assuming the colonies are independent with no dispersal between them and no changes due to selection; Fig. 3). This is indirectly supported by studies that demonstrate inbreeding depression in individual replicate colonies but a restoration of fitness with intercolony matings (Morjan et al. 1999).

In most cases, rearing managers want to limit the amount of genetic drift that occurs. At first glance, because the loci most affected by genetic drift are ones that are not under selection, the different alleles present would appear to have no fitness advantage over each other; and changes due to genetic drift should have little impact on the population. However, although these changes might be neutral in the laboratory environment, they might not be neutral in other environments (the field, for example). Changes in the environment in the laboratory might also occur, such as the outbreak of a new pathogen; and the alleles

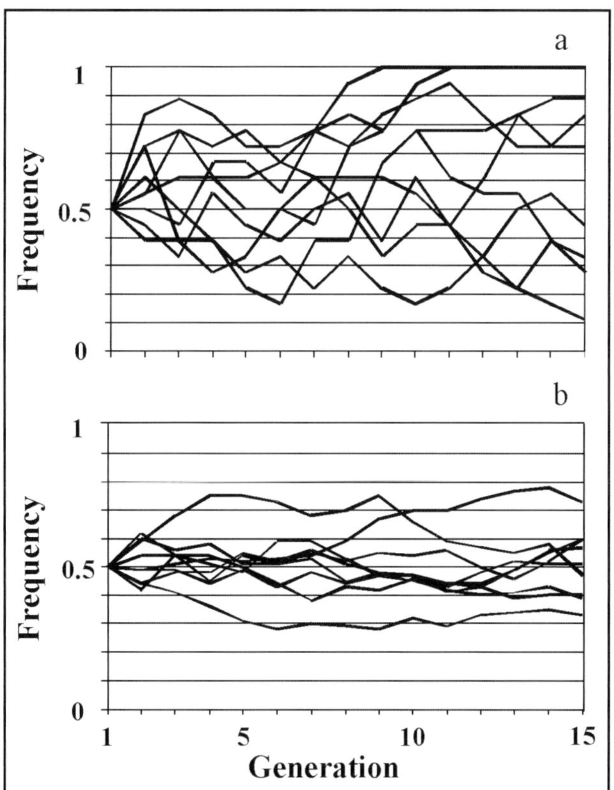

Fig. 2. Computer simulation of genetic drift for two diploid population sizes: a) 9 and b) 50.

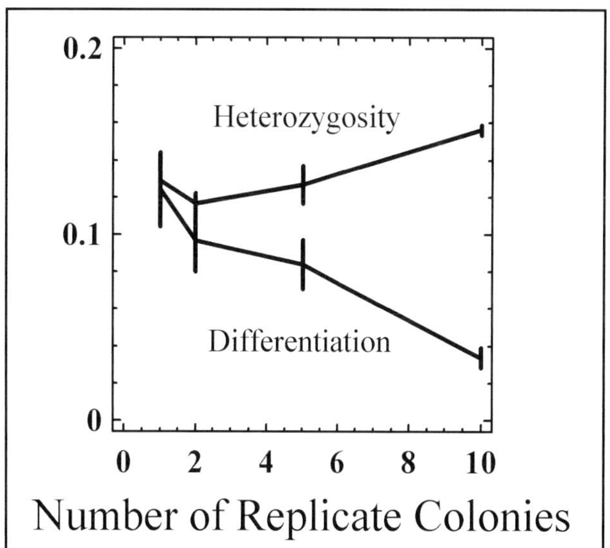

Fig. 3. The effect of maintaining the same overall number of insects divided into replicate colonies. Measures of genetic quality are given for a reconstituted colony with equal numbers from each replicate colony.

Table 1. The effect of inbreeding on population traits of the tobacco budworm, *Heliothis virescens* (Lepidoptera: Noctuidae) (Roush 1986)

Trait	Outcrossed	Sib-mated	p[a]
Successful egg hatch, %	65	58	NS
Development time, days	23.1	23.9	NS
Larval viability, %	86	52	<0.001
Successful pupal eclosion, %	75	60	<0.02
No. eggs/mated female	400	273	0.04

[a]Probability of observed difference between outcrossed and sib-mated groups under H_0: $\mu_o=\mu_s$.

might have different fitnesses in the face of this new environment. Alleles that confer resistance to a pesticide might be considered neutral in a laboratory environment (assuming they had no fitness costs associated with them), but they would certainly not be neutral in fields treated by that insecticide. Increases in homozygosity due to drift may also expose recessive lethals, alleles that are not expressed as heterozygotes but are lethal as homozygotes. Such alleles are typically maintained at low frequencies in natural populations, and the frequency of homozygotes is very low. Increases in frequency of such alleles due to drift will increase homozygosity at a rate related to the square of the frequency (i.e., much faster than the change in frequency). The expression of these alleles as recessive lethals may cause reductions in colony fitness (i.e., genetic load).

Many studies have demonstrated that inbreeding depression such as that caused by genetic drift can reduce colony fitness. Roush (1986) examined the effects of inbreeding on laboratory populations of the tobacco budworm, *Heliothis virescens* (Lepidoptera: Noctuidae), and found significant reductions in three of five fitness related traits (Table 1). Oosterhout et al. (2000) found evidence of inbreeding depression in laboratory metapopulations of *Bicylus anynana* (Lepidoptera: Satyridae), while Saccheri et al. (1998) found inbreeding depression in field metapopulations of the Glanville fritillary, *Melitaea cinxia* (Lepidoptera: Nymphalidae). Hughes (1995) found declines in male fitness components of *Drosophila melanogaster* (Diptera: Drosophilidae) associated with inbreeding. Inbreeding in the beetle *Propylea quatuordecimpunctata* (Coleoptera: Coccinellidae) resulted in large decreases in fitness characters such as egg hatch, clutch size, and first instar to adult survivorship (Morjan et al. 1999). Intercrossing the inbred strains restored most of the viability. Geden et al. (1992) found a rapid deterioration of behavioral and fitness traits in the parasitoid *Muscidifurax raptor* (Hymenoptera: Pteromalidae) when reared in culture and attributed at least some of the changes to inbreeding. Even in cases of parasitic Hymenoptera where one would expect reduced inbreeding depression because males are functionally haploid (exposing and theoretically purging recessive mutations), some cases of inbreeding depression have been identified (Luna and Hawkins 2004).

B. Selection

The second mechanism by which allele frequencies in laboratory colonies can be altered is selection. By definition, loci under selection are not neutral; so genetic drift becomes less important. Changes due to selection are directional and, for a given environment, will always occur in the same direction. In contrast to drift, selection affects only a portion of the genome. When insects are moved from a field environment into a laboratory colony, they are exposed to a novel environment; and genetic changes will occur. The process has aptly been described as domestication (Bartlett 1984). These changes may or may not interfere with the intended purpose of the colony. It is important to evaluate quality in the environment in which the colony insects are to be used. If they are to be released into the field, then quality must be evaluated in the field. There are a large number of examples in which laboratory-adapted insects did not perform well in the field, often for reasons that might not be obvious beforehand (White et al. 1977, Waldvogel et al. 1982, Raulston et al. 1987, Lance et al. 1988, Proshold et al. 1993). On the other hand, although changes were observed in a laboratory colony of pink bollworm, *Pectinophora gossypiella* (Lepidoptera: Gelechiidae), those changes apparently did not interfere with use of the colony as a source for sterile males (A. C. Bartlett, unpublished data).

It must be remembered that the fitness of individuals can vary with the environment and that different genotypes can react differently to each environment. In a recent example with the nematode *Caenorhabditis elegans*, Walker et al. (2000) examined the fitness of a strain with increased longevity. They hypothesized that the gene for increased longevity did not exist in nature because strong fitness costs were associated with it. When they examined the population for fitness costs under normal rearing conditions, none were found. However, when reared under simulated field conditions with cyclical periods of starvation, the long-lived colony had a large reduction in fitness. Clearly, what had appeared to be a highly fit strain under normal rearing conditions actually exhibited relatively poor fitness under conditions it would likely experience in the field. It is important to again emphasize that fitness must be evaluated in the environment in which one intends to use the reared insects.

III. Quality

Ultimately, a colony manager must produce quality insects while at the same time meeting financial, space, and human resource limitations. Rearing will always be a compromise. In order to rear insects of high genetic quality, one must first define *genetic quality*. There is no single, absolute metric of genetic quality. Genetic quality is best seen as the degree to which a derivative population remains similar to the source population. One cannot measure genetic quality without reference to the original source population. For example, some species of insects naturally have less genetic variability than others so that it is impossible to determine genetic quality based only on an absolute level of heterozygosity. While no universal metric exists, one can evaluate genetic quality by comparing the laboratory colony with the original source population for some relevant metric. Several metrics are possible, including polymorphism, heterozygosity, and differentiation (*Fst*, genetic distance).

Polymorphism is the proportion of loci for which more than one allele is present in the population. This statistic is a property of the population and is likely to change very slowly. It is not a particularly sensitive metric of quality.

Heterozygosity is defined as the average proportion of loci expected to have two different alleles in the average diploid individual. This metric depends on variation at the level of the individual and is therefore a more sensitive measure of genetic quality than polymorphism. It is well known that heterozygosity tends to decline as genetic drift leads towards the fixation of neutral loci (Fig. 4). Heterozygosity for loci under selection, however, can increase, though in most cases that increase is only temporary. Also, founding effects associated with colony formation may also temporarily increase heterozygosity.

Measures of genetic differentiation such as *Fst* estimate the degree of differentiation that has built up in a colony relative to a reference population. If the original source population may be referred to as the goal of rearing quality insects, then measures of differentiation would seem to be the ideal metric. The only limiting factor is that DNA from the source population must be preserved because wild populations can also change over time. Three solutions, at least one of which should be used for any established colony, are as follows: 1) storage of whole insects at –80°C, 2) extraction of DNA and storage at –20°C, or 3) storage of tissues in 70% ethanol.

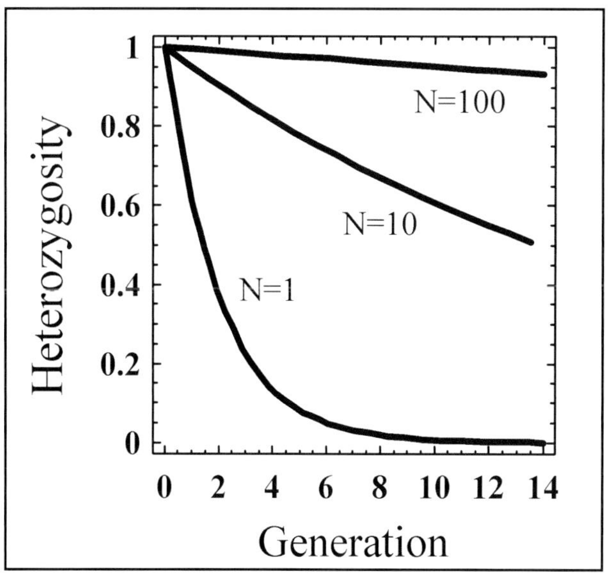

Fig. 4. The loss of heterozygosity over time due to genetic drift for various population sizes.

IV. Maintaining Genetic Quality

A. Amelioration of Genetic Drift

As noted above, the effect of drift is inversely related to population size. As population size increases, the effects of drift become less important (Fig. 3a, b). This contrasts with the effect of selection, which is invariant with regard to population size. The most effective method to reduce the impact of genetic drift is to maintain a large effective population size (N_e). The effective population size of a colony, however, may be much smaller than the actual number of individuals, i.e., census size, of the colony. Formally, N_e is the size of an ideal population showing the same level of drift/inbreeding as the non-ideal population. There are a number of factors that can affect N_e. These include the following: temporal variance in population size, sex ratio, proportion of males mating, and variance in female fecundity.

1. Temporal Variance in Population Size. The effective population size over time is determined by the harmonic mean population size. The harmonic mean is given by the following:

$$N_e = \frac{n}{\frac{1}{N_1} + \frac{1}{N_2} + \cdots + \frac{1}{N_i} + \cdots + \frac{1}{N_n}} \qquad \text{Eq. 1}$$

for n, number of generations, and N_i, population size in generation i.

While the common arithmetic mean is an unbiased estimator, the harmonic mean is biased towards smaller numbers, i.e., generations with small population size will have a disproportionately large effect on the effective population size. Consider a population maintained for two generations with population sizes of one hundred in the first generation and ten in the second. The arithmetic mean (average) would be 55, but the harmonic mean is (2/[1/100+1/10]) or 18.2. The harmonic mean is often less and never greater than the arithmetic mean. Notice that the larger the denominator, the smaller the effective population size; and the size of the denominator in this example depends primarily on the generation of population size 10, i.e., the generation with the smallest size. The implication of this result is that one should try to maintain a consistently large population size and avoid bottlenecks—generations with dramatically smaller population size. Single generations where problems with the colony (disease, hardware malfunctions, human error, etc.) reduce population size can have a long-term effect on the colony. This result also suggests that considerable effort should be made to establish a colony with a reasonably large sample relative to the maintenance size (Fig. 5). McElfresh and Millar (1993) suggested that reductions in fecundity in a colony of the peach twig borer, *Anarsia lineatella* Zeller (Lepidoptera: Gelechiidae), were due to inbreeding that occurred during colony establishment.

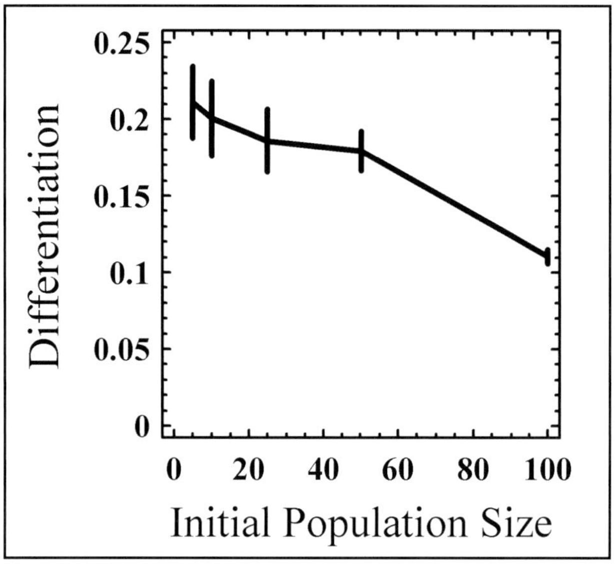

Fig. 5. The effect of initial population size on the level of differentiation (*Fst*) between the source population and the colony population after 100 simulated generations of rearing.

2. Sex Ratio. Deviations from a 1:1 sex ratio in a diploid species reduce the effective population size. This is reasonable because the gene pool from which alleles are drawn for the next generation is reduced for the sex in lower abundance. The effective population size is given as:

$$N_e = \frac{4N_mN_f}{N_m + N_f} \qquad \text{Eq. 2}$$

for N_m, number of males, and N_f, number of females.

Sex ratios can often become distorted, particularly when rearing hymenopterous parasitoids, though in this case the genetics are complicated by their haplodiploid genetic system (males are haploid and females are diploid) so that the optimal sex ratio is biased towards females.

3. Proportion of Males Mating. This is actually equivalent to variation in sex ratio, but it illustrates that even when the sex ratio is apparently unbiased, prob-

Table 2. Probability of the loss of an allele when a large population goes through a bottleneck size of N = 10 (20 copies of each gene)

Frequency	Probability
0.5	0.000001
0.1	0.12
0.01	0.82
0.001	0.98

lems can still occur. In this case, unequal probabilities of males obtaining mates can reduce population size. For example, in *Drosophila* species, studies have frequently shown that some males are "studs" and mate frequently, while other males, "duds", are less likely to secure access to females. In a simple setting with 100 females and 100 males, if only 10 of the males account for all of the mating, the effective population size would be just 36 ($N_f = 100$, $N_m = 10$).

4. Variance in Female Fecundity. In population genetic terms, an ideal population is one in which female fecundity, the number of eggs oviposited per female, follows a Poisson distribution. For the Poisson distribution, the mean is equal to the variance; and while some females by chance lay more eggs than others, the expected number of eggs per female is the same. If the variance in female fecundity is less than the mean, the effective population size is increased. In particular, if all individuals contribute equally to the next generation, the effective population size is increased to $2N - 1$, so that the effective population size is almost doubled. Arranging such an equitable distribution of female fecundity, however, is unlikely to be cost effective for insects. On the other hand, if the variance in female fecundity is greater than the mean, the effective population size is decreased. One can imagine that when a colony is established, some males and females will be more adapted to the laboratory environment than others and will produce more offspring, increasing the variance in fecundity. In addition to the effects of selection at loci associated with the laboratory adaptation, this process reduces effective population size at precisely the time when the colony size is likely to be low. This example again emphasizes the critical nature of colony establishment.

As mentioned previously, eliminating genetic bottlenecks from colony maintenance will reduce the effects of genetic drift and will decrease genetic differentiation between the source population and a colony. Bottlenecks are also responsible for the rapid loss of rare alleles (Table 2). The loss of such alleles could be important if they are involved in adaptation to novel environments.

5. Infusion. Although efforts to maintain a high effective population size can ameliorate the effects of genetic drift, eventually a laboratory colony will become differentiated from the source population. Once this has occurred, infusion with insects from the source population is a useful technique to reduce differentiation. Because wild insects may not be competitive with laboratory-adapted insects in the laboratory environment, it is probably best to mate lab insects of one sex with wild insects of the other sex. For diploid insects, either sex may be used from the source population; but for haplo-diploid insects, such as many parasitoids, males contribute only one-third of the alleles to the ensuing generations (rather than one half) so that infusion will be less effective when males are used. If one desires to infuse at all loci, then the sex of the source population should be the heterogametic sex. For example, in many lepidopterous species, females are the heterogametic sex (WZ), while males are the homogametic sex (ZZ). Infusion with males would have no impact on allele frequencies on the W chromosome.

In avoiding population bottlenecks, one should be careful not to infect the colony with disease from the source population. Frequently, it is advisable to main-

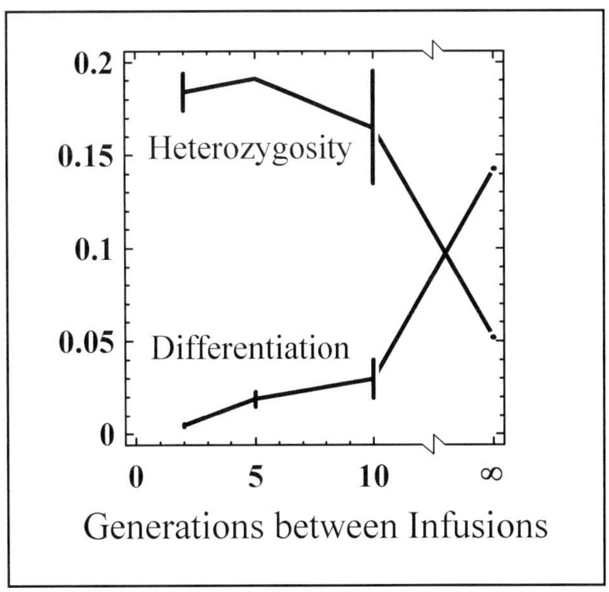

Fig. 6. Computer simulation of the effects of infusion of colony heterozygosity and differentiation from the source colony. The colony was assumed to be 100 individuals in size and was reared for 100 generations.

tain a backup colony that is not infused in case the newly infused colony suffers adverse effects. Computer simulation of the effect of infusion suggests that it is a very effective method to counteract the effects of genetic drift (Fig. 6). The frequency of infusion required for any given colony is inversely related to the colony effective population size; larger colonies require infusion less often.

B. Amelioration of Selection

The response of the colony manager to laboratory adaptation will vary with the intended use of the colony and may range from ignoring such changes (after all, a laboratory-adapted colony is easier to rear) to infusion with wild insects, to complete replacement of the colony periodically. While infusion is a good solution to negate the effects of genetic drift, it is unlikely to be as effective in ameliorating the effects of selection unless the magnitude of selection is small. If one assumes a normal infusion process (mating wild individuals of one sex with laboratory-reared insects of the opposite sex), that the colony is fixed for laboratory traits, and that the source population is nearly fixed for field-adapted traits, then a generation of infusion will only reduce the frequency of laboratory-adapted traits to 50 percent. Heritability, the rate at which a population responds to selection, is greatest for *additive-effect alleles* (i.e., loci without dominance) when allele frequencies are near 0.50 (Fig. 7). In this most extreme case, infusion moves allele frequencies for laboratory-adapted traits precisely into that range where the colony will respond most rapidly to selection. In most situations, the effect will not be this extreme; but the general process of removal of field-adapted traits by laboratory selection and replenishment by infusion will remain the same, and a sawtooth pattern of allele frequencies is likely to be observed (Fig. 8). It should be obvious that infusion cannot maintain traits against which there is strong selection. Infusion will, however, prevent those field-adapted alleles from being lost from the population. It might be suspected that if one were rearing insects that mate in culture and must perform under field conditions (such as an inundative release of a biological control agent), infusion would be a less acceptable solution than in the case of insects that are released to mate with field insects (inoculative release).

In general, however, a colony manager must accept that a laboratory colony will be different from the source field population, particularly at loci under strong selection. Even complete replacement of a colony will not entirely eliminate the effects of selection. On the other hand, a laboratory-adapted colony may still meet quality-control standards (A. C. Bartlett, unpublished data). In short, a quality insect need not necessarily be identical to a field insect.

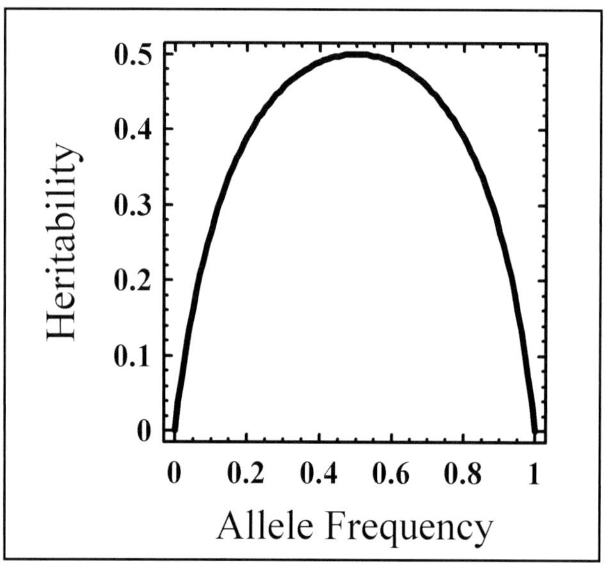

Fig. 7. The dependence of heritability on allele frequency in the case of additive genetic variance. A moderate level of environmental variance (V_e) is assumed: V_e / (phenotypic variance between genotypes) = 0.125 (Roughgarden 1979).

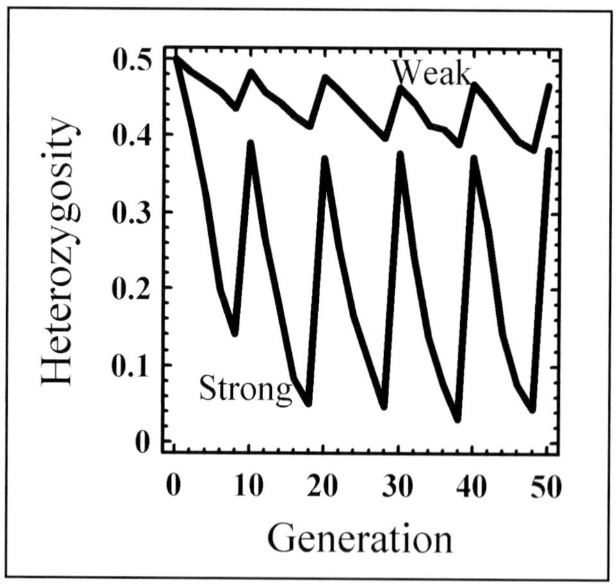

Fig. 8. The effect of infusion and selection on a trait initially present at a frequency of 0.5. The selection coefficient for strong selection is 0.5, for weak selection 0.1.

V. Conclusions

As an insect colony manager, one must be aware of the types of genetic changes that may occur in colonies and how one might ameliorate those changes. In general, changes due to genetic drift will affect the entire genome; but these changes are relatively easily ameliorated by maintaining large populations sizes and through the use of infusion from source or field populations. Selection will cause laboratory adaptation in one's colony, but the genetic changes will only occur at specific loci. Colonies will become laboratory adapted. Infusion is also an important technique for counteracting selection, although not as effective as for genetic drift. Quality assessment for loci under selection must focus on field performance, and relevant measures will depend upon the specifics of one's situation and insect. In general, however, it is extremely important to maintain a genetic record of one's population. Only with such a record can one hope to determine the cause of problems that may arise concerning the genetic quality of a colony. A genetic record should be considered an essential part of good laboratory recordkeeping for the colony. Finally, colony maintenance will of necessity be a compromise between quantity and quality, between rearing in order to harvest large numbers and rearing to maintain a high effective population size in order to minimize the risk of population bottlenecks.

References

Bartlett, A. C. 1984. Genetic processes of domestication, pp. 2–8. *In* E. G. King and N. C. Leppla (eds.), Advances and challenges in insect rearing. USDA, ARS, U.S. Government Printing Office, Washington, DC.

Geden, C. J., L. Smith, S. J. Long, and D. A. Rutz. 1992. Rapid deterioration of searching behavior, host destruction, and fecundity of the parasitoid *Muscidifurax raptor* (Hymenoptera: Pteromalidae) in culture. Ann. Entomol. Soc. Am. 85: 179–187.

Hughes, K. A. 1995. The inbreeding decline and average dominance of genes affecting male life-history characters in *Drosophila melangaster*. Genet. Res. Camb. 65: 41–52.

Lance, D. R., T. M. Odell, V. C. Mastro, and C. P. Schwalbe. 1988. Temperature-mediated programming of activity in male gypsy moths (Lepidoptera: Lymantriidae): implication for the sterile male technique. Environ. Entomol. 17: 649–653.

Luna, M. G., and B. A. Hawkins. 2004. Effects of inbreeding versus outbreeding in *Nasonia vitripennis* (Hymenoptera: Pteromalidae). Environ. Entomol. 33: 765–775.

Mangan, R. L. 1992. Evaluating the role of genetic change in insect colonies maintained for pest management, pp. 269–288. *In* T. E. Anderson and N. C. Leppla (eds), Advances in insect rearing for research and pest management. Westview Press, Boulder, CO.

Mason, L. J., D. P. Pashley, and S. J. Johnson. 1987. Insects in altered environments. Fla. Entomol. 70:49–58.

McElfresh, J. S., and J. G. Millar. 1993. Establishment and rearing of *Anarsia lineatella* (Lepidoptera: Gelechiidae) on a meridic diet. J. Econ. Entomol. 86: 1399–1404.

Morjan, W. E., J. J. Obrycki, E. S. Krafsur. 1999. Inbreeding effects of *Propylea quatuordecimpunctata* (Coleoptera: Coccinellidae). Ann. Entomol. Soc. Am. 92: 260–268.

Oosterhout, C.-van, W. G. Zijlstra, M. K. Heuven,-van, P. M. Brakefield. 2000. Inbreeding depression and genetic load in laboratory metapopulations of the butterfly *Bicyclus anynana*. Evol. 54: 218–225.

Proshold, F. I., V. C. Mastro, and G. L. Bernon. 1993. Sperm transfer by gypsy moths (Lepidoptera: Lymantriidae) from irradiated males: implication for control by inherited sterility. J. Econ. Entomol. 86:1104–1108.

Raulston, J. R., H. M. Graham, P. D. Lingren, and J. W. Snow. 1987. Mating interaction of native and laboratory-reared tobacco budworms released in the field. Environ. Entomol. 5:195–198.

Roughgarden, J. 1979. Theory of population genetics and evolutionary ecology: an introduction. Macmillan Publishing, New York, NY.

Roush, R. T. 1986. Inbreeding depression and laboratory adaptation in *Heliothis virescens* (Lepidoptera: Noctuidae). Ann. Entomol. Soc. Amer. 79: 583–587.

Saccheri, I., M. Kuussaari, M. Kankare, P. Vikman, W. Fortelius, and I. Hanski. 1998. Inbreeding and extinction in a butterfly metapopulation. Nature 392: 491–494.

Waldvogel, M. G., V. C. Mastro, C. H. Collison, and E. A. Cameron. 1982. Evaluation of pheromone-mediated responsiveness of laboratory-reared irradiated, laboratory-reared nonirradiated, and feral male gypsy moths. Environ. Entomol. 11: 35–354.

Walker, D. W., G. McColl, N. L. Jenkins, J. Harris, and G. J. Lithgow. 2000. Evolution of lifespan in *C. elegans*. Nature 405: 296–297.

White, L. D., M. Koslinska, and Z. W. Suski. 1977. Codling moth: field-cage mating competitiveness of radiosterilized males. J. Econ. Entomol. 70: 64–65.

Chapter 6

Environmental Biology of Insect Rearing

JOHN C. SCHNEIDER
Professor of Entomology
Department of Entomology & Plant Pathology
Mississippi State University
Mississippi State, MS

I. Introduction

Most biotic and abiotic environmental factors affect the biology of insects during rearing. Biotic factors include crowding and intraspecific competitive effects such as cannibalism, nutritional influences, and the pathological and beneficial effects of a wide array of microorganisms. Nutrition and the effects of microorganisms in insect rearing are considered in Chap. 7 *Insect Feeding, Nutrition, and Artificial Diets*, Chap. 8 *Microbial Contamination and Insect Rearing*, and Chap. 9 *Entomopathogens and Insect Rearing*. Abiotic environmental factors include light, temperature, humidity, and substrate. This review of environmental biology as it relates to insect rearing is restricted to abiotic factors and among abiotic factors to light, temperature, and humidity.

I have attempted to apply biological generalizations to the practical endeavor of insect rearing. At times, the starting point for the discussion of biological principles may be somewhat removed from insect rearing, but in each case the discussion is concluded with suggested applications.

II. Light

Light can be a very powerful environmental influence in insect rearing. For example, in the Handbook of Insect Rearing, Vol. II (HIRII), Baumhover (1985) reports that a low wattage light must be used at night in the cage with tobacco hornworm, *Manduca sexta* (Lepidoptera: Sphingidae), adults or else they will not feed and oviposit. Furthermore, if the intensity of light outside the cage is greater than inside, the moths will be attracted away from feeders and oviposition substrates.

Light has the properties of intensity, wavelength, and degree of polarization. The absolute level of these properties and their temporal and spatial variation can affect the physiology, behavior, and ecology of insects in various ways including the following: phototaxis, circadian rhythms, diapause, and population growth. Each of these effects of light on insects will be considered in turn.

A. Phototaxis

Directed movement of an animal in response to light is referred to as "phototaxis" (Matthews and Matthews 1978). It should be distinguished from "phototropism," which is directional growth of a plant in response to light. Phototaxis can be toward (+), away (–), or at a constant angle with respect to direction of the light. Undirected movements, i.e., changes in activity level with no preference in direction, can occur in response to light and is referred to as "photokinesis."

HIRII contains at least 22 examples of phototaxes of insects in culture. In some cases, taxis is a problem to be prevented. For example, Robertson (1985) reports that (+) phototaxis by neonate spruce budworm larvae, *Choristoneura fumiferana* (Lepidoptera: Tortricidae), must be avoided by maintaining darkness at the tops of feeding containers so that they do not leave the food and spin hibernacula at the tops of the containers. In other cases, phototaxis is an opportunity to be exploited. For example, Ankersmit (1985) reports that light can be used to attract adult summer fruit tortrix, *Adoxophyes orana* (Lepidoptera: Tortricidae), out of mass rearing containers into bags for transfer to oviposition containers.

1. Intensity & Wavelength. The light-gathering power of the eyes of insects depends on their morphology and optics (Land 1981). Insects have four main types of eyes: the stemmata of larvae and the ocelli and apposition or superposition ommatidia of adults. The function of *ocelli* is not well understood; but like *stemmata*, they are simple, single-lens eyes. The effective area over which light is gathered per photoreceptor is given in Table 1 for representatives of the main types of insect eyes and for humans. There is a 300-fold range in sensitivity among the insects. The photoreceptors in an *apposition ommatidium* (one facet of the compound eyes of most diurnally active adult insects) of the honey bee, *Apis mellifera* (Hymenoptera: Apidae), receive light through an aperture of only 25 µm and thus are relatively insensitive. Nevertheless, note that apposition ommatidia are more sensitive light detectors than is the *fovea* (the high-resolution, color-sensitive, light-adapted, central part of the retina of humans). Larval stemmata have a single, large lens, which is ca. 200 µm in diameter in a sawfly, *Perga* sp. (Hymenoptera: Pergidae). *Superposition ommatidia* (facets of the compound eyes of nocturnally active adult insects) pool light from neighboring ommatidia to produce an effective aperture (10 mm) much greater than that of a single ommatidium (340 µm) in the Mediterranean flour moth, *Ephestia kuhniella* (Lepidoptera: Pyralidae). The eyes of nocturnal moths such as *E. kuhniella* are more sensitive in terms of light-gathering power per photoreceptor than is the peripheral retina of the dark-adapted eyes of humans. Thus, the human eye is inadequate to determine whether light levels are too low for such insects to detect light.

Phototactic response of insects can vary with light intensity. For example, phantom midge larvae, *Chaoborus punctipennis* (Diptera: Chaoboridae), exhibit positive phototaxis if the light is low intensity but negative phototaxis if it is high intensity (Swift and Forward 1980). These responses contribute to vertical movements of this aquatic stage of growth toward the water's surface at dusk and away at dawn. There is also a circadian rhythm (Sec. II. B) component in phototactic responsiveness in this insect.

Most insects have three kinds of photoreceptors in their eyes, and these differ in the wavelength to which they have peak sensitivity: 340 nm (UV), 440 nm (violet/blue), and 530 nm (green) (Briscoe and Chittka 2001). Some sawflies, Lepidoptera, and Coleoptera have an additional photoreceptor maximally sensitive to light of 600 nm (orange/red). Although receptors sensitive to a wide range of wavelengths may be present in the eye, light from various directions is not detected by the same set of photoreceptor types because they do not generally occur in all ommatidia. For example, the dorsal regions of the compound eyes of many diurnal insects have a higher incidence of ommatidia with UV-sensitive photoreceptors than does the rest of the eye.

Table 1. Effective area over which light is gathered per photoreceptor for a variety of organisms (Land 1981)

Organism	Sensitivity, mm^2	Type of Eye
Ephestia kuhniella	83	Superposition ommatidia
Homo sapiens (rods)	37	Camera
Perga spp.	34	Stemma (camera)
Apis mellifera	0.32	Apposition ommatidia
H. sapiens (cones)	0.023	Camera

Many insects displaying phototaxis are most sensitive to either UV (300–400 nm) or green light (492–572 nm) (Burkhardt 1964, Menzel 1979, Gilbert 1994). Chrysomelid beetles generally are more sensitive to green light than to UV. However, honey bees are 16-fold more sensitive to UV light. The fruit fly, *Drosophila melanogaster* (Diptera: Drosophilidae), is about equally responsive to UV and green light at low light intensities but 45-fold more responsive to UV than green light at high light intensities.

TIP 1: Try using UV and green lights of various intensities to manipulate the movement of insects in culture.

The threshold light intensity triggering phototactic response can be very low for larvae with stemmata or for adults with superposition eyes. Stewart et al. (1969) found that the threshold distance for attraction of adult corn earworm, *Helicoverpa zea* (Lepidoptera: Noctuidae), and *M. sexta* to a 10-watt blacklight is 75–120 m. At a distance of 1 m, a lamp of only ca. 10^{-3} watt would be required to give the same response as that at 100 m. Larvae of the silkworm, *Bombyx mori* (Lepidoptera: Bombycidae), have a threshold for phototactic response to green light that is similar to the threshold for light detection by the dark-adapted human eye (Kitabatake et al. 1983).

TIP 2: Use a light meter—rather than one's eyes—to measure the degree of darkness to which insects are exposed in culture.

2. Polarization. Due to the fundamental structure of the photoreceptive cells of insects (as well as of other arthropods), insects generally can detect the plane of polarization of polarized light; and many have been demonstrated to do so (Waterman 1981). Polarized light has been found to be used—or hypothesized to be used—by insects for direction finding, location of water bodies, station keeping, enhancement of visual contrast, and glare reduction. The first two behavioral responses are concerned with movement and have been referred to by the term *polarotaxis*

There are two common natural processes by which light is polarized: scattering in the atmosphere and reflection (Jenkins and White 1957). Light in the sky is polarized by scattering off small particles and is most pronounced at 90° with respect to the direction of the sun. This results in spatial patterns of degree of polarization and angle of polarization of light that change with time of day. The direction of movement and level of activity of many species of adult and larval insects have been shown to depend on the polarization of light, especially near the zenith (Wellington 1974, Doane and Leonard 1975). For example, Doane and Leonard (1975) found that large gypsy moth, *Lymantria dispar* (Lepidoptera: Lymantriidae), larvae moving across a clearing in a forest tended to move parallel to the angle of polarization of light but did not move at all near noon when the degree of polarization of light near the zenith was at a minimum.

TIP 3: Try rotating a polarizing filter between the light source and insects to manipulate their direction of movement.

Light reflecting off smooth surfaces, including natural surfaces such as a body of water or plant leaves (Grant et al. 1993) and artificial surfaces such as glass or Plexiglas, is polarized. Maximum polarization is produced when light is incident on these surfaces at ca. 55°.

TIP 4: Consider the possibility that inappropriate aggregation or movement of insects housed in smooth-surfaced chambers and illuminated by point sources of light may result from patterns of reflection polarization.

3. Spatial Pattern. *L. dispar* larvae in the open move toward vertical silhouettes (Doane and Leonard 1975), and passion flower butterflies, *Heliconius* spp. (Lepidoptera: Nymphalidae), appear to recognize their passion vine host plants by leaf shape (Gilbert 1975). These behaviors suggest that both immature and adult insects have visual systems of sufficiently high resolution that spatial pattern can be very important. Estimates of the visual acuity of adult insects, based on interommatidial angle, range from 0.24° for the common green darner, *Anax junius* (Odonata: Aeshnidae), to 7.5° for a walking stick, *Dixippus morosus* (Phasmida: Heteronemiidae), with a median value of 1.7°

(Land 1997). An insect with a visual acuity of 1° can distinguish details ca. 0.5 cm in size at a distance of 30 cm. Behavioral observations of the visual acuity of holometabolous immatures (e. g., caterpillars) indicate that they have the ability—in spite of the sparseness of their widely divergent stemmata—to distinguish differences in angular size of 2–5° by scanning their surroundings and integrating this information in the central nervous system (Gilbert 1994).

Some insects can estimate distances, whether by stereoscopic vision (Frantsevich and Pichka 1976, Rossel 1983) or by velocity parallax (Eriksson 1979, Horridge 1986). Others cannot but are attracted to preferred shapes that subtend the greatest apparent angle. For example, larvae of the chequered swallowtail butterfly, *Papilio demoleus* (Lepidoptera: Papilionidae), respond as strongly to a closer, small host plant leaf as they do to a farther, large leaf (Saxena and Khattar 1977). Similarly, fall cankerworm, *Alsophila pometaria* (Lepidoptera: Geometridae), adult females, which are wingless, emerge from pupation sites in litter on the forest floor to walk to the tree trunk whose silhouette subtends the greatest angle—even if it is not the closest one (Schneider 1980).

TIP 5: Try to increase the degree of attractiveness of spatial patterns to insects in culture by presenting the patterns close to the insects.

4. Periodicity. The effects of temporal variation in light intensity on movement behavior of insects will be illustrated at two rather widely different frequencies of variation in intensity. First, day length can influence the strength of the phototactic response. A water strider, *Gerris lacustris* (Heteroptera: Gerridae), has been found to have a stronger phototactic response when adults experience long-day photoperiods (Harada 1991). Second, *A. mellifera* has a slower flight speed, increased wing beat frequency, and reduced ability to avoid obstacles under lights powered by 60-Hz AC current because it can perceive the 120 Hz flickering in intensity of these lights (Praagh 1975). Fast-flying, diurnal insects, e.g., the tsetse fly, *Glossina morsitans* (Diptera: Glossinidae), have light-adapted, flicker fusion frequencies of up to 300 Hz while slow-flying and/or nocturnal insects, e.g., the American cockroach, *Periplaneta americana* (Blattodea: Blattidae), have dark-adapted, flicker fusion frequencies of less than 100 Hz (Miall 1978). Thus, the latter are indifferent to the flickering of normal, artificial lighting, but the former are potentially affected.

TIP 6: Consider using three-phase lighting, which has a 360-Hz flicker, when rearing fast-flying, diurnal insects.

B. Circadian Rhythms

The physiological condition and the incidence of behaviors of insects vary with time of day. Cole and Adkisson (1964) reported that mortality of the boll weevil, *Anthonomus grandis* (Coleoptera: Curculionidae), due to exposure to an insecticide varied with a period of 6 h and was tied to the light:dark (LD) cycle such that mortality was as low as 10% at the beginning of *photophase* (the light phase of a LD cycle) but as high as 90% three hours later. Almost all oviposition by the European corn borer, *Ostrinia nubilalis* (Lepidoptera: Crambidae), occurs within the first 4 h after the beginning of *scotophase* (the dark phase of a LD cycle) (Skopik and Takeda 1980). However, it is not the start of scotophase per se that triggers oviposition. If females are transferred from a LD cycle to continuous darkness (DD), periods of oviposition activity will *free run*, occurring at 24.2-h intervals with a gradual increase in temporal variability, until death of the females. Apparently, oviposition in *O. nubilalis* is linked to a *pacemaker* (an internal clock) that persists independently of the LD cycle.

Each of the above is an example of a *circadian rhythm*, the cycling of a physiological or behavioral activity under the control of a pacemaker whose period is approximately 24 h and independent of temperature (Saunders et al. 2002)[1]. For behaviors or physiological processes that are expressed repeatedly during the lifetime of an individual, such as locomotion, feeding, and sometimes mating and oviposition, the circadian rhythm is expressed as a periodicity of occurrence of these behaviors or processes within individuals. For behaviors or physiological events that are expressed once in the lifetime of an individual, such as hatch, pupation, eclosion, and sometimes mating, the circadian rhythm is expressed as a periodic "gating" of the time of day these events occur within a mixed-age population.

[1]Unless otherwise indicated, information presented in the sections on circadian rhythms and diapause is taken from Saunders et al. (2002).

In some insects, the behaviors and physiological processes mentioned above are expressed periodically on a ca. 24-h cycle but are not circadian because they are under immediate control of exogenous factors. For example, mating in the Queensland fruit fly, *Bactrocera tryoni* (Diptera: Tephritidae), takes place in a narrow range of absolute light intensity occurring only at dusk, mating does not free run under DD conditions, and the timing of mating with respect to an LD cycle does not depend on *L* (duration of the photophase period) (Sec. II. B. 2).

1. Continuous Dark. If a population has been reared in continuous darkness (DD) for its entire life cycle or at least since the last photosensitive stage, e.g., early egg stage in the pink bollworm, *Pectinophora gossypiella* (Lepidoptera: Gelechiidae), circadian rhythms are not apparent; and life history events that are typically gated are temporally random. This reduces the maximum rate at which individuals enter a particular developmental stage and consequently the maximum number of individuals that can be harvested from a colony within a given time period. For example, concentration of egg hatch within a few hours per day allows the collection of many more hatchlings within a 2-h window than if egg hatch occurs uniformly over time.

Circadian rhythms of individuals in a population reared under DD can be initiated—all at the same phase—by exposure to a single pulse of light (Fig. 1). Following initiation/synchronization, a given circadian rhythm free runs at the intrinsic period of the pacemaker, τ (generally between 22 and 27 h). A subsequent pulse of light can cause a sudden shift in the phase of the circadian rhythm with respect to external clock time (Fig. 1). The magnitude and direction of the shift depend on the duration of the pulse and the phase of the circadian cycle at which the second pulse of light occurs (Fig. 2). Short pulses of low intensity can cause shifts of a few hours, while a 15-min period of high intensity light can cause a shift of up to 12 h. Exposure to light in the subjective early night causes a delay in the rhythm and exposure in the late night an advance. Shorter wavelengths of visible light, including blue to violet, and UV light are most effective in causing phase shifts. This latter characteristic is explained by the fact that a vitamin B_2-based chromophore, which is most sensitive to shorter wavelengths, is responsible for light reception in circadian rhythms (Veerman 2001).

TIP 7: To maximize availability of individuals of a particular age within stage reared in DD, expose them to a pulse of light rich in the shorter wavelengths.

TIP 8: To work with insects reared in DD without affecting their circadian rhythms, use red light.

2. Light:Dark Cycles. Because the period of a free-running circadian rhythm is generally not precisely 24 h, the rhythm drifts with respect to external clock time. A *Zeitgeber* is an external cyclic factor that can entrain a free-running rhythm so that it maintains

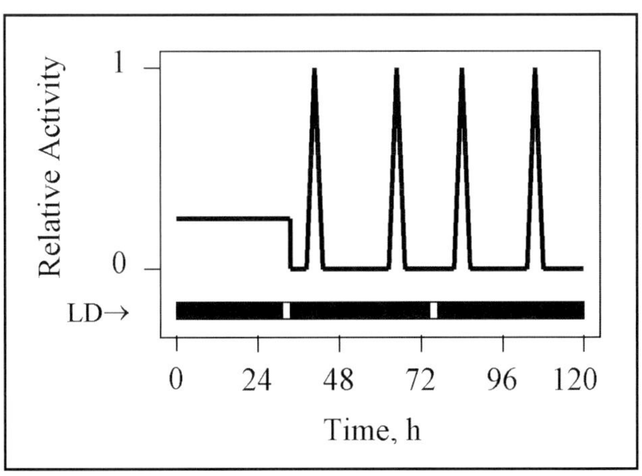

Fig. 1. Synchronization of a circadian activity in continuous darkness due to a pulse of light and a subsequent shift in the rhythm due to a second pulse.

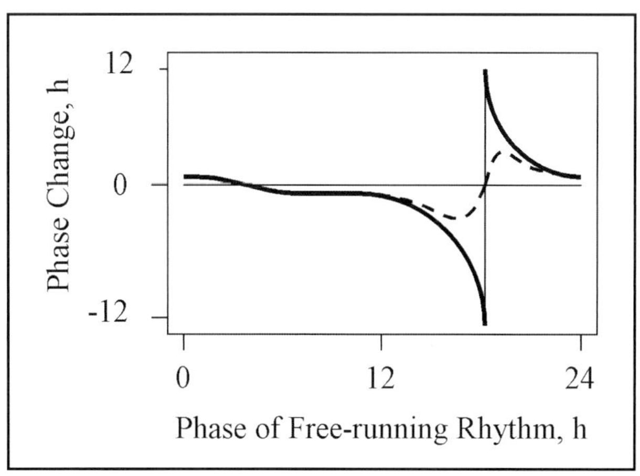

Fig. 2. Phase response curves for circadian rhythms in response to strong (solid curve) and weak (dashed curve) light pulses as a function of the phase of the free-running rhythm at which the pulse occurs.

a constant phase relationship with respect to the external cycle. The most common Zeitgeber is an LD cycle; and if $L+D = T = 24$ h, for D (duration of the scotophase period), the event/behavior/state of interest will occur at the same external clock time each day (Fig. 3). Entrainment results from a combination of phase-dependent advances and delays in the free-running rhythm at lights out, D_{out}, and lights on, D_{on}, such that $\tau + D_{out} + D_{on} = T$ is satisfied. Additional characteristics of circadian rhythms arise from the entrainment phenomenon: 1) A single pulse of light during scotophase causes a transient shift of the circadian rhythm for a few to several days before relaxation back to the previous entrained condition (Fig. 4). 2) The phase relationship of a circadian event/behavior/state with respect to the Zeitgeber depends on the combination of L and D used. For example, eclosion of the fruit fly *Drosophila pseudoobscura* (Diptera: Drosophilidae) occurs 2 h after lights on under 12:12 LD but 4 h after lights on under 16:8 LD. 3) A pulse of light that occurs regularly during the early scotophase effectively increases L (Fig. 5). A light pulse occurring regularly during the late scotophase has a similar effect. In the transition zone, near the middle of the night, small differences in the timing of a periodic light pulse can dramatically shift the circadian activity rhythm by up to 12 h (Fig. 2).

TIP 9: Avoid exposing insects reared in LD to light during the scotophase—especially the middle of scotophase.

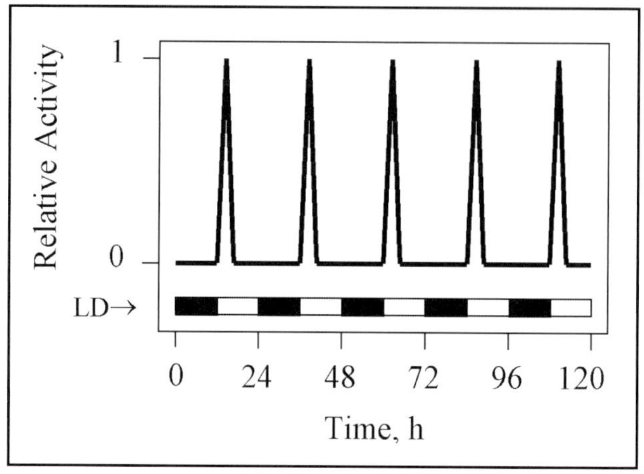

Fig. 3. Circadian rhythm entrained to a light:dark (LD) cycle.

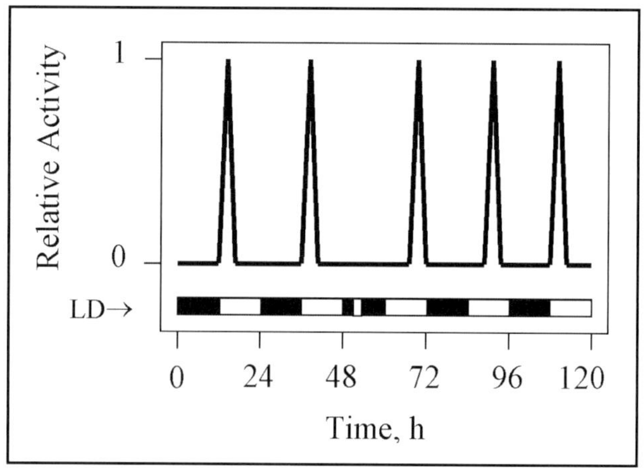

Fig. 4. Transients in a circadian rhythm entrained to a light:dark (LD) cycle following a nocturnal light pulse.

3. Continuous Light. Many population-level events gated by circadian rhythms dampen rapidly in continuous light (LL) so that they occur randomly with respect to time of day. Transfer of arrhythmic insects from LL to DD or to LD reinitiates circadian rhythms at the phase characteristic for the beginning of scotophase.

Individual, locomotory circadian rhythms, although typically not lost under LL, may assume a τ different from that under DD. *Aschoff's Rule* (Saunders et al. 2002) states that transfer from DD to LL of diurnally active species results in a decrease in τ, and the reverse occurs in nocturnally active species; however, a number of exceptions to this rule are noted.

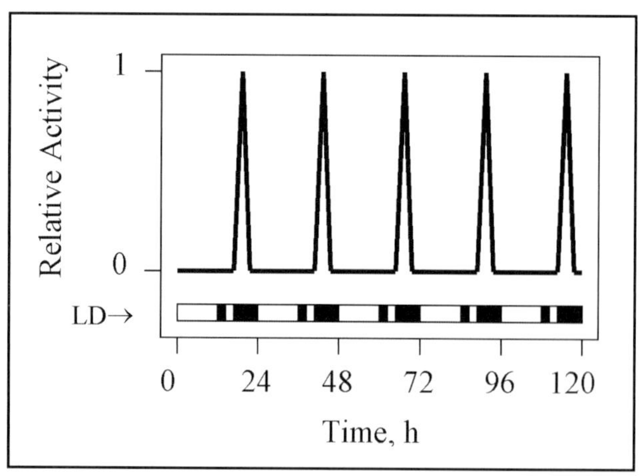

Fig. 5. Shift in circadian rhythm due to a periodic pulse of light during the scotophase.

Rearing under LL can result in decreased survival (Sec. II. D. 1) and decreased fecundity (Sec. II. D. 2).

These effects may explain why only 9/70 of the rearing procedures for Lepidoptera, Coleoptera, and Diptera in HIRII specify LL conditions.

TIP 10: Do not initiate an insect rearing program using LL lighting, unless one has specific reasons for doing so.

C. Diapause

Consider the following scenario: thousands of tobacco budworm, *Heliothis virescens* (Lepidoptera: Noctuidae), pupae were reared for a time-sensitive experiment but did not eclose as adults at the expected time. A review of rearing conditions showed that an inadvertent combination of short photophase duration and low temperature induced the insects to enter *diapause,* a type of dormancy (Tauber et al. 1986). The diapause state required weeks to terminate, thus rendering the insects useless. As this example shows, diapause can be a problem to be avoided or terminated as quickly as possible; however, in various situations, diapause can be used to one's advantage (Sec. III. B. 3).

When wide-area deterioration in habitat quality occurs predictably, e.g., during winter for most species at higher latitudes, a common adaptive response for insects has been diapause. Diapause consists of a persistent reduction in metabolic rate due to neuroendocrine changes that are triggered by a token stimulus—usually photoperiod but also thermoperiod or other factors. Diapause may occur in adults, in which case it is called *reproductive diapause*. The stage of development of the insect sensitive to the induction cue is generally not the same stage in which diapause is spent, e.g., the early instar larva may detect the induction cue, but diapause is spent as a pupa.

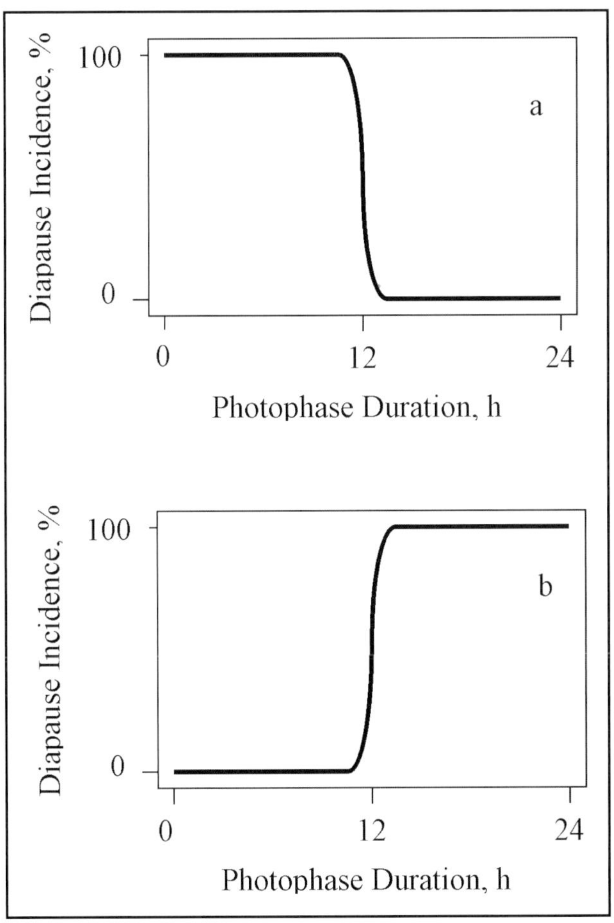

Fig. 6. a) Effect of photophase duration on incidence of diapause that is associated with the winter. b) Effect of photophase duration on incidence of diapause that is associated with the summer.

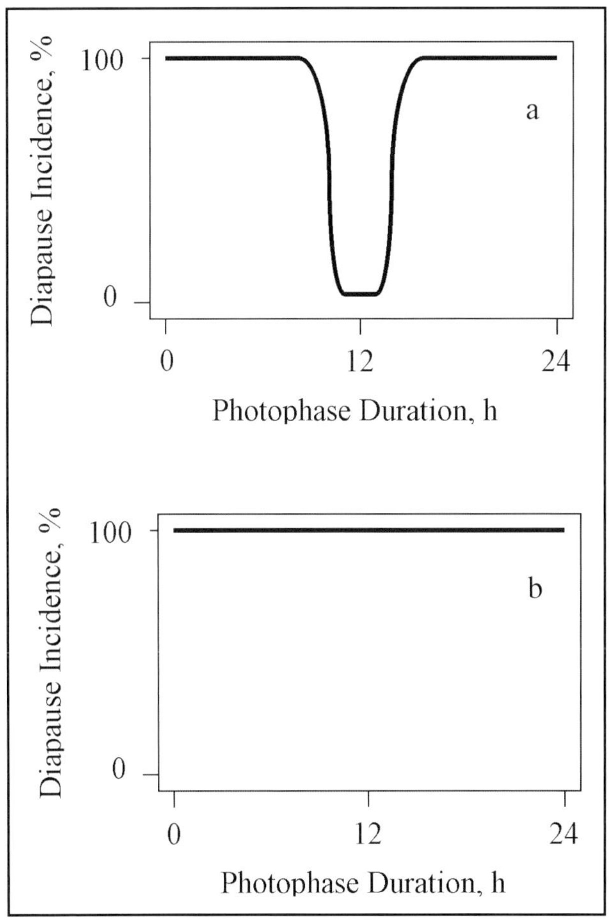

Fig. 7. a) Effect of photophase duration on incidence of diapause associated with winter and summer but not spring and fall. b) Lack of effect of photophase duration on incidence of obligatory diapause.

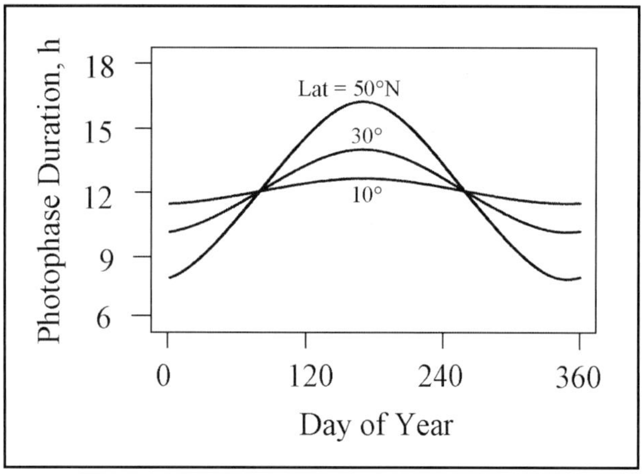

Fig. 8. Photophase duration by date for three selected latitudes.

Typically, temperature is the factor that determines the timing of diapause termination, but photoperiod is important in some species (Sec. II. C. 2 and Sec. III. B. 2).

Diapause initiated during the period of declining day length near the beginning of winter is called *hibernal diapause* or *winter diapause*, and species exhibiting hibernal diapause are called *long-day species* because they are active during times of the year with long days (Fig. 6a). Conversely, diapause initiated during the period of increasing day length near the beginning of summer is called *aestival diapause* or *summer diapause*, and species exhibiting aestival diapause are called *short-day species* (Masaki 1980; Fig. 6b). Some species of insect (e.g., the yellow-tail moth, *Euproctis similis* (Lepidoptera: Lymantriidae) are induced by shorter days to enter hibernal diapause and by longer days to enter aestival diapause, thus avoiding diapause only at intermediate photophase durations (Saunders, et al. 2002) (Fig. 7a). Diapause exhibited by univoltine species is referred to as *parapause*. Parapause is obligatory at a particular stage of development and consequently does not depend on the LD cycle (Fig. 7b; Thiele 1973) Finally, some species do not exhibit diapause under any normal environmental conditions.

The ecological function of diapause is to reduce metabolic maintenance costs when ambient temperatures are high enough for growth/development or reproduction but when for other reasons, e.g., unavailability of food, such activities are not possible or will soon be impossible. This type of situation occurs throughout aestival diapause and during the beginning of hibernal diapause. For example, insects enter hibernal diapause well before conditions unfavorable for growth/development or reproduction are likely to occur because of the inherent unpredictability of phenological events and because they must reach the stage of development in which diapause is spent before growth/development or reproduction become impossible.

Quiescence is a diapause-like state that results directly from adverse conditions such as desiccation or low temperatures. There is an immediate resumption of activity by quiescent individuals when favorable conditions return. For example, most species of insect become quiescent when placed in a refrigerator at 5°C. The use of low temperatures to store insects in quiescence will be considered in Sec. III. C. 3. b.

1. Induction. Photoperiod is the most reliable indicator of time of year on which insects can base initiation of diapause. L is maximum at the summer solstice around 21 June and reaches a minimum at the winter solstice around 21 December (Fig. 8). At the equinoxes around 21 March and 21 September, the durations of the photophase and scotophase are equal: $L = D = 12$ h. The maximum and minimum values of L are greater at higher latitudes.

In spite of the regularity of photoperiod, there are several potential complications with using the annual LD cycle as a calendar. First, the magnitude of seasonal variation in day length declines toward the equator. Although insects at low latitude are exquisitely sensitive to variation in day length, they must use a cue other than photoperiod (e.g., host plant condition) to induce diapause at less than 10° N and 10° S (Sec. III. B. 1). Second, there is an inherent ambiguity in the date of occurrence of a given value of L due to the fact that each value of L occurs twice annually. This uncertainty is resolved by one of two mechanisms: the stage sensitive to photoperiod occurs during only one of the two times of year at which diapause-inducing photoperiods occur; or the insect is sensitive to the direction of change in photoperiod, positive in winter and spring and negative in summer and fall, e.g., *H. zea* (Wellso and Adkisson 1966). Third, daily variation in cloud cover introduces variation in effective day length. However, this variation is not thought to be biologically important for most species because diapause induction involves a summation process that effectively averages day length over periods several days in duration.

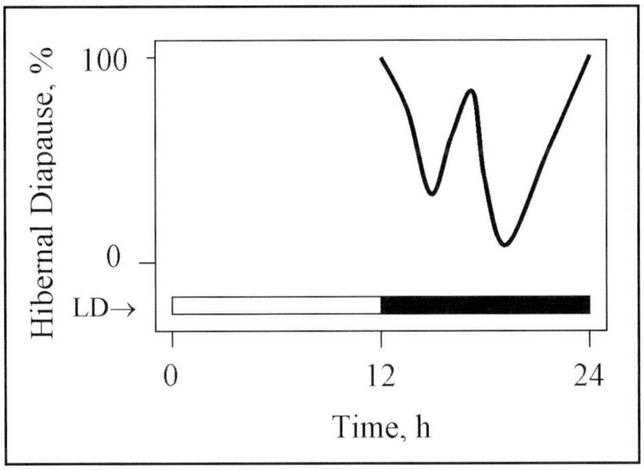

Fig. 9. Effect of recurrent light pulses during scotophase on the incidence of diapause.

The period of sensitivity to photoperiod generally occurs in the same generation as the individuals entering diapause. However, the environment of the previous generation can determine the diapause status of the next generation in some cases. In some strains of *B. mori*, exposure of eggs and young larvae to a long photoperiod results in eggs of the next generation entering diapause (Saunders et al. 2002).

Diapause induction is affected by light pulses that occur repeatedly at the same time during the scotophase of the LD cycle (Saunders, et al. 2002). For hibernal diapause, the incidence of diapause is reduced by light pulses (as short as 5 min in some species) during the scotophase of a normally diapause-inducing LD cycle, e.g., southwestern corn borer, *Diatraea grandiosella* (Lepidoptera: Crambidae) (Takeda 1985). In some species, pulses during mid-early or mid-late scotophase are most effective (Fig. 9); and in other species, pulses near the middle of the scotophase are most effective. Most species are not sensitive to UV or red light, but some species, e.g., the Colorado potato beetle, *Leptinotarsa decemlineata* (Coleoptera: Chrysomelidae) and *P. gossypiella*, are sensitive to red. A vitamin A–based chromophore appears to be responsible for light reception in diapause (Veerman 2001). Illumination sufficient to affect diapause induction can be as low as 0.025 lux, which is less than 4% of the intensity of full moonlight. The range of sensitivities observed in various species is quite broad and may reflect the fact that species vary in the intensity of light to which they are exposed during the stage of development sensitive to photoperiod. For example, some species have burrowed into plant or animal host tissues and need to be much more sensitive to light than are species that feed in the open.

TIP 11: To achieve a high incidence of hibernal diapause, avoid interruption of the scotophase with periodic light pulses—even pulses of dim, red light in the case of some species of Coleoptera and Lepidoptera.

Although little information appears to be available, the incidence of aestival diapause should be increased by light pulses periodically interrupting the scotophase of a normally diapause-averting LD cycle.

TIP 12: To prevent inadvertent induction of aestival diapause, avoid interruption of the scotophase with periodic light pulses—even pulses of dim, red light in the case of some species of Coleoptera and Lepidoptera.

In the case of parasitoids, photoperiod is the primary factor resulting in the induction of diapause in some species, while in others the diapause status of the host or, in the case of polyphagous parasitoids, the identity of the host species may affect the incidence of diapause (Saunders et al. 2002). Examples of diapause of the host causing diapause in the parasitoid include the minute egg parasite, *Trichogramma minutum* (Hymenoptera: Trichogrammatidae), (Laing and Corrigan 1995) and the blowfly parasitoids *Aphaereta minuta* and *Alysia manducator* (Hymenoptera: Braconidae) (Vinogradova and Zinovjeva 1972).

2. Termination. Although temperature determines the rate of diapause termination in most species of insects (Sec. III. B. 2), some species terminate diapause in response to photoperiod. Diapause initiated and terminated by photoperiod has been termed "oligopause" (Thiele 1973). In oligopause, the stage of development sensitive to diapause-terminating photoperiods is the same stage sensitive to diapause-initiating photoperiods. For example, in laboratory observations on *D. grandiosella* collected in Missouri (Takeda 1985), hibernal diapause is initiated as a last instar larva in response to $L < 15.5$ h and is terminated in response to $L > 16$ h. This results in larvae entering diapause in the summer when L is decreasing and terminating diapause in the spring when L is increasing.

D. Population Growth

Light can affect developmental rates and therefore generation time (T_G) of insects in culture. It can also affect survival (*S*) and fecundity (*F*), which together with T_G determine population growth rate and potential productivity of a colony. How these three factors combine to determine the rate of population growth and potential productivity is discussed below (Sec. III. C. 4).

1. Survival. Survival of adults can be reduced by frequent shifts in phase of the LD cycle. Aschoff et al. (1971) observed a 22% reduction in the 90th percentile for longevity of the black blowfly *Phormia terranovae* (Diptera: Colliphoridae) when the phase of the LD cycle was shifted by 6 h at weekly intervals.

Survival of adults can be reduced under LL conditions. The median longevity of *D. melanogaster* was reduced by 14% under LL compared to 12:12 *L:D* (Pittendrigh and Minis 1972). The 90th percentile for longevity of *P. terranovae* was reduced by up to 48% under LL (300 lux) compared to 12:12 *L:D* (Saint Paul and Aschoff 1978). The reduction in longevity of *P. terranovae* also depends on light intensity: the reduction in 90th percentile for longevity was only 25% at 3 lux. If Aschoff's Rule (Sec. II. B. 3) holds for these diurnal species, then transfer to LL results in more circadian cycles being completed in a given time period.

2. Fecundity. Some behaviors such as oviposition can be suppressed under LL. For example, oviposition is under circadian control in *O. nubilalis* and becomes arrhythmic in LL (Skopik and Takeda 1980). In addition, the oviposition rate in LL is suppressed to less than 5% of that in a free-running oviposition rhythm in DD or for that in an entrained oviposition rhythm in LD.

3. Developmental Time. The LD cycle to which insects are exposed affects developmental time (*d*) in some insect species but not others. An example of the latter is the lady beetle *Coccinella novemnotata* (Coleoptera: Coccinellidae) for which changes in the LD cycle did not affect *d* (McMullen 1967). However, *d* for larval development of the flesh fly *Sarcophaga argyrostoma* (Diptera: Sarcophagidae) was 25% longer in 12:12 *L:D* than in 16:8 at 17°C (Saunders 1972). For *H. zea* and *H. virescens*, there was a 14 % increase in *d* of the larval stage at 10:14 *L:D* compared to 15:9 at 24°C (Benschoter 1968). For *H. virescens*, a 13–15% increase in *d* of the larval stage was observed at 10:14 *L:D* compared to 14:10 at 21.2–26.7°C (Henneberry et al. 1993). In each of these cases, shorter *L* resulted in longer *d*. This effect was independent of differential induction of diapause at the different photoperiods.

TIP 13: To reduce the developmental time of insects reared under an LD cycle, try increasing photophase duration.

Benschoter (1968) also observed an effect of intensity of illumination on *d*: 26% increase for *H. zea* and 11% increase for *H. virescens* under the lower illumination level of 25 foot-candles compared to 400 foot-candles at 21°C and 10:14 *L:D*.

III. Temperature

Temperature is an extremely important environmental factor for insects because of their lack of internal temperature regulation. Temperature has the property of magnitude, which can vary temporally and spatially. This variation can affect the physiology and ecology of insects in many different ways including thermotaxis, circadian rhythm, diapause, and population growth. Temperature can influence all three of the demographic factors determining population growth rates and productivity: generation time, survival, and fecundity. Temperature interacts strongly with photoperiod with respect to diapause induction and is usually the primary environmental factor determining rate of diapause termination. Moreover, thermoperiodic cycles can substitute for photoperiodic cycles in entrainment of circadian rhythms and in diapause induction.

The effects of temperature on circadian rhythms, diapause, and population growth and the consequences for insect rearing will be considered in turn.

A. Circadian Rhythms

1. Constant Temperature. As discussed above (Sec. II. B. 1), circadian rhythms are free-running under a DD environment with a period $\tau \approx 24$ h. However, to the condition of DD must be added "constant temperature." Just as a light pulse can cause an advance or delay in a free-running circadian rhythm depending on the phase of the rhythm at which the pulse occurs, so can a temperature pulse or step change result in a shift in circadian rhythms (Beck 1983, Saunders et al. 2002; Fig. 10). For example, the eclosion rhythm of *D. pseudoobscura* is advanced up to 5 h by a temperature step up (20→28°C) and is delayed by up to 2.5 h by a temperature step down (28→20°C). In general, much longer temperature pulses are required to achieve phase shifts of a given magnitude than is the case for light pulses.

2. Warm:Cool Cycles. Given that changes in temperature cause phase shifts in circadian rhythms, it is not surprising that daily temperature cycles can entrain the same individual-level and population-level activities that can be entrained by LD cycles (Fig. 11). For example, *H. virescens* pupae held in DD and a warm:cool (WC) cycle with a *thermophase* (*W*, warmer phase of a WC cycle) of 14 h duration at 30°C and a *cryophase* (*C*, cooler phase of a WC cycle) of 10 h duration at 22°C, i.e., 14(30):10(22) *W*:*C*, displayed a gated emergence pattern for adults: peak emergence occurred 3 h after the onset of the cryophase (Roush and Schneider 1985). Mating behavior of *O. nubilalis* provides an additional example: mating occurs only during the cryophase of a 16(26.6):8(18.3) *W*:*C* and DD environment (Loughner 1972).

Use of a WC cycle may be desirable. For example, one may use a thermoperiodic environment to avert hibernal diapause or to induce aestival diapause under DD conditions (Sec. III. B. 1). However, as in the case of photoperiodically induced diapause, periodic temperature disturbances can affect the incidence of diapause.

In nature, temperature varies with time of day such that minimum air temperature occurs near dawn and maximum temperature in midafternoon. If one artificially shifts the phase relationship of a WC cycle with respect to a co-occurring LD cycle, circadian rhythms are shifted with respect to the LD cycle up to some limit (Figs. 12a, b). For example, peak eclosion of *D. pseudoobscura* adults occurs 2 h after the beginning of photophase in 12:12 *L*:*D* at constant temperature or in 12:12 *L*:*D* plus a sinusoidally varying WC cycle with a minimum temperature (19°C) at lights on and a maximum temperature (29°C) at lights off. If the WC cycle is shifted later with respect to the LD cycle, eclosion occurs later during photophase. However, when the minimum point of the WC cycle falls closer than ca. 2 h from the end of photophase, peak eclosion is advanced about 12 h to the latter part of scotophase.

TIP 14: To avoid disturbance of circadian rhythms, one should not inadvertently or haphazardly expose insects reared under DD to temperature pulses or shifts.

TIP 15: To manipulate the timing of circadian rhythms with respect to a fixed LD cycle, try imposing a WC cycle that is shifted with respect to the "natural" phase relationship between WC and LD cycles.

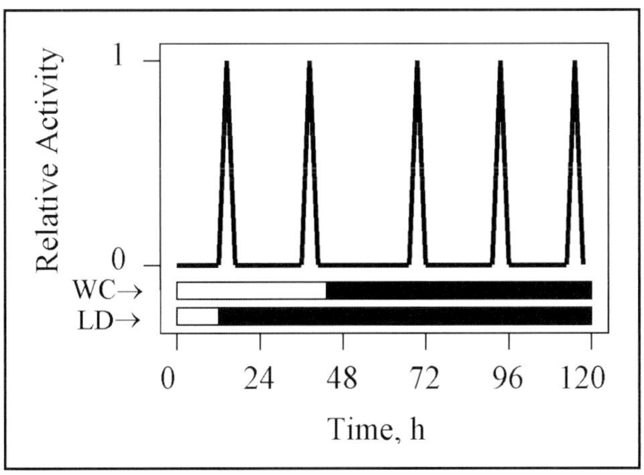

Fig. 10. Shift in the phase of a free-running circadian rhythm in response to a temperature step-down.

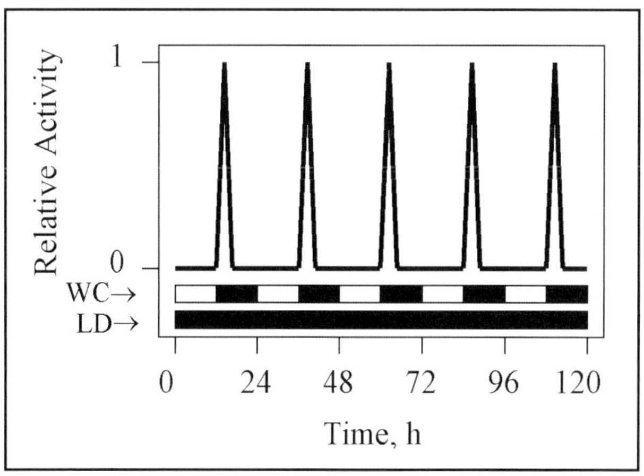

Fig. 11. Entrainment of a circadian rhythm by a warm:cool (WC) cycle under continuous dark (DD) conditions.

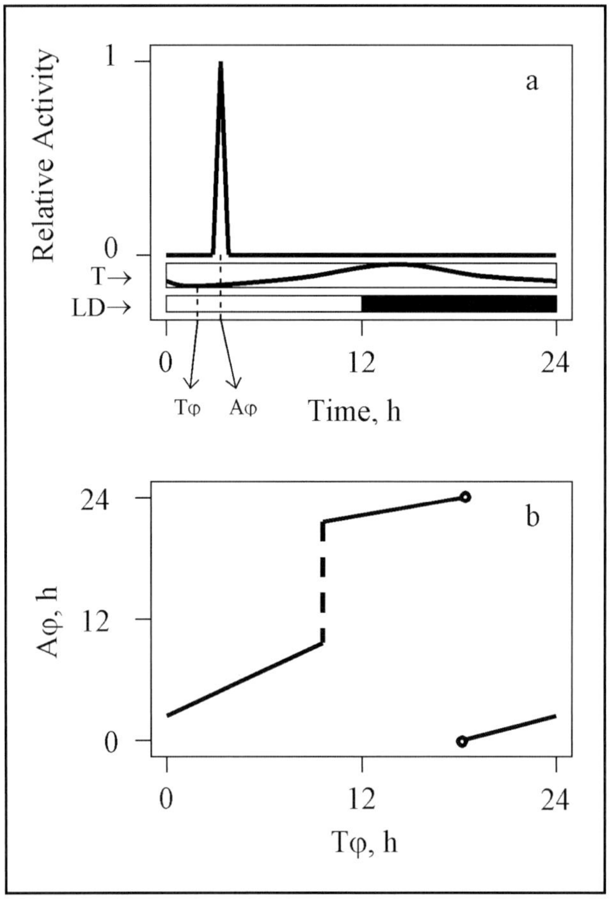

Fig. 12. a) Lag (A_{φ}) in the peak of a circadian rhythm with respect to lights on in the presence of a warm:cool cycle (WC), the minimum of which lags lights-on by T_{φ}. b) A_{φ} as a function of T_{φ}.

B. Diapause

1. Induction. Some species of insects occur in situations where photoperiodic cues are not available, e. g., stem borers and species within 10° of latitude of the equator. In such situations, WC cycles may be the best seasonal cues available. However, perhaps surprisingly, hibernal diapause in most insects—not just those in aphotoperiodic sites—is affected by WC cycles under DD conditions (Saunders, et al. 2002, Beck 1983). As expected from the natural association of thermophase with photophase in WC and LD cycles, cycles with shorter *W* induce hibernal diapause and those with longer avert it (Fig. 13). For example, the large white, *Pieris brassicae* (Lepidoptera: Pieridae), subjected to DD and a 16(21):8(13) *W*:*C* experiences <10% diapause, but incidence of diapause jumps to >70% when the insects are subjected to a 14(21):10(13) *W*:*C*. Frequently, thermophase duration that induces 50% diapause in a population is similar to the photophase duration that induces 50% diapause. There appear to have been few studies of the effect of WC cycles on induction of aestival diapause although it would be expected that shorter *W* would avert diapause and longer *W* induce it.

TIP 16: To avert hibernal diapause or to induce aestival diapause under DD conditions, try using a WC cycle.

For most species of insects, diapause is determined by a mechanism that results in the average temperature across the entire 24-h cycle affecting the incidence of diapause. Higher temperatures work in concert with longer photophases to avert hibernal diapause and promote aestival diapause (Fig. 14a), and lower temperatures have the opposite effect (Fig. 14b).

For some species of insects, diapause appears to depend on measurement of *D*, e.g., *O. nubilalis*, *D. grandiosella*, and the vetch aphid, *Megoura viciae,* (Homoptera: Aphididae). If these species are maintained simultaneously under LD and WC cycles with coincident photophase and thermophase, the incidence of hibernal diapause is affected by the temperature during scotophase but is generally not affected by the temperature during photophase. For example, *O. nubilalis* experiences 15% diapause when reared under 15(31):9(31) *W*:*C* but 96% whether reared under 15(31):9(21) or 15(21):9(21) *W*:*C*.

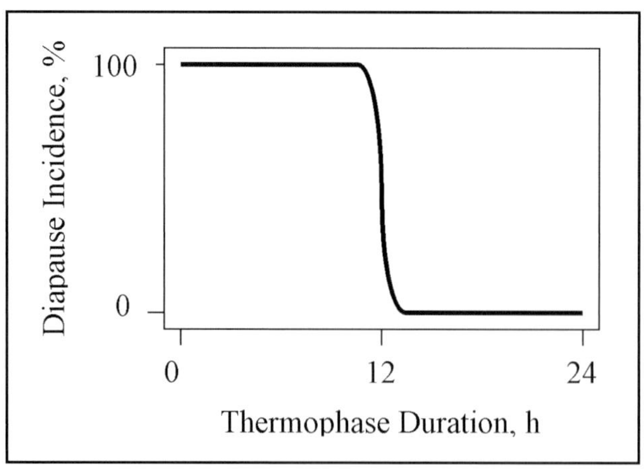

Fig. 13. Effect of thermophase duration on incidence of hibernal diapause under continuous darkness.

Some additional effects of temperature on diapause have been observed. First, for at least some insects when reared under an LD cycle that averts hibernal diapause, periodic chilling to 5°C at the beginning or end of photophase results in a high incidence of diapause.

TIP 17: When immobilizing insects using cold for weighing or other measurements, chill them during the middle of the photophase to avoid induction of hibernal diapause.

Second, a step change in temperature soon after the end of the period of sensitivity to photoperiod for induction of hibernal diapause can increase (step-down) or decrease (step-up) the incidence of diapause. For example, transfer of *S. argyrostoma* from 10:14 *L:D* at 16°C to 10:14 *L:D* at 28°C immediately after puparium formation causes the incidence of diapause to drop from 100% to near 50% (Gibbs 1975).

TIP 18: To avert hibernal diapause in individuals recently exposed to diapause-inducing conditions, e.g., late summer field collections made for the purpose of genetic rejuvenation of a colony, transfer them to temperatures in the lab higher than those experienced in the field.

2. Termination. Temperatures near those prevailing at the time of diapause induction tend to maintain the diapause state. Termination of diapause occurs in most species following exposure, over a period of time, to temperatures lower than those prevailing when diapause was induced. In the case of hibernal diapause, termination frequently occurs during the middle of the winter, e.g., the cabbage root fly, *Delia radicum* (Diptera: Anthomyiidae) (Collier and Finch 1983) and *H. virescens* (Lopez and Hartstack 1985). This fact reinforces the point made above (Sec. II. C) that the ecological function of diapause is to reduce metabolic costs when ambient temperatures are high enough for growth/development or reproduction but when such activities are not possible for other reasons, e.g., unavailability of food.

The time of resumption of activity may be delayed following the termination of diapause. In the case of termination of hibernal diapause during the middle of the winter, the insects remain quiescent until temperatures allow post-diapause development to proceed. For some species, once diapause has terminated, a specific cue may be required for resumption of activity (Tauber et al. 1986). As a result, variation in developmental stage or physiological state within a population is frequently much reduced following diapause.

TIP 19: To increase the level of developmental and physiological synchrony of a group of insects in culture, try leaving the population continuously at temperatures well below the level allowing appreciable post-diapause development until all individuals have terminated diapause.

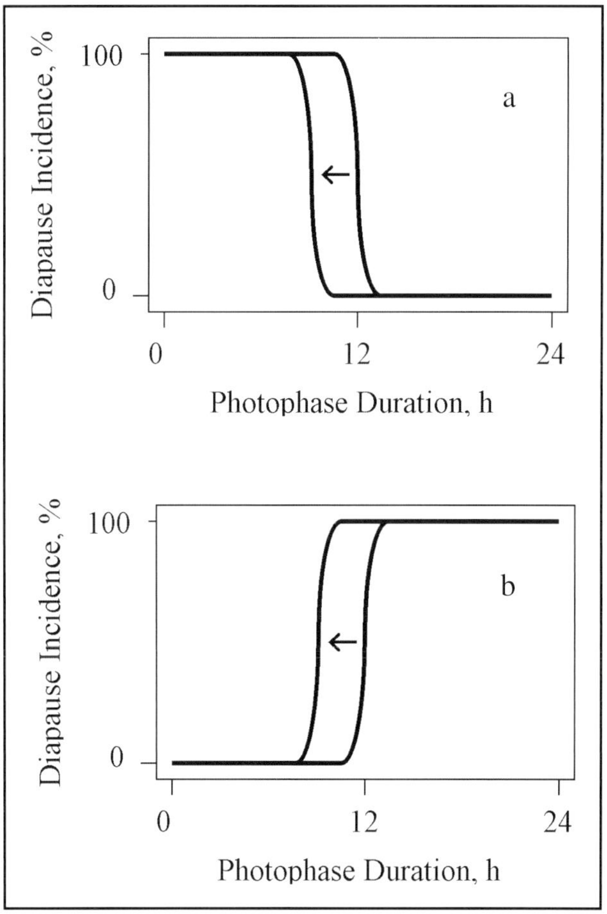

Fig. 14. a) Effect of higher constant temperatures on the incidence of hibernal diapause as a function of photophase duration. b) Effect of higher constant temperatures on the incidence of aestival diapause as a function of photophase duration.

3. Stockpiling and Maintenance. Several biological control agents, including predators and parasitoids in at least four orders, have been induced to enter hibernal diapause for the purpose of accumulating large numbers of individuals for mass release (Whistlecraft et al. 1985, Gilkeson 1990, Tauber et al. 1993, Laing and Corrigan 1995). These species were held at low temperatures with high survival rates for periods of 6–10 mo. Various herbivorous insects have been stockpiled at low temperatures including the codling moth, *Cydia pomonella* (Lepidoptera: Tortricidae) (Singh et al. 1986) and *H. virescens* (Mulrooney et al. 1989) to use in sterile release programs and to preserve the genetic integrity of the colony during the "off season."

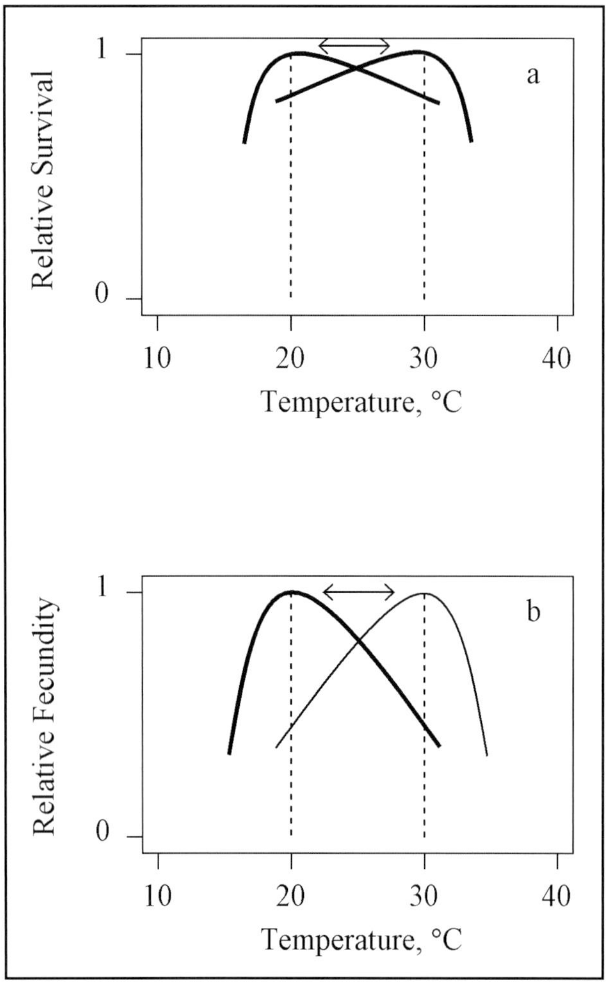

Fig. 15. a) Effect of rearing temperature on survival for hypothetical species having temperature optima either near the lower or upper end of the range of typical rearing temperatures. b) Effect of rearing/adult holding temperature on the fecundity of hypothetical species having either a low temperature optimum (more common; thick line) or a higher optimum (less common; thin line).

C. Population Growth

1. Survival. The survival of insects during development from egg to adult can vary significantly over a range of temperatures but to different degrees or even in opposite directions for different species. For example, Tolman et al. (1985) showed that survival of the onion maggot dropped from 76% at 20°C to 24% at 30°C. In contrast, survival of *Catolaccus grandis* (Hymenoptera: Pteromalidae), a parasitoid of *A. grandis*, increased from 19% at 20°C to 30% at 30°C (Morales-Ramos et al. 1996). One possible explanation for such variation among species is as follows: survival is a convex function of temperature, so a positive or negative relationship will be observed depending on whether the temperature range of observation falls to the left or right, respectively, of the optimum temperature for that species (Fig. 15a). For a few species, observations of survival of immatures have been made over a very wide range of temperatures, and a convex relationship has been revealed. For example, the flesh fly *Parasarcophaga ruficornis* (Diptera: Sarcophagidae) was reared at nine temperatures ranging from 13 to 37°C, and survival of immatures ranged from 0% at both ends of the range to a maximum of 90% at 28°C (Amoudi et al. 1994).

2. Fecundity. Fecundity of insects can vary with the temperature at which the adults are maintained. Data on the relationship between adult maintenance temperature and female fecundity was found for 12 species. These 12 include 9 for which fecundity was higher at lower temperature over at least part of the temperature range tested [one Hymenoptera (Hentz et al. 1998), two Diptera (Tolman et al. 1985, Yang et al. 1994), and six Lepidoptera (Fye and McAda 1972)] and three species for which the reverse was true [two Hymenoptera (Morales-Ramos et al. 1996, Lyons 1996) and one Homoptera (Savopoulou-Soultani 1997)]. An example of the majority response is Tolman et al. (1985): fecundity of *D. radicum* nearly tripled from 71 per female at 30°C to 211 per female at 20°C. Similar to the relationship between temperature and survival, the general relationship between temperature and fecundity might be described by Fig. 15b.

An increase in fecundity at lower temperature is "explained" by two additional effects of temperature on adults. First, rearing at lower temperatures generally results in larger adult size. Atkinson (1994) collated information on 66 species of insects of which 55 showed a larger adult size at lower temperature over at least part of the range of temperatures tested. Second, lower temperature typically results in increased adult life span. Data on the relationship between adult maintenance temperature and adult life span was found for 10 species. In all 10 cases, lower temperature resulted in increased adult life span: two Hymenoptera (Lyons 1996, Hentz et al. 1998), two Diptera (Yang et al. 1994, Nunney and Cheung 1997), and six Lepidoptera (Fye and McAda 1972). Given the above, it is not surprising that larger, longer-lived adults produced at lower temperatures have a higher fecundity than the smaller, shorter-lived adults produced at higher temperatures. However, the existence of species that have a lower fecundity when maintained at lower temperatures (3/12 of species surveyed), urges caution in generalization.

3. Developmental Rate. A very large number of studies have been done on the effect of temperature on the developmental rate of insects. This is not surprising given the large effects observed. For example, *H. virescens* larvae reared in 14:10 *L:D* take about half as much time to develop from egg hatch to adult emergence at 32.2°C as at 21.1°C (Henneberry et al. 1993). Models quantifying the relationship between developmental rate and temperature are the basis of two related approaches for the prediction of average developmental times: day-degrees and rate summation. Both will be described below. In addition, a generalization of the relationship between variance and average developmental time will be used to estimate the availability of individuals from a cohort that fall within a desired range of developmental ages on a given future date.

a. Day-degrees. In general, developmental time (d) is shorter at higher temperature (T) (Fig. 16a). This hyperbolic-shaped relationship can be partially linearized by transforming d to developmental rate ($R = 1/d$): AVE[R] increases linearly with temperature over some range of temperatures (T_{lindev}) (Fig. 16b). Within T_{lindev}, a convenient method called *day-degrees* is available for estimating d even when T is variable. The reciprocal of the slope of AVE[R] vs. T is called the *day-degree requirement* (K). The x-intercept of AVE[R] vs. T (i.e., T at $R = 0$) is the *lower developmental threshold* (τ). Given these assumptions and definitions, the general day-degree relationship when temperature is variable can be stated as follows: $dd_t = K$ at $t =$ AVE[d], for dd_t, the cumulative sum over time (t) of the area bounded by T and τ (Fig. 17). If t is measured in units of days and T in units of °C, then the units of dd_t and K are d•°C, i.e., day-degrees. The general day-degree relationship for a constant temperature (T_c) is simpler:

$$K = \text{AVE}[d] \cdot (T_c - \tau) \qquad \text{Eq. 1.}$$

K and τ are species specific (Table 2) and also depend on diet and photoperiod (Sec. II. D. 3).

TIP 20: To accurately predict developmental times of a given insect species, do not use values of K and τ obtained from a different species or even from the same species reared on a different diet or under different conditions.

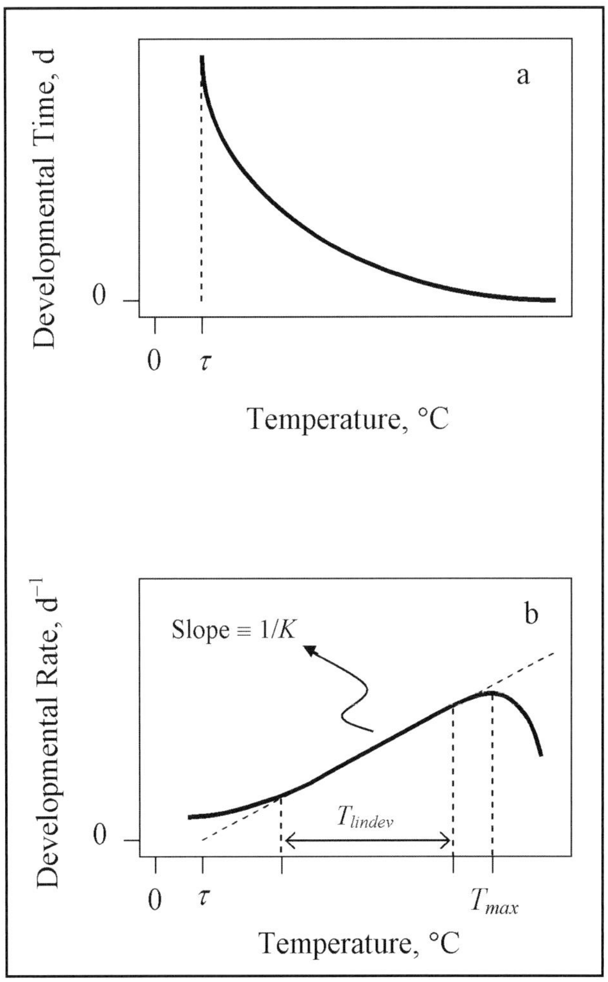

Fig. 16. a) Typical, concave relationship between developmental time and temperature. b) Typical, partially linear relationship between developmental rate and temperature.

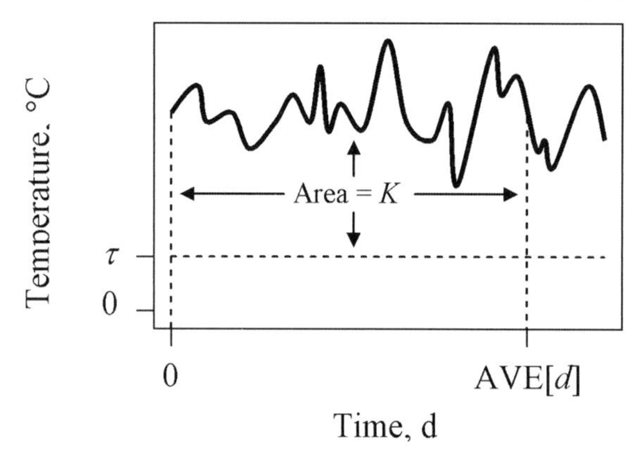

Fig. 17. Relationship between day-degree requirement (*K*) and temperature (*T*) experienced over time: area between *T* and the lower developmental threshold (τ) sums to *K* at AVE[*d*] for developmental time (*d*).

b. Rate Summation. "Developmental zero" is a misnomer for τ. Relative to the trend line between *R* and *T* for *T* within T_{lindev}, *R* is higher than expected for $T < T_{lindev}$. As described in Sec. III. C. 1 above, developing insects suffer 100% mortality at sufficiently low temperatures; however, before this point is reached, *R* appears to decline asymptotically toward zero. Thus, even theoretically, there is no temperature at which $R = 0$, and certainly, at $T = \tau$, $R \neq 0$. Consequently, the fact that insects commonly terminate hibernal diapause during the winter (Sec. III. B. 2) means that most species experience $T < T_{lindev}$ during post-diapause development in the diapausing generation. Thus, using day-degrees to model post-diapause development should tend to result in prediction of a later date of emergence from overwintering than the actual date. In addition, to hold insects in developmental stasis, low temperature quiescence may not be adequate because developmental rate is not zero at even the lowest temperatures tolerated by the insect.

Generally, there exists a temperature at which *R* peaks (T_{max}, Fig. 16b). Thus, the generalization that *d* is shorter at higher temperatures is not true for temperatures higher than T_{max}. Interestingly, T_{max} can be low enough that $T > T_{max}$ is experienced in the field. For example, $T_{max} = 32°C$ for *H. virescens* and the daily average maximum temperature in central Mississippi in July is 33.6°C (Moghal and Schneider 1998). The fact that *R* is lower than expected for $T > T_{lindev}$ suggests that such temperatures are stressful and that quality of the insects produced for certain uses may be reduced.

TIP 21: To avoid stressing insects during rearing, do not subject them to temperatures near T_{max}.

If temperatures outside of T_{lindev} are experienced by an insect and if *T* varies with time [$T(t)$], then *d* is estimated using rate summation. For R_t, the cumulative sum over *t* of the area under $R(T(t))$, $R_t = 1.0$ at $t =$ AVE[*d*] (Fig. 18). For a constant temperature, T_c, outside of T_{lindev}, AVE[*d*] = $1/R(T_c)$, i.e., one must make direct observations of *d* at T_c or estimate a value of $R(T_c)$ by interpolation from adjacent values of *T* or by using a model of $R(T)$ (e.g., Ali et al. 1990).

Table 2. Developmental characteristics of various insect pests of cotton

Species[a]	τ, °C	*K*, d°C	*d* @ 28°C, d	Literature Citation
TBW (cotton)	5.6	509	22.7	Butler and Hamilton (1976)
TBW (art. diet)	12.6	200	13.0	Butler et al. (1979), Nadgauda and Pitre (1983)
CEW	7.5	438	21.4	Butler (1976)
FAW	12.9	270	17.9	Ali et al. (1990)
BAW	13.4	159	10.9	Ali and Gaylor (1992)
TPB	11.8	259	15.9	Fleischer and Gaylor (1988)
BW	10.6	124	7.1	Bacheler et al. (1975)
TSSM	11.6	111	6.8	Gergis and El-Duweini (1992)
CA	8.0	98	4.9	Akey and Butler (1989)

Developmental characteristics including the following: lower developmental threshold, τ; day-degree requirement, *K*; and developmental time, *d*. A diet of appropriate cotton plant part(s) was fed except for "TBW (art. diet)," which consisted of a wheat germ-based artificial diet. [a]TBW, tobacco budworm (*Heliothis virescens*); CEW, corn earworm (*Helicoverpa zea*); FAW, fall armyworm (*Spodoptera frugiperda*); BAW, beet armyworm (*Spodoptera exigua*); TPB, tarnished plant bug (*Lygus lineolaris*); BW, boll weevil (*Anthonomus grandis*); TSSM, twospotted spider mite (*Tetranychus urticae*); CA, cotton aphid (*Aphis gossypii*).

It has been claimed for various species of insects that developmental rate is accelerated or decelerated by a variable temperature regime relative to the rate observed at a constant temperature equal to the average for the variable regime. However, almost all of these differences can be explained by the nonlinearity of the R vs. T relationship: variable temperature regimes with AVE[T] within T_{lindev} and minimum $T < T_{lindev}$ (R above the linear trend) result in apparent rate accelerations, and variable regimes with AVE[T] within T_{lindev} and maximum $T > T_{lindev}$ (R below the linear trend) result in decelerations (Howe 1967). Consequently, a variable temperature regime generally has no beneficial effect on insects in culture.

A few insects do exhibit a residual accelerating effect of variable temperature on developmental rate relative to that expected from the nonlinear rate summation effect (Liu et al. 1995). These exceptions involve LL lighting conditions. As discussed above, LL can result in a damping of circadian rhythms and a decline in general vigor (Sec. II. B. 3) and a WC cycle can induce and entrain circadian rhythms (Sec. III. A. 2).

TIP 22: Although a variable temperature regime generally has no effect on developmental rate, if an insect is reared under LL, providing a WC cycle may result in an increase in developmental rate compared to that at a constant temperature equal to the average temperature over the WC cycle.

c. Variance. If one monitors a *cohort* of insects, an even-aged group, variation in developmental stage among individuals will develop over time. This can reduce yields in rearing operations. For example, if one requires a given number of 0- to 2-day-old pupae 21 days from now, i.e., individuals that pupate 19–21 days from now, and one has a supply of neonate larvae, day-degrees can be used to calculate the temperature at which AVE[d] = 20 d. However, some of the larvae will develop too quickly and some too slowly to fall within the desired developmental range. In general, in order to determine the size of the cohort necessary to produce the desired number of useable individuals, one must estimate the fraction of the cohort that falls into an age-restricted window some time in the future.

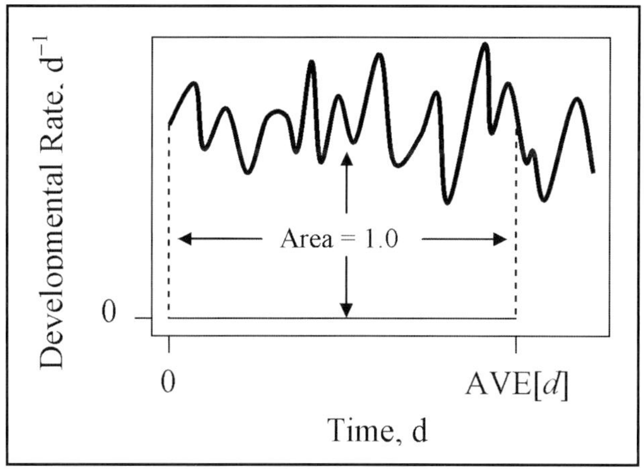

Fig. 18. Relationship between developmental rate (*R*) and temperature (*T*) where *T* is a function of time (*t*): area under *R* sums to 1.0 at *t* = AVE[*d*] for developmental time (*d*).

The frequency distribution for d is typically asymmetrically skewed to the right, i.e., there is a long tail to the right representing individuals that take longer than the average to develop. In contrast, the frequency distribution for R tends to be *normal*, i.e., symmetrically bell-shaped (Wagner et al. 1984). One can use standard statistical tables for the normal probability distribution to describe the frequency distribution of R. Normal distributions depend on the two statistics mean (AVE) and variance (VAR), but these statistics appear to be simply related in the case of developmental rate of insects:

$$\text{SD[R]} \cong 0.10 \cdot \text{AVE[R]} \qquad \text{Eq. 2.}$$

for $d \geq 6$ d and SD[R] = √(VAR[R]) (Shaffer 1983). Thus, one can reduce the dependence of the frequency distribution of R to a single statistic, say AVE[R]; and given that AVE[d] = 1/AVE[R], one can estimate the percentage of a cohort that falls within a given range of d as a function of AVE[d]. For example, for AVE[d] = 20 d, an age-restricted window 2 d wide requires a cohort ca. 2.5-fold larger than the required number of individuals, and a 1-d-wide window requires a ca. 5-fold larger cohort (Fig. 19). These should be considered minimum estimates because mortality is ignored.

4. Models. How generation time, T_G, survival, S, and fecundity, F, combine to determine rate of population growth depends on the age structure of the population. If there is no age structure and reproduction occurs at intervals of T_G, then population growth is

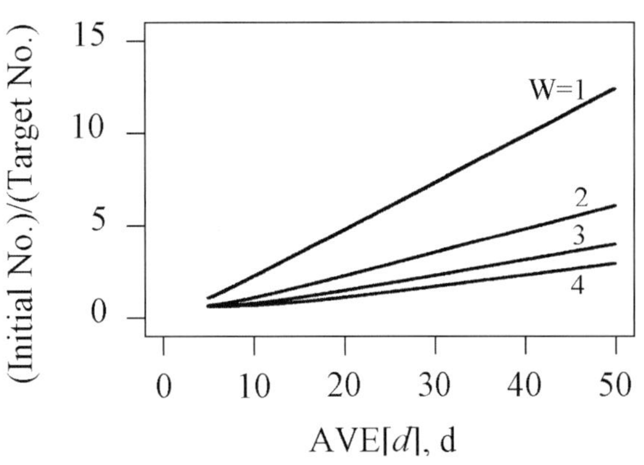

Fig. 19. Number of insects that must be produced per individual harvested within an age window of a given width (*W*) centered on AVE[*d*] as a function of AVE[*d*] for developmental time (*d*).

said to be *discrete*, has units of *generation*$^{-1}$, and is calculated as the finite rate of increase (R_0):

$$R_0 = \frac{F \cdot S}{(SR + 1)} \qquad \text{Eq. 3.}$$

where the effect of sex ratio (*SR*, ♂:♀) has been included. For example, for $F = 100$, $S = 0.5$, and $SR = 1$, one has $R_0 = 25$, a typical value for pest insects in culture.

Assuming that R_0 remains constant and availability of males does not limit population growth, a population initially containing N_0 females will contain N_G females after G generations:

$$N_G = N_0 \cdot R_0^{\,G} \qquad \text{Eq. 4.}$$

If reproduction occurs continuously at the population level and the population has a stable age structure, then population growth is said to be *continuous*, has units of *time*$^{-1}$, and is calculated as the instantaneous rate of increase (r):

$$r = \frac{\ln(R_0)}{T_G} \qquad \text{Eq. 5.}$$

Assuming that r remains constant, the female population will attain a size N_t after a time period (t) given by the following:

$$N_t = N_0 \cdot e^{r \cdot t} \qquad \text{Eq. 6.}$$

For a given population size, there is a maximum sustainable harvest rate (H) that allows the population to replace itself with the excess eggs being harvested. For discrete generations and a production colony of size N females, H has units of *generation*$^{-1}$ and is calculated as follows:

$$H = N \cdot F \cdot \left[\frac{(R_0 - 1)}{R_0}\right] \qquad \text{Eq. 7.}$$

For $R_0 >> 1$, $H \cong N \cdot F$, the total number of eggs produced. For example, for a colony of $N = 1{,}000$ females, $F = 100$, and $R_0 = 25$, one could sustainably harvest $H = 96{,}000$ eggs each generation (out of a total of $N \cdot F = 100{,}000$ eggs produced) for uses other than colony propagation.

TIP 23: The sustainable rate of harvest of eggs from a typical pest insect species in culture is approximately equal to the culture's total rate of egg production.

IV. Humidity

Insects, like other organisms, are typically composed of ca. 75% water by weight. Moreover, because the water molecule is small relative to enzymes, structural proteins, and carbohydrate polymers, approximately 99 of every 100 molecules in live insects are water (Wharton and Richards 1978). The primary route by which most insects acquire water is through ingestion, and the nutritional effects of water are considered in Chap. 7 *Insect Nutrition, Feeding, and Artificial Diets*. The effects of ambient humidity on insects will be considered here.

The concentration of water vapor in the air most strongly affects small species of insects or species that feed on foods of low water content. Also, early developmental stages (e.g., eggs) and nonfeeding stages of larger species can be strongly affected. The effects of ambient humidity on survival, fecundity, and developmental rate, the demographic factors that determine population growth rate, will be considered in turn. First, however, the choice of units for measurement of humidity will be discussed.

A. Units of Measurement

Of the 19 studies examined for this review, 15 measured the amount of water vapor in the air as *relative humidity* (*RH*); and the remainder used *vapor*

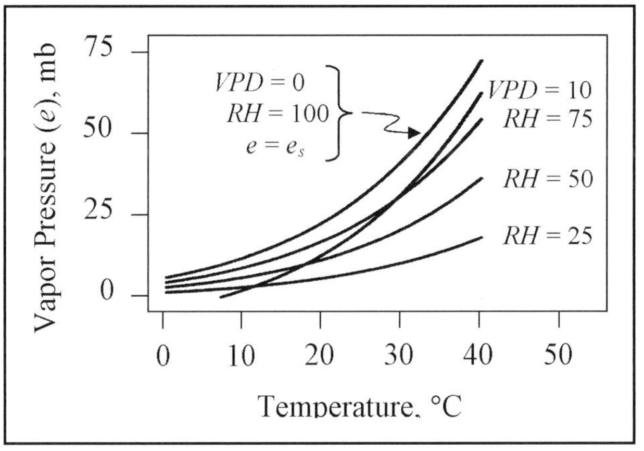

Fig. 20. Vapor pressure of water as a function of temperature for various constant levels of relative humidity (*RH*) and for a fixed vapor pressure deficit (*VPD*) of 10 mb.

pressure deficit (*VPD*). *VPD* is defined as the difference between the saturation vapor pressure of water (e_s) and the actual vapor pressure of water in the air (e): $VPD \equiv e_s - e$, where the concentration of a gas in a mixture of gases is proportional to the partial vapor pressure of the gas. *RH* is defined as the ratio of e to e_s on a percentage basis: $RH \equiv (e/e_s) \cdot 100\%$.

Although less commonly used, *VPD* may be preferable to *RH* when working with biological systems. The rate of diffusion of water vapor is proportional to the slope of its concentration gradient, which is equal to the change, i.e., difference, in vapor pressure per unit distance. In still air, the vapor pressure of water at the surface of an insect is close to e_s, and at a distance it is e, so the drying power of the ambient air is proportional to *VPD* rather than to *RH* (Kucera 1954, Ferro and Chapman 1979). If temperature is fixed, then (100% – *RH*) is proportional to *VPD* and also may be used as a measure of the drying power of the air. However, if temperature increases or decreases with *RH* fixed, then *VPD* and the drying power of the air increases or decreases, respectively (Fig. 20). Consequently, when temperature is not fixed, one should report humidity in units of mb[1] of *VPD* rather than percentage of *RH*.

One method to manipulate humidity is to confine a saturated salt solution in a closed container. For a given temperature, the equilibrium water vapor pressure in the air above the salt solution varies with type of salt to produce an equilibrium *RH* falling in the range of ca. 10–98% (CRC 1970). For a given salt, *RH* is nearly constant over the biologically important range of 15–30°C. Thus, it is much easier to maintain *RH* at a constant level in the face of a variable temperature regime than is the case for *VPD*. This fact may explain the historical preference of *RH* over *VPD* in spite of the limitations of *RH* discussed above.

Hygrometers frequently measure humidity in units of *RH*. However, *RH* can be converted to *VPD* using the following relationship (Monteith and Unsworth 1990):

$$VPD \cong (1 - RH/100) \cdot (23.1 \text{ mb}) \cdot e^{18.00\,[(T(K)-293.0)/T(K)]} \qquad \text{Eq. 8.}$$

for T(K), temperature in degrees Kelvin: $T(K) = T(C) + 273.2$, for T(C), temperature in degrees Celsius.

The rate of air movement across the surface of an insect is another important factor determining the rate of water loss. In still air and $e < e_s$, a gradient in water vapor pressure develops with higher vapor pressure at the surface of the insect. In moving air, this gradient becomes steeper, and the rate of water loss can be increased 2–3X even for air velocities of <3 m/s (Kucera 1954, Bursell 1964).

B. Population Growth

1. Survival. Published data for nine species of arthropods for which survival was reported to be a function of humidity were examined. Eight of the studies involved survival of the eggs, and one was concerned with survival of the egg parasitoid *Trichogramma pretiosum* (Hymenoptera: Trichogrammatidae) (Gross 1988). Of the eight species for which eggs were studied, four were mites (Acarina: Phytoseiidae): *Neoseiulus fallacis* and *Metaseiulus occidentalis* (Mangini and Hain 1991); the spruce spider mite *Oligonychus ununguis* (Kramer and Hain 1989); and the twospotted spider mite, *Tetranychus urticae* (Ferro and Chapman 1979). Another species was *O. nubilalis* (Godfrey and Holtzer 1991), and three others were species of citrus root weevils (Coleoptera: Curculionidae): *Artipus floridanus*, *Pachnaeus opalus*, and *Pantomorus cervinus* (Tarrant and McCoy 1989). Decrease in survival with increase in *VPD* varied by a wide range: 2.1 to 13.3% per mb *VPD*. This high level of variation was not explained by differences in size because the eggs of some mite species were less susceptible to higher *VPD*s than were the eggs of *O. nubilalis* even though the former

[1] 1 millibar = 0.75 g H_20/m^3 = 0.75 mm Hg = 0.99 X 10^{-3} atm

are much smaller than the latter. Thus, there appears to exist among insects a wide range in the degree of permeability of the integument to water vapor.

Two types of results illustrate the point made above that *RH* is not a satisfactory measure of humidity with respect to its effects on biological systems when temperature varies. First, four of the species showed lower survival at higher temperatures with *RH* held constant. This effect was apparent even though developmental rates were higher at higher temperatures so that the insects lost water at the higher *VPD*s for shorter periods of time. Second, the higher the temperature, the greater is the effect of a given difference in *RH* on survival. Gross (1988) reported that at 20% *RH* survival of *T. pretiosum* was 46% lower compared to that at 80% *RH* when the temperature was 27°C but 68% lower when the temperature was 32°C. Tarrant and McCoy (1989) found a similar interaction between temperature and *RH* for the citrus root weevils they studied.

Mortality is typically high at *VPD* = 0 mb, but the pattern of change in mortality close to saturation varies among species. In the case of eggs of *N. fallacis* and *M. occidentalis*, mortality increased gradually up to saturation over a wide range of *VPD* (Mangini and Hain 1991). In contrast, mortality of eggs of *T. pretiosum* (Gross 1988) appeared to display a threshold response to *VPD*, with mortality increasing only close to *VPD* = 0 mb.

It is not surprising that insect and mite eggs and minute parasitoids should be a focus for research on the effects of ambient humidity on survival. The fraction of water loss per unit time (*WL*) for organisms of similar shape and composition is proportional ($\propto$) to the ratio of their surface area (*SA*) and volume (*V*): $WL \propto SA/V$ (Thompson 1961). Because, for any linear dimension (L), $SA \propto L^2$ and $V \propto L^3$, $WL \propto SA/V \propto L^2/L^3 = L^{-1}$, $WL \propto L^{-1}$. Thus, the smaller the organism, the higher is the potential fraction of water loss per unit time. In addition, most eggs cannot replenish the water they lose through uptake of water vapor.

Bursell (1964) presents information on mortality of the pupae of three species of tsetse fly, *Glossina brevipalpis*, *G. pallidipes*, and *G. swynnertoni* (Diptera: Muscidae), in response to ambient humidity. For *G. brevipalpis*, mortality is nearly 100% at a *VPD* of ca. 12 mb. For *G. swynnertoni*, mortality is only slightly elevated at 20 mb *VPD*, and the response of *G. pallidipes* is intermediate. Although in this case the organisms are not particularly small, pupae—like eggs—cannot replenish lost water. In addition, these results again illustrate the point that there exists a wide range among insects in the degree of permeability of the integument to water vapor.

2. Fecundity. Egg production is typically lower at low humidity. Ambient humidity can reduce fecundity in two ways: lower oviposition rate and/or shorter longevity of the females. Throne (1994) reported that the number of eggs oviposited per day by the maize weevil, *Sitophilus zeamais* (Coleoptera: Curculionidae), declines strongly with a decrease in *RH* from 75% to 45% at both 25 and 30°C. Cline (1991) showed that fecundity of the flat grain beetle, *Cryptolestes pusillus* (Coleoptera: Cucujidae), feeding on cracked corn was reduced at lower *RH*s at 27°C because of a combination of reduced oviposition rate and reduced female longevity.

In these examples, the already low moisture content of the grain on which the beetles fed was further reduced at low ambient humidity (Throne 1994). As discussed below (Sec. IV. B. 3), lower moisture content of grain can reduce fitness of stored product pests. This may explain the lower tendency of females to oviposit under such circumstances.

3. Developmental Rate. Throne (1994) observed a 46% slower rate of development for the immatures of *S. zeamais* at 13.3 mb compared to 5.8 mb *VPD* at 20°C. At 25°C, Ferro and Chapman (1979) observed a reduction in the absolute rate of development of eggs of *T. urticae* at 29.9 mb relative to 6.6 mb *VPD* that was ca. 4X the reduction observed at 30°C. Thus, lower ambient humidity can result in slower developmental rate, and the magnitude of this effect is greater at lower temperatures. The latter effect can be explained by the supposition that, for a given difference in *VPD*, a greater percentage of water should be lost during the longer period of development at lower temperatures.

When an insect uses food with much higher water content than grain, the effect of ambient humidity on developmental rate may be lost. Parajulee et al. (1995) found no effect of ambient humidity on the developmental rate of the larger pirate bug, *Lyctocoris campestris* (Hemiptera: Anthocoridae), which occurs in the same environments as the stored product pests and is approximately the same size (ca. 4 mm length) but which feeds on the stored product pests themselves. However, when the moisture content of an insect's food is low and varies with ambient humidity, such as in the case of stored product pests, reduced developmental rate at higher *VPD*s is readily explained. The products of oxidative catabolism are CO_2 and H_2O. For example, the stoichiometric relationship for catabolism of carbohydrate is as follows:

$$6O_2 + C_6H_{12}O_6 \rightarrow 6CO_2 + 6H_2O \quad \text{Eq. 9.}$$

About 0.55 g of water is produced for each 1.00 g of carbohydrate catabolized (Fraenkel and Blewett 1944). Thus, an organism can obtain *metabolic water* by diverting nutritional substrates from anabolic to catabolic pathways. The lower the moisture content of the food, the greater the percentage of the ingested material that is diverted from growth to the production of metabolic water to prevent desiccation. Some stored product pests apparently partially compensate for this loss by increasing their feeding rate at lower *VPD*s.

V. Conclusions

The numbers of ways for inadequacies in the physical environment of insects to prevent their successful rearing are myriad. For a species never before reared, one can examine the natural environment in which it persists and translate these conditions into a rearing protocol. Unfortunately, identification of the crucial elements and interactions of the natural environment to include in the necessarily simplified rearing environment is generally not straightforward. For a species with an established rearing protocol, one must remain vigilant for inadvertent or uncontrolled, deleterious changes in the physical environment. The premise of this chapter is that an understanding of the general principles governing the effects of the physical environment on rearing insects is preferable to the alternative—trial and error.

Placing the environmental factors light, temperature, and humidity at the highest organizational level in this chapter facilitates comparison of the effects of each on the several biological processes considered: circadian rhythms, taxes, diapause, and population growth rate (developmental rate, survival, and fecundity). Of these biological processes, light frequently has its strongest effect on circadian rhythms, taxes, and diapause initiation. Temperature frequently has its strongest effect on diapause termination and population growth rate. Humidity can also have very strong effects on population growth rate but only for small species and growth stages or when water intake per os is limited.

The organization followed in this chapter does not facilitate consideration of the interactions of environmental factors on biological processes. The most important such interactions appear to be between temperature and humidity with respect to population growth, between temperature and light cycles with respect to diapause induction, and between temperature cycles and light cycles with respect to circadian rhythms.

Taxes in response to spatial variation in temperature (thermotaxis) and humidity (hygrotaxis) have not been considered here nor have the effects of humidity on diapause. It is anticipated that these topics will be covered in future editions of this chapter.

References

Akey, D. H., and G. D. Butler, Jr. 1989. Developmental rates and fecundity of apterous *Aphis gossypii* on seedlings of *Gossypium hirsutum*. Southwest. Entomol. 14: 295–299.

Ali, A., and M. J. Gaylor. 1992. Effects of temperature and larval diet on development of the beet armyworm (Lepidoptera: Noctuidae). Environ. Entomol. 21: 780–786.

Ali, A., R. G. Luttrell, and J. C. Schneider. 1990. Effects of temperature and larval diet on development of the fall armyworm (Lepidoptera: Noctuidae). Ann. Entomol. Soc. Am. 83: 725–733.

Amoudi, M. A., F. M. Diab, and S. S. M. Abou-Fannah. 1994. Development rate and mortality of immature *Parasarcophaga* (Liopygia) *ruficornis* (Diptera: Sarcophagidae) at constant laboratory temperatures. J. Med. Entomol. 31: 168–170.

Ankersmit, G. W. 1985. *Adoxyphes orana*, pp. 165-175. *In* P. Singh and R. F. Moore, (*eds.*), Handbook of insect rearing, vol. II. Elsevier, New York, NY.

Aschoff, J., U. v. Saint Paul, and R. Wever. 1971. Die Lebensdauer von Fliegen unter dem Einfluss von Zeit-Verschiebungen. Naturwiss. 58: 574.

Atkinson, D. 1994. Temperature and organism size—a biological law for ectotherms?, pp. 1–58. *In* M. Begon and A. H. Fitter (eds.), Advances in ecological research: vol. 25. Academic Press, New York, NY.

Bacheler, J. S., J. W. Jones, J. R. Bradley, Jr., and H. D. Bowen. 1975. The effect of temperature on development and mortality of boll weevil immature stages. Environ. Entomol. 4: 808–810.

Baumhover, A. H. 1985. *Manduca sexta*, pp. 387–400. *In* P. Singh and R. F. Moore (eds.), Handbook of insect rearing, vol. II. Elsevier, New York, NY.

Beck, S. D. 1983. Insect thermoperiodism. Ann. Rev. Entomol. 28: 91–108.

Benschoter, C. A. 1968. Diapause and development of *Heliothis zea* and *H. virescens* in controlled environments. Ann. Entomol. Soc. Am. 61: 953–956.

Briscoe, A. D., and L. Chittka. 2001. The evolution of color vision in insects. Annu. Rev. Entomol. 46: 471–510.

Burkhardt, D. 1964. Colour discrimination in insects, pp. 131–173. *In* J. W. L. Beament, J. E. Treherne, and V. B. Wigglesworth (eds.), Advances in insect physiology: vol. 2. Academic Press, New York, NY.

Bursell, E. 1974. Environmental aspects: humidity, pp. 43–84. *In* M. Rockstein (ed.). The physiology of Insecta, vol. I. Academic Press, New York, NY.

Butler, G. D., Jr. 1976. Bollworm: development in relation to temperature and larval food. Environ. Entomol. 5: 520–522.

Butler, G. D., Jr., and A. G. Hamilton. 1976. Development time of *Heliothis virescens* in relation to constant temperature. Environ. Entomol. 5: 759–760.

Butler, G. D., Jr., A. G. Hamilton, and F. I. Proshold. 1979. Developmental times of *Heliothis virescens* and *H. subflexa* in relation to constant temperature. Ann. Entomol. Soc. Am. 72: 263–266.

Cline, L. D. 1991. Progeny production and adult longevity of *Cryptolestes pusillus* (Coleoptera: Cucujidae) on broken and whole corn at selected humidities. J. Econ. Entomol. 84: 120–125.

Cole, C. L., and P. L. Adkisson. 1964. Daily rhythm in the susceptibility of an insect to a toxic agent. Science 144: 1148–1149.

Collier, R. H., and S. Finch. 1983. Completion of diapause in field populations of the cabbage fly (*Delia radicum*). Entomol. Exp. & Appl. 34: 186–192.

(CRC) Chemical Rubber Company. 1970. Constant humidity, p. E-40. *In* R. C. Weast (ed.), Handbook of chemistry and physics, 51st ed. Chemical Rubber Company, Cleveland, OH.

Doane, C. C., and D. E. Leonard. 1975. Orientation and dispersal of late-stage larvae of *Porthetria dispar* (Lepidoptera: Lymantriidae). Can. Entomol. 107: 1333–1338.

Eriksson, E. S. 1979. Movement parallax and distance perception in the grasshopper (*Phaulacridium vittatum* (Sjöstedt)). J. Exp. Biol. 86: 337–340.

Ferro, D. N., and R. B. Chapman. 1979. Effects of different constant humidities on twospotted spider mite egg hatch. Environ. Entomol. 8: 701–705.

Fleischer, S. J., and M. J. Gaylor. 1988. *Lygus lineolaris* (Heteroptera: Miridae) population dynamics: nymphal development, life tables, and Leslie matrices on selected weeds and cotton. Environ. Entomol. 17: 246–253.

Fraenkel, G., and M. Blewett. 1944. Utilization of metabolic water in insects. Bull. Entomol. Res. 35: 127–139.

Frantsevich, L. I., and V. E. Pichka. 1976. [Size of the binocular zone of the field of vision of insects] (Russian). Zh. Evol. Biokhim. Fiziol. 12: 461–465.

Fye, R. E., and W. C. McAda. 1972. Laboratory studies on the development, longevity, and fecundity of six lepidopterous pests of cotton in Arizona. USDA/ARS Tech. Bull. No. 1454. Washington, D.C.

Gergis, M. F., and F. K. El-Duweini. 1992. Temperature-dependent development and prediction models for simulating the use of acaricides to control *Tetranychus urticae* Koch in cotton fields in Egypt, pp. 771–776. *In* D. J. Herber and D. A. Richter (eds.). Proceedings of the Beltwide Cotton Conference. National Cotton Council, Memphis, TN.

Gibbs, D. 1975. Reversal of pupal diapause in *Sarcophaga argyrostoma* by temperature shifts after puparium formation. J. Insect Physiol. 21: 1179–1186.

Gilbert, C. 1994. Form and function of stemmata in larvae of holometabolous insects. Annu. Rev. Entomol. 39: 323–349.

Gilbert, L. E. 1975. Ecological consequences of a coevolved mutualism between butterflies and plants, pp. 210–240. *In* L. E. Gilbert and R. H. Raven (eds.), Coevolution of animals and plants, symposium V, First International Congress of Systematic and Evolutionary Biology, Boulder, Colorado, August 1973. University of Texas Press, Austin, TX.

Gilkeson, L. A. 1990. Cold storage of the predatory midge *Aphidoletes aphidimyza* (Diptera: Cecidomyiidae). J. Econ. Entomol. 83: 965–970.

Godfrey, L. D., and T. O. Holtzer. 1991. Influence of temperature and humidity on European corn borer (Lepidoptera: Pyralidae) egg hatchability. Environ. Entomol. 20: 8–14.

Grant, L., C. S. T. Daughtry, and V. C. Vanderbilt. 1993. Polarized and specular reflectance variation with leaf surface features. Physiologia Plantarum 88: 1–9.

Gross, H. R. 1988. Effect of temperature, relative humidity, and free water on the number and normalcy of *Trichogramma pretiosum* Riley (Hymenoptera: Trichogrammatidae) emerging from eggs of *Heliothis zea* (Boddie) (Lepidoptera: Noctuidae). Environ. Entomol. 17: 470–475.

Harada, T. 1991. Effects of photoperiod on phototaxis in *Gerris lacustris latiabdominis* (Heteroptera: Gerridae). Environ. Entomol. 20: 1149–1154.

Henneberry, T. J., G. D. Butler, Jr., and D. L. Coudriet. 1993. Tobacco budworm (Lepidoptera: Noctuidae): effects of temperature and photoperiod on larval and pupal development, larval mortality and induction of pupal diapause. Southwest. Entomol. 18: 269–279.

Hentz, M. G., P. C. Ellsworth, S. E. Naranjo, and T. F. Watson. 1998. Development, longevity, and fecundity of *Chelonus* sp. nr. *curvimaculatus* (Hymenoptera: Braconidae), an egg-larval parasitoid of pink bollworm (Lepidoptera: Gelechiidae). Environ. Entomol. 27: 443–449.

Horridge, G. A. 1986. A theory of insect vision: velocity parallax. Proc. Roy. Soc. Lond., Ser. B, Biol. Sci. 229: 13–27.

Howe, R. W. 1967. Temperature effects on embryonic development in insects. Annu. Rev. Entomol. 12: 15–42.

Jenkins, F. A., and H. E. White. 1957. Fundamentals of optics. McGraw-Hill, New York, NY.

Kitabatake, S., I. Shimizu, and M. Kato. 1983. Wavelength-dependent properties of phototaxis in larvae of *Bombyx mori*. Photochem. & Photobiol. 37: 321–327.

Kramer, D. A., and F. P. Hain. 1989. Effect of constant- and variable-humidity and temperature regimes on the survival and developmental periods of *Oligonychus ununguis* (Acarina: Tetranychidae) and *Neoseiulus fallacis* (Acarina: Phytoseiidae). Environ. Entomol. 18: 741–746.

Kucera, C. L. 1954. Some relationships of evaporation rate to vapor pressure deficit and low wind velocity. Ecology 35: 71–75.

Laing, J. E., and J. E. Corrigan. 1995. Diapause induction and post-diapause emergence in *Trichogramma minutum* Riley (Hymenoptera: Trichogrammatidae): the role of host species, temperature, and photoperiod. Can. Entomol. 127: 103–110.

Land, M. F. 1981. Optics and vision in invertebrates, pp. 471–592. *In* H. Autrum (ed.). Handbook of sensory physiology VII/6B. Springer Verlag, New York, NY.

Land, M. F. 1997. Visual acuity in insects. Annu. Rev. Entomol. 42: 147–177.

Liu, S.-S., G.-M. Zhang, and J. Zhu. 1995. Influence of temperature variations on rate of development in insects: analysis of case studies from entomological literature. Ann. Entomol. Soc. Am. 88: 107–119.

Lopez, J. D., Jr., and A. W. Hartstack, Jr. 1985. Comparison of diapause development in *Heliothis zea* and *H. virescens* (Lepidoptera: Noctuidae). Ann. Entomol. Soc. Am. 78: 415–422.

Loughner, G. E. 1972. Mating behavior of the European corn borer, *Ostrinia nubilalis*, as influenced by photoperiod and thermoperiod. Ann. Entomol. Soc. Am. 65: 1016–1019.

Lyons, D. B. 1996. Oviposition and fecundity of pine false webworm (Hymenoptera: Pamphiliidae). Can. Entomol. 128: 779–790.

Mangini, A. C., and F. P. Hain. 1991. Vapor pressure deficit differentially affects laboratory populations of *Metaseiulus occidentalis* and *Neoseiulus fallacis* (Acarina: Phytoseiidae) reared together. Environ. Entomol. 20: 823–831.

Masaki, S. 1980. Summer diapause. Annu. Rev. Entomol. 25: 1–25.

Mathews, R. W., and J. R. Mathews. 1978. Insect behavior. John Wiley & Sons, New York, NY.

McMullen, R. D. 1967. The effects of photoperiod, temperature, and food supply on rate of development and diapause in *Coccinella novemnotata*. Can. Entomol. 99: 578–586.

Menzel, R. 1979. Spectral sensitivity and color vision in invertebrates, pp. 503–580. *In* H. Autrum (ed.). Handbook of sensory physiology VII/5A. Springer Verlag, New York, NY.

Miall, R. C. 1978. The flicker fusion frequencies of six laboratory insects and the response of the compound eye to mains fluorescent 'ripple.' Physiol. Entomol. 3: 99–106.

Moghal, M. S., and J. C. Schneider. 1998. Effects of high temperature on development of *Heliothis virescens* (F.) on cotton fruiting structures in the laboratory. Southwest. Entomol. 23: 123–126.

Monteith, J. L., and M. H. Unsworth. 1990. Principles of environmental physics, 2nd ed. Edward Arnold (Division of Hodder & Stoughton), London, England.

Morales-Ramos, J. A., S. M. Greenberg, and E. G. King. 1996. Selection of optimal physical conditions for mass propagation of *Catolaccus grandis* (Hymenoptera: Pteromalidae) aided by regression. Environ. Entomol. 25: 165–173.

Mulrooney, J. E., W. L. Parrott, F. M. Davis, and J. N. Jenkins. 1989. Storage of diapausing tobacco budworm pupae (Lepidoptera: Noctuidae). J. Econ. Entomol. 82: 1490–1493.

Nadgauda, D., and H. Pitre. 1983. Development, fecundity, and longevity of the tobacco budworm (Lepidoptera: Noctuidae) fed soybean, cotton, and artificial diet at three temperatures. Environ. Entomol. 12: 582–586.

Nunney, L., and W. Cheung. 1997. The effect of temperature on body size and fecundity in female *Drosophila melanogaster*: evidence for adaptive plasticity. Evolution 51: 1529–1535.

Parajulee, M. N., T. W. Phillips, J. E. Throne, and E. V. Nordheim. 1995. Life history of immature *Lyctocoris compestris* (Hemiptera: Anthocoridae): effects of constant temperatures and relative humidities. Environ. Entomol. 24: 889–897.

Pittendrigh, C. S., and D. H. Minis. 1972. Circadian systems: longevity as a function of circadian resonance in *Drosophila melanogaster*. Proc. Nat. Acad. Sci. USA 69: 1537–1539.

Praagh, J. P. van. 1975. Light-ripple and visual acuity in a climate room for honeybees (*Apis mellifera* L.). Netherlands J. Zool. 25: 506–515.

Robertson, J. L. 1985. *Choristoneura occidentalis* and *Choristoneura fumiferana*, pp. 227–236. *In* P. Singh and R. F. Moore (eds.). Handbook of insect rearing, vol. II. Elsevier, New York, NY.

Rossel, S. 1983. Binocular stereopsis in an insect. Nature 302: 821–822.

Roush, R. T., and J. C. Schneider. 1985. Thermoperiod and photoperiod as temporal cues for adult eclosion of *Heliothis virescens* (Lepidoptera: Noctuidae). Ann. Entomol. Soc. Am. 78: 514–517.

Saint Paul, U. von, and J. Aschoff. 1978. Longevity among blowflies *Phormia terraenovae* R.D. kept in non-24-hour light-dark cycles. J. Comp. Physiol. 127: 191–195.

Saunders, D. S. 1972. Circadian control of larval growth rate in *Sarcophaga argyrostoma*. Proc. Nat. Acad. Sci. USA 69: 2738–2740.

Saunders, D. S, C. G. H. Steel, X. Vafapoulou, and R. D. Lewis. 2002. Insect clocks, 3rd ed. Elsevier, Boston, MA.

Savopoulou-Soultani, M. 1997. Laboratory rearing of euonymus scale (Homoptera: Diaspididae) at different temperatures. J. Econ. Entomol. 90: 955–960.

Saxena, K. N., and P. Khattar. 1977. Orientation of *Papilio demoleus* larvae in relation to size, distance, and combination pattern of visual stimuli. J. Insect Physiol. 23: 1421–1428.

Schneider, J. C. 1980. The role of parthenogenesis and female aptery in microgeographic, ecological adaptation in the fall cankerworm, *Alsophila pometaria* Harris (Lepidoptera: Geometridae). Ecology 61: 1082–1090.

Shaffer, P. L. 1983. Prediction of variation in development period of insects and mites reared at constant temperatures. Environ. Entomol. 12: 1012–1019.

Singh, P., M. D. Ashby, and P. Singh. 1986. Production and storage of diapausing codling moth larvae. Entomol. Exp. & Appl. 41: 75–78.

Skopik, S. D., and M. Takeda. 1980. Circadian control of oviposition activity in *Ostrinia nubilalis*. Am. J. Physiol. 239: R259–R264.

Stewart, P. A, J. J. Lam, Jr., and J. L. Blythe. 1969. Influence of distance on attraction of tobacco hornworm and corn earworm moths to radiations of a blacklight lamp. J. Econ. Entomol. 62: 58–60.

Swift, M. C., and R. B. Forward, Jr. 1980. Photoresponses of *Chaoborus* larvae. J. Insect Physiol. 26: 365–371.

Takeda, M. 1985. An "hourglass-like" feature in the photoperiodic time measurement in *Diatraea grandiosella* (Pyralidae). J. Insect Physiol. 31: 397–406.

Tarrant, C. A., and C. W. McCoy. 1989. Effect of temperature and relative humidity on the egg and larval stages of some citrus root weevils. Fla. Entomol. 72: 117–123.

Tauber, M. J., C. A. Tauber, and S. Masaki. 1986. Seasonal adaptations of insects. Oxford University Press, New York, NY.

Tauber, M. J., C. A. Tauber, and S. Gardescu. 1993. Prolonged storage of *Chrysoperla carnea* (Neuroptera: Chrysopidae). Environ. Entomol. 22: 843–848.

Thiele, H. U. 1973. Remarks about Mansingh's and Müllers's classifications of dormancies in insects. Can. Entomol. 105: 925–928.

Thompson, D. W. 1961. On growth and form, abridged. J. T. Bonner (ed.). Cambridge University Press, London, England.

Throne, J. E. 1994. Life history of immature maize weevils (Coleoptera: Curculionidae) on corn stored at constant temperatures and relative humidities in the laboratory. Environ. Entomol. 23: 1459–1471.

Tolman, J. H., J. W. Whistlecraft, and C. R. Harris. 1985. *Delia antiqua*, pp. 49–57. *In* P. Singh and R. F. Moore (eds.). Handbook of insect rearing, vol. II. Elsevier, New York, NY.

Veerman, A. 2001. Photoperiodic time measurement in insects and mites: a critical evaluation of the oscillator-clock hypothesis. J. Insect Physiol. 47: 1097–1109.

Vinogradova, E. B., and K. B. Zinovjeva. 1972. Experimental investigation of the seasonal aspect of the relationship between blowflies and their parasites. J. Insect Physiol. 18: 1629–1638.

Wagner, T. L., H.-I. Wu, P. J. H. Sharpe, and R. N. Coulson. 1984. Modeling distributions of insect development time: a literature review and application of the Weibull function. Ann. Entomol. Soc. Am. 77: 475–487.

Waterman, T. H. 1981. Polarization sensitivity, pp. 281-469. *In* H. Autrum (*ed.*), Handbook of sensory physiology, vol. VII/6B. Springer Verlag, New York, NY.

Wellington, W. G. 1974. Changes in mosquito flight associated with natural changes in polarized light. Can. Entomol. 106: 941–948.

Wellso, S. G., and P. L. Adkisson. 1966. A long-day short-day effect in the photoperiodic control of the pupal diapause of the bollworm, *Heliothis zea* (Boddie). J. Insect Physiol. 12: 1455–1465.

Wharton, G. W., and A. G. Richards. 1978. Water vapor exchange kinetics in insects and acarines. Annu. Rev. Entomol. 23: 309–328.

Whistlecraft, J. W., C. R. Harris, J. H. Tolman, and A. D. Tomlin. 1985. Mass-rearing technique for *Aleochara bilineata* (Coleoptera: Staphylinidae). J. Econ. Entomol. 78: 995–997.

Yang, P., J. R. Carey, and R. V. Dowell. 1994. Temperature influences on the development and demography of *Bactrocera dorsalis* (Diptera: Tephritidae) in China. Environ. Entomol. 23: 971–974.

Chapter 7

Insect Nutrition, Feeding, and Artificial Diets

MUHAMMAD F. CHAUDHURY
Research Entomologist
USDA-ARS, Screwworm Research Unit
Mexican-American Commission for Screwworm Eradication
Tuxtla Gutierrez, Chiapas, Mexico

I. Introduction

Insects have been a subject of interest to humans particularly because of the pest status of some species and the beneficial nature of others. Humankind competes with pest insects throughout the world for food resources. Some insects are vectors of human diseases. The beneficial nature of insects includes their importance in the natural food chain, their use as feed for other animals, their role as biological control agents, as a source of honey and other products beneficial to humans, and the aesthetic quality of some species. Consequently, insects have been the subject of study in nature as well as in the laboratory. To facilitate these studies and research, it became necessary to rear insects so that a large number was available at all times. Moreover, many species of insects are reared in large quantities for various purposes, such as to use in biological control and sterile insect release programs. Reliability of these programs depends on the quality of the insects used, and the quality of insects produced depends on the quality of the diet used to rear them. Therefore, an important first step in rearing quality insects is to develop a successful diet. Success in developing a quality diet requires thorough knowledge of the biology of the insect to be reared.

Insects that feed primarily on animals, plants and decaying organic materials are referred to as *zoophagous*, *phytophagous*, and *saprophagous*, respectively. Some insects are *polyphagous*, and others are more specific feeders being restricted to a particular type of plant or animal, i.e., *monophagous*. Also, insects themselves constitute an important link in the food chain. In spite of being constantly preyed upon by other animals, insects thrive well in all possible environmental conditions and places on the earth. One of the reasons for this is that they have abundant food materials.

In general, insects may be reared on either natural or artificial diets. However, tremendous progress has been made in the last few decades in formulating artificial diets for rearing insects. This technology has contributed to the development of research in both economic and basic entomology but particularly in mass rearing programs to eradicate or manage various pest species. Consideration of diets in this chapter will be restricted to artificial diets.

It is important to be aware of the limitations of artificial diets. For example, a chemical added to a diet may adversely affect the diet by interacting with other dietary ingredients in a manner different from that in the natural food. In this case, the artificial diet may not mimic the natural food. A diet may be rejected by the insect because one or more *phagostimulants* essential for normal feeding are absent or low or because one or more *phagodeterrents* are present. A diet may be eaten poorly due to inadequate moisture content, or it may be eaten adequately but fail to support development or fecundity, probably due to poor digestive capability resulting from the lack of requisite enzymes.

This chapter will discuss briefly the various aspects of insect diets and artificial diet preparation with special reference to insects' nutritional requirements, feeding behavior, and processes; food consumption and utilization methods; and the processes of digestion, absorption, and metabolism. Valuable sources of information on insect diets are available in the following references that have been consulted in writing this chapter: Singh (1977), King and Leppla (1984), Reinecke (1985), Singh and Moore (1985), Anderson and Leppla (1992), and Cohen (2004). Aspects of insect feeding have been reviewed by Chapman (1985), Smith (1985), Chapman and deBoer (1995), and Cohen (2004). Various aspects of insect digestion, absorption and nutrition have been reviewed by Applebaum (1985), Dadd (1985), Turunen (1985), and Chapman (1998).

II. Nutrition and Dietetics

Nutrition can be defined as the nature and processing of materials taken in from the environment by organisms to provide energy and substance to support growth, development, and reproduction. In short, nutrition is the study of the requirements and function of food components. *Dietetics* is the practical application of nutritional information. This section will deal with the nutritional requirements of insects and the various categories of ingredients of insect diets.

A. Nutritional Requirements

Although the basic nutritional requirements of insects have been studied extensively and are well understood, developing a suitable diet for a specific insect can be a challenging task. Sometimes it can be very difficult to determine appropriate concentrations and ratios of required nutrients and nonnutritive components such as inert texturizers or *phagostimulants* (feeding stimulants). One has to be sure about the presence of adequate nutrients that are essential for proper growth. Sometimes excess nutrients can be detrimental or just simply wasted.

Diets for zoophagous insects have high proportions of protein and amino acids relative to carbohydrates, with the amino N/carbohydrate ratio being normally more than 2.0 (Singh 1977). For the flesh-eating screwworm larvae, *Cochliomyia hominivorax* (Diptera: Calliphoridae), and black blow fly, *Phormia regina* (Diptera: Calliphoridae), this ratio is infinity. These larvae are adversely affected by any dietary sugar and can satisfy their energy needs entirely through metabolism of amino acids. Singh (1977) named several phytophagous insects that feed on various plant tissues in nature, having diets in which protein or amino acid mixtures make up about a third of the total dietary components with sugar or carbohydrate in an approximately equal proportion, giving about a 1:1 ratio of amino acid:sugar. Amino acid:sugar ratio in the artificial diets for the green peach aphid, *Myzus persicae* (Homoptera: Aphididae), and the leafhopper, *Macrosteles fascifrons* (Hemiptera: Deltocephalidae), is very low (0.2 to 0.3), which simulates the composition of their natural diet of phloem sap.

Various diets for stored-product insects contain neither water nor nonnutrient components, which reflects the composition of their natural diets of dry and normally low-fiber foods, e.g., seeds. For other diets, the proportion of both water and inert texturizers roughly mimics the conditions of their natural foods. Artificial diets for phytophagous insects are usually based on an agar gel which incorporates an optimal amount of water in a semisolid texture similar to the plant parts normally eaten because for phytophagous insects the water content of the diet can be very critical (Reese and Beck 1978). Artificial diet for the black twig borer, *Xyleborus ferrugineus* (Coleoptera: Curculionidae), contains a very high proportion of nutritionally inert filler because in nature the beetle bores in wood and thus ingests large amounts of indigestible cellulose. An endosymbiotic fungus digests the cellulose in the gut of the beetle to provide its host with carbohydrate (Sec. IV. A).

The literature on insect nutrition is enormous and has been reviewed numerous times by various authors. The following key reviews should be consulted to access the original literature: Trager (1947, 1953), Lipke and Fraenkel (1956), House (1958, 1961, 1972, 1974a), Dadd (1970, 1973, 1977, 1985), Rodriguez (1972).

B. Dietary Components

1. Carbohydrates. Generally, carbohydrates serve as energy and carbon sources and are often found to be phagostimulatory. Carbohydrates as energy sources are an essential part of most insects' diet in their larval stage. Insects generally survive better when carbohydrates are in their diet. Carbohydrates can be used in the form of poly- to monosaccharides and some sugar alcohols. Use of these components other than monosaccharides is dependent upon the digestive enzymes present in the insect to degrade the larger molecule to smaller subunits for absorption. In artificial diets, polysaccharides such as dextrin, starch, and glycogen may be supplied as purified compounds or as semipurified plant extracts such as corn or potato starch, corn syrup, and molasses. Carbohydrates are frequently incorporated into diet in the form of crude plant products that may also contain other nutritional components. Such additives include the following: wheat germ, soybean meal, corn meal, and other bean or vegetable powders.

Several insects absolutely require carbohydrate or sugars in their diets (Dadd 1963, Rock and Sharma 1974, Thomson 1981a). A few other insects studied were found to utilize triglyceride, fatty acids, or glycerol in place of carbohydrate as the energy source. Greater wax moth, *Galleria mellonella* (Lepidoptera: Pyralidae), and hide beetle, *Dermestes maculates* (Coleoptera: Dermestidae), larvae show optimal growth on diet with beeswax and triglyceride, respectively, as the main energy source. Both species, however, grow nearly as well with carbohydrate entirely replacing the lipid components (Dadd 1964, 1966).

Larvae of flesh flies such as the screwworm and black blow fly do not require any carbohydrate in their diets and probably derive their energy by the oxidation of suitable amino acids. The dietary amino acids presumably are metabolized to tricarboxylic acids that can enter the Krebs cycle and ultimately release energy similar to glycolysis of carbohydrate, sugars, or acetylcoenzyme A from the β-oxidation of fatty acids (Dadd 1966). There are other species that can complete development without carbohydrate but require it to grow at a maximum rate. For example, the spruce budworm, *Choristoneura fumiferana* (Lepidoptera: Tortricidae), develops to the adult stage on meridic diet without carbohydrate but with a prolonged larval period and low-weight pupae, compared to normal development with an optimum level of sucrose in the diet (Harvey 1974). Larvae of the yellowfever mosquito, *Aedes aegypti* (Diptera: Culicidae), show similar facultative dependency on sugar to maintain normal growth rate (Sneller and Dadd 1981).

Carbohydrate requirements have been studied in both immature and adult insects. In most species, the requirement of carbohydrate is uniform throughout larval growth; however, in some insects the carbohydrate requirement may vary from one instar to the next. For example, the last two larval instars of locusts require twice the dietary sugar concentration required by the earlier instars (Dadd 1963). European corn borer larvae, *Ostrinia nubilalis* (Lepidoptera: Crambidae), have been found to change their location on the host plant during feeding by moving to parts with higher sugar concentration during late instars (Beck 1956). The preference for higher carbohydrate concentration during later larval instars probably results from the need to accumulate fat reserves, which can be synthesized from sugars.

In nature, many adult Lepidoptera, Diptera, and Hymenoptera feed on nectars and other sugary fluids to achieve maximum longevity. Those dipterous insects that require no carbohydrate in their diet during the larval stage, such as the screwworm, need sugar during the adult stage for longevity. Usefulness of dietary carbohydrate in adult insects has been studied extensively and has been reviewed by Dadd (1977).

For both larval growth and adult longevity, the sugars that are utilized best are those available in plant and animal tissues that constitute natural food. These are the hexose monosaccharides glucose and fructose and the disaccharide sucrose. Most insects can also utilize hexoses like galactose and mannose; the disaccharides such as maltose, trehalose and melibiose; and the trisaccharides melezitose and raffinose. Disaccharides such as lactose and cellobiose are normally not utilized. However, the insects that can digest cellulose can use cellobiose, which is produced by the action of the enzyme cellulase on cellulose. Pentose sugars such as ribose, xylose, rhamnose, and arabinose do not have any nutritive value and are considered toxic to some insects (Nettles 1972, Barker and Lehner 1974 a, b).

Starch in plants and glycogen in animals are polysaccharide polymers of sugars. Most widespread of plant polysaccharides is cellulose. However, cellulose is only important to those insects possessing cellulose-digesting symbionts. Various starches and dextrins, all polymers of glucose in α-1, 4 and α-1, 6 linkages, are utilized by most stored-product insects, cockroaches, and many phytophagous species of the orders Orthoptera, Coleoptera, and Lepidoptera. However, many leaf-eating lepidopteran larvae requiring dietary sugar cannot utilize starch or do so very poorly (see Dadd 1985). Starches and other polysaccharides generally can not replace sugars for the maintenance of adult Diptera and Hymenoptera, although certain plant-sucking Heteroptera can utilize soluble starch (Dadd 1985). Many carbohydrates, particularly sugars, act as phagostimulants (Sec. III. B).

House (1972) listed 33 carbohydrates that are used by 17 species in their diets. Dadd (1977) listed 23 species that make use of carbohydrates. The summary of this review shows that generally the D-forms of glucose, fructose, sucrose, maltose, trehalose, melezitose, and raffinose are utilized by Diptera, Coleoptera, Lepidoptera, Orthoptera, Hemiptera, and Hymenoptera. Compounds such as xylose, L-arabinose, L-sorbose, and inulin are not used. Diptera are generally unable to utilize glycogen, starch, and cellobiose. A diet for melon fly maggot, *Bactrocera (Dacus) cucurbitae* (Diptera: Tephritidae), contained as many as 18 carbohydrates (Srivastava et al. 1978); however, only fructose, sucrose, ribose, or maltose supported adequate growth. The sugar melezitose was toxic. A number of sugars act as phagostimulants in the diet. Therefore, sometimes it is difficult to assess the actual dietary value of a sugar in the diet.

Sugars are often used to maintain adult insects. A solution of 5% sucrose in water can sustain a number of Lepidoptera, Diptera, and Hymenoptera as the only source of nutrition. For some insects, honey and water is better than sucrose solution. For example, colony growth rate, vigor, and queen survival of red imported fire ant, *Solenopsis invicta* (Hymenoptera: Formicidae), was significantly increased by supplementing the standard colony diet with a 1:1 ratio of honey and water used *ad libitum* (Williams et al. 1980). Generally, most-liquid feeding female adults must have a more complex diet for egg development. The diet should contain some source of amino acids, and often various vitamins, other growth factors, and minerals. For example, adult screwworms can produce some eggs during the first reproductive cycle with a diet of sucrose solution only but require additional protein in the diet, such as spray-dried powdered chicken egg, to accelerate and maximize egg production during the first reproductive cycle (Chaudhury et al. 2000). Alternatively, many other adults, such as many moths, can exhibit reproductive function without any feeding. In such cases, the nutritional requirement for egg production has to be satisfied during the larval period.

Sometimes carbohydrates can be nutritionally inhibitory. For example, larval black blow fly can not tolerate any sugar without adverse growth effects (Cheldelin and Newburgh 1959). Sucrose has been found to act as a feeding inhibitor in the tea tortrix, *Homona coffearia* (Lepidoptera: Tortricidae) (Sivapalan and Gnanapragasam 1979). Nonnutritive carbohydrates such as agar and other gelling agents, bran, cellulose, and other fibers have been widely used in insect diet to provide texture and firmness (Sec. II. C. 1).

2. Proteins. Proteins serve as important structural components and as enzymes to regulate all biochemical functions of living cells. In all insects, both proteins and amino acids are required for growth and maintenance. The requirement is higher in growing immature stages and lower in adults except for females with developing eggs or young. Adult insects that take nonproteinaceous food normally utilize protein stored during the larval to stage to meet their demand. The suitability of food protein depends on how easy it is to break down into amino acids during digestion and absorption. Also, one has to consider how well the absorbed amino acids are utilized. Many indices have been developed to assess and compare suitability of proteins, such as "protein efficiency ratio," "net protein utilization," and the "protein score." Application of these measures to insect studies, especially protein score, is discussed by McGinnis and Kasting (1972) and Rock (1972). On the basis of these evaluations, food protein can be classified as superior or inferior, though these rankings differ for different species.

For the development of artificial diets for rearing or nutritional studies, a high-grade animal protein is assumed to be most suitable. Early studies with protein-based diets found that plant-derived proteins (zein, gluten, edestin) were inadequate for insects unless augmented by particular amino acids (Dadd 1985).

During the early years of insect diet research, egg albumen, lactalbumin, and especially casein, sometimes supplemented with one or more amino acids to correct imbalances indicated by amino acid analysis, were found to be effective amino N sources for most insects studied. In such partially defined insect diets, amino acids are usually provided by casein, chosen largely because of its availability as highly purified fat- and vitamin-free preparations, a useful property for the study of the lipid and vitamin requirements. Sometimes assessments of protein value may be difficult because of the presence of trace impurities of other essential nutrients or variation in digestibility due to different processing methods used for the same protein as, for example, in the case of the common fruit fly, *Drosophila melanogaster* (Diptera: Drosophilidae) (Dadd 1985). For some plant-feeding insects, proteins derived from cottonseed or soy meal were found superior to casein (Geer 1963, Ito and Arai 1965). For some insects, mixtures of proteins and peptones (soluble substances into which proteins are converted by partial hydrolysis) have been found superior to single proteins (Dadd 1963, Vanderzant 1973).

Besides digestibility, the major factor that determines the suitability of a protein is the complement of amino acids it provides, especially the number and proportion of essential amino acids. Generally, there are 10 essential amino acids for insects: L-isomers of arginine, histidine, isoleucine, leucine, lysine, methionine, phenylalanine, threonine, tryptophan, and valine. Understanding of the nutritional requirements of amino acids comes from the development of synthetic diets using amino acid mixtures to replace proteins. The amino acid most commonly needed in addition to the 10 essentials is proline. Omission of any one of the essential amino acids normally causes growth to stop entirely within the first or early stadia, whereas omission of proline causes a more gradual cessation of growth and development, with a proportion of individuals reaching a delayed adulthood, such as in the northern house mosquito, *Culex pipiens* (Diptera: Culicidae) (Dadd 1978).

Several insects studied require additional essential amino acids, such as glycine, glutamic or aspartic acids. For the black blow fly and the silkworm, *Bombyx mori* (Lepidoptera: Bombycidae), either glutamic acid or aspartic acid is essential; and for the screwworm, glutamic acid specifically is essential (Gingrich 1964; Dadd 1977, 1978).

Nonessential amino acids often have important nonspecific functions in the diet. All nonessential amino acids can theoretically be derived through pathways involving aminations, transaminations, deaminations, carboxylations, and oxidations from essential amino acids. In a particular insect, if the metabolic pathways available for derivation of a nonessential amino acid are weak, this may be a limiting factor on growth. For most insects studied, tyrosine is nutritionally nonessential and is derived, so far as is known, entirely by hydroxylation of phenylalanine, an essential amino acid. Conversion of phenylalanine to tyrosine has been demonstrated in various insects (Ito and Arai 1967, Davis 1968). A requirement for asparagine that cannot be satisfied by aspartic acid was detected in the northern house mosquito and the mosquito *Culiseta incidens* (Diptera: Culicidae); so this amide is a true essential for these mosquitoes, and unless it is provided in the diet, the first instar larvae will not grow. This requirement has been found only for these few species of mosquito, indicating that they have secondarily lost the ability to aminate aspartic acid to asparagine. In contrast, the yellow fever mosquito does not need asparagine or aspartic acid (Dadd 1978, 1980).

Insects generally acquire amino acids from ingested proteins. Proteins are enzymatically degraded to amino acids and absorbed through the gut wall. Full complements of amino acids are provided in most complex diets. A few particular amino acids may be used as supplements in diets that contain a protein source that may be deficient in those amino acids. Insects that do not possess proteases, such as plant-sucking hemipteran insects, need free amino acids in their liquid diet.

There are a number of protein sources that have become popular over the years for insect diets. Some of the well-known proteins are casein, soybean protein, cottonseed protein, gelatin, glycinin, lactalbumin, peanut protein, wheat gluten, brain extract, various milk products, egg products, animal whole blood and blood cells, and zein. These protein sources may be interchangeable, which may be advantageous because of availability, cost, and simplicity in use or new research findings. Generally, different proteins and different sources for the same protein have different amino acid profiles. In addition, a crude protein source may also contain active enzymes, lipids, vitamins, minerals and carbohydrates. On the other hand, it may contain unwanted contaminating components that may be toxic, such as pesticide residues, or that act as feeding inhibitors.

Casein is one of the most commonly used protein source of insect diets. Casein is a phosphoprotein normally obtained from bovine milk and is also available as a fat- and vitamin free preparation. Casein contains 15 amino acids but may have an insufficient amount of certain amino acids, such as cystine, glycine, and glutamic acid to meet the dietary requirements of some species. In such cases, the diet must be supplemented with these amino acids. Caseins are generally water insoluble, but some are soluble and may be used in some liquid diets. Highly soluble casein was used in a diet to rear larvae of the northern house mosquito (Dadd et al. 1973).

Yeast and yeast products were used very early in the history of insect diet development and were found to be a useful source of nutrients. Yeast is the most common crude nutrient source used in diets of both phytophagous and zoophagous insects. Yeast can vary depending on growth conditions and species, and thus it may not be consistent in composition between sources or even between batches from the same source. Yeast is a good source of amino acids, vitamins, phospholipids, glycerides, and sterol (mainly ergosterol). However, certain useful nutrients may be lacking or present in low concentrations. Therefore, yeast is usually combined with other materials such as casein, wheat germ, and specific vitamins and amino acids. Yeast may be used whole, as an extract, or as a hydrolysate.

Wheat germ was first used in the diet of pink bollworm, *Pectinophora gossypiella* (Lepidoptera: Gelechiidae). Whole wheat germ and defatted wheat germ are popular additions to insect diets. Wheat germ contains about 25% protein, which contains most of the amino acids including cystine, glycine, and glutamic acid that may be deficient in casein. Raw wheat germ has a large number of enzymes that are denatured easily by heat.

Soybean and its several products are regularly used for insect diets. Products such as soybean meal and defatted soybean meal are most commonly used. Soybean is high in protein (40–50%) and contain other nutrients such as carbohydrates and lipids (Wolf and Cowan 1971, Smith and Circle 1978). Soybean products may not be uniform and can vary with respect to variety of the bean, growth environment, processing, and storage (Ivy et al. 1971, Wolf and Cowan 1971). Soybean products are normally used in combination with other protein sources to compensate for any deficiency in amino acids. However, soy pulp, a by-product of tofu (bean curd) production, was the only protein source used in a diet for southwestern corn borer, *Diatraea grandiosella* (Lepidoptera: Crambidae) (Yin and Peng 1981). Soybean products are most commonly used in diets for Lepidoptera but also for Coleoptera, Diptera, and Neuroptera.

Algae have been used as a protein source for insect diets with some promising results. The spiral blue-green algae have been used for larvae of the silkworm. A commercially developed species of algae, *Spirulina geitleri*, was used as a substitute for casein, soy flour, and wheat germ in the diet of the larvae of tobacco budworm, *Heliothis virescens* (Lepidoptera: Noctuidae) (Guerra and Garza 1978). The authors also reared bollworm, *Helicoverpa zea* (Lepidoptera: Noctuidae), and boll weevil, *Anthonomus grandis grandis* (Coleoptera: Curculionidae), on diets containing *S. geitleri*.

Other crude protein sources for artificial insect diets include a number of legumes such as pinto, kidney, lima, and horse beans. Raw beans frequently have toxic components that can significantly retard insect growth and even cause high mortality when used in diet. These beans can be detoxified for use in insect diets with heat, such as by autoclaving, or through solvent extraction. Beans are used ground, soaked and mashed, and even sprouted.

Diets prepared for some zoophagous insects use whole eggs, powdered eggs, and egg albumen, whole and skimmed milk powder, milk substitute, fish meal, or cow and other animal blood. The choice of these protein sources, particularly for diets used in mass rearing, depends not only on their suitability as insect diet components but also on their cost and availability (Chaudhury et al. 1998, 2000).

3. Lipids. It is now well established that all insects, particularly the immature stages, need certain lipids in their diets. Although insects may use a wide variety of lipids as an energy source alternative to carbohydrates, only a few are frequently essential, and only the sterols are universally required (Dadd 1985). The only exceptions to this generalization are insects in which symbiotic microorganisms synthesize the necessary sterols. Insects can utilize both animal and plant sterols from their diet. The range of dietary sterols that satisfy the requirement has been documented in more than 40 species studied (Dadd 1977). Certain dermestid beetles that feed preferentially on animal products utilize the animal sterol, cholesterol, but none of the common plant sterols. In contrast, many phytophagous insects are able to develop well using a variety of phytosterols. Phytophagous insects may prefer plant sterols to cholesterol (Svoboda and Thomson 1983). The screwworm, an extremely zoophagous insect, cannot utilize sitosterol or ergosterol in place of cholesterol (Gingrich 1964); but other blow fly larvae, seemingly just as zoophagous, are able to utilize certain phytosterols very well (Clayton 1964). Certain parasitoid Hymenoptera grow best with cholesterol in synthetic media, but phytosterols such as β-sitosterol can also be utilized by these insects (Thomson 1981b). Normally the dietary phytosterols are dealkylated to cholesterol, which is then converted to α-ecdysone and to 20-hydroxyecdysone. Not all insects dealkylate phytosterols to cholesterol. Examples are the house fly, *Musca domestica* (Diptera: Muscidae), and honey bee, *Apis mellifera* (Hymenoptera: Apidae). In artificial diets, cholesterol fulfills the sterol requirement for most insects. However, it is useful when developing a diet for plant feeding insects to have a sterol profile of the host. This may avoid a subsequent problem with the diet.

Insects use fatty acids as an energy source. However, given that they generally are able to synthesize the common fatty acids, they can be reared on artificial diets with reduced lipid content. An example is the greater wax moth, which has a natural diet of spent honeycomb but which can be reared on a high carbohydrate diet in the laboratory (Marston et al. 1975). Lepidopterans show fatty acid deficiency by failure of pupal to adult ecdysis, as in the classic study of the Mediterranean flour moth, *Anagasta (Ephestia) kuehniella* (Lepidoptera: Pyralidae) (Dadd 1985). A critical physiological need for fatty acid during metamorphosis has been reported by Grau and Terriere (1971). This need can be alleviated using linolenic acid during late larval development in some Lepidoptera (Rock et al. 1965, Kato 1978). A similar problem in some parasitic Hymenoptera can also be prevented using linolenic acid in the diet (Yazgan 1972). Acridid grasshoppers sometimes show retarded nymphal growth and deformed adults due lack of a fatty acid (Nayar 1964). The Colorado potato beetle, *Leptinotarsa decemlineata* (Coleoptera: Chrysomelidae), shows slow larval development and decreased fecundity with fatty acid deficient diet (Wardojo 1969).

Polyunsaturated fatty acids are essential for a number of insects. The polyunsaturated di- and trienoic C-18 fatty acids linoleic and linolenic are found to be most essential, particularly in Lepidoptera and Orthoptera, which apparently cannot synthesize them. Adult eclosion of moths is affected by an inadequate level of linoleic and linolenic acids. Arachidonic acid, a C-20 polyenoic acid, was found to be essential for three mosquito species (Dadd et al. 1980). When deprived, the mosquito larvae grew normally, but the success rate of adult eclosion was reduced.

Termites, cockroaches, and crickets are able to synthesize linoleic acid (Blomquist et al. 1982). Other fatty acids such as animal-derived phospholipids added to larval diets for the northern house mosquito improved adult viability (Dadd and Kleinjan 1978). Arachidonic acid can be substituted for phospholipids; however, this fatty acid in the free state in the diet was quite unstable, and measures must be taken for its protection (Dadd and Kleinjan 1979).

Several vegetable oils are commonly used in artificial diets to provide essential fatty acids. Examples are corn oil, cottonseed oil, linseed oil, olive oil, safflower oil, and sunflower oil. These oils differ significantly in their composition. For example, olive oil contains 4% linoleic acid whereas sunflower oil contains 66.2% (Merk 2001). Variation in fatty acid composition within a given type of oil is common.

Therefore, it is imperative that the source of oil should be considered carefully for use in the diet. In storage, the oil may become rancid due to oxidation of polyunsaturated fatty acids, which not only reduces the essential nutrient but replaces it with toxic peroxides (Reinecke 1985).

4. Nucleic acids. Most insects are capable of synthesizing nucleic acids from other dietary constituents. However, for Diptera it appears that an optimal larval diet must contain nucleic acid or certain components thereof (Friend 1968). A number of Diptera including the screwworm (Gingrich 1964), the common fruit fly (Sang 1978), the northern house mosquito, and the yellow fever mosquito (Dadd and Kleinjan 1978) have an absolute requirement for dietary nucleic acid or certain nucleotides to complete development. Presumably, the biosynthetic pathways for these components are lost or weakened in these species.

5. Minerals. Minerals have received little attention in the formulation of insect artificial diets. This is probably because insects can live without additional minerals other than what is already available from the other ingredients. Minerals cannot be biosynthesized and therefore must come from the diet. Diets containing added minerals generally use commercially available mixtures, such as Wesson salt mixture (Table 1). Some other mixtures are salt mixture No.2 U.S.P. XIII; salt mixture U.S.P. XIV; and McCollum-Davis' salt mixture No. 185 (Singh 1977). From the many insect diets listed by Singh (1977), it appears that salt mixtures rarely exceed 1% of the total diet weight. For example, lepidopteran diets have salt mixtures between 0.3 and 0.9%, and diets fed to some dipteran species have salt concentrations of 0.07-0.2%. Aphid diets contain about 0.6% salt. A few insects have a high calcium requirement, for example, certain flies that form calcareous puparia. Calcium must be added in high quantities to the diet, but other essential salt-derived ions such as sodium and chloride are generally present in other diet components (Reinecke 1985). Magnesium helps in the glycolytic pathway and in enzymatic reactions in other pathways (Lehninger et al. 1993). Trace metals such as copper, iron, manganese, and zinc are also present in various major diet components. These trace metals are known to function as cofactors in common enzymatic reactions. The pea aphid, *Acyrthosiphon pisum* (Hemiptera: Aphididae), requires boron, cobalt, and molybdenum for proper growth and development. It is possible that one or more of these trace metals benefit the symbionts rather than the insects (Dadd 1968).

Table 1. Composition of Wesson salt mixture (Singh 1977)

Ingredient	Weight, %
Calcium carbonate	21.0
Copper sulfate $5H_2O$	0.039
Ferric phosphate	1.47
Manganese sulfate (anhydrous)	0.020
Magnesium sulfate (anhydrous)	9.0
Potassium aluminum sulfate	0.0090
Potassium chloride	12.0
Potassium dihydrogen phosphate	31.0
Potassium iodide	0.0050
Sodium chloride	10.5
Sodium fluoride	0.057
Tricalcium phosphate	14.9

A chelating agent such as sodium EDTA is sometimes used to add trace elements to diets in soluble form (Reinecke 1985, Cohen 2004). However, for the green peach aphid, sodium EDTA was found to be a feeding deterrent (Mittler and Kleinjan 1970), and use of ascorbic or citric acid as a chelating agent in place of sodium EDTA was suggested.

6. Vitamins. Although we know that many insects require various vitamins in their diets, our understanding of these requirements is limited. There are two main types of vitamins depending on their solubility properties: water-soluble and lipid-soluble. The water-soluble vitamins include the B vitamins, vitamin C, and a few other compounds such as choline and carnitine that have vitamin-like properties. Requirements for B vitamins have been established for representatives of Diptera, Coleoptera, Lepidoptera, Hemiptera, Hymenoptera, Orthoptera, and Siphonaptera. Insects generally require the following B vitamins: thiamine, riboflavin, niacin, pantothenic acid, pyridoxine, folic acid and biotin. Insects get vitamins through their diet or through their symbionts. An example showing requirements for all seven B vitamins is the northern house mosquito (Kleinjan and Dadd 1977). For some of the vitamins, such as biotin and folic acid, the required amount is extremely small. Pyridoxine is essential for nearly all symbiont-independent insects studied with the exception of the screwworm, which develops to the adult better without a dietary pyridoxine (Gingrich 1964). Other B vitamins such as B_{12}, cyanocobalamin, have been used in insect diets but have not been found to be essential except for a few species (Armstrong 1978). Carnitine, though not a classic B vitamin, is needed in the diets of several insects. For all beetles of the family Tenebrionidae that have been studied nutritionally, carnitine was found to be an essential vitamin (House 1963). The sawtoothed grain beetle, *Oryzaephilus surinamensis* (Coleoptera: Silvanidae), developed better with carnitine in its diet (Davis 1964).

Initially, vitamin C, ascorbic acid, was not considered important in the nutrition of insects. However, a clear dietary need for ascorbic acid was demonstrated in the desert locust. *Schistocerca gregaria* (Orthoptera: Acrididae), and the migratory locust, *Locusta migratoria* (Orthoptera: Acrididae). When reared on semisynthetic diet without ascorbic acid, these locusts grew poorly with high mortality and poor adult emergence (Dadd 1960). Later investigations with other species belonging to Lepidoptera, Coleoptera, Hemiptera, and Hymenoptera indicated that ascorbic acid is essential for most insects that feed on fresh plant tissue (Kramer and Seib 1982).

Insects in general do not seem to need any of the fat-soluble vitamins such as A (ß-carotenes and other carotenoids), D (calciferol), E (α-tocopherol), and K (menadione), at least during the larval growth period. The parasitoid, *Agria affinis* (Diptera: Sarcophagidae), shows slightly retarded larval growth in the absence of vitamin A (House 1966). Beta-carotene may have a phagostimulatory role in the synthetic diet for larvae of the Indian meal moth, *Plodia interpunctella* (Lepidoptera: Pyralidae). Vitamin D has been studied as a possible substitute for cholesterol in insects but with negative results. Nutritional requirement for vitamin E has been demonstrated in a few insects, particularly in connection with reproductive function, including improved fecundity in certain Lepidoptera (David and Ellaby 1975) and Coleoptera (Atallah and Newsom 1966). In *A. affinis*, vitamin D was found to accelerate larval growth (House 1966). Vitamin K has been tested with insects as a possible growth-stimulatory factor but without positive results.

A number of diets are prepared with commercially available vitamin premixes such as Vanderzant vitamin mixture (Table 2; Singh 1977). Other vitamin mixtures that have been used in insect diets are vitamin mixture U.S.P. XIV, Ferbelan (BDH) vitamin mixture, vitamin premix No. 26862 by Hoffman-LaRoche, Deca-Vi-SOL, and Viterra (multivitamins and minerals) (Singh 1977).

Table 2. Composition of Vanderzant vitamin mixture (Singh 1977)

Constituent	Weight, %
Tocopherol	0.80
Ascorbic acid	27.0
Biotin	0.0020
Calcium pantothenate	0.100
Choline chloride	5.0
Folic acid (crystalline)	0.025
Inositol	2.0
Niacinamide	0.100
Pyridoxine hydrochloride	0.025
Riboflavin	0.050
Thiamine hydrochloride	0.025
Vitamin B_{12}	0.20
Glucose	64.7

7. Water. Water is an indispensable ingredient in most insect diets. Failure to maintain proper water content in diets has often caused problems in rearing insects. Moist diets tend to lose water by evaporation, and this may lead to adverse changes in texture and palatability. Evaporation can be avoided by spraying water on the surface of the diet or placing a thin film of paraffin over the diet (Howell 1967).

Water used in the preparation of artificial diets is commonly described in the literature as tap or distilled water. In many publications, the source of water is not mentioned. With growing concerns regarding water contamination due to agricultural runoff, tap water should be questioned for use in insect diet. Use of water from a purification device should be a priority for all insect rearing programs. A combination of processes used simultaneously should provide desired water purity. These may include activated charcoal filters, particle removal filters, reverse osmosis, water softeners, and sterilizing systems. Although the occurrence of a disaster in the insectary due to water impurities is highly unlikely, the use of high quality water should at least eliminate one potential rearing problem.

C. Nonnutritive Dietary Additives

Besides nutrients, an artificial diet requires the addition of various other ingredients that are nutritionally inert. These additives improve the stability, texture, palatability, and, indirectly, overall nutritional quality of the diet. A summary of the most important additives is given below. For a detailed discussion of these additives see Cohen (2004).

1. Gels and Bulking Agents. Gelling agents are used to make the artificial diet moist and semisolid in consistency. The most popular gelling agent is agar. Agar is generally considered nonnutritional; however, it may cause problems when used in nutritional studies, because of the metal ion components and other contaminants trapped in it during production. Agar can also be used at a low concentration (0.05%) to optimize the viscosity of diets for mosquito larvae (Dadd 1975, 1978). Nettles et al. (1980) used a combination of 1.5% agar liquid diets for rearing larvae of the endoparasitoid *Eucelatoria* sp. (Diptera: Tachinidae). Absorption/desorption of liquid at the gel/liquid interface maintained the diet consistency necessary for feeding as well as proper respiratory function in the absence of the host's tracheal system.

Agar is a costly ingredient and may be prohibitively expensive for mass rearing programs. Therefore, various gelling agents have been developed as agar substitutes. One substitute known as *carrageenan* was developed from seaweeds. It is a polysaccharide composed of galactose, dextrose and levulose. It exists in two forms: one is extracted with cold water to give a viscous solution and the other is extracted with hot water to form a gel. The latter form is used in some insect diets (Shapiro and Bell 1982). Recently sodium polyacrylate or starch-grafted polyacrylate gels were found to be useful for screwworm larval diet (Taylor 1988, Chaudhury and Alvarez 1999).

Other gelling agents that have been tried in artificial insect diets include starch, gelatin, guar gum, locust bean gum, glutin, soya lecithin, and carboxymethylcellulose (CMC). These agents were tried in various combinations by Howell (1977) to find adequate and inexpensive agar substitutes for rearing the codling moth, *Cydia pomonella* (Lepidoptera: Tortricidae).

There are a number of liquid or moist diets that do not use gelling agents but rely on other bulking agents such as sawdust or sugarcane pulp to solidify the diet sufficiently to provide a suitable texture. Recently, cellulose fibers recycled from newsprint have been successfully used as a bulking agent for screwworm larval diet (Chaudhury and Skoda 2007, Chaudhury et al. 2002). However, in such gel-free diets, poor retention of moisture can be a problem. Repeated spraying of the medium with water or a more viscous liquid such as syrup may be necessary. For example, the trays of cellulose fiber–based diet for screwworm larval rearing tend to lose moisture during the 5-day period of larval development. These diet trays are provided with adequate moisture by spraying them on a daily basis (Chaudhury and Skoda 2007). Insects reared on liquid diet may require a physical support such as an inert mat, sponge, or cotton that can be saturated with the diet. Such a system has been used for mass rearing screwworm (Taylor 1992). The olive fruit fly, *Bactrocera* (*Dacus*) *oleae* (Diptera: Tephritidae), was reared on liquid soaked cotton toweling (Mittler and Tsitsipis 1973).

2. Diet Stabilization. Problems of diet degradation and microbial contamination are common in insect artificial diets. Approaches to managing microbial contamination are considered in detail in Chap. 8 *Microbial Contamination and Insect Rearing*. A number of methods are available to stabilize prepared insect diets. These methods include heating to deactivate unwanted chemicals, chilling for extended storage, and adding antioxidants to preserve unstable compounds.

In addition to beneficial effects, heating insect diets may have some ill effects. Besides killing microbial contaminants, heating may help activate some gelling agents, denature some harmful enzymes such as lipo-oxygenases and phenol oxidases, destroy some digestive inhibitors, and accelerate desirable chemical reactions. On the other hand, heating causes destruction of certain nutrients such as ascorbic acid and unsaturated lipids and decreases protein digestibility. Therefore, one must find a balance between under and overheating (Cohen 2004).

Chilling or cold storage of prepared diet is a procedure frequently used in many rearing situations to prevent degradation of diets and their components. Reactions between interactive compounds causing deterioration in the diet can be prevented by freezing. However, cold storage, particularly freezing, may cause desiccation, oxidation, and formation of ice crystals, reducing nutritional quality and palatability of the diet.

Artificial diets may contain some chemicals that are added to serve as antioxidants, such as ascorbic acid, tocopherols, and carotenes. The importance of various vitamins in insect diets is well known (Gilmour 1961); however, their function as antioxidants in insects has only recently been recognized (Cohen 2004).

III. Insect Feeding

Feeding is of enormous practical concern in insect nutrition because the feeding mechanisms of insects are so diverse. The biology of insect feeding includes the study of mouthparts of both adult and immature forms, feeding behavior and mechanisms, and the physical and chemical factors that influence feeding. Many insect species feed poorly in the absence of specific positive stimuli or in the presence of various deterrent stimuli.

A. Insect Mouthparts

In general, there are three types of insect mouthparts: biting/chewing, piercing/sucking, and lapping/sponging (Fig. 1). These three types can further be divided into several subtypes: chewing (grasshopper, beetle, and many immature insects), chewing/lapping (bee), chewing/raptorial (dragonfly), piercing/sucking (cicada, flea, and mosquito), sucking/mouth hooks (maggots), piercing/sponging (biting fly), rasping/sucking (thrips), sponging (housefly), and sucking/siphoning (butterflies and moths). Mouthparts are adapted to feed on liquid diets, solid diets, and a mixture of solid and liquid diets. Most insects are solid feeders and have chewing type mouthparts that include mandibles (the main biting apparatus), labrum, labium, and maxillae. They also possess gustatory sensory structures such as labial and maxillary palps that contain sensory hairs and pegs. Liquid-feeding insects are less common. These are sap- and nectar-feeding species and also those that feed on vertebrate blood. Whether the diet is solid, liquid, or in between, it is important that the diet possesses the appropriate texture and viscosity.

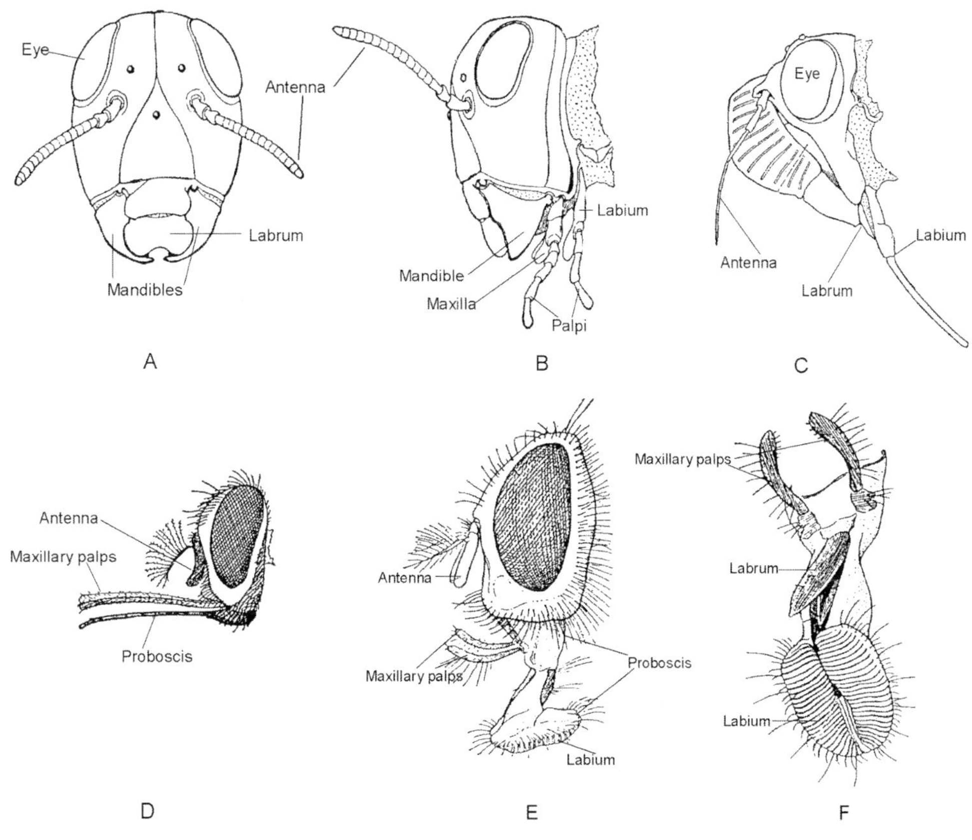

Fig. 1. Types of insect mouthparts presented diagrammatically (not to scale) to show various structures used in manipulating food materials. (A) Biting/chewing, frontal view (grasshopper). (B) Biting/chewing, lateral view (grasshopper). (C) Piercing/sucking type (cicada). (D) piercing/sucking type (tsetse fly). (E) Sponging type (house fly). (F) Sponging type showing the proboscis (house fly) [adapted from Snodgrass (1935)].

B. Feeding Stimulants and Deterrents

Insects may fail to develop on a certain diet because of abnormally low food intake due to the absence or deficiency of a feeding stimulant or due to the presence of a feeding deterrent. Chemoreception is very important in initiating food intake in animals, including insects. Various ablation experiments on chemical sensory organs such as maxillary palps indicate that all the major groups of chemoreceptors are involved in the regulation of food intake. For example, in grasshoppers, all the chemoreceptors tested neurophysiologically respond to many feeding stimulants and deterrents. Phagostimulants, are essential for feeding to be initiated and to continue, while phagodeterrents may inhibit feeding even in the presence of phagostimulants (Bernays and Simpson 1982).

Several compounds showing phagostimulatory activity are listed in Table 3. A more extensive list is provided by Reinecke (1985). Sugars, particularly sucrose and fructose, are the best known phagostimulants. Bees, wasps, butterflies, moths, and many Diptera, including blood sucking species, feed on sugar solutions. For example, in the black blow fly, maltose, fructose, and sucrose are very effective phagostimulants. In many phytophagous insects such as grasshoppers, beetles, caterpillars, aphids, and planthoppers, sucrose stimulates feeding. The concentration causing maximum intake normally lies between 0.1 and 1.0 M.

In most phytophagous species, a variety of nutrients in addition to sugars serve as phagostimulants. Among these are amino acids, although their effects vary with species. For example, only a few amino acids such as serine, valine, and threonine are even weakly stimulating in locusts, whereas a number of amino acids including glycine, D-alanine, alanine, and methionine are strongly stimulating in the large milkweed bug, *Oncopeltus fasciatus* (Hemiptera: Lygaeidae). Additional nutrients that may stimulate feeding include phospholipids, certain fatty acids and sterols, ascorbic acid, and various salts (Dadd 1970, Cook 1977).

In some instances, an amino acid or a secondary plant chemical stimulates feeding only in the presence of another compound such as sucrose (Ma 1972). In such cases where the amount of food eaten when both com-

Table 3. Compounds showing phagostimulatory activity in insects

Chemical group	Compound	Species (Order)	Reference
Amino acids	Aspartic acid	*Gastrophysa atrocyanea* (Coleoptera)	Matsuda (1981)
	Glutamic acid	*Spodoptera littoralis* (Lepidoptera)	Ascher et al. (1976)
	Lysine	*Gastrophysa atrocyanea* (Coleoptera)	Matsuda (1981)
	Methionine	*Sitona lineatus* (Coleoptera)	Havlickova (1976)
	Proline	*Locusta migratoria* (Orthoptera)	Cook (1977)
	Serine	*Locusta migratoria* (Orthoptera)	Cook (1977)
Sugars	Fructose	*Choristoneura fumiferana* (Lepidoptera)	Albert et al. (1982)
		Gastrophysa atrocyanea (Coleoptera)	Matsuda (1981)
		Oxyrhachis tarandus (Hemiptera)	Gandhi (1977)
	Glucose	*Choristoneura fumiferana* (Lepidoptera)	Albert et al. (1982)
		Empoasca fabae (Hemiptera)	Dahlman et al. (1981)
		Gastrophysa atrocyanea (Coleoptera)	Matsuda (1981)
	Sucrose	*Camponotus compressus* (Hymenoptera)	Shetty (1982)
		Choristoneura fumiferana (Lepidoptera)	Albert et al. (1982)
		Culiseta inornata (Diptera)	Friend (1978)
		Empoasca fabae (Hemiptera)	Dahlman et al. (1981)
		Spodoptera exempta (Lepidoptera)	Ma & Kubo (1977)
Salts	KCl	*Gibbium psylloides* (Coleoptera)	El-Sawaf & El-Sayes (1978)
	KH_2PO_4	*Gibbium psylloides* (Coleoptera)	El-Sawaf & El-Sayes (1978)
	NaCl	*Gibbium psylloides* (Coleoptera)	El-Sawaf & El-Sayes (1978)
Sterols	Sitosterol	*Bombyx mori* (Lepidoptera)	Tsai et al. (1978)
		Spodoptera littoralis (Lepidoptera)	Abbassy et al. (1977)
Vitamins	Ascorbic acid	*Gastrophysa atrocyanea* (Coleoptera)	Matsuda (1981)
	Niacin	*Gastrophysa atrocyanea* (Coleoptera)	Matsuda (1981)
	Tocopherol	*Agrotis segetum* (Lepidoptera)	Charpentier (1979)
Miscellaneous	ATP	*Culiseta inornata* (Diptera)	Friend (1978)
		Rhodnius prolixus (Hemiptera)	Smith (1979)
	Ethanol	*Xyleborus ferrugineus* (Coleoptera)	Norris & Baker (1969)
	Flavonoids	*Aphis fabae* (Hemiptera)	Jordens-Rottger (1979)
	Phenol	*Aphis fabae* (Hemiptera)	Jordens-Rottger (1979)
	Sinigrin	*Schistocerca gregaria* (Orthoptera)	Aspirot & Lauge (1981)

pounds are present is greater than that expected under a model of independent effects, the chemicals are said to be *synergistic* in their action. Other examples of these effects are given by Dadd (1970).

Some chemicals have been shown to have phagostimulatory effects, even though they are neither required nor utilized. For example, the sugar fucose is strongly stimulating to flies and causes increased feeding, yet it is not utilized metabolically (Bernays and Chapman 1994). These chemicals are sometimes called *token stimuli* or *sign stimuli* to differentiate them from "real" feeding stimuli such as nutrients. Additionally, nonnutrient materials that are specific to host plants stimulate feeding in some oligophagous species. These token stimuli indicate high nutritional suitability. For example, a number of species feed on cruciferous plants that contain glucosinolates such as sinigrin. This type of compound is normally phagostimulatory for these insects (Schoonhoven 1972).

There are many substances that act as phagodeterrents, showing inhibitory effects on insect feeding. As is the case for phagostimulants, phagodeterrents act on specific chemoreceptors. For example, strychnine-HCl is a feeding deterrent for adult western corn rootworm beetles, *Diabrotica virgifera virgifera* (Coleoptera: Chrysomelidae), as shown by the electrophysiological responses of chemoreceptors located on the galea (Chyb et al. 1995). Feeding of the American cockroach, *Periplaneta americana* (Blattodea: Blattidae), is inhibited by 1, 4-naphthoquinone; and ammonium nitrate inhibits feeding by the sweetclover weevil, *Sitona cylindricollis* (Coleoptera: Curculionidae). An extensive list of phagodeterrents is given by Schoonhoven (1972).

The balance of chemical phagostimulants and deterrents is very important in determining the palatability of food. Sometimes a nutrient commonly used in artificial diets has a phagostimulatory effect on one insect and an inhibitory effect on another. For example, sucrose, a phagostimulant for many phytophagous insects, is a phagodeterrent for the tea tortrix,(Sivapalan and Gnanapragasam 1979). In acridids, low levels of deterrents are ignored if sufficient levels of phagostimulants are available (Bernays and Chapman 1994). In flea beetles that feed on Cruciferae, the amount eaten depends largely on the combinations and amounts of phagostimulatory glucosinolates and phagodeterrent flavonoids (Bernays and Chapman 1994). Other examples are known from Lepidoptera (Schoonhoven 1972).

C. Physical Characteristics of Diets

Physical characteristics of diets are important factors affecting diet acceptability (Vanderzant 1969). These characteristics include hardness, texture, moisture content, temperature, and homogeneity. It is therefore necessary to create and maintain the appropriate physical characteristics in the diet. For example, the screwworm larval diet is most acceptable when it is prepared using warm water (~40°C) and viscosity is maintained at around 5,000 cP (Chaudhury, unpublished). Optimum viscosity provides adequate moisture and proper texture. To achieve the optimum consistency, various nonnutritive ingredients such as cellulose fiber and gelling agents may be added to the diet (Sec. II. C).

Hardness and toughness are the properties of food that are of particular relevance to phytophagous and wood-boring insects. Hardness can reduce feeding or increase time taken to ingest a given amount of food. Seed feeders may be unable to feed on particular seeds, or they may eat less of harder specimens (Smith 1985). Physical form can also affect the rate of ingestion and the total amount ingested. For example, particle size affects ingestion rate in filter feeders such as certain mosquito larvae (Dadd 1971).

In the case of certain fluid feeders, viscosity, fluid pressure, and other physical factors can affect the rate of food intake. The rate of feeding on high concentrations of sugar by flies is relatively low due to high viscosity (Bernays and Simpson 1982). Mosquitoes that insert their stylets into the capillaries of their vertebrate hosts feed faster than those that feed from small local hemorrhages caused in the host tissues (Friend 1978). Other physical factors influencing probing and ingestion in blood-feeding insects are temperature gradients; humidity; the thickness, texture, and hairiness of the host skin; and osmotic pressure and pH of the food medium and host tissues (Friend and Smith 1977, Friend 1978). For example, many sucking insects can be induced to feed when the diet is presented under a membrane (Friend and Smith 1977). In the case of the mosquito *Culiseta inornata* (Diptera: Culicidae) feeding was induced by heating a membrane-covered diet to 37°C (Friend 1978). The use of host-plant preparations, as fresh or dried plant parts, are sometimes used to induce feeding in fastidious insects.

Another important physical property of artificial diets is *dispersion*. An insect diet is a heterogeneous mixture of ingredients consisting of an organized and compartmentalized matrix that assures availability of each component of the diet to the insect (Cohen 2004). Normally the liquid matrix can be water and is solidified or gelled. For example, the screwworm larval diet is prepared with dietary ingredients such as spray-dried bovine blood, spray-dried poultry eggs, and a milk substitute dispersed in water and finally solidified by adding a sodium polyacrylate gelling agent (Chaudhury et al. 1998).

D. Evaluating Diet Quality

Normally, measurement of growth rate has been used to determine the nutritional quality of a diet for the immature stages. Larval and pupal weights, time between molts, and time to adult emergence have also been used as indicators of growth. Number or percentage of pupae produced from a given number of eggs or larvae can also serve as the criterion for evaluating a diet. Diets for adult insects are normally evaluated by recording the number or mass of eggs produced and the percentage of eggs that hatch. To compare two diets differing in performance profiles, e.g., $S(5\%\uparrow)/F(10\%\downarrow)/T_G(3\%\uparrow)$ (*S*, survival; *F*, fecundity; T_G, generation time), one can compute a composite index of quality such as *r*, instantaneous rate of increase (Eq. 5, Chap. 6 *Environmental Biology of Insect Rearing*). A diet needs to be evaluated for a number of generations because some nutritional deficiencies may not be manifested during the first or second generations (Chaudhury et al. 2002).

E. Measures of Food Utilization: Nutritional Indices

Because of the increasing cost of mass-rearing programs, maximizing efficiency of food utilization by the insects produced is of utmost important. Measures of food intake and subsequent utilization of ingested food provide the information necessary to design diets that are efficiently utilized and to determine the quantity of diet that is adequate but not in excess of needs.

Individual performance is frequently quantified in part through the calculation of *nutritional indices* (Slansky and Rodriguez 1987): growth rate (*GR*), consumption rate (*CR*), approximate digestibility (*AD*), and efficiency of conversion of digested food (*ECD*). Based on measurements of food consumption or ingestion (*I*), feces output or egestion (*E*), and biomass production (*B*) during a particular time period (*T*), the nutritional indices are defined and related to one another as follows:

$$GR = B/T \qquad \text{Eq. 1}$$
$$CR = C/T \qquad \text{Eq. 2}$$
$$AD = (I-E)/I \qquad \text{Eq. 3}$$
$$ECD = B/(I-E) \qquad \text{Eq. 4}$$
$$GR = CR \times AD \times ECD \qquad \text{Eq. 5}$$

Efficiencies are expressed generally as percentages, while rates are expressed as milligrams dry mass per day. Frequently, the weight-specific rates relative consumption rate ($RCR = CR/W$) and relative growth rate ($RGR = GR/W$) for average weight during the observation period (*W*) are used instead of *CR* and *GR*, respectively. Rates of metabolism and excretion can also be calculated.

Unfortunately, these rates and efficiencies interact in complex ways that are not yet fully understood. For example, a change in temperature may affect each rate and efficiency directly; but in addition, each change in rate or efficiency may in turn affect other rates and efficiencies. For example, if an increase in temperature results in an increase in CR and a decrease in AD, it will be difficult to determine to what extent the decline in AD is due to the increase in CR resulting in a shorter occupancy-time of food in the gut or to a direct effect of temperature on the digestive and assimilative process. One can establish baseline data for interpreting the rate X efficiency interactions by conducting a controlled consumption study in which the amount of food offered is quantitatively altered by the researcher. Waldbauer (1968) and Scriber and Slansky (1981, 1985) provide interpretations for rate-efficiency interactions and discuss the factors that influence these parameters.

Sources of error in nutritional index studies on artificial diets have been investigated by Schmidt and Reese (1985) in order to refine the nutritional index technique proposed by Waldbauer (1968). Their findings suggest that unless certain experimental conditions are met, obtaining useful results with artificial diets is difficult. An improved approach to nutritional indices that makes use of analysis of covariance and a geometric analysis of feeding and nutrition has been provided by Raubenheimer and Simpson (1992, 1994).

IV. Digestion, Absorption, and Metabolism of Nutrients

Insects are diverse in food and feeding habits. The alimentary canal, normally called the *gut*, is diversified and modified in special ways in insects to deal with different types of food they consume. However, the basic plan of the various regions of the gut including *foregut*, *midgut*, and *hindgut* and associated organs remains the same in all insects (Figs. 2 and 3). Major changes in the structure of digestive organs occur where the larval and adult food is different, such as insects with complete metamorphosis (Fig. 4).

Digestive enzymes break down the complex substances in the food into simpler substances that can be absorbed and later assimilated. Various parts of the gut are concerned with different aspects of digestion and absorption of foodstuffs, but the midgut is the principal site for secretion of digestive enzymes, digestion of food, and absorption of nutrients in most insects. Digestion in insects is reviewed by House (1965) and Applebaum (1985), and absorption is reviewed by Turunen (1985). For a description of a generalized insect digestive system, consult Chapman (1998) and Cohen (2004).

A. Digestion

Enzymes that are used in digestion are present in the saliva and in secretions of the midgut. Digestion in different insects is appropriately adapted to the nutritional composition of hosts upon which the specific insect feeds. The gut of insects that feed on solid food is normally a straight tube, such as the lepidopterous caterpillars. These are phytophagous insects and normally feed frequently or almost continuously. The undigested or partially digested food can easily and rapidly pass through the straight alimentary canal. On the other hand, many liquid-feeding insects have a convoluted gut through which their food can nevertheless pass. Insects that feed on dilute liquid food have the ability to get rid of the excess water using Malpighian tubules. Plant sap feeders, such as Hemipteran plant bugs, have a morphological modification of the alimentary canal known as a filter chambers (Fig. 5) where a loop of the hindgut comes in close contact with the foregut, thus allowing a large volume of water to bypass the midgut, moving directly from the foregut into the hindgut. For example, aphids must ingest large volumes of plant sap and excrete the excess water and carbohydrates, i.e., "honeydew," in order to concentrate the remaining nutrients in the plant sap.

Digestive hydrolysis of food macromolecules occurs with a wide range of efficiencies in different insects and developmental stages of the same insect depending on factors such as the types of digestive enzymes produced, antimetabolites, and various condi-

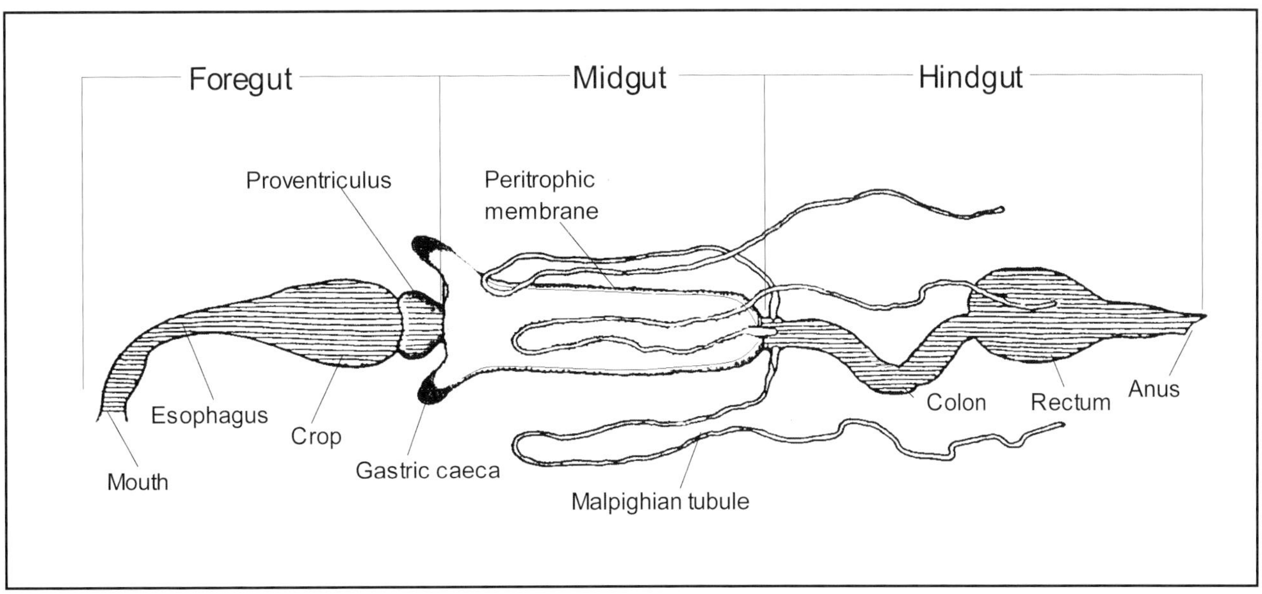

Fig. 2. Generalized, diagrammatic view of the insect alimentary canal [adapted from Snodgrass (1935)].

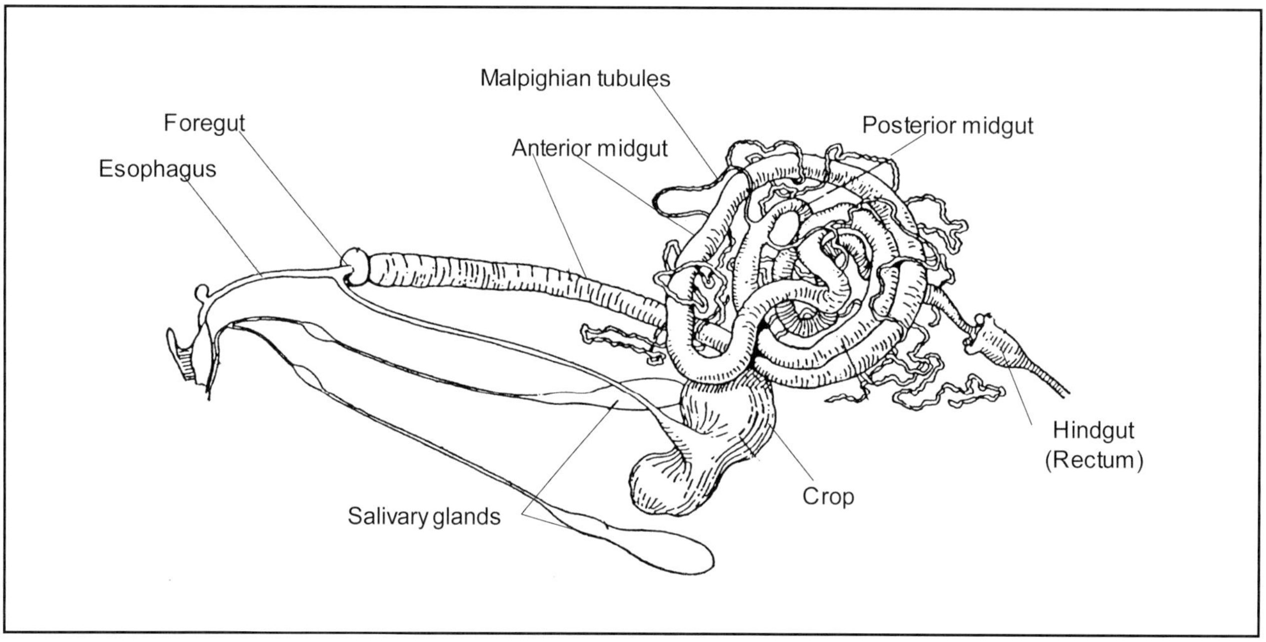

Fig. 3. Alimentary canal and salivary glands of a fruit fly, the apple maggot, *Rhagoletis pomonella* (Diptera: Tephritidae) [adapted from Snodgrass (1935)].

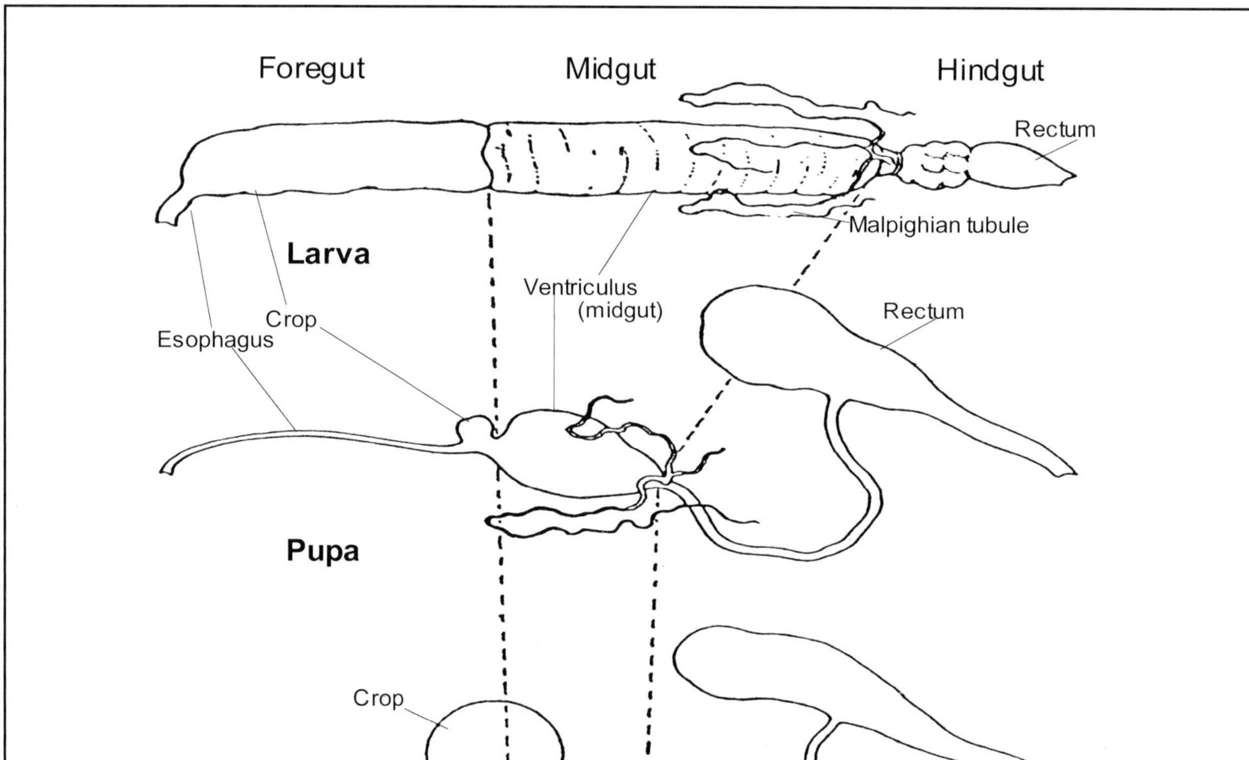

Fig. 4. Transformation of the alimentary canal of a moth, the eastern tent caterpillar, *Malacosoma americanum* (Lepidoptera: Lasiocampidae), from the larva through the pupa to the adult [adapted from Snodgrass (1935)].

tions governing rates of enzyme activity. Gut pH influences the rate of digestion because digestive enzymes exhibit maximal activity at a particular pH. Insects using trypsin as a major proteolytic enzyme must maintain a gut pH above 7.0 and sometimes as high as 10.0 to have optimal enzymatic activity. The pH of the foregut varies with the diet since there is no appropriate buffering of the foregut contents. For example, with a cockroach fed on a diet high in protein, the foregut pH is 6.3, fed with maltose it is 5.8, and with glucose, 4.5–4.8. The more acidic pH with the sugars results from microorganisms producing organic acids. The pH values in various parts of the gut of selected insects have been reported by Nation (2002).

1. External Digestion. Digestion may start, or even be substantially completed, outside the body of the insect through injection of digestive enzymes into the food. For example, in some zoophagous insects with piercing/sucking mouthparts, salivary enzymes are injected into the host. In the suborder Heteroptera (Hemiptera) and the robber flies (Diptera: Asilidae), the contents of the prey are completely histolysed before ingestion. Extraintestinal digestion also occurs in diving beetle larvae, *Dytiscus* spp. (Coleoptera: Dytiscidae), which have no salivary glands so that the midgut enzymes must be employed. These are injected through the mandibles, which are perforated by a narrow tube; and in a short period, when the contents of the prey have been digested, the resulting fluid is taken up through the same route. The salivary enzymes responsible for digestion of starch in the plant bug *Lygus disponsi* (Heteroptera: Miridae) are strongly activated by glutamine and to a lesser extent by other amino acids and by nitrate and chloride ions in host plant tissues into which the enzymes are injected (House 1974b).

Larvae of the common cattle grub, *Hypoderma lineatum* (Diptera: Oestridae), burrow beneath the skin of cattle and feed on the connective tissue of the host. The larvae produce a mixture of salivary proteases including *collagenase*, which is specific to digestion of connective tissue. The resulting mixture of degraded protein and proteases is then ingested, with proteolysis continuing in a midgut reservoir. To prevent contamination of their closed environment, common cattle grub larvae do not excrete feces during their whole period of development.

During pharate adult development, *Antheraea pernyi* (Lepidoptera: Saturniidae), a silkmoth, synthesizes an inactive zymogen precursor of *cocoonase*, a proteolytic enzyme, in specialized giant epidermal cells of the maxillary galea. Two days before adult emergence, this precursor is secreted as a highly concentrated solution that dries as a film on the outer sur-

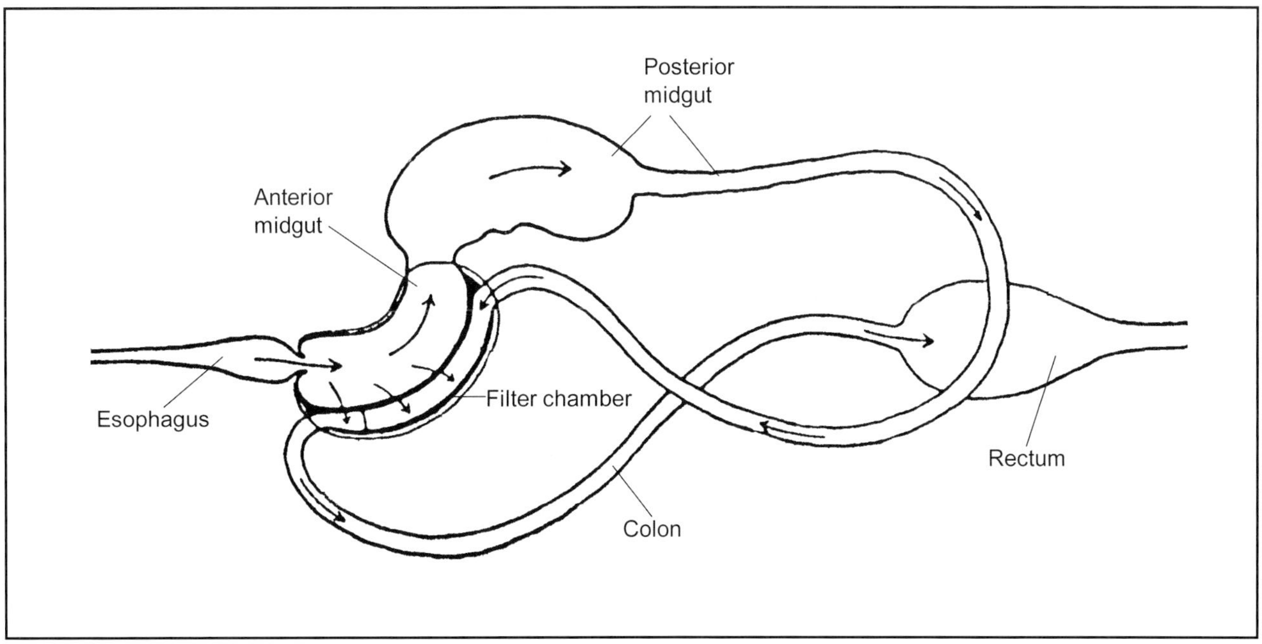

Fig. 5. Alimentary canal of a hemipteran showing a simple type of filter chamber in which two extremities of the midgut and the anterior portion of the colon are bound in a sheath (filter chamber) [adapted from Snodgrass (1935)].

face of the galea. Following adult ecdysis from the pupa within the cocoon, the adult labial glands secrete a solution of $KHCO_3$ (0.15 M) that activates the cocoonase. Application of the activated cocoonase to the inner surface of the cocoon weakens its structure and permits emergence of the adult moth.

2. Internal Digestion. Most digestion occurs in the midgut where the digestive enzymes are secreted; but through regurgitation of midgut fluids, some can also take place in the crop. In Orthoptera, the bulk of digestion occurs in the crop. Little digestion occurs in the hindgut apart from cellulose digestion in a few insects. In these insects, microorganisms produce the cellulases responsible for digestion of cellulose.

Enzymes present in the midgut can vary facultatively with the diet of the insect. For example, when larvae of the black carpet beetle, *Attagenus megatoma* (Coleoptera: Dermestidae), were fed on nutrient-supplemented wool, total midgut protein content was nearly three times that observed when larvae were fed on a Purina Laboratory Chow diet (Dadd 1985). Protease activity was somewhat higher on the wool diet, while sucrase activity was 5-fold higher on the Purina diet. Substrate specificity of enzyme secretion results in different ratios of proteases to carbohydrases, depending on the presence and predominance of protein and carbohydrate in the diet (see Dadd 1985).

3. Digestion of Carbohydrates. Carbohydrates are generally absorbed as *monosaccharides* (e.g., glucose, fructose, and galactose), so disaccharides and polysaccharides must be broken down to their component monosaccharides by the enzymatic action of *glycosidases*. Different glycosidases are usually necessary to hydrolyze different series of sugars (Lehninger et al. 1993).

The common disaccharides like maltose, trehalose, and sucrose all contain a glucose residue that is linked to a second sugar residue by an α-linkage. All these are hydrolyzed by the enzymatic invertase α-glucosidase, the most common invertase in insects (Gilmour 1961). The naturally occurring β-glucosides (e.g., salicin, arbutin, and cellobiose) are usually of plant origin, and the highest β-glucosidase activity is found in phytophagous insects. The glucosidases are the most common glycosidases; but α-galactosidase, hydrolyzing substances such as melibiose, is recorded from various Diptera and the desert locust; and β-galactosidase, hydrolyzing lactose, is also present in the desert locust.

In the hydrolysis of carbohydrates, water is the typical acceptor for the sugar residues; but other sugars may also act as acceptors with the formation of oligosaccharides, which can serve as excellent energy sources. For example, in the hydrolysis of sucrose, other sucrose molecules may act as acceptors to form the trisaccharides glucosucrose and melezitose. These in turn may accept another glucose molecule to form tetrasaccharides. This process is known as *transglucosylation*.

Amylase catalyzes the hydrolysis of α-1,4 glycosidic linkages in *starch*, the most common energy storage polysaccharide in plants, to maltose; and those in *glycogen*, the energy storage polysaccharide in animals, to glucose. Actually, there are two types of amylases. An *exoamylase* splits off maltose residues from the end of the molecule leading to a rapid increase in concentration of maltose, while an *endoamylase* attacks bonds internal to the starch molecule so that the initial rate of increase in the concentration of maltose is slower. The products are then ultimately digested to glucose by an α-glucosidase.

Only a minority of plant- and wood-eating insects have a *cellulase*, an enzyme capable of hydrolyzing cellulose. Many such insects use symbionts to solve this problem (Sec. IV. A. 6).

4. Digestion of Proteins. Insects possess a series of *proteases* to digest proteins. A trypsin-like protease is produced in the midgut that hydrolyzes proteins to *peptones*, the longer chain products of initial protein digestion, and to *polypeptides*, shorter chains of amino acids. These in turn are acted on by *peptidases* some of which occur in the gut lumen but most of which are found in the epithelial cells. There are various types of peptidase: *carboxypolypeptidase* attacks the peptide chain from the carboxyl (COOH) end provided that tyrosine or other specific amino acids are present in the chain, *aminopolypeptidase* attacks the chain from the amino (NH_2) end, and *dipeptidase* hydrolyzes all dipeptides.

Some insects are able to digest the generally inert animal proteins keratin and collagen. *Keratin* is the protein found in wool, hair, and feathers; and *collagen* is a primary constituent of connective tissues. Keratin consists of polypeptide chains, including sulfur-containing amino acids, which are linked together by disulphide bonds. Only Mallophaga using birds as hosts, larvae of some dermestid beetles, and the larvae of some tineid moths are able to digest it. These insects depend on symbionts for their ability to utilize keratin. It also appears that the very low *redox* (reduction/oxidation) potential (–300 mV) in the midgut of the wool-eating species is necessary for their ability to digest keratin (Gilmour 1961).

5. Digestion of Lipids. Most of the fat eaten by insects consists of *triacylglycerols*. Many insects produce *lipases*, which hydrolyze triacylglycerols to fatty acids and glycerol. Lipases are water-soluble enzymes while their substrates are not, so catalysis occurs only at the oil-water interface, and catalytic efficiency depends on suitable dispersion and stabilization of the oil droplets.

The digestion of beeswax by larvae of the greater wax moth is a specialized case of lipid digestion. Honeycomb normally forms a large part of the diet of these larvae although they can survive without it. Beeswax is a mixture of esters, fatty acids, and aliphatic hydrocarbons. These larvae are able to utilize ca. 50% of the beeswax, mainly the fatty acids but also some of the unsaponifiable material and aliphatic hydrocarbons, possibly with the assistance of symbiotic bacteria.

6. Symbionts. Many insects possess microorganisms, referred to as *symbionts*, that allow them to digest otherwise inert substances. Most symbiont-dependent insects can be placed into the following categories: 1) wood feeders; 2) plant sap suckers; 3) blood feeders; 4) feeders on dry seeds, cereals, or other dry foods; 5) scavengers on feathers, hair, old honeycomb, and similar materials; and 6) saprophytic feeders. These symbionts reside in the insect midgut in specialized cells called *mycetocytes* or in specialized gastric caeca called *mycetomes*. They can be bacteria, fungi, or protozoa. They may have an obligatory relationship with the insect host and provide a wide range of nutrients including vitamins, sterols, amino acids, and various digestive enzymes.

Wood is potentially a very abundant source of energy-producing carbohydrate. However, insects lack the necessary digestive enzymes such as *cellulases*, *lignases*, and *cellobiases* (hydrolyzing cellulose, lignin, and cellobiose, respectively) and can only digest wood through the mediation of symbionts. Without these enzymes, phytophagous insects must mechanically disrupt cell walls to extract nutrients from cell contents. Among the larvae of wood-boring beetles, various methods of digestion are observed. For example, the powder post beetles (Coleoptera: Lyctidae), having no cellulase, feed only on cell contents. Death watch beetles (Coleoptera: Anobiidae) and long-horned beetles (Coleoptera: Cerambycidae) ingest cellulases and can digest cell walls along with the cell contents. There are other wood-boring insects whose permanent gut flora or fauna assist in the digestion of cellulose. For example, the wood roaches, *Cryptocercus* spp. (Blattodea: Cryptocercidae), possess various flagellates that are responsible for cellulose digestion. This is also true of the larvae, soldiers, and workers of most termites. The flagellates occur in an expansion of the hindgut; and in *Zootermopsis* spp. (Isoptera: Termopsidae), the mass of flagellates is about a third that of the insect.

Insects of various groups that feed only on blood, e.g., kissing bugs (Hemiptera: Reduviidae), bedbugs (Hemiptera: Cimicidae); lice (Phthiraptera: Anoplura), and tsetse flies (Diptera: Glossinidae), or only on plant sap, e.g., cicadas, leafhoppers, treehoppers, planthoppers, and froghoppers (Hemiptera: Auchenorrhyncha), and aphids, whiteflies, and scale insects (Hemiptera: Sternorrhyncha), possess symbionts to provide essential B vitamins and amino acids.

Some of the clearest evidence of insect nutritional dependence on symbionts has been demonstrated with various stored-product beetles that normally contain yeasts located in mycetomes. Since transmission between generations occurs by external contamination of the eggs, aposymbiotic insects may be obtained by external asepsis of the eggs before hatching. With the cigarette beetle, *Lasioderma serricorne* (Coleoptera: Anobiidae), and the drugstore beetle, *Stegobium paniceum* (Coleoptera: Anobiidae), comparisons of development of normal and aposymbiotic larvae on semisynthetic diets deficient in specific nutrients show that symbionts supply a variety of B vitamins, sterols, and several essential amino acids (Dadd 1985).

B. Absorption

All nutrients must be digested and absorbed in solution. No *phagocytosis*, engulfment of particles, occurs; but *pinocytosis*, engulfment of extracellular fluid, may be involved. Absorption in some cases is a passive process, but in other instances active transport occurs. Molecules may pass through the membrane by diffusion, carrier-facilitated transport, or pinocytosis. Furthermore, the molecules must be transported from the absorptive cell across the lateral or basal membrane, the basement membrane, and a sheath of muscle and connective tissue before reaching the hemolymph.

Simple diffusion depends on the presence of a chemical gradient: substances move from a higher concentration outside the cell to a lower concentration inside without utilization of metabolic energy. For an electrolyte, the rate also depends on the electrical potential prevailing across the cell membrane. In active transport, amino acids, monosaccharides, free fatty acids, minerals, and vitamins are absorbed at the appropriate sites on the surface of the *microvilli*, hair-like extensions of the gut wall, which increase gut surface area. There are various types of receptor sites on the microvilli that facilitate the absorption process. The absorption of water is particularly important in terrestrial insects, and the rectum plays an important part in removing water from the feces in a process that involves active transport.

Factors such as gut pH, temperature, and redox potential play major roles in the absorption processes. Many phytophagous insects have very high pH in the midgut (10.0–11.0). Various aspects of absorption processes have been discussed in detail by Applebaum (1985), Turunen (1985) and Chapman (1998).

C. Metabolism

Digested and absorbed materials are transported by the hemolymph to the target cells and tissues. Water-soluble components such as monosaccharides, amino acids, and many minerals and vitamins are easily transported by hemolymph; but the lipid-soluble components are transported by special molecules such as lipoproteins. These components are transported from the gut into the various tissues such as fat body, muscles, and reproductive tissues. When the nutrients arrive at their target cells, receptors on the cell surface can bind them and take them into the cells, where they participate in various specialized functions known collectively as *metabolism*. There are two types of metabolic processes: *anabolic* (i.e., synthetic) and *catabolic* (i.e., degradative) (Lehninger et al. 1993). One example of anabolism is egg production, where fat body cells produce yolk protein vitellin, which is transported again via hemolymph and taken up into the developing eggs. Organisms use catabolic processes to produce energy by breaking down the ingested dietary components. For example, insects can use the carbohydrate glycogen stored in the fat body to produce trehalose and glucose and eventually *ATP*, the energy-transfer molecule used by all multicellular organisms. There is an inherent tradeoff between anabolism and catabolism. For example, if an immature insect must detoxify a toxic material in its food using an energy-intensive process, then it will have less material available for growth.

V. Developing an Artificial Diet

Artificial diets may be crude, synthetic, semi-synthetic, or chemically defined. However, these terms are considered ambiguous and not systematic. Dougherty (1959) described artificial diets by degree of chemical definition. This system classified the artificial diets into three basic groups: 1) *oligidic*, consisting mainly of crude natural materials such as wheat germ or other bulk plant material and animal products such as liver; 2) *meridic*, composed mostly of defined chemicals but with one or more undefined or ill-defined components (such as casein); and 3) *holidic*, consisting entirely of known, pure chemicals. Oligidic diets are used for mass-rearing projects. The ingredients needed for these diets have to be easily available and inexpensive. These diets are normally easy to prepare in large quantities so that the cost of production is kept minimal. Examples of oligidic diets are egg powder, milk powder, and spray-dried blood mixed with gel and water for screwworm larvae, and casein and yeast for flour beetles. Sometimes an oligidic diet can hardly be considered artificial because the dietary ingredients used are very similar to the insect's natural diet. A meridic diet can contain components such as vitamin-free casein or yeast extract in combination with salt and vitamins, cholesterol, agar, and water; or they can be more complex, containing specific amino acids, sugars, sterols, vitamins, minerals, agar, and water. Although casein (or yeast extract) is the only ingredient that is not chemically defined, the diet is classified as meridic. Most artificial diets for insects are meridic. A completely holidic diet probably does not exist, because trace elements and/or contaminants are present even in the "pure" components that are used for the diet. Also, additional compounds may form during formulation and use of a diet rendering it chemically impure. Nearly holidic diets are most useful for determining nutritional requirements.

To develop an artificial diet for an insect, one must have a thorough knowledge and understanding of the biology of the insect in question including its feeding behavior and mechanisms, growth and developmental processes, and the natural dietary components on which it feeds. One must also have ingenuity, time, patience, and good luck.

A. Starting Diet

Normally, an artificial diet of a closely related species may serve as a starting point to develop an artificial diet. Diets that have been used successfully for species of similar feeding habits may also serve as a starting point. The pink bollworm diet (Adkisson et al. 1960) is an example of a diet that has been modified to rear similar species. Because there is no certainty as to which of these diets will give the best result, more than one diet should be tested. Following the initial testing, the diet showing the best results should be selected as the "test diet" for further testing and modification needed to produce quality insects.

B. Testing

Once a diet is ready to be tested, a well-designed set of procedures should be followed. Generally, performance on the test diet may be compared to performance on the insect's natural diet with respect to several growth and fitness variables such as body weight and size of larva, pupa, and adult; adult emergence success rate; longevity of adults; reproductive characteristics such as fecundity, egg hatch, and adult sex ratio; and survivorship.

These tests have to be run for several generations to determine that no chronic effects are manifest in later generations. Unfortunately, a diet may support growth for one or two generations but fail gradually thereafter due to the lack or deficiency of one or more essential nutrients. A successful diet is one that supports normal growth and development, normal mating behavior, and normal fecundity for at least 10 generations.

Great care and patience are needed while conducting the initial tests using selected artificial diets. It may be necessary to select and test several diets before a suitable one is obtained. As mentioned earlier, it is advisable to first select three or four existing diets and through comparative tests, select the one with best results. This may give some degree of initial success but not result in a completely adequate diet. Therefore, the diet selected will require some modifications to obtain satisfactory results, such as changing dietary ingredients to manipulate protein, fat, and carbohydrate content.

C. Troubleshooting

The simplest way to determine the factors responsible for the poor quality of a diet is to follow a systematic troubleshooting procedure. A prepared artificial diet, if totally acceptable to the target insect, should induce robust feeding resulting in good growth and reproduction for many generations. If the diet induces no feeding, it is possible that a feeding stimulant is missing, a feeding deterrent is present, or the texture is completely wrong for the species. If a diet induces robust feeding but results in poor growth and reproduction, the diet may not have all the proper nutrients, the nutrients may not be in proper proportions, or one or more toxic compounds may be present.

D. Modification

Moore (1985) has suggested a method for systematic modification of insect diets. The strategy is to modify one dietary ingredient each in turn. He suggests that a single change be made at each step because multiple, simultaneous changes make interpretation difficult. He illustrated his approach using the general purpose diet of Singh (Singh 1983) as the starting diet. This diet consists of the following ingredients (total = 100 g): casein (3.5 g) and wheat germ (3 g) for protein; sucrose (3 g), glucose (0.5 g), and dextrose (see vitamins) for carbohydrate; linoleic acid (0.25 g) for lipid; cholesterol (0.05 g) for sterol; Wesson salts (1 g) for minerals; Vanderzant vitamin mix (0.7 g in 1.3 g of dextrose) for vitamins and carbohydrate; agar (2.5 g) as a gelling agent; cellulose (10 g) as a bulking agent; streptomycin (0.015 g), benzylpenicillin sodium (0.015 g), methyl p-hydroxybenzoate (0.112 g), and sorbic acid (0.3 g) as microbial inhibitors; and water (72.17 g). The diet modifications (excluding those dealing with microbial contamination) that Moore (1985) recommended trying are as follows:

1. Increase the amount of casein and wheat germ in proportionate amount to a maximum of 15% to determine if there is enough protein.

2. Replace casein and wheat germ with another protein that could contain a feeding stimulant or a different balance of amino acids.
3. Incorporate some processed host materials in choice tests to determine if the insect exhibits a feeding preference.
4. Take precautions to ensure that there are no adverse reactions between the proteins and other diet ingredients (e.g., the Maillard reaction between amino acids and glucose) that affect the availability of protein.
5. Increase specific amino acids to achieve a balance in the diet. Casein contains most essential amino acids, but it may need augmentation with glycine, glutamic acid, or cystine (Dadd 1977). Analysis of amino acids in the insect carcass has been used as the basis for determining the optimum amino acid content of a diet (Rock 1972).
6. Vary the amount of sucrose to determine the most effective level. Sucrose appears to be beneficial in most insect diets though many dipterous larvae do not require it and it may even be detrimental to them.
7. Substitute other forms of sugars or carbohydrates, preferably ones that would be related to the natural host. Refer to Sec. II. B. 1 for details on the nutritive value of carbohydrates.
8. Substitute wheat germ oil for linoleic acid because of its feeding stimulant qualities and its completeness in fatty acid composition.
9. Vary the amount of lipid (between 0.1 and 1%). Insects generally require a source of polyunsaturated fatty acids, primarily linoleic and linolenic, especially in Lepidoptera (see discussion in Sec. II. B. 3).
10. Substitute other crude lipids such as plant or animal lecithin that may be more acceptable to the insect.
11. Add antioxidants such as tocopherol, butylated hydroxytoluene (BHT) or butylated hydroxyanisole (BHA) to prevent oxidation of the essential polyunsaturated fatty acids.
12. To ensure uniform distribution of the sterol, dissolve it in acetone or ether, add the solution to the dry ingredients, and allow the solvent to evaporate. Cholesterol and lipids are normally dissolved in acetone and added to the protein and other dry ingredients to achieve uniform distribution.
13. Substitute a plant sterol or a mixture of sterols for the cholesterol. Some insects have special requirements for sterols, and the diet might be improved by substituting other forms such as sitosterol, stigmasterol, and campesterol (Sec. II. B. 3).
14. Substitute a salt mixture made for insects for the Wesson salt, which is a vertebrate mix. Refer to Sec. II. B. 5 for other salt mixtures and also volumes 1 and 2 of the handbook by Singh and Moore (1985).
15. Vary the amount of vitamin mixture. The general purpose diet contains 2% Vanderzant vitamin mix, but about 65% of this is dextrose.
16. Substitute a vitamin mixture used for a closely related species.
17. To change the firmness of the diet, try various concentrations of agar.
18. Substitute other gelling agents for agar. Try using gelling agents that do not need heating, such as sodium polyacrylate polymers used in screwworm diets (Chaudhury et al. 1998).
19. Vary the amount of cellulose or add some other bulking agent to modify the texture of the diet.

E. Additional Considerations

1. Sources of Diet Ingredients. Most suppliers of dietary ingredients include material safety data sheets (MSDS) and complete chemical analyses of the components. The chemical analysis is especially useful for diet formulation. The MSDS provides instructions on handling and disposal of the materials and also information regarding any human toxicity issues.

Ingredients should always be obtained from a supplier who will deliver chemically and physically identical ingredients every time an order is placed. After developing an acceptable diet, if it becomes necessary to change suppliers, ingredients from the new suppliers must first be tested to see that acceptable results are obtained before placing a bulk order.

2. Mixing. Utmost care should be taken in weighing the ingredients and allowing the ingredients to mix completely, especially the ingredients that are used in minute quantities. Sometimes these minute quantities may be incorporated into the mixture by trituration with sucrose or by dissolving them in an organic solvent such as alcohol or ether, wetting the solid components with this solution followed by evaporation of the solvent. When agar is used as the gelling agent, the diet will require heating. Heating may cause breakdown of some ingredients, especially vitamins. Therefore, heating should be at the lowest practical temperature and for the shortest possible time.

The order of mixing ingredients may be of importance due to interaction of various ingredients. Some ingredients are difficult to mix and may need further processing or addition in a certain order. For example, in preparing screwworm larval diet, the gelling agent sodium polyacrylate polymer can be mixed with the solid dietary ingredients and then mixed with water. In contrast, a starch-grafted gel has to be mixed with water first and then added to the solid ingredients (Chaudhury and Alvarez 1999).

3. Storage. Ideally, diets should be prepared fresh daily for testing and, if possible, for normal feeding as well. However, if storing of a prepared diet is absolutely necessary, it should be stored in a refrigerator for the short term (week) or in a freezer for the long term (month). It should be noted that diet stored in a refrigerator or freezer may deteriorate over time, and therefore the maximum storage time for each diet should be determined and followed.

VI. Conclusions

Because of the many species possessing various feeding habits and nutritional requirements, developing artificial diets for insects will always be a challenging task. I have tried to provide some basic and general information on insect feeding and nutritional requirements and some general guidelines and procedures on the development of artificial diets for insects. Information provided in this chapter may not prove sufficient in solving all problems in insect diet development, but I hope that this information will at least provide a starting point and a direction towards successful development of a new diet or modification of an old one.

References

Abbassy, M. A., F. El-Gayar, and A. El-Shazli. 1977. Phagostimulants from *Acokanthera spectabilis* Hook. tested on *Spodoptera littoralis* Boisd. (Lepidoptera: Noctuidae). Z. Ang. Ent. 82: 441–444.

Adkisson, P. L., E. S. Vanderzant, D. L. Bull, and W. E. Allison. 1960. A wheat germ medium for rearing the pink bollworm. J. Econ. Entomol. 53: 759–762.

Albert, P. J., C. Cearley, F. Hanson, and S. Perisella. 1982. Feeding responses to eastern spruce budworm, *Choristoneura fumiferana*, larvae to sucrose and other carbohydrates. J. Chem. Ecol. 8: 233–240.

Anderson, T. E., and N. C. Leppla (eds.). 1992. Advances in insect rearing for research and pest management. Westview Press, New Delhi, India

Applebaum, S. W. 1985. Biochemistry of digestion, pp. 279–311. *In* G.A. Kerkut and L.I. Gilbert (eds.), Comprehensive insect physiology, biochemistry and pharmocology, Vol. 4. Pergamon Press, Oxford, U.K.

Armstrong, E. 1978. The effect of vitamin deficiencies on the growth and mortality of *Tribolium castaneum* infected with *Nosema whitei*. J. Invert. Pathol. 31: 303–306.

Ascher, K. R. S., J. Meiner, and H. M. Flowers. 1976. Effect of amino acids on the feeding behavior of larva of the Egyptian cotton leafworm, *Spodoptera littoralis*. Phytoparasitica 4: 85–92.

Aspirot, J., and G. Lauge. 1981. Experimental study of stimulatory effect of feeding sucrose and sinigrin in *Schistocerca gregaria* (Orthoptera: Acrididae) and regulation phenomena. Reprod. Nutr. Dev. 21: 695–704.

Atallah, Y. H., and L. D. Newsom. 1966. Ecological and nutritional studies on *Coleomagilla maculata* De Geer (Coleoptera: Coccinellidae) I. The development of an artificial diet and a laboratory rearing technique. J. Econ. Entomol. 59: 1173–1179.

Barker, R. J., and Y. Lehner. 1974a. Acceptance and sustenance value of naturally occurring sugars fed to newly emerged adult workers of honey bees (*Apis mellifera* L.). J. Exp. Zool. 187: 277–285.

Barker, R. J., and Y. Lehner. 1974b. Influence of diet on sugars found by thin-layer chromatography in thoraces of honey bees, *Apis mellifera* L. J. Exp. Zool. 188: 157–164.

Bernays, E. A., and R. F. Chapman. 1994. Host-plant selection by phytophagous insects. Chapman and Hall, New York, NY.

Bernays, E. A., and S. J. Simpson. 1982. Control of food intake. Adv. Insect Physiol. 16: 59–118.

Beck, S. D. 1956. European corn borer, *Pyrausta nubilalis* (Hubn.) and its principal host plant. II. The influence of nutritional factors on larval establishment and development on the corn plant. Ann. Entomol. Soc. Amer. 49: 582–588.

Blomquist, G. J., L. A. Dwyer, A. J. Chu, R. O. Ryan and M. de Renobales. 1982. Biosynthesis of lenoleic acid in a termite, cockroach and cricket. Insect Biochem. 12: 349–353.

Chapman, R. F. 1985. Structure of the digestive system, pp. 165–211. *In* G. A. Kerkut and L. I. Gilbert (eds.), Comprehensive insect physiology, biochemistry and pharmacology, vol. 4. Pergamon Press, Oxford, U.K.

Chapman, R. F. 1998. The insects: structure and function. Cambridge University Press, Cambridge, U.K.

Chapman, R. F., and G. DeBoer (eds.). 1995. Regulatory mechanisms in insect feeding. Chapman and Hall, New York, NY.

Charpentier, R. 1979. Rearing and biology of the turnip moth, *Scotia segetum* (Lepidoptera: Noctuidae) on artificial diets. Zoon 7: 95–120.

Chaudhury, M. F., and L.A. Alvarez. 1999. A new starch-grafted gelling agent for screwworm (Diptera: Calliphoridae) larval diet. J. Econ. Entomol. 92: 1138–1141.

Chaudhury, M. F. and S. R. Skoda. 2007. A cellulose fiber-based diet for screwworm (Diptera: Calliphoridae) larvae. J. Econ. Entomol. 100: 241-245.

Chaudhury, M. F., **L. A. Alvarez, and J. B. Welch. 1998.** An alternative source of blood protein for screwworm (Diptera: Calliphoridae) larval diet. J. Econ. Entomol. 91: 1397–1400.

Chaudhury, M. F., **L. A. Alvarez, and L. L. Velazquez. 2000.** A new meatless diet for adult screwworm (Diptera: Calliphoridae). J. Econ. Entomol. 93: 1398–1401.

Chaudhury, M. F., **L. A. Alvarez, and L. L. Velazquez. 2002.** Recycled paper products as substitutes for gelling agent in screwworm (Diptera: Calliphoridae) larval diet. J. Econ. Entomol. 95: 1337–1341.

Cheldelin, V. H., and R. W. Newburgh. 1959. Nutritional studies on the blow fly. Ann. New York Acad. Sci. 77: 373–383.

Chyb, S., H. Eichenseer, B. Hollister, C. A. Mullin, and J. L. Frazier. 1995. Identification of sensilla involved in taste mediation in adult western corn rootworm (*Diabrotica virgifera virgifera* LeConte). J. Chem. Ecol. 21: 313–329.

Clayton, R. B. 1964. The utilization of sterols by insects. J. Lipid Res. 5: 3–19.

Cohen, A. C. 2004. Insect diets: science and technology. CRC Press, Boca Raton, FL.

Cook, A. G. 1977. Nutrient chemicals as phagostimulants for *Locusta migratoria*. Ecol. Entomol. 2: 113–122.

Dadd, R. H. 1960. Some effects of dietary ascorbic acid on locusts. Proc. Roy. Soc. B 153: 128–143.

Dadd, R. H. 1963. Feeding behaviour and nutrition in grasshoppers and locusts. Adv. Insect Physiol. 1: 47–109.

Dadd, R. H. 1964. A study of carbohydrate and lipid nutrition in the wax moth, *Galleria mellonella* (L.) using partially synthetic diets. J. Insect Physiol. 10: 161–178.

Dadd, R. H. 1966. Beeswax in the nutrition of the wax moth, *Galleria mellonella* (L.). J. Insect Physiol. 12: 1479–1492.

Dadd, R. H. 1968. Problems connected with inorganic components of aqueous diets. Bull. Entomol. Soc. Amer. 14: 22–26.

Dadd, R. H. 1970. Arthropod nutrition, pp. 35–95. *In* M. Florkin and B.T. Scheer (eds.), Chemical zoology, vol. 5. Academic Press, New York, NY.

Dadd, R. H. 1973. Insect nutrition: current developments and metabolic implications. Annu. Rev. Entomol. 18: 381–420.

Dadd, R. H. 1975. Ingestion of colloid solutions by filter-feeding mosquito larvae: relationship to viscosity. J. Exp. Zool. 191: 395-406.

Dadd, R. H. 1977. Qualitative requirements and utilization of nutrients: insects, pp. 305–348. *In* M. Recheigl (ed.), Handbook series in nutrition and food, vol. 1. CRC Press, Cleveland, OH.

Dadd, R. H. 1978. Amino acid requirements of the mosquito, *Culex pipiens*: asparagine essential. J. Insect Physiol. 24: 25-30.

Dadd, R. H. 1980. Essential fatty acids for the mosquito, *Culex pipiens*. J. Nutrition 110: 1152-1160.

Dadd, R. H. 1985. Nutrition: organisms, pp. 313–390. *In* G. A. Kerkut and L. I. Gilbert (eds.), Comprehensive insect physiology, biochemistry and pharmocology, vol. 4. Pergamon Press, Oxford, U.K.

Dadd, R. H., and J. E. Kleinjan. 1978. Dietary nucleotide requirements of the mosquito, *Culex pipiens*. J. Insect Physiol. 23: 333–341.

Dadd, R. H., and J. E. Kleinjan. 1979a. Essential fatty acid for the mosquito, *Culex pipiens*. J. Insect Physiol. 25: 495–502.

Dadd, R. H., I. Gomze, and M. Nambe. 1973. Requirements for ribonucleic acid in a semi-synthetic larval diet for the mosquito *Culex pipiens*. J. Med. Entomol. 10: 47-52.

Dadd, R. H., W. G. Friend, and J. E. Kleinjan. 1980. Arachidonic acid requirement for two species of *Culiseta* reared on synthetic diet. Canad. J. of Zool. 58: 1845-1850.

Dahlman, D. L., L. A. Schroeder, R. H. Tomhave, and E. T. Hibbs. 1981. Inhibition and survival response of the potato leafhopper, *Empoasca fabae* to selected sugars in agar media. Entomol. Exp. Appl. 29: 228–233.

David, W. A. L., and S. Ellaby. 1975. The viability of the eggs of the African armyworm, *Spodoptera exempta* in laboratory cultures. Entomol. Exp. Appl. 18: 269–280.

Davis, G. R. F. 1964. The importance of carnitine in the diet of larvae *Oryzaephilus surinamensis* (L.) (Coleoptera: Silvanidae). Archiv. Int. Physiol. Biochem. 72: 70–75.

Davis, G. R. F. 1968. Quantitative dietary requirements of *Oryzaephilus surinamensis*. (L.) (Coleoptera: Silvanidae) for phenylalanine and tyrosine. Canad. J. Zool. 46: 469–473.

Dougherty, E. C. 1959. Introduction to axenic culture of invertebrate metazoan: a goal. Ann. N.Y. Acad. Sci. 77: 27–54.

El-Sawaf, B. M., and S. M. El-Sayes. 1978. Feeding behavior of the hump beetle, *Gibbium psylloides* (Coleoptera: Ptinidae). Z. Angew. Entomol. 86: 46–52.

Friend, W. G. 1968. The nutritional requirements of Diptera, pp. 41–57. *In* Radiation, radioisotopes and rearing methods in the control of pest insects. Proceedings International Atomic Energy Association, Vienna.

Friend, W. G. 1978. Physical factors affecting the feeding responses of *Culiseta inornata* to ATP, sucrose and blood. Ann. Entomol. Soc. Amer. 71: 935–940.

Friend, W. G., and J. J. B. Smith. 1977. Factors affecting feeding by blood sucking insects. Annu. Rev. Entomol. 22: 309–331.

Funke, B. R. 1983. Mold control for insect rearing media. Bull. Entomol. Soc. Amer. 29: 41–44.

Gandhi, J. R. 1977. Some factors governing feeding behavior of Homopteran, *Oxyrhachis tarandus*. Indian J. Entomol. 39: 12–16.

Geer, B. W. 1963. A ribonucleic acid-protein relationship in *Drosophila* nutrition. J. Exp. Zool. 154: 353–364.

Gilmour, D. 1961. The biochemistry of insects. Academic Press, New York, NY.

Gingrich, R. E. 1964. Nutritional studies on screwworm larvae with chemically defined media. Ann. Entomol. Soc. Amer. 57: 351–360.

Grau, P. A. and L. C. Terriere. 1971. Fatty acid profile of the cabbage looper, *Trichoplusia ni*, and the effect of diet and rearing conditions. J. Insect Physiol. 17: 1637–1649.

Guerra, A. A., and A. Garza. 1978. Algae, an economical protein source for insect diets. Southwest Entomol. 3: 76–79.

Harvey, G. T. 1974. Nutritional studies of eastern spruce budworm (Lepidoptera: Tortricidae). I. Soluble sugars. Canad. Entomol. 106: 353–365.

Havlickova, H. 1976. Effect of free amino acid and sugars on feeding rate of pea leaf weevil, *Sitona lineatus*. Ochr. Rostl. 12: 71–78.

House, H. L. 1958. Nutritional requirements of insects associated with animal parasitism. Exp. Parasitol. 7: 555–609.

House, H. L. 1961. Insect nutrition. Annu. Rev. Entomol. 6: 13–26.

House, H. L. 1963. Nutritional diseases, pp. 133–160. *In* E. A. Steinhaus (ed.), Insect pathology, an advanced treatise, vol. 1. Academic Press, New York, NY.

House, H. L. 1965. Effects of vitamin A acetate and structurally related substances on growth and reproduction of *Agria affinis* (Fallen) (Diptera: Sarcophagidae). J. Insect Physiol. 10: 1039–1045.

House, H. L. 1966. Effects of vitamin E and A on growth and development, and the necessity of vitamin E for reproduction in the parasitoid *Agria affinis* (Fallen) (Diptera: Sarcophagidae). J. Insect Physiol. 12: 409–417.

House, H. L. 1972. Insect nutrition, pp. 513-573. *In* R. N. Fiennes (ed.), Biology of nutrition, vol. 18. Pergamon Press, New York, NY.

House, H. L. 1974a. Nutrition, pp. 1–62. *In* M. Rockstein (ed.), The physiology of Insecta, vol. 5. Academic Press, New York, NY.

House, H. L. 1974b. Digestion, pp. 63–117. *In* M. Rockstein (ed.), The physiology of Insecta, vol. 5. Academic Press, New York, NY.

Howell, J. F. 1967. Paraffin films to control dehydration of an artificial rearing medium for codling moth. J. Econ. Entomol. 60: 289–290.

Howell, J. F. 1977. Codling moth: Agar substitutes in artificial diets. USDA ARS Bull. W-43, pp. 1–9.

Ito, T., and N. Arai. 1965 Nutrition of the silkworm, *Bombyx mori*. VIII. Amino acid requirements and nutritive effects of ascorbic acid. Bull. Sericult. Exp. Station Tokyo 20: 1–19.

Ito, T., and N. Arai. 1967. Nutritive effects of alanine, cystine, glycine, serine, and tyrosine on the silkworm, *Bombyx mori*. J. Insect Physiol. 13: 1813–1824.

Ivy, C. A., D. B. Bragg, and E. L. Stephenson. 1971. The availability of amino acids from soybean meal for the growing chick. Poultry Sci. 50: 408–410.

Jordens-Rottger, D. 1979. The role of phenolic substances for host selection behavior of the black bean aphid, *Aphis fabae*. Entomol. Exp. Appl. 29: 49–54.

Kato, M. 1978. Phenols as indispensable components of the synthetic diet of the silkworm, *Bombyx mori*. Entomol. Exp. Appl. 24: 485–490.

King, E. G., and N. C. Leppla (eds.). 1984. Advances and challenges in insect rearing. USDA ARS, New Orleans, LA.

Kleinjan, J. E., and R. H. Dadd. 1977. Vitamin requirements of the larval mosquito, *Culex pipiens*. Ann. Entomol. Soc. Amer. 70: 541–543.

Kramer, K. J., and P. A. Seib. 1982. Ascorbic acid and the growth and development of insects, pp. 275-291. *In* P. A. Seib and B. M. Tolbert (eds.), Ascorbic acid: chemistry, metabolism and uses. Advances in Chemistry Series No. 200.. American Chemical Society, Washington, DC.

Lehninger, A. L., D. L. Nelson, and M. M. Cox. 1993. Principles of biochemistry. Worth, New York, NY.

Lipke, H., and G. Fraenkel. 1956. Insect nutrition. Annu. Rev. Entomol. 1:17–44.

Ma, W. -C. 1972. Dynamics of feeding responses in *Pieris brassicae* Linn. as a function of chemosensory input: a behavioural, ultrastructural and electrophysiological study. Mededelingen Landbouwhogeschool Wageningen 72: 1–162.

Ma, W. -C., and I. Kubo. 1977. Phagostimulants for *Spodoptera exempta* identification of adenosine from *Zea mays*. Entomol. Exp. Appl. 22: 107–112.

Marston, N., B. Campbell, and P. E. Boldt. 1975. Mass producing eggs of greater wax moth, *Galleria mellonella* (L.). USDA ARS Tech. Bull. 1510.

Matsuda, K. 1981. Feeding stimulation of nutrient chemicals in *Gastrophysa atrocyanea* (Coleoptera: Chrysomelidae). Jap. J. Appl. Entomol. Zool. 25: 84–88.

McGinnis, A. J., and R. Kasting. 1972. Quantitative nutrition and evaluation of protein in foods of phytophagous insects, pp. 57–71. *In* J. G. Rodriguez (ed.), Insect and mite nutrition: significance and implications in ecology and pest management. Elsevier–North Holland, Amsterdam, Netherlands.

Merck. 2001. Merck index. M. J. O'Neil, A. Smith, P. E. Heckelman, and S. Budavari, eds. John Wiley & Sons, Whitehall, NJ.

Mittler, T. E., and J. E. Kleinjan. 1970. Effect of artificial diet composition on wing production by the aphid, *Myzus persicae*. J. Insect Physiol. 16: 833–850.

Mittler, T. E., and J. A. Tsitsipis. 1973. Economical rearing of larvae of the olive fruit fly, *Dacus oleae*, on a liquid diet offered on cotton toweling. Entomol. Exp. Appl. 16: 292–293.

Moore, R. F. 1985. Artificial diets: Development and improvement, pp. 67–84. *In* P. Singh and R. F. Moore (eds.), Handbook of insect rearing, vol. 1. Elsevier, Amsterdam, Netherlands.

Nation, J. L. 2002. Insect physiology and biochemistry. CRC Press. Boca Raton, FL.

Nayar, J.K. 1964. The nutritional requirements of grasshoppers: I. rearing of the grasshopper, *Melanoplus bivittatus* (say), on a completely defined synthetic diet and some effects of different concentrations of B-vitamin mixture, linoleic acid and ß-carotene. Canad. J. Zool. 42: 11-22.

Nettles, W. C., Jr. 1972. Toxic sugars and longevity of *Anthonomus grandis*. J. Insect Physiol. 18: 1107–1113.

Nettles, W. C., Jr., C. M. Wilson, and S.W. Ziser. 1980. A diet and methods for the in vitro rearing of the tachinid, *Eucelatoria* sp. Ann. Entomol. Soc. Amer. 73: 180–184.

Norris, D. M., and J. M. Baker. 1969. Nutrition of *Xyleborus ferrugineus*. I. Ethanol in diets as a tunneling (feeding) stimulant. Ann. Entomol. Soc. Amer. 62: 592–594.

Raubenheimer, D., and S. J. Simpson. 1992. Analysis of covariance: an alternative to nutritional indices. Entomol. Exp. Appl. 62: 221–231.

Raubenheimer, D., and S. J. Simpson. 1994. The analysis of nutrient budgets. Functional Ecology 8: 783–791.

Reese, J. C., and S. D. Beck. 1978. Interrelationships of nutritional indices and dietary moisture in the black cutworm (*Agrotis ipsilon*) digestive efficiency. J. Insect Physiol. 24: 473–479.

Reinecke, J. P. 1985. Nutrition: artificial diets, pp. 391-419. *In* G. A. Kerkut and L. I. Gilbert (eds.), Comprehensive insect physiology, biochemistry and pharmacology, vol. 4. Pergamon Press, Oxford, U.K.

Rock, G. C. 1972. Optimal proportions of dietary amino acids, pp. 183–197. *In* J. G. Rodriguez (ed.), Insect and mite nutrition: significance and implications in ecology and pest management. Elsevier-North Holland, Amsterdam, Netherlands.

Rock, G. C., and G. K. Sharma. 1974. Comparison of the nutrition values of carbohydrates to *Argyrotaenia velutinana*. Ann. Entomol. Soc. Amer. 57: 617–621.

Rock, G. C., R. L. Patton, and E. H. Glass. 1965. Studies of the fatty acid requirements of *Argerotaenia velutinana* (Walker). J. Insect Physiol. 11: 91-101.

Rodriguez, J. G. (ed.). 1972. Insect and mite nutrition: significance and implications in ecology and pest management. Elsevier-North Holland, Amsterdam, Netherlands.

Sang, J. H. 1978. The nutritional requirements of *Drosophila*. *In* M. Ashburner and T. R. F. Wright (eds.), The genetics and biology of *Drosophila*, vol. 2a. Academic Press, New York, NY.

Schmidt, D. J., and J. C. Reese. 1985. Sources of error in nutritional index studies of insects on artificial diet. J. Insect Physiol. 32: 193–198.

Schoonhoven, L. M. 1972. Some aspects of host selection and feeding in phytophagous insects, pp. 557–566. *In* J. G. Rodriguez (ed.), Insect and mite nutrition: significance and implications in ecology and pest management. Elsevier-North Holland, Amsterdam, Netherlands.

Scriber, J. M., and F. Slansky. 1981. The nutritional ecology of immature insects. Annu. Rev. Entomol. 26: 183–211.

Shapiro, M., and R. A. Bell. 1982. Production of the gypsy moth, *Lymantria dispar*, nuclear polyhedrosis virus using carrageenan as dietary gelling agents. Ann. Entomol. Soc. Amer. 75: 43–45.

Shetty, P. S. 1982. Gustatory preferences of ants *Camponotus compessus* for urea and sugars. Experientia 38: 259–260.

Sikorowski, P., and R. H. Goodwin. 1985. Contaminant control and disease recognition in laboratory colonies, pp. 85–105. *In* P. Singh and R. F. Moore (eds.), Handbook of insect rearing, vol. 1. Elsevier, Amsterdam, Netherlands.

Singh, P. 1977. Artificial diets for insects, mites and spiders. Plenum Press, New York, NY.

Singh, P. 1983. A general purpose laboratory diet mixture for rearing insects. Insect Sci. Applic. 4: 357–362.

Singh, P., and R. F. Moore (eds). 1985. Handbook of insect rearing, vol. 1 & 2. Elsevier, Amsterdam, Netherlands.

Sivapalan, P., and N. C. Gnanapragasam. 1979. The influence of linoleic and linolenic acid on adult moth emergence of *Homona coffearia* from meridic diets in vitro. J. Insect Physiol. 25: 393–398.

Slansky F., Jr., and J. G. Rodriguez (eds). 1987. Nutritional ecology of insects, mites, spiders, and related invertebrates. Wiley-Interscience, New York, NY.

Slansky, F., Jr. and J. M. Scriber. 1985. Food consumption and utilization, pp. 87–163. *In* G. A. Kerkut and L. I. Gilbert (eds.), Comprehensive insect physiology, biochemistry and pharmocology, vol. 4. Pergamon Press, Oxford, U.K.

Smith, A. K., and S. J. Circle. 1978. Chemical composition of the seed, pp. 61–92. *In* A. K. Smith and S. J. Circle (eds.), Soybeans: chemistry and technology, vol.1. AVI Publishing, Westport, CT.

Smith, J. J. B. 1979. The feeding response of *Rhodnius prolixus* to blood fractions and the role of ATP. J. Exp. Biol. 78: 225–232.

Smith, J. J. B. 1985. Feeding mechanisms, pp. 33–85. *In* G. A. Kerkut and L. I. Gilbert (eds.), Comprehensive insect physiology, biochemistry and pharmocology, vol. 4. Pergamon Press, Oxford, U.K.

Sneller, V. -P., and R. H. Dadd. 1981. Interaction of amino acids and glucose on growth of *Aedes aegypti* (Diptera: Culicidae) in a synthetic rearing medium. J. Medical Entomol. 18: 235–239.

Snodgrass, R. E. 1935. Principles of insect morphology. McGraw-Hill, New York, NY.

Srivastava, B. G., N. C. Pant, and H. S. Choudhary. 1978. Carbohydrate requirement of *Dacus cucurbitae* maggots under aseptic conditions. Indian J. Entomol. 40: 150–155.

Svoboda, J. A., and M. J. Thomson. 1983. Comparative sterol metabolism, pp. 1–16. *In* T. E. Mittler and R. H. Dadd (eds.), Metabolic aspects of lipid nutrition in insects. Westview Press, Boulder, CO.

Taylor, D. B. 1988. Comparison of two gelling agents for screwworm (Diptera: Calliphoridae) larval diet. J. Econ. Entomol. 81: 1414-1419.

Taylor, D. B. 1992. Rearing system for screwworm mass production, pp. 393–403. *In* T. E. Anderson and N. C. Leppla (eds.), Advances in insect rearing for research and pest management. Westview Press, Boulder, CO.

Thomson, S. N. 1981a. Effects of dietary carbohydrate and lipid on nutrition and metabolism of metazoan parasites with special reference to parasitic Hymenoptera, pp. 215-252. *In* G. Bhaskaran, S. Friedman, and J. G. Rodriguez (eds.), Current topics in insect endocrinology and nutrition. Plenum Press, New York, NY.

Thomson, S.N. 1981b. *Brachymeria lasus* and *Pachycrepoideus vindemiae*: sterol requirements during larval growth of two hymenopterous insect parasites reared in vitro on chemically defined media. Exp. Parasitol. 51: 220–235.

Trager, W. 1947. Insect nutrition. Biol. Rev. 22: 148–177.

Trager, W. 1953. Nutrition, pp. 350-386. *In* D. D. Roeder (ed.), Insect physiology. John Wiley & Sons, New York, NY.

Tsai, Y. M, H. L. Wang, C. L. Ho, and S. C. Li. 1978. Studies on the artificial diet of silkworm, *Bombyx mori*: the effects of dietary physiochemical factors on the feeding and growth of the silkworm. Acta Entomol. Sinica 21: 369–384.

Turunen, S. 1985. Absorption, pp. 241-277. *In* G. A. Kerkut and L. I. Gilbert (eds.), Comprehensive insect physiology, biochemistry and pharmocology, vol. 4. Pergamon Press, Oxford, U.K.

Vanderzant, E. S. 1969. Physical aspects of artificial diets. Entomol. Exp. Appl. 12: 642–650.

Vanderzant, E. S. 1973. Axenic rearing of larvae and adults of the boll weevil on defined diets: additional tests with amino acids and vitamins. Ann. Entomol. Soc. Amer. 66: 1184–1186.

Waldbauer, G. P. 1968. The consumption and utilization of foods by insects. Adv. Insect Physiol. 5: 229–288.

Wardojo, S. 1969. Some factors relating to larval growth of the Colorado potato beetle, *Leptinotarsa decemlineata* Say on artificial diets. Mededelingen Landbouwhogeschool Wageningen 16: 1–71.

Williams, D. F., C. S. Lofgren, and A. Lemire. 1980. A simple diet for rearing laboratory colonies of the red imported fire ant. J. Econ. Entomol. 73: 176–177.

Wolf, W. J., and J. C. Cowan. 1971. Soybean as a food source. CRC Crit. Rev. Food Technol. 2: 81–158.

Yazgan, S. 1972. A chemically defined synthetic diet and larval nutritional requirements of the endoparasitoid, *Itoplectis conquisitor* (Hymenoptera). J. Insect Physiol. 18: 2123–2141.

Yin, C. -M., and W. K. Peng. 1981. Simplified soy pulp wheat germ diets for rearing the southwestern cornborer, *Diatraea grandiosella* Dyar. Ann. Entomol. Soc. Amer. 74: 425–427.

Chapter 8

Microbial Contamination and Insect Rearing

G. DOUGLAS INGLIS
Research Scientist
Zoonotic Bacteriology
Agriculture and Agri-Food Canada
Lethbridge, AB Canada

PETER P. SIKOROWSKI, Deceased
Professor of Entomology
Department of Entomology and Plant Pathology
Mississippi State University
Mississippi State, MS

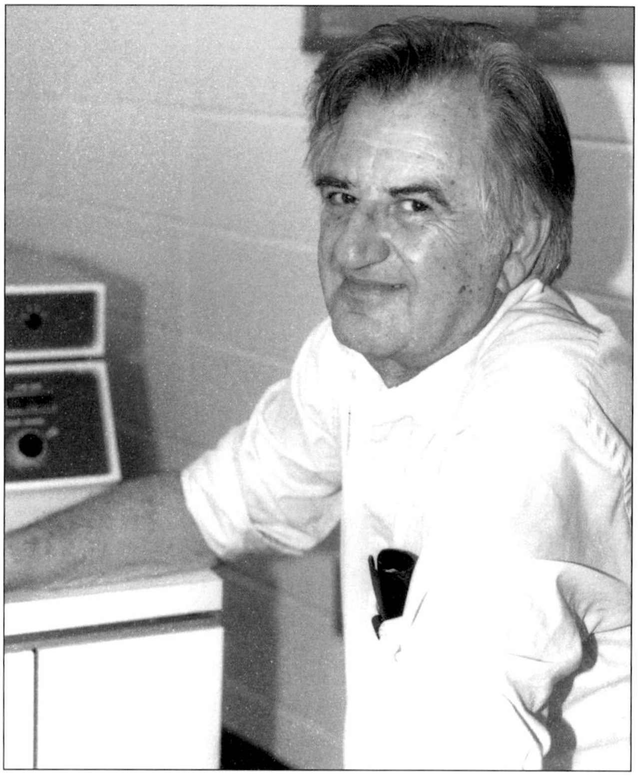

Fig. 1. Peter Paul Sikorowski (1926–2000).

Dedication

This book chapter is dedicated to Peter P. Sikorowski (Fig. 1), who unfortunately died during its preparation. Peter, more than any other person, has promoted the study and management of microbial contaminants in relation to insect rearing. This is illustrated by the frequent citations to his research within this chapter. He was an untiring proponent of the virtues of studying and managing microbial contaminants and was active in this area of research until his death. He will be sorely missed.

I. Introduction

As indicated in Chap. 9 *Entomopathogens and Insect Rearing*, the management of disease-inciting microorganisms can determine the success of a rearing program. In this chapter, we deal with microorganisms that are considered to be microbial contaminants. These spoilage microorganisms almost exclusively belong to two groups: the saprotrophic bacteria and fungi. A single, universally efficacious method that can be applied to successfully control microbial contaminants does not exist, and successful management relies almost exclusively on the application of integrated measures. Despite the obvious importance of microbial contaminants to insect rearing, research on microbial contaminants has generally not been a priority area within entomology. Few entomologists and rearing specialists possess a background in microbiology, which may explain the reluctance of many entomologists and rearing specialists to address microbial contamination issues in a substantive manner. Given the complexities of the interactions involved, the development and application of efficacious strategies for managing microbial contaminants require an interdisciplinary approach. Entomologists and rearing specialists should embrace this challenge and draw upon the expertise of microbiologists. We also believe that the application of sound, mechanistic research principles to the study of microbial contaminants is essential; but, to our knowledge, this type of research is not currently being conducted. The purpose of this chapter is to examine what microbial contamination is; what its potential consequences are; from where it comes; how it can be detected; and, most importantly, how its effects can be mitigated.

II. What Is Microbial Contamination?

Most readers of this chapter know that microbial contaminants are a serious problem in insectary settings, but defining microbial contamination is not an easy task. Some people have defined *microbial contaminants* as microorganisms that do not form mutualistic or parasitic relationships with insects. This definition is rather narrow and does not address whether microbial contaminants are pathogens or not. In Chap. 9 *Entomopathogens and Insect Rearing*, we defined *disease* as "a dynamic interaction between an organism and its environment that results in abnormal physiological and often morphological or neurological changes in the organism." Further, a *pathogen* was defined as "any agent that incites disease." Some have defined microbial pathogens as microorganisms that are capable of producing disease only under normal conditions of host resistance and that rarely live in close association with the host without producing disease (Steinhaus and Martignoni 1970). Defining "normal conditions of host resistance" is not possible, especially in rearing settings, which are inherently stressful to insects. In many cases, the dissimilarity between pathogen and microbial contaminant is narrow or nonexistent. The pathogenic potential of a given microorganism is exceptionally complex, determined by many biotic factors (some of microbial and others of host origin) as well as abiotic environmental factors (Fig. 2).

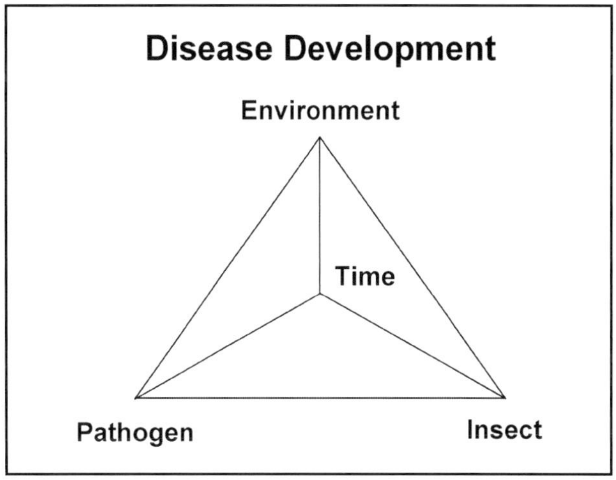

Fig. 2. The disease tetrad or "theory of predisposition" model. Disease can only occur in the presence of a susceptible host, a pathogen, and a suitable environment. Since all diseases are manifested over time, a temporal component is added.

Consider a typical scenario in which an insect captured in the field is returned to an insectary. Most insects typically harbor various types of microorganisms within their alimentary canal with no apparent harm to the insect. However, when such insects are exposed to a stressor within the insectary, they often become infected by microorganisms that are present in their guts, and a large proportion die. These microorganisms do not incite disease under normal conditions, but they may become pathogenic when conditions are changed: they are said to be *facultative* pathogens. If they did not adversely affect the health of an insect, then we would not be concerned about their management in insectaries. Furthermore, this example illustrates that disease cannot be examined independent of the environmental component of the interaction.

To illustrate this aspect of the disease tetrad (Fig. 2), consider a situation in which a microorganism present in an insect diet never enters an insect while it is living but decreases the pH of the diet to such a degree that the insect is unable to obtain adequate nutrition. In this case, the microorganism influences the environment in which the insect exists and contributes to disease resulting from inadequate nutrition. It is important to realize that the interaction between environment, microorganisms, and host is never static but rather changes over time. Furthermore, microorganisms almost always exist in communities within very complex ecosystems such as within an alimentary canal or an artificial diet. These microorganisms have evolved elaborate mechanisms to colonize substrates and outcompete other organisms, including insects, for these substrates.

When we combine the two above examples, we obtain a more realistic scenario. An insect from the field is returned to the insectary and placed on an artificial diet. Microorganisms associated with one of the diet ingredients rapidly proliferate within the diet. One species produces extracellular acids as a by-product of metabolism. Other microorganisms are adversely affected by the lowered pH, and the acid-producing microorganism predominates. As it proliferates, the pH of the diet decreases to such an extent that the insect is no longer able to utilize many nutrients present within the artificial diet and it becomes physiologically stressed. As a consequence, the insect is no longer able to apply metabolic resources to maintenance of normal barriers against microorganisms. Other microorganisms present within the midgut lumen find that they are able pass through an incomplete *peritrophic matrix* (complex, chitin-based lining of the midgut) and that they are able to come into contact with the midgut epithelium (Fig. 3) at multiple locations. These microorganisms produce toxins as part of their strategy to secure nutrients. As a consequence, these toxins damage the microvillar membranes, further contributing to the inability of the insect to take up nutrients. As a result of the damage to the epithelium, the microorganisms gain access to the cytoplasm of cells where they begin to proliferate, catabolizing the rich source of cellular nutrients. Under normal circumstances, the insect is able to defend itself against such an insult (e.g., by production of cellular and/or chemical defense responses). However, given the insect's impaired nutritional status, its defense response is inadequate; and the microorganisms continue to proliferate, eventually breeching the basement membrane and gaining access to the hemocoel. They continue to proliferate within the hemocoel, gaining access to the entire body of the insect. As with the diet, the various types of microorganisms produce a cadre of antibiotic materials (e.g., toxins) to enhance their competitive colonization of the substrates from which they are deriving nutrition. The consequence is massive damage to the insect tissues and death.

This scenario illustrates that interactions between insects and microorganisms are never simple. Given the complexity of the interaction, we broadly define *microbial contaminants* as primarily saprotrophic bacteria and fungi that rarely produce disease in healthy insects but that, under the appropriate conditions, can adversely affect the health of an insect. It also is very important to realize that not all microorganisms are detrimental to insects. To the contrary, many are beneficial (e.g., mutualistic symbionts) as discussed below (Sec. III. C).

A. Bacteria

Bacteria are a large group of phylogenetically diverse microorganisms that are *prokaryotic* (lacking a nuclear membrane). Of the two major groups of bacteria, we are almost exclusively concerned with organisms within the domain Bacteria (true bacteria) as opposed to the

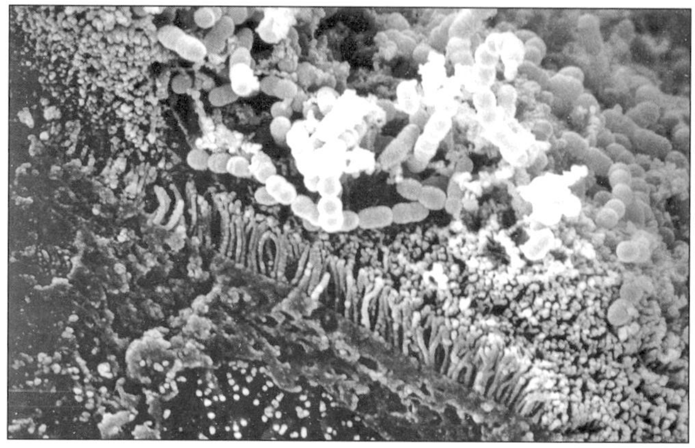

Fig. 3. Bacterial cells in contact with the midgut of a boll weevil, *Anthonomus grandis*, (Coleoptera: Curculionidae), adult. The bacteria have breeched the peritrophic matrix to gain access to the midgut epithelium.

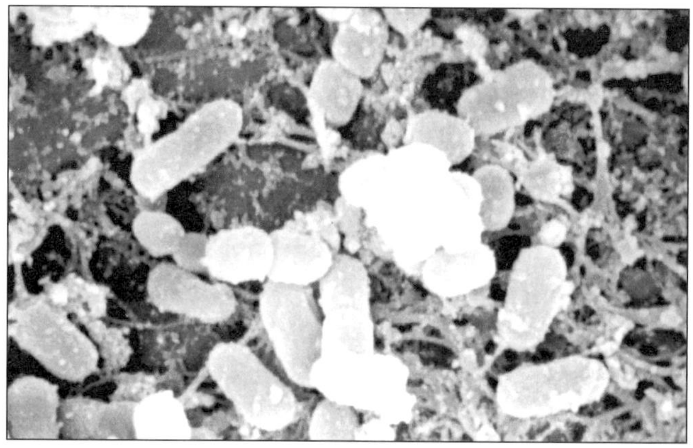

Fig. 4. Scanning electron micrograph of bacilli (rod-shaped) bacteria.

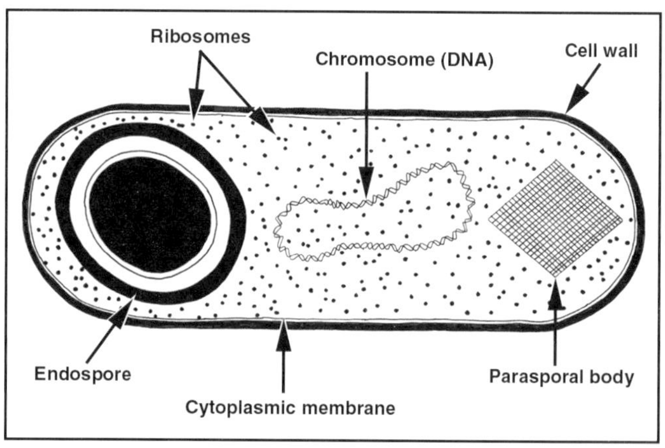

Fig. 5. Longitudinal section through a *Bacillus* cell. The cell has a well-defined, peptidoglycan cell wall enclosing the cytoplasmic membrane, which in turn contains the cytoplasm. Within the cytoplasm, there is a circular chromosome comprised of double-stranded DNA that stores the hereditary information, numerous ribosomes by means of which proteins are synthesized, and a single endospore. Endospores contain a copy of the bacterial chromosome; and once they are liberated from the cell, they are capable of germinating and producing a clone of the parent cell.

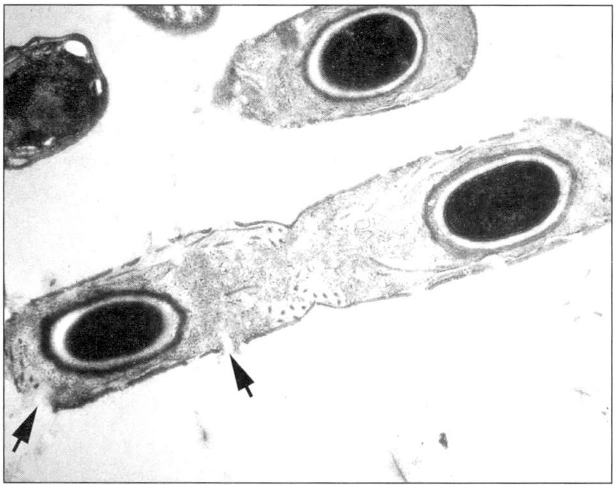

Fig. 6. Transmission electron micrograph of *Clostridium bifermentans*. Note the thick-walled endospores within the cells. In this micrograph, the cells are undergoing lysis (arrows) to release the endospores. [Image provided by J. F. Charles (Institute Pasteur, Paris, France).]

archaebacteria within the domain Archaea. The true bacteria are primarily unicellular with the exception of the *actinomycetes*, which are filamentous, 1.0–8.0 μm in length, and 0.2–2.0 μm in width. Some taxa form persistent chains, whereas others do not or form ephemeral chains. The primary cell shapes are *cocci* (spherical cells; Fig. 3) and *bacilli* (rod-shaped cells; Fig. 4). Within the bacteria, there are two major types of cell wall morphologies, and the two types can be distinguished based on their Gram stain reaction: *Gram negative* or *Gram positive*. For more information on the cell walls of bacteria and on the Gram staining procedure, the reader is referred to Chap. 9 *Entomopathogens and Insect Rearing*. Some taxa of bacteria (e.g., *Bacillus*, *Paenibacillus*, and *Clostridium*) form endospores, which has important consequences from a microbial contamination perspective (Figs. 5, 6). *Endospores* are survival structures and are some of the most heat-resistant cells known to science. As a consequence, the deactivation of endospores requires the application of special control strategies (Sec. VI. B. 2).

B. Fungi

Fungi are a phylogenetically diverse group of *eukaryotes* (possessing a nuclear membrane). They reproduce by spores, with both sexual and asexual spores being produced depending on the fungus and conditions; they are typically not motile; they are *heterotrophic* (deriving nutrition from organic substrates); and they obtain nutrients extracellularly. There are two primary growth forms produced by fungi: 1) yeasts and 2) molds. It is important to realize that growth form is not a phylogenetic character. *Yeast* refers to the nonfilamentous form of fungi adapted to a liquid environment, whereas *mold* refers to the filamentous form of fungi adapted to a terrestrial existence. Some fungi are dimorphic and grow both as yeast and mold forms. Many fungi have evolved specialized relationships with other organisms, but perhaps the most important role of fungi in nature is as primary decomposers of organic matter.

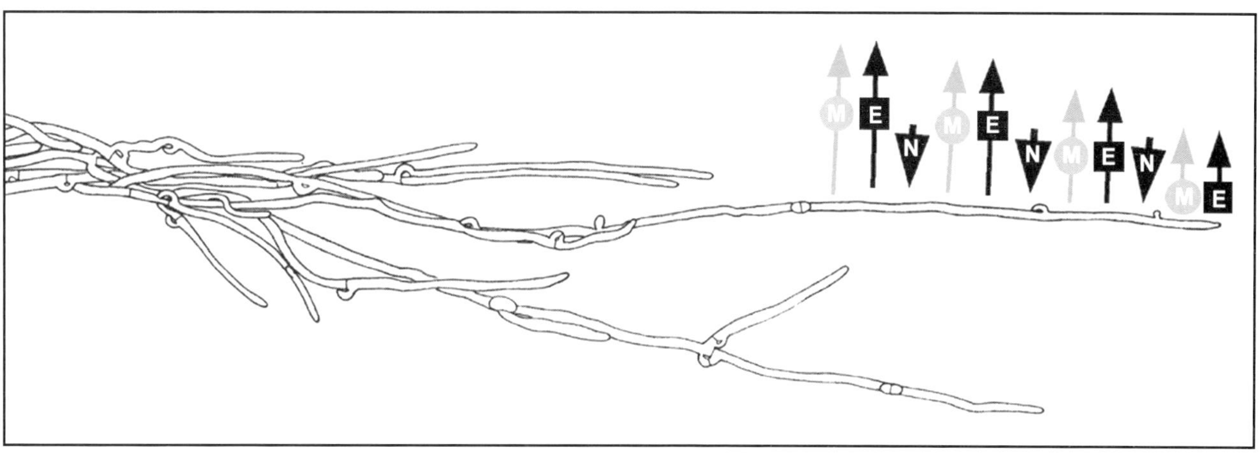

Fig. 7. Hyphal growth of a saprotrophic fungus. In the "digest then ingest" trophic strategy, enzymes (E) are exported from the hyphal cells and diffuse into the substrate where they catalyze the breakdown of complex polymeric molecules (e.g., starch) to smaller nutrient molecules (N) (e.g., glucose), which are taken up by the hyphal cells. Most fungi also produce secondary metabolites (M) such as mycotoxins to exclude competing organisms. [Webster (1980) with permission from Cambridge University Press.]

1. Molds. The filamentous nature of molds derives from their ability to produce *hyphae* (specialized threadlike structures; *mycelium* when massed; Fig. 7). Hyphae are the assimilative structure of such fungi and are key to the success of this large group of microorganisms. Hyphae grow from the tip, and thus they are able to ramify into solid substrates. Most molds produce copious numbers of asexual spores, which contributes to their success as primary decomposers. As a group, they are a classic example of *r-selected* organisms (Andrews 1992). Such organisms are characterized by the following: small size, production of many offspring, reduced energy invested per offspring, rapid maturation, and high mortality rates. The molds typically produce millions, if not billions, of asexual spores, many of which are adapted for airborne dissemination (Fig. 8). Two genera of molds that are commonly found in air in large numbers are *Aspergillus* and *Penicillium*. In particular, *Aspergillus niger* (Fig. 9A) is a common contaminant in insectaries. However other taxa, including human pathogenic species such as *Aspergillus fumigatus* (Fig. 9B), also are common.

Members of the genus *Aspergillus* are easily identified based on the way their *conidia* (asexual spores) are formed. The diagnostic features of the genus *Aspergillus* are the production of *conidiophores* (conidial stalks) with a conspicuous, apical vesicle bearing *phialides* (flask-shaped conidiogenous cells) from which conidia are produced in long, dry chains (Fig. 9). Currently, there are over 100 recognized species within the genus (Hocking 2001).

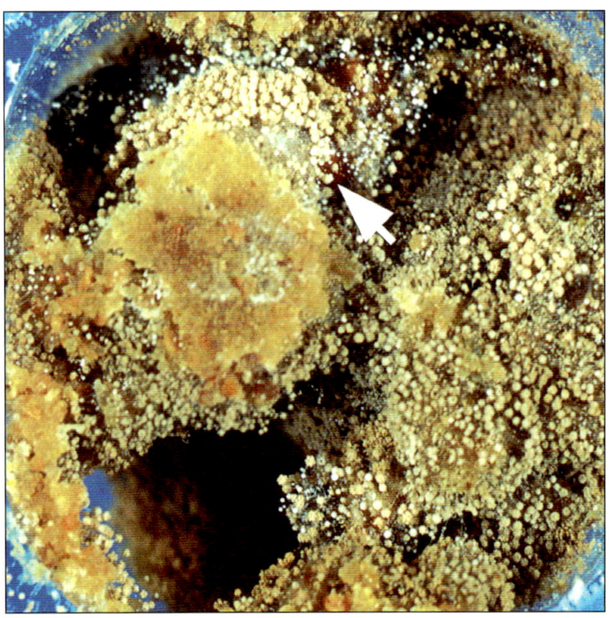

Fig. 8. Prolific sporulation by *Aspergillus ustus* on a semisynthetic artificial diet within a plastic rearing container. The arrow points to conidial heads where large numbers of asexual spores termed conidia are produced. These conidia are adapted for airborne dissemination, and opening the lid covering the rearing cell can result in the liberation of billions of conidia, each capable of contaminating a suitable substrate within an insect rearing facility.

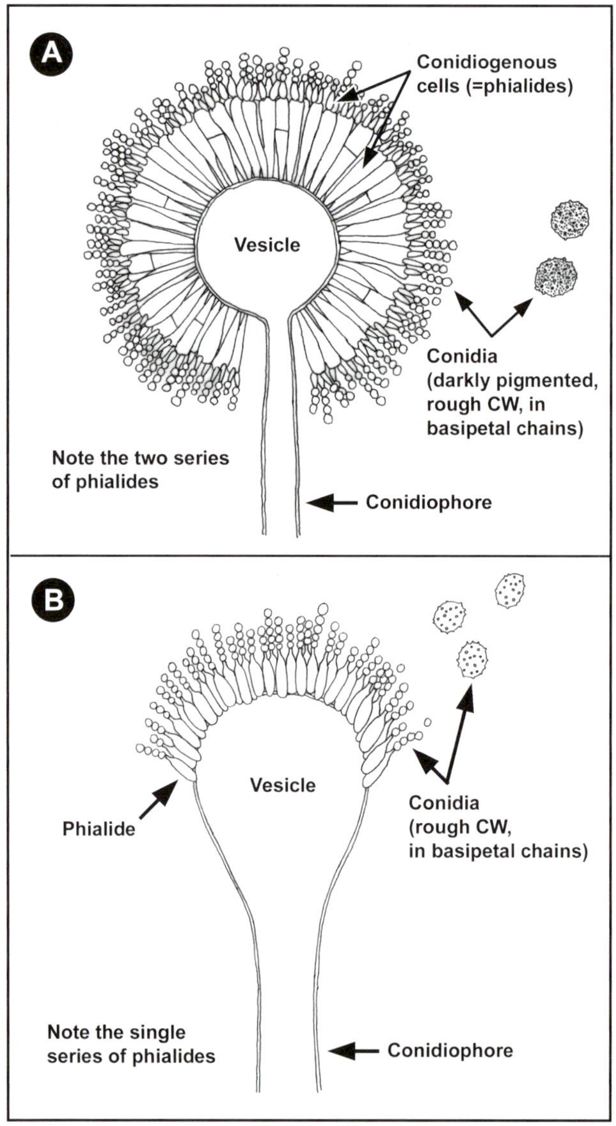

Fig. 9. Drawings of two species of *Aspergillus* encountered in insect rearing facilities. All *Aspergillus* species produce a well-defined conidiophore with an apical vesicle bearing numerous phialides, each of which produces a dry chain of conidia. (A) *Aspergillus niger*. This species is characterized by the production of copious quantities of black conidia possessing irregularly roughened cell walls (CW) and arranged in large radiating heads. (B) *Aspergillus fumigatus*. This species is a common causal agent of aspergillosis, a respiratory disease of humans. It is characterized by the production of greyish-turquoise or dark turquoise to dark green conidia possessing smooth to slightly rough cell walls (CW) and arranged in columnar heads.

Members of the genus *Penicillium* also are readily identified to the genus level based on their microscopic appearance. Like *Aspergillus*, *Penicillium* species produce long chains of conidia from phialides. In contrast to the genus *Aspergillus*, *Penicillium* species produce conidiogenous cells in *pencillate* (brushlike) heads (Fig. 10). There are 150 recognized species of *Penicillium* of which at least 50 are common (Pitt 2001).

Members of both of these genera are well-known producers of toxic secondary metabolites. Furthermore, they produce exceptionally large numbers of small (2–3 μm diameter) conidia. These often remain suspended in air for long periods of time. Furthermore, the conidia of these taxa are typically electrostatically charged, and they adhere to inanimate objects common in insectaries. While only a few conidia produced by molds will ever land on a suitable substrate and subsequently colonize it to produce offspring, the sheer numbers produced by these r-selected microorganisms make them successful pioneer colonizers of substrates such as insect diets.

Another commonly encountered group of molds in insectaries are the *bread molds* belonging to the order Mucorales. Members of this group produce *sporangiospores* (asexual spores) in a *sporangium* (spore sac; Fig. 11). The *peridium* (wall of the sporangium) is very fragile and easily ruptured, thereby releasing copious numbers of sporangiospores. Often all that is left of the specimen after it is mounted on a glass slide for microscopy are remnants of the peridium and the wide *sporangiophore* (specialized hypha subtending the sporangium) with its *columnella* (central axis and terminal vesicle of the sporangiophore). Common genera of Mucorales encountered in insectaries are *Mucor* and *Rhizopus*.

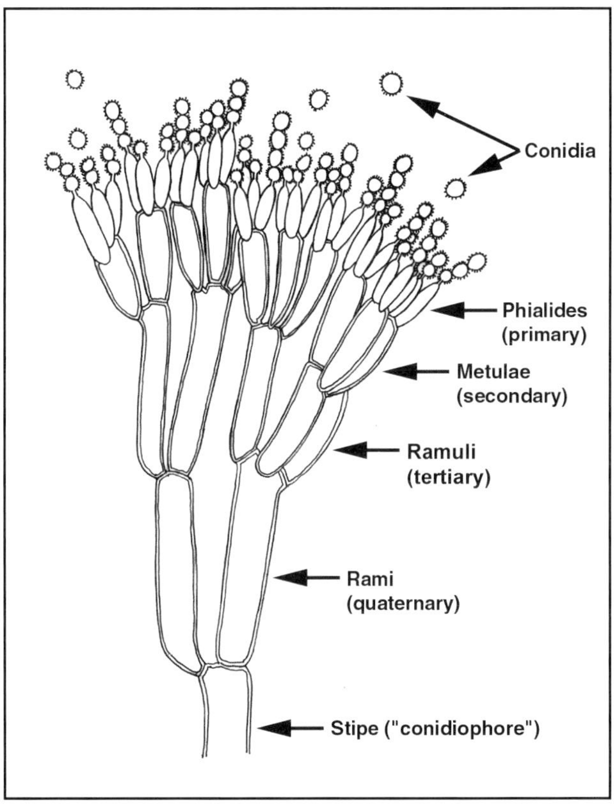

Fig. 10. *Penicillium brevicompactum*. All *Penicillium* species are characterized by the production of long chains of conidia on flask-shaped *phialides* (conidiogenous cells) in a brushlike head termed a *penicillus*. Most species produce a well-defined *stipe* or *conidiophore* (stalk) but differ in the number of series of phialides (1-4) produced. *P. brevicompactum* is a representative of the most complex form of conidiogenesis in the genus and produces four series of phialides: *phialides*, *metulae*, *ramuli*, and rami, from distal to proximal, respectively.

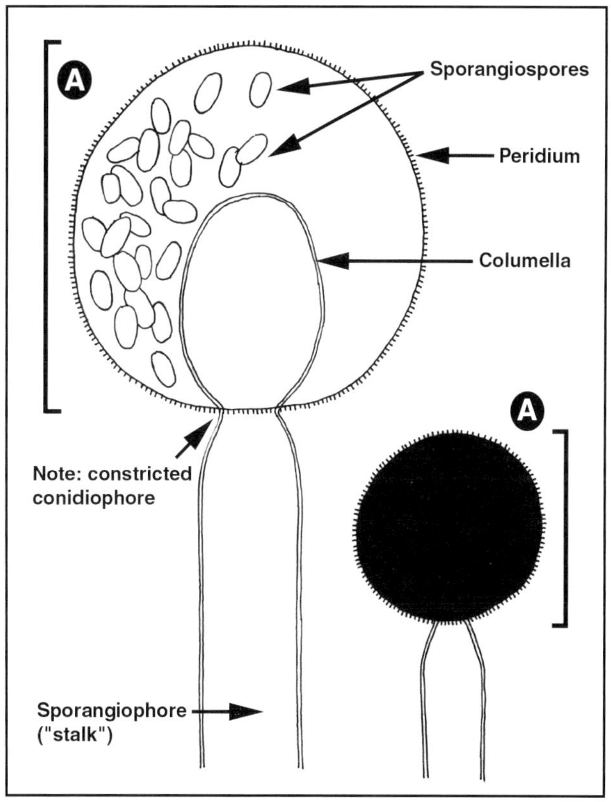

Fig. 11. *Mucor mucedo*, an example of a fungus in the order Mucorales. Members of this group of fungi are characterized by the production of *sporangiospores*, which are asexual spores within a *sporangium* [upper (A), cross section; lower (A), external), which is a saclike structure formed at the apex of a conspicuous sporangiophore. For most *Mucor* species, the sporangiophore is constricted below the sporangium.

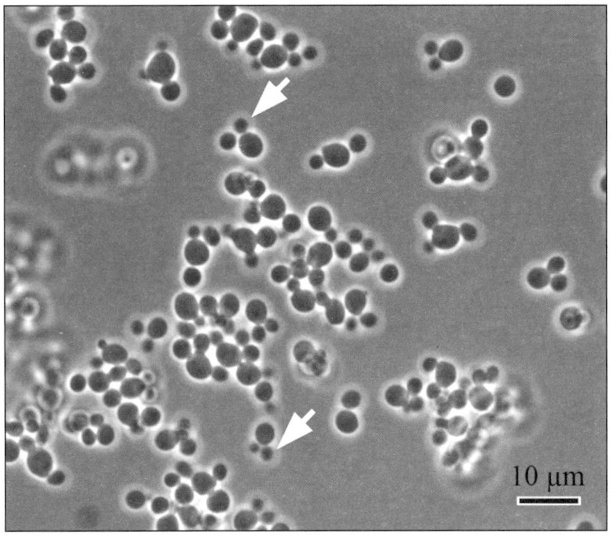

Fig. 12. Photograph of a yeast showing the characteristic *blastic conidiogenesis* (budding). The arrows point to buds or daughter cells that form from the parent cell. This yeast is *Saccharomyces dairiensis*, a commonly encountered contaminant of a meat-based entomophage diet.

2. Yeasts. As mentioned previously, the yeast form is adapted to existence within liquid environments. The yeasts are primarily unicellular and rapidly reproduce asexually by *blastic conidiogenesis* (budding; Fig. 12). The smaller cell is often referred to as the *daughter cell* and the larger the *mother cell*. The presence of blastic conidiogenesis is a character unique to the fungi, and this character is used to rapidly distinguish yeast cells from bacteria, which divide by binary fission. In some instances, yeast cells also can produce sexual spores, and these spores are often formed within the mother cell. Most people are familiar with the Brewer's yeast, *Saccharomyces cerevisiae*. However, many people do not realize the tremendous diversity of fungi that exist as yeasts. For example, Barnett et al. (1990) describe 597 species, many of which contaminate foods.

C. Diversity and Characterization of Microorganisms

There is tremendous diversity within the Bacteria: for example, a ton of soil could contain > 10^6 different taxa (Curtis et al. 2002). Identification is primarily based on physiological and molecular characters, but morphological characters (e.g., endospore formation) also are used on occasion. Basic identification methods are presented within Chap. 9 *Entomopathogens and Insect Rearing*. For detailed information on the identification of the bacteria, the reader is referred to Bergey's manuals (e.g., Holt et al. 1994). Like the bacteria, there is high diversity represented by the fungi: 10^6 species estimated (Hawksworth 2001). In contrast to the bacteria, identification of molds is based primarily on morphological characters, with *conidiogenesis* (the development of conidial structures) the primary character used. Methods used and a key to the entomopathogenic fungi are provided in Chap. 9 *Entomopathogens and Insect Rearing*. Similar to the bacteria, identification of yeasts relies primarily on physiological and molecular characters. For more information on the identification of the fungi, the reader is referred to Raper and Fennell (1965), Barron (1968), Raper et al. (1968), Ellis (1971), Domsch et al. (1980), Von Arx (1981), Pitt (1988), Barnett et al. (1990), and Kurtzman and Fell (2000).

It is important for the reader to realize that microorganisms are ubiquitous and that not all taxa are equal—some are harmful, some are beneficial, and some are benign. Furthermore, microorganisms almost always exist in communities in complex ecosystems resulting in very complex interactions. As a consequence of this complexity, it is essential that the reader understand that a panacea for controlling microbial contaminants does not exist. Efficacious management of contaminants relies on understanding the relationship between microorganisms, the insect, and the rearing conditions utilized.

D. Saprotrophic Microorganisms

Saprotrophic microorganisms degrade nonliving substrates. Both bacteria and fungi produce enzymes that catalyze the breakdown of complex substrates into smaller chemicals that then can be taken up by the microbial cell(s) and utilized for growth and reproduction (Fig. 7). In addition to altering the organic substrate by breaking down complex components, a variety of by-products of metabolism are produced (e.g., metabolic acids). Furthermore, microbial degradation of a substrate is a very difficult task. Not only do the microorganisms have to access the substrate and possess the enzyme systems necessary to degrade it, but they have to compete with other microorganisms. Competitive exclusion prevents two populations from occupying the same ecological niche; that is, they cannot play the same functional role at the same location at the same time (Atlas 1997). When two populations attempt to occupy the same niche, one will succeed and one will fail. Both bacteria and fungi have evolved specialized mechanisms to facilitate the competitive colonization of substrates. One such strategy is to produce large numbers of offspring and to flood the environment (e.g., *r*-selected fungi). If a substrate becomes available, there is a high probability that at least a few of the offspring will access that substrate and rapidly colonize it relatively free of competition in the absence of other microorganisms. However, secondary invaders almost always supplant pioneer colonizers with time so that the composition of the microflora undergoes succession.

Within a substrate, spatial separation does allow microorganisms to escape competitive pressures and permit coexistence of competing species. However, many microorganisms produce *secondary metabolites* (specialized, antagonistic chemicals) that are inhibitory to other organisms (Fig. 7). These chemicals are generally produced in a phase subsequent to growth and are not essential intermediaries of the primary metabolism. Secondary metabolites generally possess an unusual chemical structure. There is an energy cost to producing these materials, and therefore the function of these chemicals is believed to facilitate competitive colonization of substrates. Examples of secondary metabolites include toxins such as aflatoxin B and antibacterial agents such as penicillin. A *mycotoxin* is a nonenzymatic, secondary metabolite produced by a fungus that is injurious to another organism—particularly animals and especially humans poisoned by feeds and food products colonized by the fungus. While these materials can be detrimental to some insects, they can be beneficial to others. For example, many of the antibiotics currently used to manage spoilage microorganisms are produced by microorganisms as a strategy to competitively colonize substrates. The use of antibiotics and microorganisms to manage microbial contaminants of insect diets is discussed below (Sec. VI. B. 4. c).

III. Consequences of Microbial Contamination

The objective of the following discussion is to provide documentation of the consequences of microbial contaminants in insectaries. This involves their impact on the production of quality insects and on the health of insectary workers.

A. Detrimental Effects on Insects

There is no doubt that in some circumstances microorganisms can seriously compromise insect health. Bacterial and fungal contaminants can cause considerable physiological stress to insects in captivity, and they are responsible for chemical modifications to diets that adversely affect insect nutrition, which results in delayed development, decreased survival and emergence, morphological alterations (e.g., to midgut tissues), and biochemical and physiological effects such as changes in amino acids, fatty acids, blood chemistry, metabolism, pheromone production, and susceptibility to insecticides.

1. Organismal. Under certain instances, microorganisms can incite substantial insect mortality in insectaries. For example, 1- to 2-day-old boll weevil, *Anthonomus grandis* (Coleoptera: Curculionidae), larvae are extremely susceptible to *Asp. niger*; and 100% mortality is common if fungal contamination of the diet occurs at this stage of development (Sikorowski et al. 1980). *A. grandis* that ingest diet contaminated with the bacteria *Staphylococcus aureus* produce substantially fewer (≤ 57%) offspring than those consuming an uncontaminated diet (Sikorowski and Lawrence 1994), and the rate of pupation by tobacco budworm, *Heliothis virescens* (Lepidoptera: Noctuidae), larvae ingesting a variety of bacterial contaminants is severely

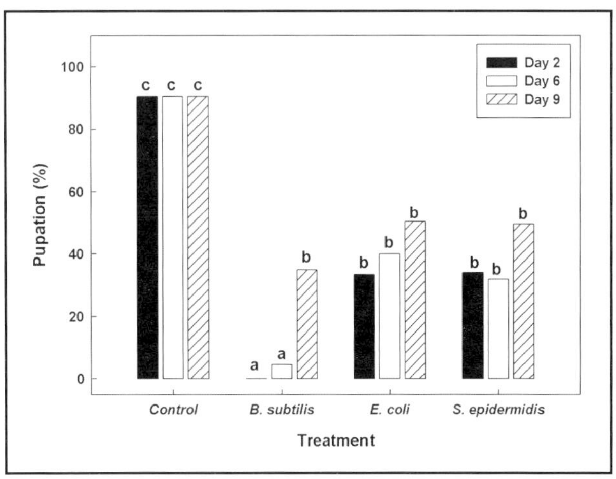

Fig. 13. Effects of ingestion of *Bacillus subtilis*, *Escherichia coli*, and *Streptococcus epidermidis* on pupation success by tobacco budworm, *Heliothis virescens* (Lepidoptera: Noctuidae), larvae. Larvae ingested the bacteria 2, 6, and 9 d after eclosion from the egg. Bars not topped by the same letter are significantly different. [Based on Sikorowski and Thompson (1984).]

affected (Sikorowski and Thompson 1984; Fig. 13). The type of microorganism present influences the impact. For example, *Stenotrophomonas maltophilia* and *Bacillus subtilis*, but not *Leuconostoc mesenteroides* and *Escherichia coli*, caused statistically significant mortality in adults of the parasitoid *Microplitis croceipes* (Hymenoptera: Braconidae) within 5 days (Sikorowski et al. 1992; Fig. 14). However, beyond this period, all bacterial species caused higher mortality than the control treatment.

Although microbial contaminants can incite acute mortality in insects, sublethal effects are a more common outcome. For example, the normal developmental time of *A. grandis* is 13 days from egg to adult, but this period is extended to ≥ 15 days on diets contaminated with some species of bacteria (Sikorowski and Lawrence 1994). *H. virescens* larvae fed diet contaminated with *Bac. subtilis*, *Esc. coli*, and *Streptococcus epidermidis* produce pupae that are 30, 12, and 8% smaller than controls, respectively (Sikorowski and Thompson 1984). These bacteria also reduce egg eclosion by 12–13% in the F_1 generation, and *Bac. subtilis* also greatly reduces adult eclosion. The adult longevity of bacterially contaminated insects is also often affected, and this is of particular significance for predators and parasitoids. For example, *M. croceipes* contaminated with *Stenotrophomonas* and *Bacillus* exhibit reduced rates of host parasitization and higher rates of parasitoid mortality (Sikorowski et al. 1992). The effects of microbial contaminants on insects are complex and influenced not only by the microbial species present but also by the stage and age of the insect at the time of contamination, temperature, pH of the diet, and many other factors.

Relatively limited work has been done to investigate morphological alterations as a consequence of microbial contamination. However, bacterial damage to the midgut can be severe in some circumstances, and damage is greatly influenced by the species of microorganism present (Sikorowski and Thompson 1980, Shapiro 1984, Bong and Sikorowski 1991, Sikorowski and Lawrence 1991, Sikorowski et al. 1992). The epithelial brush border of bacterially contaminated insects is often ulcerated and overgrown by bacteria (McLaughlin and Sikorowski 1978; Sikorowski and Thompson 1979, 1980; MacGown and Sikorowski 1980; Fig. 15).

The presence of microorganisms can influence the susceptibility of insects to insecticides. *A. grandis* contaminated with bacteria are less susceptible to methyl parathion than uncontaminated individuals (Hurej et al. 1982). The effect of contamination on the toxicity of mirex, an organochlorine insecticide, also is influenced by the presence of microorganisms; but the degree of inhibition varies according to the microbial taxon present (Hurej et al. 1982). They speculated that certain

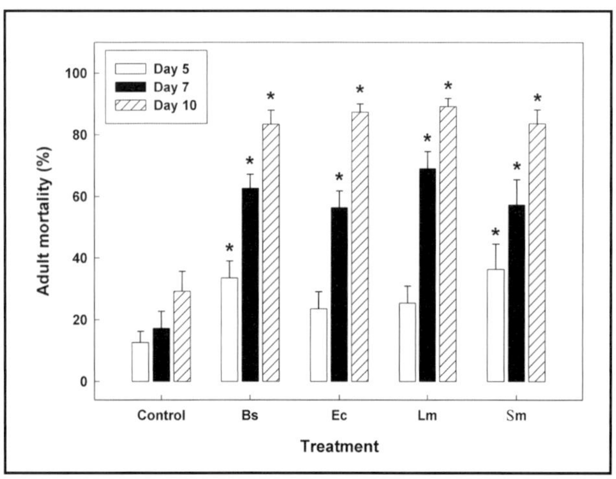

Fig. 14. Longevity of adults of the hymenopteran parasitoid *Microplitis croceipes* (Hymenoptera: Braconidae) ingesting *Bacillus subtilis*, *Escherichia coli*, *Leuconostoc mesenteroides*, and *Stenotrophomonas maltophilia* 5, 7, and 10 d after exposure in honey water relative to control adults. Bars with an "*" are significantly different from the corresponding control treatment. [Data from Sikorowski et al. (1992).]

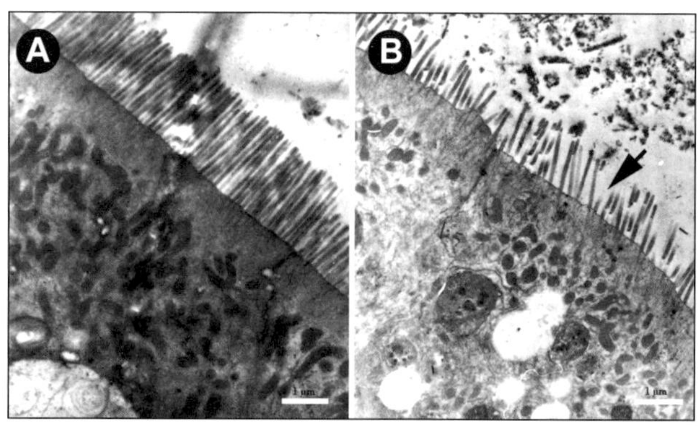

Fig. 15. Transmission electron micrographs of the brush border of the midgut epithelium of boll weevil, *Anthonomus grandis* (Coleoptera: Curculionidae), adults. (A) Healthy epithelium of a young adult. Note the luxuriant brush border. (B) An epithelium of an older adult from which considerable numbers of bacteria were isolated. Note the deteriorated brush border (arrow). Often, ruptures were detected in the epithelium, and ulcerated areas were observed where the brush border had been destroyed. [From McLaughlin and Sikorowski (1978) with permission from the Entomological Society of America.]

populations of bacteria present in the insect digestive system are able to metabolize, and thereby detoxify, insecticides within the living insect.

2. Biochemical and Physiological. Seven-fold more pheromone was isolated from the feces and body homogenates of uncontaminated *A. grandis* than from contaminated weevils (Gueldner et al. 1977, Wiygul and Sikorowski 1986). This pheromone is composed of four components: two alcohols and two aldehydes. The two alcohol components are significantly decreased in *A. grandis* exposed to a toxin from *Sta. aureus* (Wiygul and Sikorowski 1986). The reason for this reduction in pheromone production is not known; however, bacteria are known to produce various toxins that can influence a variety of metabolic processes within insects.

With the exception of tyrosine and glutamine in females and cysteine and tyrosine in males, all of the amino acids were found in greater amounts in the whole bodies of bacteria-free *A. grandis* compared to bacteria-contaminated individuals (Fig. 16; Thompson and Sikorowski 1978). However, it was not determined whether the decrease in amino acids resulted from competition by bacteria for amino acids in the diet or from the inability of insects to absorb and metabolize amino acids and proteins.

All of the fatty acids examined were decreased in the bodies of *A. grandis* contaminated with *Streptococcus* sp., *Kocuria varians*, and *Enterobacter aerogenes* (Thompson et al. 1977; Fig. 17). Comparison of individual *A. grandis* adults from the contaminated and uncontaminated treatments showed that exposure to bacteria resulted in a substantial reduction (76%) in the number of calories represented by total fatty acids (Fig. 18). This is of particular significance to the fecundity of *A. grandis*, because a healthy female must use 25% of her total fatty acids per day in producing eggs (≈ 7 eggs per day) (Thompson et al. 1977). Therefore, reductions in caloric content as a result of exposure to bacterial contaminants could severely decrease fecundity.

In adult *A. grandis* that feed on a diet contaminated with *Sta. aureus*, concentration of uric acid in the blood is substantially higher compared to controls (Thompson and Sikorowski 1982; Fig. 19). Furthermore, a substantially different effect was imparted by the two strains of *Sta. aureus* tested. Uric acid is the end product of protein and purine catabolism (Chapman 1998),

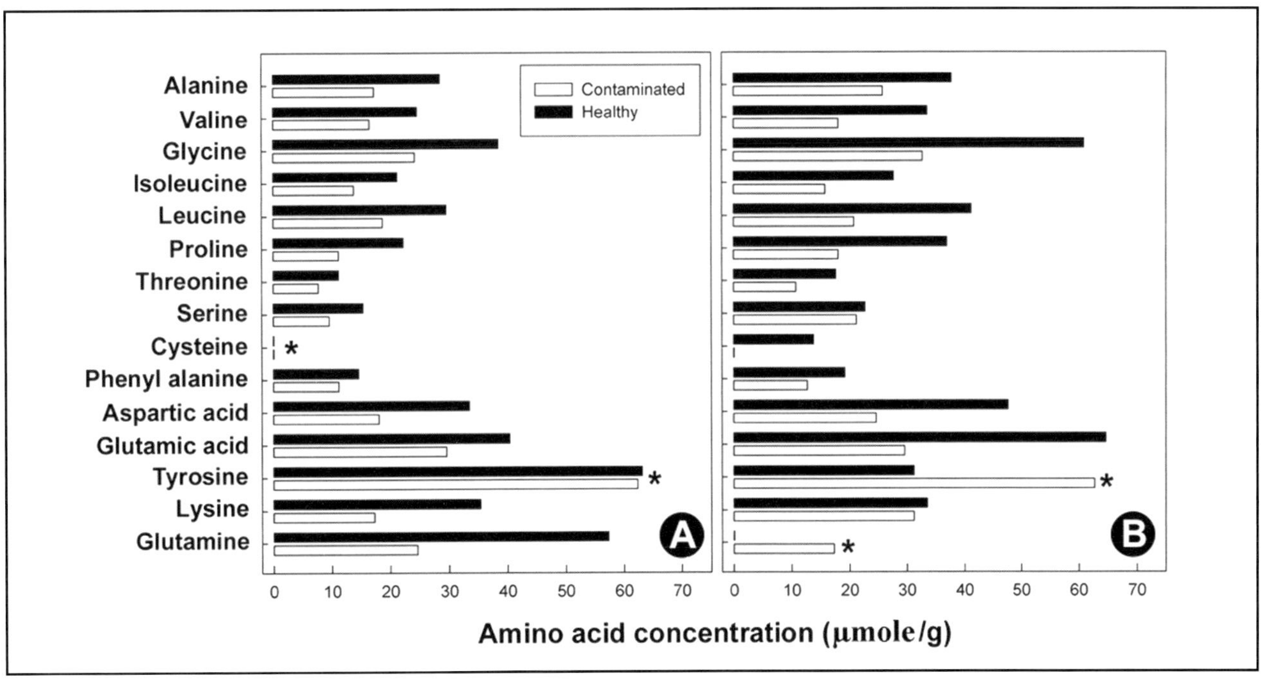

Fig. 16. Effects of simultaneous infection with *Streptococcus* sp., *Kocuria varians*, and *Enterobacter aerogenes* on the amino acids of male (A) and female (B) boll weevil, *Anthonomus grandis* (Coleoptera: Curculionidae), adults. With the exception of tyrosine and glutamine in females and cysteine and tyrosine in males (indicated by an asterisk), all of the amino acids were found in greater amounts in the whole bodies of bacteria-free *A. grandis* compared to bacteria-contaminated weevils. [Based on Thompson and Sikorowski (1978).]

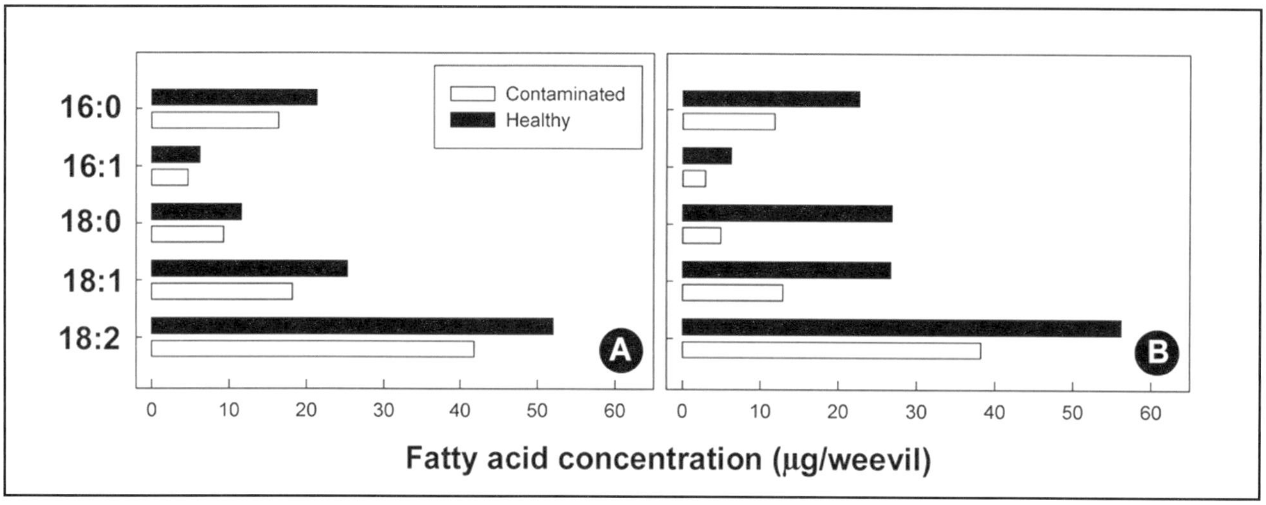

Fig. 17. Effects of simultaneous infection with *Streptococcus* sp., *Kocuria varians*, and *Enterobacter aerogenes* on C_{16}-C_{18} saturated and unsaturated fatty acids of male (A) and female (B) boll weevil, *Anthonomus grandis* (Coleoptera: Curculionidae), adults. Comparisons showed a decrease in total fatty acids of up to 75% in heavily contaminated individuals. [From Sikorowski and Thompson (1980) with permission from the Mississippi Agricultural and Forestry Experiment Station.]

and the high concentrations of uric acid observed in the blood of *A. grandis* exposed to *Sta. aureus* indicate that they were catabolizing proteins and/or purines—indicative of physiological stress. A similar observation was made for *H. virescens* (Sikorowski and Thompson 1984). In addition to elevated concentrations of uric acid, conspicuous differences in the concentrations of chloride, cholesterol, and tryglyceride in the hemolymph of *A. grandis* exposed to *Sta. aureus* were observed.

Given the effects that microbial contaminants can have on various metabolic processes, it is not surprising that differences in oxygen uptake also result. Wiygul and Sikorowski (1981) demonstrated that oxygen uptake by 1-, 5-, or 10-day-old, *Sta. aureus*–contaminated *A. grandis* tended to increase as the degree of contamination increased. In contrast, oxygen consumption tended to decrease when contaminated with *Streptococcus* sp., once again emphasizing the complexity of the interaction. The rates of use of oxygen by the bacteria were not determined in this study.

3. Dietary. As with human foods, biochemical changes caused by contaminating microorganisms would be expected to alter the palatability and nutritional value of insectary diets and, consequently, to affect the vigor of the insects being reared. For more information on the nutritional complexities of insect diets, the reader is referred to Cohen (2003). Surprisingly, very limited research has been conducted on physical and chemical changes to insect diets caused by microbial contaminants. A meat-based entomophage diet developed by Cohen and Smith (1998) is subject to microbial degradation as evidenced by the considerable accumulation of gases within the diet packets as a result of microbial respiration (Fig. 20). In the absence of antibacterial agents in the diet, populations of bacteria increase very rapidly—four orders of magnitude within 24 hours (Inglis and Cohen 2004; Fig. 21). The primary bacteria present in the diet are lactobacilli (i.e., *Lactobacillus* and *Carnobacterium*

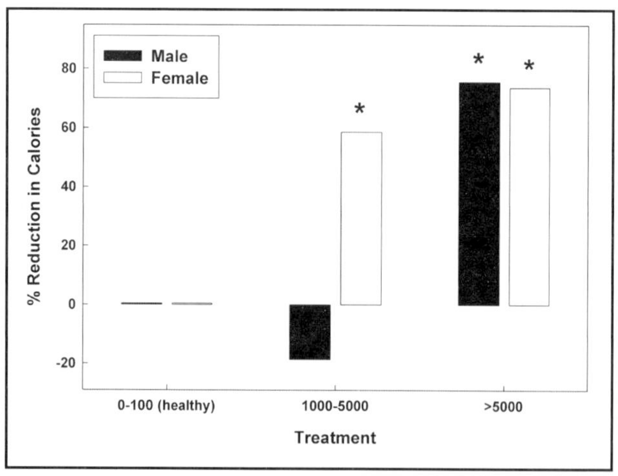

Fig. 18. Reduction in caloric content (calculated from fatty acids) of male and female adult boll weevils, *Anthonomus grandis* (Coleoptera: Curculionidae), exposed to the bacterial contaminants *Streptococcus* sp., *Kocuria varians*, and *Enterobacter aerogenes*. Bars with an "*" are significantly different from the control treatment. [Based on Thompson et al. (1977).]

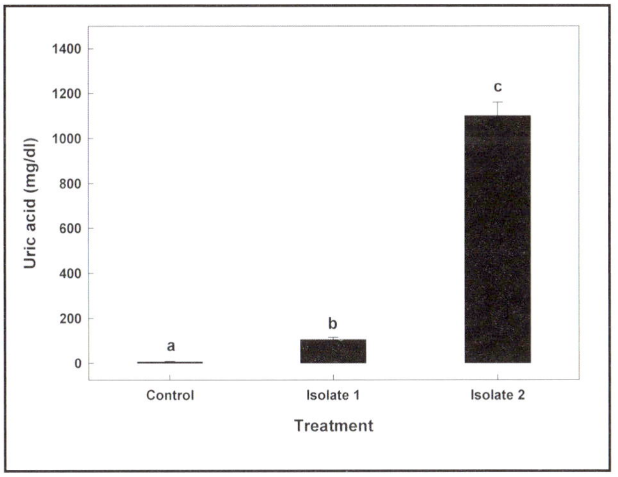

Fig. 19. Concentrations of uric acid in the hemolymph of 1-d-old adult boll weevils, *Anthonomus grandis* (Coleoptera: Curculionidae), that were bacteria-free or contaminated with one of two strains of *Staphylococcus aureus*. "Isolate 1" is American Type Culture Collection (ATCC) strain 8095, and "Isolate 2" is ATCC strain 8096. Bars not accompanied by the same letter are significantly different (P < 0.05). [Based on Thompson and Sikorowski (1982).]

spp.), and the number of lactobacilli present is highly correlated with decreases in pH of the diet. Lactobacilli are *microaerophilic* (able to grow in low oxygen environments), and they produce CO_2, ethanol, lactic acid, and/or acetic acid as the end products of glucose fermentation. Most insects will only accept diets within the relatively narrow pH range of 7 ± 2 pH units (Cohen 2003), and the observed pH of < 4.5 observed for the meat-based entomophage diet spoiled by lactobacilli would be expected to severely limit the palatability of this diet.

4. Stress-related. Overcrowding, poor ventilation, and incomplete diet have been frequently blamed for various maladies of insects in culture including the silkworm, *Bombyx mori* (Lepidoptera: Bombycidae), as early as the eighteenth century (Steinhaus 1975). Because bacteria, as a group, invade organisms with poor vigor (i.e., stressed), knowing the sources of microorganisms associated with insect rearing, the factors affecting their multiplication, and measures that can be used for their control are important to managers of rearing facilities. While contaminants can incite physiological stress in insects directly, exposure to additional stress factors (e.g., high temperature) can exacerbate the adverse effects of microbial contaminants on insect quality. The transfer of insects from a field to a laboratory setting is a stressful event in itself. If the insect is further stressed by bacterial or fungal contaminants, resulting in a microbial-modified diet or the presence of microbial toxins, the consequences can be devastating. A variety of factors, including temperature, wounding, stagnant heat and humidity, and nutrition all influence the effects of microbial contaminants on insect health.

Several workers have attributed increases in insect mortality to the presence of microbial contaminants exacerbated by increases in rearing room temperature (Steinhaus 1953, McLaughlin 1962, Stephens 1962, Bucher and Williams 1967, Howell 1971). High pupal mortality of the longtailed fruit fly parasite, *Opius longicaudatus* (Hymenoptera: Braconidae), a parasitoid of the Caribbean fruit fly, *Anastrepha suspensa* (Diptera: Tephritidae), is attributed to the action of the bacterial contaminants *Serratia marcescens* and *Pseudomonas aeruginosa* as a function of temperature (Greany et al. 1977). These bacteria are able to overwhelm both parasitized and unparasitized *A. suspensa* larvae and pupae when they are subjected to thermal stress of rearing temperatures > 30°C. Silkworms exposed to thermal shock of as low as 30°C often develop an acute and fatal bacteriosis caused by *Bacillus thuringiensis* (Vago 1963).

The effects of wounding on insects are discussed in detail by Day and Oster (1963) and Vago (1963). Most researchers conclude that wounding increases the prevalence of infections by both bacteria and fungi independent of providing a portal for their entry. Stress due to wounding is often observed in insects reared in

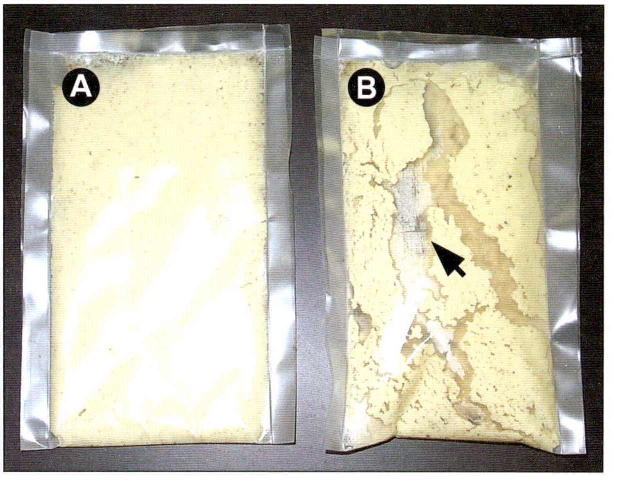

Fig. 20. A meat-based diet presented in Parafilm M™ packets. (A) A freshly prepared packet. (B) A packet that has been incubated at 27°C for ca. 3 d. Note the accumulation of gas within the diet packet (arrow) resulting in a "puffy" appearance. [Image provided by Janet Alverson, USDA-ARS, Washington State University, Pullman, WA.]

aggregations where physical injury is more common. Stress caused by crowding, independent of physical injury, also can contribute to microbial infection. For example, Hensley and Hammond (1968) reported that the incidence of *Asp. niger* contamination increased as the number of sugarcane borer, *Diatraea saccharalis* (Lepidoptera: Crambidae), larvae per rearing container increased. Heat and humidity are often associated with dysentery and the occurrence of bacteria in lepidopterous insects (Vago 1963). In addition, high humidity that condenses as free water on the surface of rearing containers as a result of a rapid change in the holding room temperature (Chap. 11 *Insect Rearing Production Systems, A Case Study: The Southwestern Corn Borer*) may create an environment that is conducive for microbial growth (e.g., *Asp. niger*). Furthermore, changing from natural food to artificial diet may lead to digestive problems, followed by an increase in intestinal bacteria. In general, any deviation from conditions present in the ecosystem in which an insect normally lives can be considered a physiological stressor; and under such stressful conditions, the insect's immune responses will be compromised, increasing its susceptibility to infection and onset of disease.

B. Detrimental Effects on Humans

Fungi such as *Aspergillus* (causative agent of *aspergillosis*), *Penicillium* (causative agent of *penicilliosis*), *Rhizopus* (causative agent of *zygomycosis*), and a variety of yeast and yeast-like organisms (causative agents for a wide variety of diseases) can colonize insect diets and may be a hazard to employees. Such fungi may be capable of infecting humans directly, they may produce secondary metabolites that can be toxic to humans if ingested, or they can act as allergens. Furthermore, fungi (e.g., *Stachybotrys chartarum*) have been implicated as a significant cause of "sick building syndrome" (e.g., Hintikka 2004). There are four classes of *mycoses* (primary infections caused by fungi): 1) *dermatophytoses* (cutaneous mycoses), 2) subcutaneous mycoses, 3) systemic mycoses, and 4) opportunistic mycoses. Fortunately, only opportunistic fungal pathogens represent a threat in most insectaries. Nonetheless, such infections are not to be taken lightly. Species of fungi commonly encountered in insectaries (e.g., *Aspergillus*, *Penicillium, Mucor*, and *Rhizopus*) are capable of infecting humans. For example, *Asp. fumigatus* and *Asp. niger* are able to infect the ears, eyes, and lungs (Beneke et al. 1984). In afflicted individuals, an *aspergilloma* (fungus ball) can form, most commonly in the lungs, and can develop into an invasive systemic infection in susceptible individuals. The use

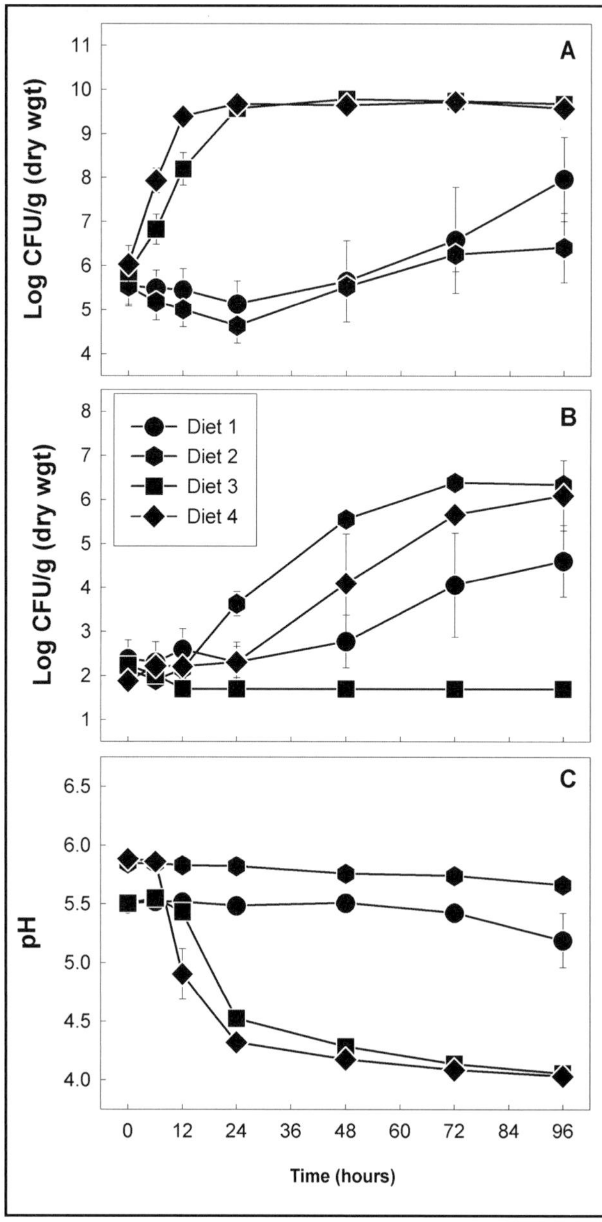

Fig. 21. Changes in pH and microbial populations recovered from a meat-based entomophage diet over a 96-h period. (A) Bacteria isolated on trypticase soy agar. (B) Fungi isolated on Sabouraud's dextrose agar. (C) pH. Diet treatments consisted of the following: (1) diet containing the antibacterial agents chloramphenicol and tetracycline and the antifungal agents sorbic acid and propionic acid, (2) diet containing antibacterial agents but no antifungal agents, (3) diet containing antifungal agents but no antibacterial agents, and (4) diet containing neither antibacterial or antifungal agents. The minimum detection level for both fungi and bacteria was 50 CFU/g, and bars represent SEM (n = 3). SEM bars not visible are within the point symbols. [From Inglis and Cohen (2004) with permission from the Entomological Society of America.]

of antifungal antibiotics such as amphotericin B is the usual treatment for invasive aspergillosis, but the fungus often responds poorly to treatment and rapidly proves fatal (Beneke et al. 1984). Diseases caused by fungi in the order Mucorales (e.g., *Mucor* and *Rhizopus*) are typically referred to as *mucormycoses*. These fungi can invade the nasal sinuses, orbital tissues, central nervous system, pulmonary tissue, cornea, and subcutaneous tissues in susceptible individuals. For more information on human mycoses, the reader is referred to Doctor-Fungus (http://www.doctorfungus.org/mycoses?human/human_index.htm).

Inhalation of airborne fungal propagules can cause a variety of human allergies such as allergic rhinitis or sinusitis, hypersensitivity pneumonitis due to sensitization to fungal spores, and organic dust syndrome caused by inhalation of large quantities of toxin-containing microbial particles (Sigler et al. 1996). About 20% of the human population is sensitized by exposures of $< 10^6$ mold spores/m^3 and are said to be *atopic*. The remainder of the population will develop allergies to molds only after long-term exposure to higher spore concentrations (Kendrick 2000). Such exposures can lead to allergic alveolitis, typified by breathlessness. Repeated exposures can lead to a much more serious allergic disease commonly referred to as *farmer's lung*, which can cause death in severe cases. Quite often, repeated exposure to *Penicillium* and *Aspergillus* spp. can lead to sensitization and produce acute or chronic symptoms. The acute stage is often seen in individuals who are briefly exposed to overwhelming spore numbers. Although they experience chills, fever, and malaise, they do recover. The chronic stage is a much more serious situation. This is typically experienced in workers who are repeatedly exposed to moderate spore levels. Such an affliction causes degenerative changes in the respiratory tract, which can lead to obstructions. This may lead to emphysema and possibly death. Furthermore, bronchial asthma attacks can be triggered by exposure to spores in air.

Bacteria encountered in insectaries also can represent a serious health risk to humans. *Serratia* spp. and *Pseudomonas* spp. causing *septicemia* (blood infections) are encountered with increasing frequency in hospital patients, and some species of *Pseudomonas* are commonly found in wounds, sputum, pus, blood exudate, urine, stool, and biopsy and autopsy tissue samples (e.g., Lindberg 1974).

Table 1. Mutualistic symbioses formed between insects and microorganisms

Insect group	Bacteria	Fungi	Protozoa
Anoplura (sucking lice)	+	-	-
Coleoptera (beetles)	+	+	-
Dictyoptera (cockroaches)	+	-	+
Diptera (flies)	+	-	-
Hemiptera (true bugs)	+	-	-
Homoptera (aphids/hoppers)	+	+	-
Hymenoptera (bees/wasps)	+	+	-
Isoptera (termites)	+	+	+
Mallophaga (chewing lice)	+	-	-
Neuroptera (lacewings)	-	+	-
Thysanoptera (thrips)	+	-	-

C. Beneficial and Benign Effects

As mentioned earlier, microorganisms are not necessarily detrimental to insects. In Chap. 9 *Entomopathogens and Insect Rearing*, we discussed mutualistic symbioses. In some instances, the beneficial relationship between microorganisms and insects is very specific (Table 1). In other circumstances, the relationship is less specific. We have previously mentioned that many microorganisms (e.g., *Lactobacillus* species) produce antimicrobial compounds that can be used to prevent spoilage of foods. In addition, microorganisms naturally occurring within the alimentary canals of animals can provide a degree of protection from pathogens. A large effort is now focused on the development of *prebiotics* and *probiotics* (foods or preparations of microorganisms, respectively, that result in beneficial alterations of gut microflora) to protect humans and other mammals and birds from harmful microorganisms within the gastrointestinal tract. Very limited research in this area has been conducted with insects. The use of broad-spectrum antibacterial agents is a common cause of *iatrogenic* (treatment-induced) disease in humans: the antibacterial agents kill the indigenous microflora of the gastrointestinal tract allowing harmful microorganisms to access and infect tissues. Iatrogenic disease, in conjunction with the rapid development of resistance to antimicrobial compounds, has provided an impetus to develop alternative strategies to manage microorganisms in human and veterinary medicine. Entomologists and rearing specialists should proactively develop similar strategies.

IV. Sources of Microbial Contaminants

The major sources of microbial contaminants in insectary settings are: 1) insects, 2) diets and diet ingredients, 3) inanimate objects including processing equipment and rearing containers, 4) personnel, and 5) air within the insectary. Arguably the two most significant sources of contamination are air and insectary personnel. As indicated previously, fungi are primary decomposers of vegetation and other suitable substrates (e.g., insect diets), and the propagules produced by many of these fungi are adapted for airborne dissemination. Therefore, air within the insectary is the most important source of fungi. Many bacteria also are disseminated in air either directly or indirectly. For example, spores produced by the actinomycetes readily become airborne. However, most other bacteria are only disseminated in air as *aerosols* (suspensions of particles or droplets in a gas; Sec. IV. A). The primary source of most bacteria within insectaries is directly or indirectly from insectary personnel. It is important to emphasize that microbial contaminants affect all aspects of the insectary (Fig. 22). Even if microorganisms are effectively removed from a diet, there is a tremendous potential for subsequent recontamination of the diet. Not all processes are equally at risk or affected by microbial contaminants. The relative importance of contamination from air, personnel, and insects on key aspects of the insect production continuum is illustrated in Fig. 23. Once a particular point in the production process becomes affected by contaminants, it can serve in turn as a source of contamination if appropriate measures are not implemented.

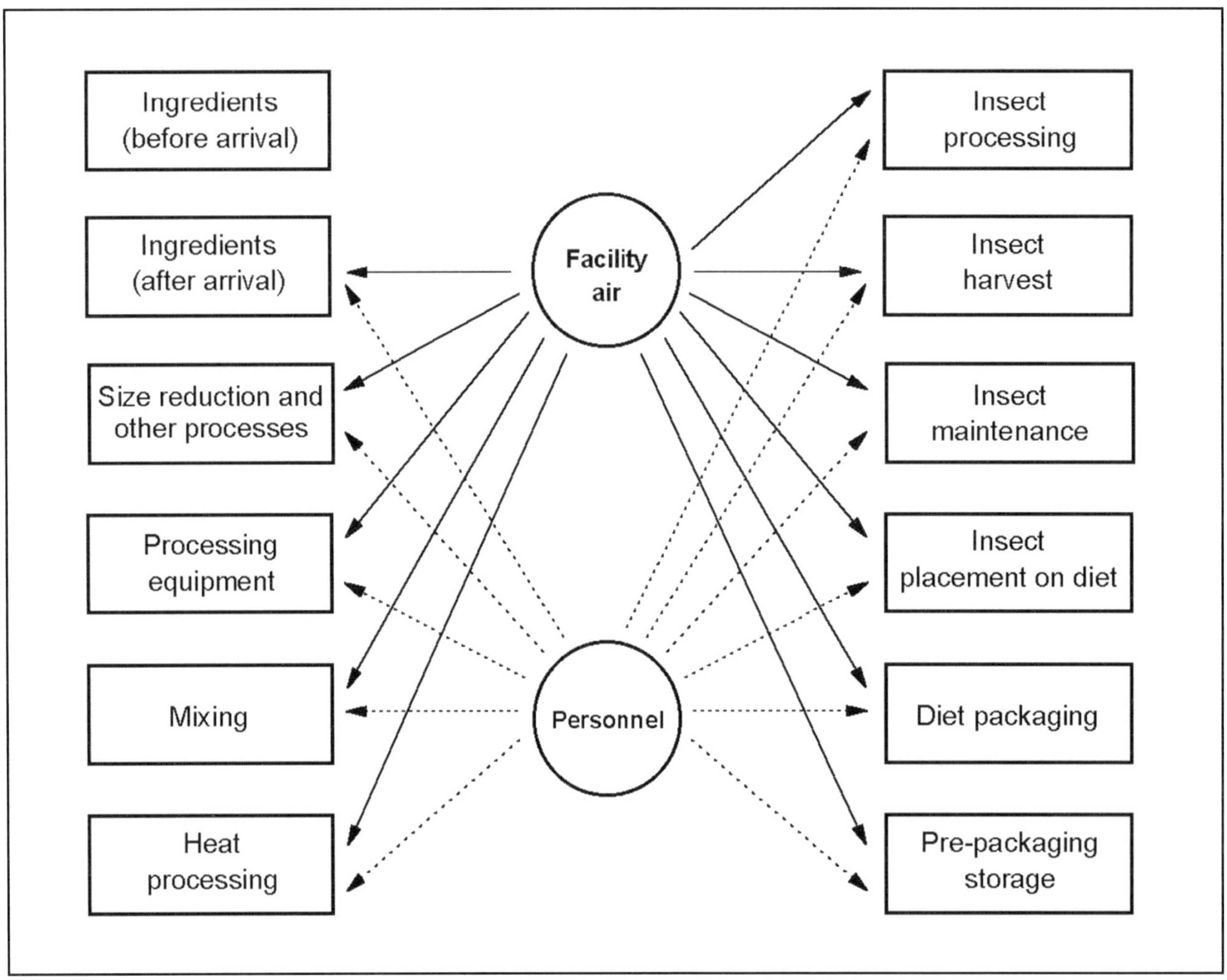

Fig. 22. Rearing processes affected by microbial contaminants transferred in air and by or from insectary personnel.

A. Air

Microorganisms are omnipresent in air, but the composition of microflora in the air varies by geographical location and time of year due to variation in the environment, including vegetation. The majority of bacteria in air occur in association with aerosols. When aerosols are generated from wet suspensions, as in a cough or sneeze or during certain laboratory procedures (e.g., pipetting, blending, or transferring liquids and mixing diets), the air-suspended droplets dehydrate rapidly. Conversely, when aerosols are made from dry or solid materials (e.g., diet ingredients), the air-suspended particles rapidly imbibe water from the environment. *Microbial aerosol* refers to air-suspended particles containing microorganisms. All aerosols carry an electric charge, and the charge of the particles affects their activity and causes them to accumulate on various surfaces including walls and equipment. The magnitude of the electrical charge of an aerosol depends upon the origin and age of the particulate component (Whitfield 1963, Phillips and Runkle 1973).

A cough produces from a few to a dozen aerosol droplets, an individual speaking for a few minutes produces several hundred, and a sneeze produces 10^3–10^4 (Jennison 1942). Another source of airborne contamination is from the normal skin-shedding

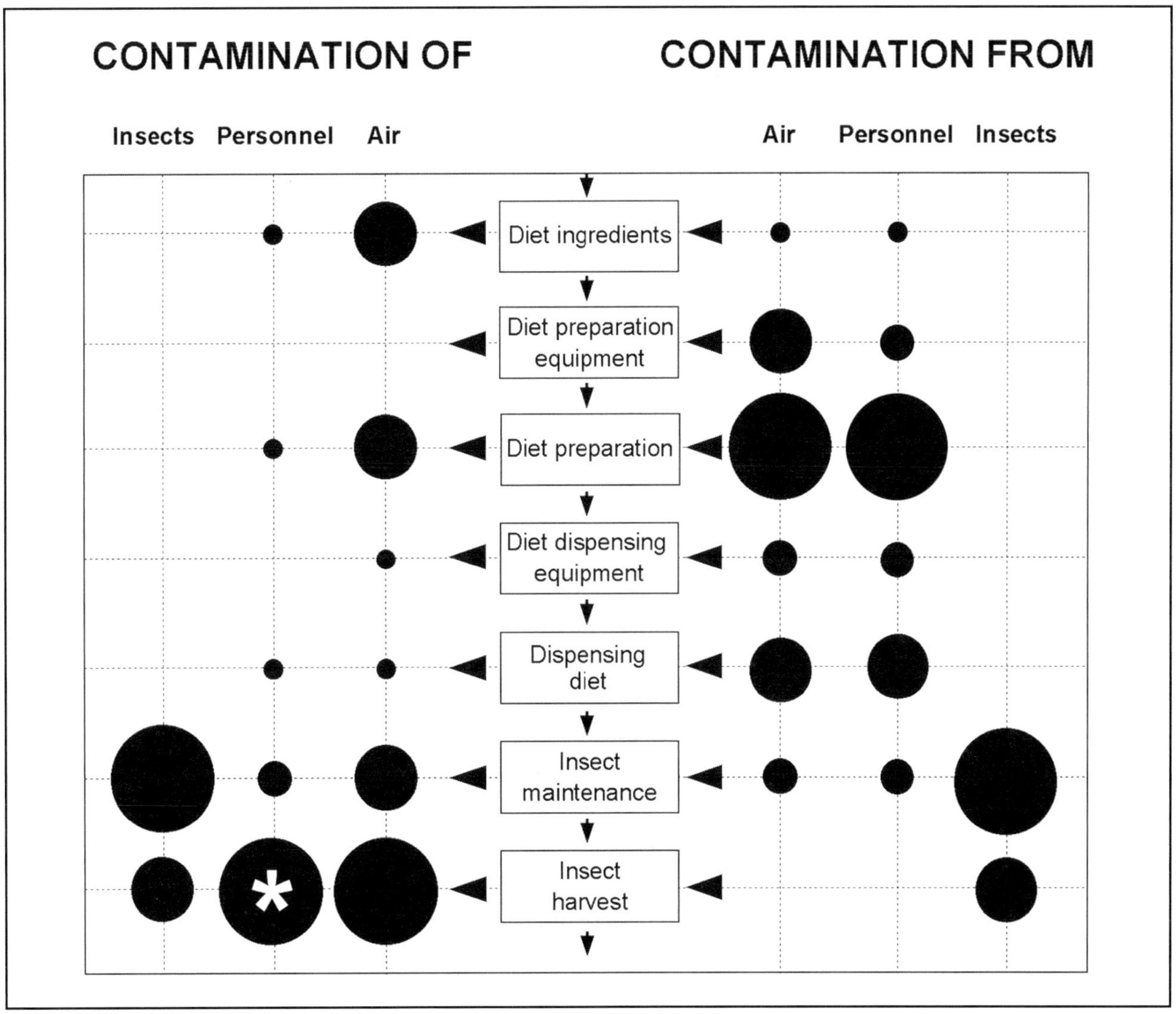

Fig. 23. The relative importance of facility air, personnel, and insects as sources of microbial contaminants in an insect rearing facility. The area of the circle indicates the degree of importance. The circle with an "*" is of paramount importance from a worker health perspective. [See Chap. 4 *Health and Safety Issues in the Rearing of Arthropods* for additional detail.]

process. An average individual with normal skin flora sheds approximately 10^4 bacterial cells each minute (Phillips and Runkle 1973).

Fungi from a variety of sources (e.g., decomposing organic materials) are also a common component of the air's microflora, and the composition of this mycoflora varies temporally. As mentioned previously, the inhalation of fungal propagules can cause a number of adverse reactions in humans. However, because of the complexities associated with assigning human health risks to inhalation of molds, there has not been universal adoption of numerical guidelines for safe levels of exposure. However, the World Health Organization (WHO) of the United Nations has recognized that 150 fungal *CFU* (colony forming units) per m^3 of air is cause for concern, especially if potentially toxic or pathogenic taxa are present (WHO 1988). Few studies have qualitatively or quantitatively examined the mycoflora of air within insectaries. However, in a study in which airborne fungi were quantified in honeybee, *Apis mellifera* (Hymenoptera: Apidae), overwintering and equipment cleaning facilities, spore counts reached 1.4×10^3 CFU/m^3 prior to disturbance by workers and increased 10X while workers swept up debris. Even the former level is higher than what is considered safe for workers (Sigler et al. 1996).

B. Personnel

Humans are carriers of many microorganisms, and Jawetz et al. (1978) divided microorganisms into two groups: 1) the resident microflora consisting of specific types of microorganisms regularly found on a given area of the human body and on persons of a given age and 2) the transient microflora consisting of various microorganisms that inhabit the skin, hair, and mucous membranes sporadically for hours, days, or weeks following exposure to them in the environment. A healthy human body harbors millions of microorganisms, mainly bacteria, on the skin and in the mouth, respiratory tract, genitourinary tract, and intestines (Fig. 24). In the absence of appropriate sanitation measures, bacteria in human fecal material can easily be transferred to diet ingredients, preparatory equipment, or the diet itself. Bacteria also can be liberated from personnel via air. Smith and Bruch (1969) monitored healthy individuals who exercised naked for 30 min and found that such individuals released 2–6 x 10^6 viable microorganisms into the air. The level of contamination in housing space is directly related to activity and density of the persons using them (Favero et al. 1966; Runkle and Phillips 1969). Consequently, bacterial composition of insectary environments and insects may vary from day to day or among areas within the insectary, depending on variation in the personnel. Furthermore, there is tremendous diversity within bacterial taxa associated with humans. For example, the American Type Culture Collection (ATCC) lists > 800 strains of *Esc. coli* representing hundreds of different serotypes (http://www.atcc.org). This bacterium is part of the normal intestinal flora of humans, but only a few strains (e.g., verotoxigenic serotypes such as O157:H7) are pathogenic to humans (Freeman 1979). However, the presence of *Esc. coli* is frequently used as an indicator of fecal contamination, and this taxon is commonly isolated from diet ingredients and prepared diets. The tremendous diversity within human-associated microbial taxa complicates predictions of their impact on reared insects.

C. Insects

1. Feral. The alimentary canal of many feral insects harbors microorganisms, which are usually obtained from food. Feral insects may be grouped based on the presence or absence of microorganisms as follows: 1) insects with a microflora normally residing within the gut or intracellularly in various tissues and 2) insects with a microflora casually present in the gut (i.e., transients) or free of a microflora. The most common, internally harbored microorganisms in insects are bacteria or bacteria-like, mutualistic or commensalistic symbionts found in Blattaria, Isoptera, Homoptera, Hemiptera, Anoplura, Mallophaga, Coleoptera, Hymenoptera, and Diptera (Table 1). Yeast and yeast-like symbionts are found in Homoptera, Coleoptera, and some Neurospora; and protozoan flagellates are common in wood-feeding insects (Steinhaus 1949, Chapman 1998). Most symbionts are intracellular in various parts of the body, but in some insects they are free living in the gut lumen (Chapman 1998).

For insects forming obligate symbiotic relationships with microorganisms, their absence usually has serious consequences on the ability to rear them. The majority of insects currently reared in laboratory set-

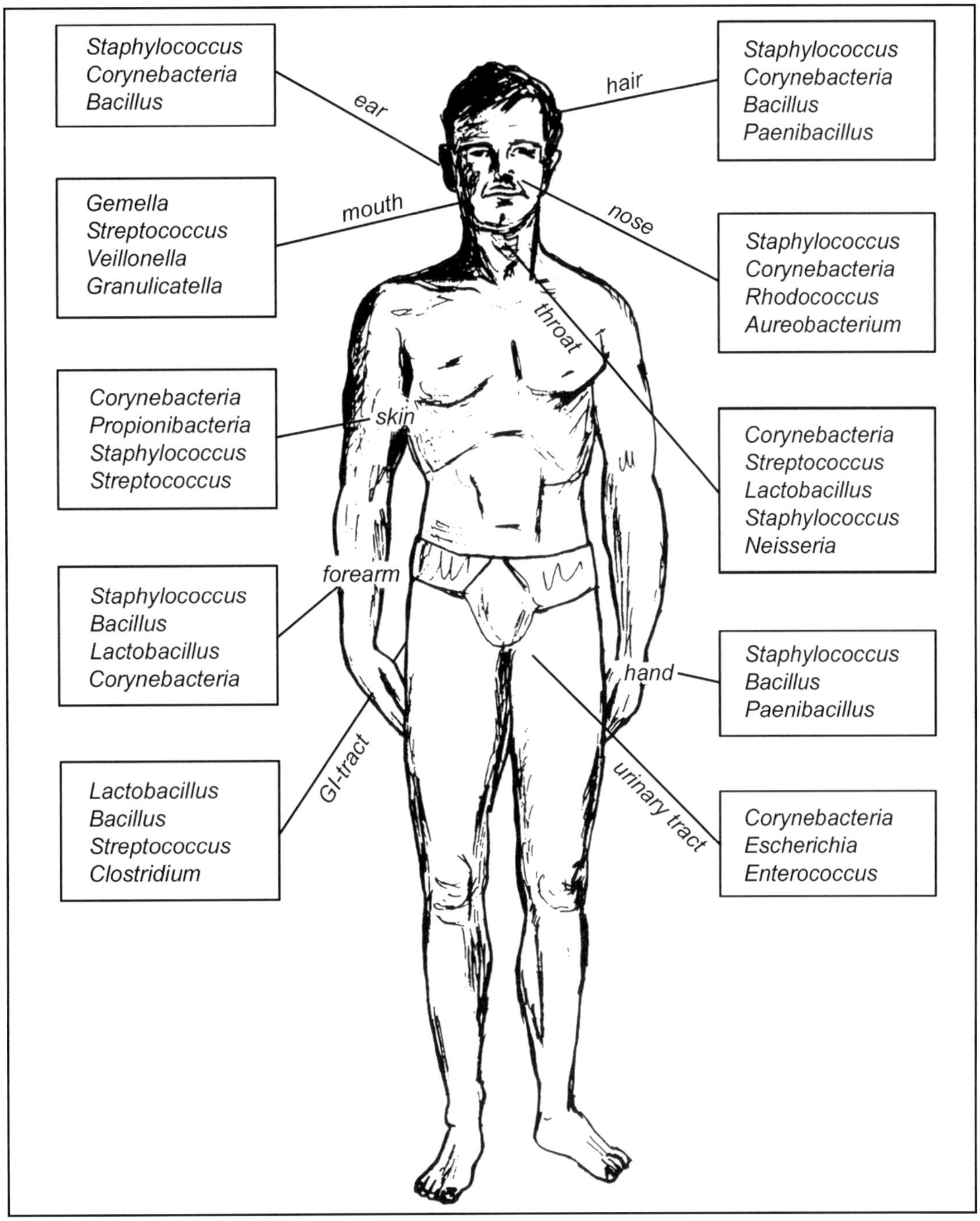

Fig. 24. Prevalent microflora associated with humans. [From Sikorowski and Thompson (1980) with permission from the Mississippi Agricultural and Forestry Experiment Station.]

tings do not require an obligate symbiosis with microorganisms. In such insects, the vast majority of the microorganisms present in the alimentary canal are cosmopolitan taxa, or they are infrequently isolated and considered to be transients. Steinhaus (1941) studied the bacterial flora of 30 species of insects and isolated only 93 taxa of bacteria and two species each of yeasts and filamentous fungi.

In some insects, surprisingly small numbers of microbial cells and consequently a limited diversity of microorganisms are found within their gut. For example, more than 75% of field-collected, adult *A. grandis* of various ages had 100 or fewer bacterial cells per insect (Sikorowski 1984). Reasons for the low populations and limited diversity can be complex. Insects frequently exhibit short residence times for food in the gut, and the peritrophic matrix provides a physical barrier to the gut wall. In addition, antimicrobial compounds released from diet components can severely inhibit microorganisms within the alimentary canal. For example, a number of materials within cotton leaves (e.g., gossypol, caryophyllene, gallic acid, and tannins) are thought to be responsible for the small numbers of bacteria in the gut of *A. grandis* (Hedin et al. 1974). In other insects, the reduced diversity of microorganisms in the alimentary canal has been attributed to competitive exclusion by certain microorganisms. The predominant bacterium present in the gut of greater wax moth, *Galleria mellonella* (Lepidoptera: Pyralidae), larvae is *Enterococcus faecalis*; and this bacterium is able to suppress the growth of other bacterial species ingested with food through the production of antibacterial agents (Jarosz 1979).

Interestingly, recently eclosed adults of some insects possess a very limited microflora. For example, > 70% of house flies, *Musca domestica* (Diptera: Muscidae), and > 50% of stable flies, *Stomoxys calcitrans* (Diptera: Muscidae), emerging from the puparium are devoid of microorganisms, although larvae possess a variety of bacteria (Greenberg 1962). The author concluded that metamorphosis tends to sterilize the digestive tract, and subsequent research indicated that maggots possess autosterilization mechanisms that limit their capacity to carry various bacteria into the adult stage (Greenberg and Klowden 1972).

Although insects often have a small number of bacteria and limited microbial diversity within their guts, a variety of factors (e.g., change in temperature and moisture content of diet) favor the rapid multiplication of certain gut bacteria (Sikorowski 1985, Krieg 1987). For reviews of the literature concerning the microorganisms of healthy insects, the reader is referred to Steinhaus (1947), Brooks (1963), and Tanada and Kaya (1993).

2. Insectary-reared. Insects in insectaries are often contaminated with diverse species of aerobic and anaerobic bacteria and fungi (e.g., Goodwin 1968, Sikorowski 1975, Greany et al. 1977, Habib 1978, McLaughlin and Sikorowski 1978, Smalley and Ourth 1979, Sikorowski and Thompson 1980, Sikorowski et al. 1980). Perhaps one of the major differences between microflora of feral and insectary-reared insects is the composition of the flora. Feral insects may harbor microorganisms that are part of the insect's ecosystem and may represent an ancient association. Insectary-reared insects usually harbor microorganisms that are part of a normal human flora. The utilization of artificial diets has substantially reduced the prevalence of infectious pathogens affecting reared insects, but artificial diets represent a rich substrate for saprotrophic microorganisms. Furthermore, antimicrobial compounds present in natural foods are largely absent in artificial diets. As a result, reared insects are frequently exposed to large numbers of microorganisms that they would not normally encounter in nature. Sanitation is important where diets with a low level of antimicrobials are used (Sec. VI. C).

D. Diets and Diet Ingredients

Many components of artificial diets for insects contain various bacteria and fungi, and these components provide the nutrients essential for growth not only of the insects but also of the microorganisms (McLaughlin and Sikorowski 1978). Common constituents of artificial diets for phytophagous insects, including vitamins, wheat germ, casein, gelling agents, and tap water, are all potential sources of microorganisms (e.g., Ignoffo 1965). Populations of these microorganisms can explode following diet preparation. For example, bacterial populations in one such artificial diet reached 2.9 x 10^8 cells/g at pH 6.5 after 7 days (Gifawesen et al. 1975). The constituents of artificial diets for entomophagous insects are also a source of microbial contaminants. Inglis and Cohen (2004) isolated large populations of bacteria from liver and hamburger and to a lesser extent eggs. The composition of the microflora of the diet ingredients and prepared diet was very complex, including at least 59 taxa of bacteria (Fig. 25) and 25 taxa of fungi (Fig. 26; Inglis and Cohen 2004). In the absence of antibacterial agents, bacterial populations exploded from 10^6 cells/g to 10^9 cells/g with a concomitant decrease in diet pH from 6 to 5 within 24 hours (Fig. 21).

Natural diets such as leaves, fruits, flowers, and pollen possess a substantial microflora, particularly on their surfaces (Fig. 27). Consequently, it is not surprising that corn cob grits and wheat germ, common constituents or adjuncts to artificial diets, are typically contaminated with fungi and bacteria (Fig. 28). Microorganisms can be found within plant tis-

Taxon	Eggs	Liver	GB	Diet	Taxon	Eggs	Liver	GB	Diet
Acetobacter sp.	○	❶	○	○	*Lactococcus garvieae*	○	❷	❶	❷
Acinetobacter iwoffii	○	❶	○	❷	*Lactococcus lactis*	○	○	❶	❶
Acinetobacter sp.	○	○	○	❶	*Lactococcus plantarum*	○	○	❶	○
Ancyclobacter sp.	○	○	○	❶	*Leuconostoc mesenteroides*	○	❷	❶	❸
Arthrobacter agilis	○	○	○	❶	*Leuteococcus japonicus*	○	○	○	❶
Arthrobacter atrocyaneus	○	❶	❶	❶	*Microbacterium liquefaciens*	○	❶	○	○
Arthrobacter ilicis	○	○	❶	❶	*Microbacterium terregens*	○	❶	○	❶
Arthrobacter sp.	○	❶	○	❶	*Microbacterium* sp.	❶	○	○	❶
Bacillus cereus	○	○	❶	❶	*Micrococcus luteus*	○	❷	○	❶
Bacillus megaterium	○	❶	❶	❷	*Moraxella* sp.	○	❶	❶	❶
Bacillus sphaericus	❶	❶	❶	❶	*Neisseria* sp.	○	❶	○	○
Bacillus subtilis	○	○	○	❶	*Ochrobactrum grignonense*	○	○	○	❶
Brevibacterium sp.	○	○	○	❶	*Paenibacillus polymyxa*	○	○	○	❶
Brochothrix thermosphacta	○	○	❶	○	*Paracoccus* sp.	❶	○	○	○
Carnobacterium divergens	○	❸	❸	❸	*Pediococcus penotsaceus*	○	❶	○	○
Carnobacterium piscicola	○	❸	❷	❸	*Planococcus okeanokoites*	○	○	○	❶
Carnobacterium sp.	○	○	❶	❸	*Pseudomonas fragi*	○	○	❶	❶
Citrobacter sp.	○	○	○	❶	*Pseudmonas putida*	○	○	❷	❷
Cornyebacterium callunae	○	○	○	❶	*Pseudomonas* sp.	○	○	❶	❷
Corynebacterium sp.	○	○	○	❶	*Rhodococcus equi*	❶	○	○	❶
Curtobacterium albidum[c]	○	○	❶	○	*Rhodococcus erythropolis*	○	○	○	❶
Enterobacter cloacae	○	○	○	❶	*Rhodococcus* sp.	○	❶	○	○
Enterobacter intermedius	○	○	○	❶	*Sphingobacterium multivorum*	❶	○	○	○
Enterococcus faecalis	○	❶	○	❶	*Staphylococcus aureus*	○	❶	○	○
Enterococcus faecium	○	❶	○	○	*Staphylococcus epidermidis*	❸	○	❶	❶
Enterococcus hirae	○	○	○	❶	*Staphylococcus hominis*	❶	○	○	○
Enterococcus sulfureus	○	○	○	❷	*Staphylococcus pasteuri*	❷	○	○	○
Enterococcus sp.	○	❷	❷	❸	*Staphylococcus sciuri*	○	○	❶	○
Klebsiella terrigena	○	○	○	❶	*Staphylococcus warneri*	❶	○	○	○
Kocuria kristinae	○	○	○	❶	*Stenotrophomonas maltophilia*	○	○	○	❶
Kocuria varians	❶	○	❶	❷	*Streptococcus pyogenes*	○	❶	❶	❷
Kurthia gibsonii	○	○	○	❶	*Terrabacter* sp.	○	❶	○	○
Lactobacillus curvatus	○	❶	❶	❸	*Yersinia intermedia*	○	○	○	❶
Lactobacillus sakei	❶	❷	❷	❸	Unidentifed	❷	❸	❷	❸

[a]Empty circles indicate that the corresponding bacterium was not isolated.

[b]Dark circles indicate positive isolation, and the number within the symbol indicates the number of replicates from which the taxon was isolated.

[c]Identifications indefinite.

Fig. 25. Bacteria isolated from a meat-based entomophage diet and some of its ingredients: chicken eggs, beef liver, and ground beef (GB). The experiment consisted of three replicates conducted on separate occasions. Additional ingredients of the diet are sucrose, Brewers yeast, honey, acetic acid, and tap water. [From Inglis and Cohen (2004) with permission from the Entomological Society of America.]

Group/Taxon	Eggs	Liver	GB	Diet
Filamentous fungi				
Aspergillus ustus	○[a]	○	○	❶[b]
Cunninghamella elegans	○	❶	○	❶
Fusarium sp.	❶	❶	❶	❶
Mucor hiemalis	○	○	○	❶
Mucor sp.	○	○	○	❶
Penicillium citrinum	○	❷	❷	❸
Penicillium janthinellum	○	❶	○	○
Penicillium oxalicum	○	○	○	❶
Penicillium simplicissimum	○	❶	○	○
Penicillium sp.	○	❶	○	❸
Trichoderma sp.	○	○	○	❶
Unidentified	○	❶	○	❶
Yeasts				
Candida albicans	○	❶	○	○
Candida glabrata	○	○	○	❷
Candida parapsilosis	○	○	○	❷
Candida pararugosa	○	○	❶	❸
Candida zeylanoides	○	○	❸	❸
Cryptococcus curvatus	○	○	❷	❸
Cryptococcus marinus	○	○	❶	❷
Leucosporidium scottii[c]	○	○	○	❶
Pichia pastoris[c]	○	○	○	❶
Rhodotorula acheniorum[c]	○	○	❶	❷
Rhodotorula minuta	○	○	❶	○
Saccharomyces dairensis	○	❶	○	❶
Schizoblastosporon starkeyi-henricii[c]	○	❶	○	○
Torulaspora globosa	○	○	○	❸
Trichosporon brassicae[c]	○	○	○	❶
Yarrowia lipolytica	○	❷	❷	❸
Unidentified	❶	❶	❷	❸

[a]Empty circles indicate that the corresponding fungus was not isolated.

[b]Dark circles indicate positive isolation, and the number within the symbol indicates the number of replicates from which the taxon was isolated.

[c]Identification indefinite.

Fig. 26. Fungi isolated from a meat-based entomophage diet and some of its ingredients: chicken eggs, beef liver, and ground beef (GB). The experiment consisted of three replicates conducted on separate occasions. Additional ingredients of the diet are sucrose, Brewers yeast, honey, acetic acid, and tap water. [From Inglis and Cohen (2004) with permission from the Entomological Society of America.]

sues, and many plants possess such *endophytic* microorganisms (e.g., Saikkonen et al. 1998). Also, when plant tissues senesce, a variety of microorganisms gain entry to internal tissues. When plant substrates are placed in containers under conditions of suitable moisture (e.g., provided by a moist paper towel), they are rapidly colonized by saprotrophic microorganisms. It should be realized that paper towels also are frequently contaminated with bacteria such as *Ser. marcescens* and with fungi such as *Asp. niger* (Sikorowski and Lawrence 1991; Fig. 29). Surface sanitation is not efficacious against microorganisms within plant tissues. Furthermore, even after plant materials have been surface-sanitized, microorganisms are readily reintroduced from insects (e.g., from feces); and this is often exacerbated through the action of insect feeding. Microbial contaminants from plant materials can easily contaminate an insectary with a large number of undesirable microorganisms including *Asp. niger* and *Aspergillus falvus*, which grow on almost any type of organic matter.

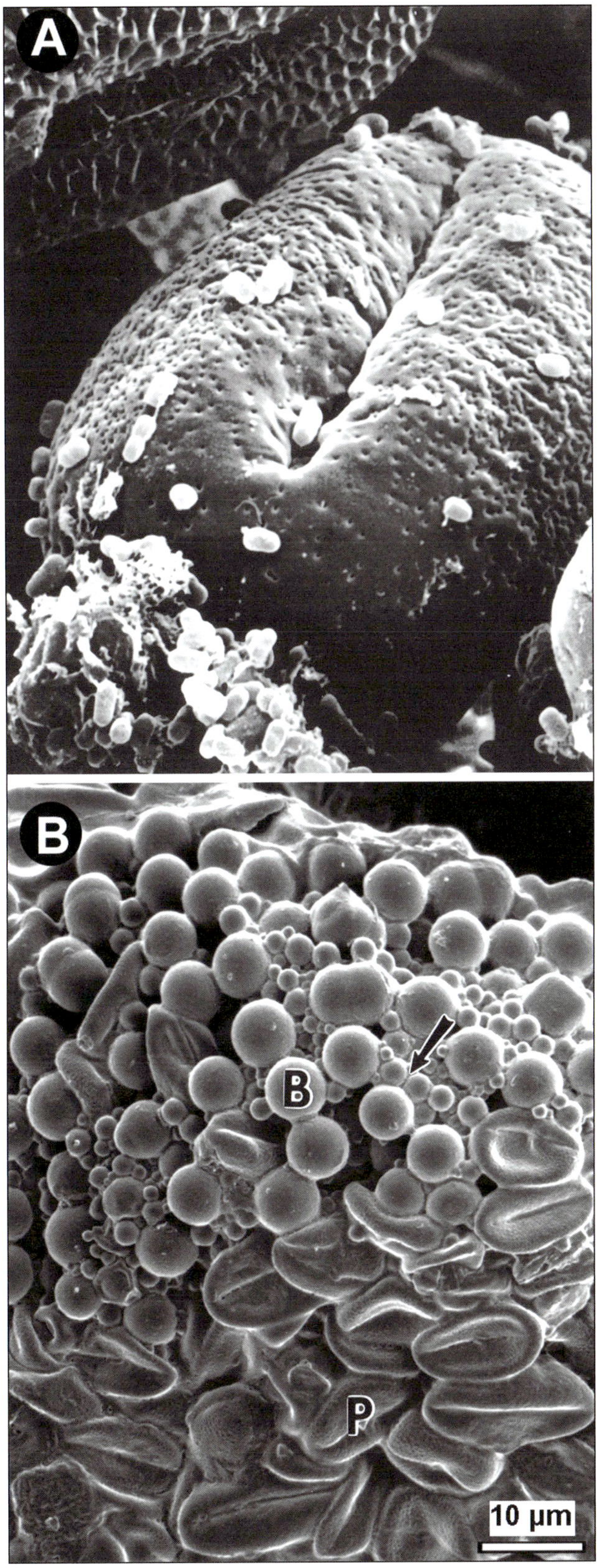

Fig. 27. Microorganisms associated with the alimentary canal of alfalfa leafcutter bee, *Megachile rotundata* (Hymenoptera: Megachilidae), adults. (A) Bacteria associated with a pollen grain. (B) The black yeast *Trichosporonoides megachiliensis*, pollen grains (P), blastic cells (B), and blastic cells produced in chains (arrow).

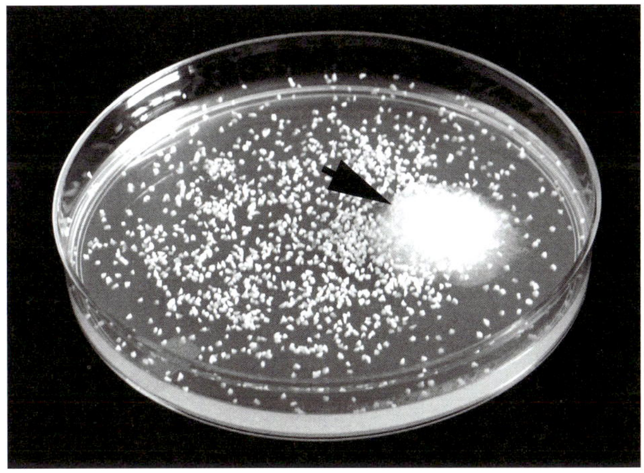

Fig. 28. Corn cob grits sprinkled on a mycological solid medium and incubated at 25°C for 3 d. Note the white mycelial growth (arrow).

Fig. 29. Pieces of paper toweling were treated and incubated to demonstrate microbial growth under various conditions. (A) Unautoclaved toweling on bacteriological medium. (B) Autoclaved toweling on bacteriological medium. (C) Unautoclaved toweling on mycological medium. (D) Autoclaved toweling on mycological medium. Bacterial cultures were incubated at 30°C for 3 d, and fungal cultures were incubated at 25°C for 5 d. Note the luxuriant growth of bacteria (arrow) and fungi from unautoclaved paper towel pieces.

V. Monitoring for Microbial Contamination

Routine monitoring is essential to properly manage microbial contamination. Air, diets and associated materials, insects, workrooms, equipment, and personnel should all be considered as potential sources of contamination. Monitoring microbial contamination is an *SOP* (standard operating procedure) for all food and many other industries, yet it has not been extensively used in most insect rearing applications. Monitoring microbial contamination primarily involves the use of basic microbiological techniques, although more specialized equipment may be used in some circumstances. Many of these techniques are relatively simple and inexpensive to implement, and rearing personnel should be educated to make use of them.

A. Air

Air sampling is an excellent indicator of microbial contamination in a rearing facility and should be conducted as an SOP. Air sampling quantifies the number of particles per m^3 of air, and typical total airborne particulate counts in occupied areas range from several hundred thousand to many million particles per m^3 of air depending on the type of room and activity within that room (Phillips and Runkle 1973). Development of an effective sampling program should include consideration of the following: 1) number and movement of personnel, 2) potential levels of contamination, 3) airflow patterns around equipment and other objects, 4) time of sampling, 5) effects of gravity, and 6) effects of electrostatic forces. For better determination of airflow patterns such as turbulence, dead space, and areas of back streaming, smoke methods should be used (Phillips and Runkle 1973). As mentioned previously, the composition and numbers of fungi in air are influenced by season and to a lesser extent by environmental conditions within the facility. Because humans are the primary source of bacteria, the greatest number of bacteria in the air normally occurs during peak human activity; and the lowest number occurs in the morning before the beginning of the workday. There are two basic sampling strategies that are used to assess microorganisms in the air: 1) sampling for *microbial propagules* (particles containing one or more viable microorganisms), and 2) sampling to determine the frequency distribution of particles as a function of particle size. Here we describe the most common methods used for each strategy and their relative merits and disadvantages.

1. Propagule Sampling. *a. Sedimentation Plates.* The simplest method of air sampling is the sedimentation plate method (also referred to as the *settle plate* method). For this method, gravity causes microbial propagules in the air to be deposited onto a suitable solid medium, usually within a Petri dish. The Petri dish is situated on an appropriate surface, and the lid is removed for a defined period of time. The lid is then replaced, the culture is incubated at an appropriate temperature, and colonies are enumerated after 2 to 5 days. This method permits determination of the numbers of bacteria or fungi or both that settle per unit area in a given time.

The advantages of the sedimentation plate method are its low cost, simplicity, and the fact that the count is based on particles with viable microorganisms only. In spite of these advantages, we do not advocate its use because of a number of limitations. These include the inability to provide accurate quantitative counts, the long waiting period for settling of small particles, and a bias towards larger propagules, which have higher sedimentation rates. Small-spored genera such as *Aspergillus* and *Penicillium* are always underestimated using sedimentation plates. Despite these limitations, sedimentation plates are still widely used, especially in preliminary assessments.

b. Vacuum Samplers. There are two constraints on all viable particle samples: 1) the necessity of separation of the particles from the air and 2) maintenance of the viability of the particles during the collection procedure. Vacuum samplers are divided into the following three types: 1) inertial samplers that direct a stream of air onto a solid medium; 2) samplers that trap propagules from an air stream in a viscous fluid, which can then be cultured on a solid culture medium; and 3) filter samplers that trap propagules on a membrane filter, which is subsequently cultured on a solid medium.

Inertial samplers include Andersen-type samplers, slit samplers, and centrifugal impactors. In most studies, the Andersen and slit samplers have been shown to be the most precise (e.g., Smid et al. 1989). Andersen "Impactor" series samplers are the industrial standard and have been used to quantify airborne microorganisms for over 35 years (http://www.thermo.com/). Several models are currently marketed by Andersen Instruments Inc,

Smyrna, GA, and have been used in many air quality studies (e.g., domiciles, pharmaceutical production facilities, animal care laboratories, wastewater treatment plants, and filter and clean room efficiency studies).

All inertial samplers work on the same basic principle. Air is drawn through an orifice at the top of the sampler, accelerating the air (Fig. 30). The accelerated air containing microbial propagules is then forced through each of 400 holes in a separation plate. A vacuum pump is used to draw air through the sampler, and the flow rate is set at 28.3 L/min (1 ft^3/min). The plate separates the air into equal aliquots and directs the air at an agar medium in a Petri dish situated below it. Propagules then strike the medium and are deposited there. In multiple-stage models, up to six plates occur in a stack. Each of the plates contains 400 equal-sized holes, but the hole size is smaller in each successive stage. Consequently, the velocity of the air is uniform among holes within a stage but increases in each succeeding stage. When the velocity imparted to a particle is sufficiently great, its inertia will overcome its aerodynamic drag, and the particle will strike the surface. Thus, each stage collects smaller particles than the preceding one. For a six-stage sampler, the size range of particles collected on the Petri dish under successive separation plates is as follows: > 7.1 μm, 4.7–7.1 μm, 3.3–4.7 μm, 2.1–3.3 μm, 1.1–2.1 μm, and 0.65–1.1 μm. Particles are deposited on the agar medium situated below the separation plate for a defined period of time. The Petri dishes containing a solid medium with or without antibacterial or antifungal agents are removed from the instrument, incubated at an appropriate temperature, colonies counted, representative colonies aseptically collected, and the microorganisms characterized, if desired. The number of CFU per unit volume of air is then calculated for each size category as a function of time and flow rate. However, as the number of particles impinging on the agar medium rises, the probability that two particles will go through the same hole and produce a single colony increases. Therefore, a statistical correction should be applied based on observed colony counts to obtain *P* (the statistically corrected propagule count) (Table 2).

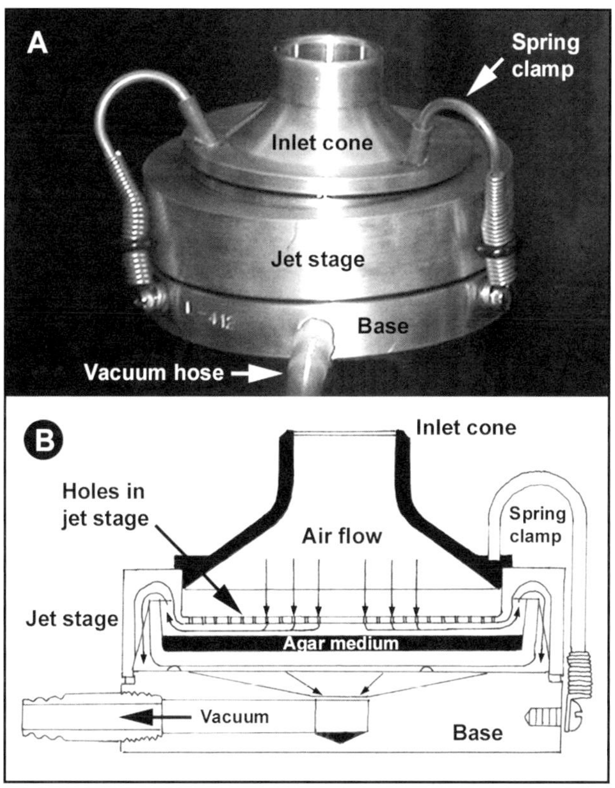

Fig. 30. An Anderson, single-stage inertial air sampler. (A) External view of the sampler showing the aluminum construction. (B) Cross-section drawing of the sampler showing the components including a vacuum hose; a base upon which the Petri dish is placed; the jet stage, which contains 400 precision-drilled holes; and the inlet cone through which air is drawn. The base and the inlet cone are held together by three spring clamps and sealed with an O-ring gasket.

Fig. 31A shows the results of sampling a rearing room at Mississippi State University with a single-stage Andersen air sampler. Twenty-six fungal colonies developed on a general purpose, mycological medium containing antibacterial agents that was exposed for 5 min [*SP* (exposure period) = 5 min] at an airflow rate of 28.3 L/min [*F* (airflow rate) = 28.3 L/min) (1 x 10^{-3} m^3/L) = 2.83 x 10^{-2} m^3/min)] following incubation at 25°C for 4 days. Using Table 2, a statistically corrected total of 27 propagules was indicated (*P* = 27 propagules). To calculate *AD* (the aerial density of microorganisms), the following equation is applied as indicated:

$$AD = P/(F \cdot SP) \qquad \text{Eq. 1}$$

$$AD = [27 \text{ propagules}/[(2.83 \times 10^{-2} \text{ m}^3/\text{min}) \cdot (5 \text{ min})] = 191 \text{ propagules/m}^3$$

The 2-min and 1-min samples taken in the same rearing room indicated the presence of 247 and 353 viable propagules/m^3 of air, respectively. This variability (1.3 to 1.8 times greater than indicated by the 5-min sample) illustrates the need to conduct multiple samples, in multiple locations, at multiple times within an insectary. Also of note, the above example was of air in a rearing room in which the air was filtered. The significance of this will be discussed (Sec. VI. A).

The single-stage Andersen or equivalent air samplers can be used with excellent results. The advantages of the inertial samplers include: 1) quantification of propagules per unit volume of air, 2) quantification of the size distribution of propagules, 3) collection of very small propagules, 4) ease of operation, 5) ease of calibration, 6) versatility in the media that can be used, 7) ease of sanitation using wipes, and 8) high reproducibility of results. The primary disadvantages of the Andersen samplers are cost, and some evidence suggests that particles can be broken, resulting in inaccurate particle separation. Although, the single stage sampler does not provide a size distribution and is one-sixth the cost of the six-stage sampler, it is still very expensive relative to sedimentation sampling. The Andersen air samplers operate with an external vacuum, which must be calibrated periodically. Inertial samplers of a similar de-

Table 2. Positive-hole correction table for a 400-hole Andersen-type inertial air sampler (Feller 1950)

r[a]	P[b]	r	P	r	P	r	P	r	P	r	P	r	P	r	P
1	**1**	51	**55**	101	**116**	151	**190**	201	**279**	251	**395**	301	**559**	351	**840**
2	**2**	52	**56**	102	**118**	152	**191**	202	**281**	252	**398**	302	**563**	352	**848**
3	**3**	53	**57**	103	**119**	153	**193**	203	**283**	253	**400**	303	**567**	353	**857**
4	**4**	54	**58**	104	**120**	154	**194**	204	**285**	254	**403**	304	**571**	354	**865**
5	**5**	55	**59**	105	**122**	155	**196**	205	**287**	255	**406**	305	**575**	355	**874**
6	**6**	56	**60**	106	**123**	156	**198**	206	**289**	256	**409**	306	**579**	356	**883**
7	**7**	57	**61**	107	**125**	157	**199**	207	**292**	257	**411**	307	**584**	357	**892**
8	**8**	58	**63**	108	**126**	158	**201**	208	**294**	258	**414**	308	**588**	358	**902**
9	**9**	59	**64**	109	**127**	159	**203**	209	**296**	259	**417**	309	**592**	359	**911**
10	**10**	60	**65**	110	**129**	160	**204**	210	**298**	260	**420**	310	**597**	360	**921**
11	**11**	61	**66**	111	**130**	161	**206**	211	**300**	261	**423**	311	**601**	361	**931**
12	**12**	62	**67**	112	**131**	162	**208**	212	**302**	262	**426**	312	**606**	362	**942**
13	**13**	63	**69**	113	**133**	163	**209**	213	**304**	263	**429**	313	**610**	363	**952**
14	**14**	64	**70**	114	**134**	164	**211**	214	**306**	264	**432**	314	**615**	364	**963**
15	**15**	65	**71**	115	**136**	165	**213**	215	**308**	265	**434**	315	**620**	365	**974**
16	**16**	66	**72**	116	**138**	166	**214**	216	**311**	266	**437**	316	**624**	366	**986**
17	**17**	67	**73**	117	**138**	167	**216**	217	**313**	267	**440**	317	**629**	367	**998**
18	**18**	68	**75**	118	**140**	168	**218**	218	**315**	268	**443**	318	**634**	368	**1010**
19	**19**	69	**76**	119	**141**	169	**220**	219	**317**	269	**447**	319	**639**	369	**1023**
20	**21**	70	**77**	120	**143**	170	**221**	220	**319**	270	**450**	320	**644**	370	**1036**
21	**22**	71	**78**	121	**144**	171	**223**	221	**322**	271	**453**	321	**649**	371	**1050**
22	**23**	72	**79**	122	**146**	172	**225**	222	**324**	272	**456**	322	**654**	372	**1064**
23	**24**	73	**81**	123	**147**	173	**227**	223	**326**	273	**459**	323	**659**	373	**1078**
24	**25**	74	**82**	124	**148**	174	**228**	224	**328**	274	**462**	324	**664**	374	**1093**
25	**26**	75	**83**	125	**150**	175	**230**	225	**331**	275	**465**	325	**670**	375	**1109**
26	**27**	76	**84**	126	**151**	176	**232**	226	**333**	276	**468**	326	**675**	376	**1125**
27	**28**	77	**86**	127	**153**	177	**234**	227	**335**	277	**472**	327	**680**	377	**1142**
28	**29**	78	**87**	128	**154**	178	**263**	228	**338**	278	**475**	328	**686**	378	**1160**
29	**30**	79	**88**	129	**156**	179	**237**	229	**340**	279	**478**	329	**692**	379	**1179**
30	**31**	80	**89**	130	**157**	180	**239**	230	**342**	280	**482**	330	**697**	380	**1198**
31	**32**	81	**91**	131	**159**	181	**241**	231	**345**	281	**485**	331	**703**	381	**1219**
32	**33**	82	**92**	132	**160**	182	**243**	232	**347**	282	**488**	332	**709**	382	**1241**
33	**34**	83	**93**	133	**162**	183	**245**	233	**349**	283	**492**	333	**715**	383	**1263**
34	**36**	84	**94**	134	**163**	184	**246**	234	**352**	284	**495**	334	**721**	384	**1288**
35	**37**	85	**96**	135	**165**	185	**248**	235	**354**	285	**499**	335	**727**	385	**1314**
36	**38**	86	**97**	136	**166**	186	**250**	236	**357**	286	**502**	336	**733**	386	**1341**
37	**39**	87	**98**	137	**168**	187	**252**	237	**359**	287	**506**	337	**739**	387	**1371**
38	**40**	88	**99**	138	**169**	188	**254**	238	**362**	288	**508**	338	**746**	388	**1403**
39	**41**	89	**101**	139	**171**	189	**256**	239	**364**	289	**513**	339	**752**	389	**1438**
40	**42**	90	**102**	140	**172**	190	**258**	240	**367**	290	**516**	340	**759**	390	**1476**
41	**43**	91	**103**	141	**174**	191	**260**	241	**369**	291	**520**	341	**766**	391	**1518**
42	**44**	92	**105**	142	**175**	192	**262**	242	**372**	292	**524**	342	**772**	392	**1565**
43	**45**	93	**106**	143	**177**	193	**263**	243	**374**	293	**527**	343	**779**	393	**1619**
44	**47**	94	**107**	144	**179**	194	**265**	244	**377**	294	**531**	344	**786**	394	**1681**
45	**48**	95	**108**	145	**180**	195	**267**	245	**379**	295	**535**	345	**793**	395	**1754**
46	**49**	96	**110**	146	**182**	196	**269**	246	**382**	296	**539**	346	**801**	396	**1844**
47	**50**	97	**111**	147	**183**	197	**271**	247	**384**	297	**543**	347	**808**	397	**1961**
48	**51**	98	**112**	148	**185**	198	**273**	248	**387**	298	**547**	348	**816**	398	**2127**
49	**52**	99	**114**	149	**186**	199	**275**	249	**390**	299	**551**	349	**824**	399	**2427**
50	**53**	100	**115**	150	**188**	200	**277**	250	**392**	300	**555**	350	**832**	400	**–**[c]

[a]Observed positive hole count
[b]Estimated positive hole count corrected for probability that two propagules enter the same hole
[c]Quantitative limit for correction exceeded (ca. 2,628 particles)

sign but produced by different manufacturers possess a built-in, battery-powered vacuum pump featuring a convenient calibration mode.

As mentioned previously, propagules of many microorganisms are adapted for airborne dissemination, and such propagules do not readily settle. As a result, the sedimentation plate method often provides a false sense of security with respect to airborne contamination in an insectary. For example, 26 colonies were obtained with the Andersen air sampler run for 5 min but only 2 colonies with the sedimentation plate after 2 h (Fig. 31). The sedimentation plate method almost always underestimates the level of contamination of air in insectaries. In heavily contaminated environments, Andersen samplers may suffer from overexposure, with multiple propagules being counted as one due to colony overgrowth after impaction via the same sieve hole. The correction factors (see above) are only effective for moderately overexposed Petri dishes but are inadequate for heavily overexposed dishes. Reducing the sampling time is one strategy that can be employed in heavily contaminated environments, but it should be realized that spatial/temporal discontinuities in airborne microorganisms may bias results. For example, a 30-sec exposure may fortuitously sample a current of relatively clean air from a duct draft not representative of a contaminated room. As a result, in heavily contaminated rooms, devices that trap propagules in liquid may be more accurate.

General purpose air filtration equipment can be used to collect airborne microorganisms if more specialized equipment is lacking. The size of most bacteria is 2.0–8.0 μm in length and 0.2–2.0 μm in width (Tortora et al. 1989). Because most bacteria are associated with dust particles as aerosols, the pore size of the membrane filter required to remove microorganisms from the air should have a diameter no greater than 0.45–0.60 μm. Some researchers have devised filtering methods to monitor microorganisms in air (e.g., Runkle and Phillips 1969, Dark and Harper 1972, Sikorowski 1975). The air filtration equipment used by Sikorowski (1975) is pictured in Fig. 32.

An alternative to the above method is to dislodge propagules from the filter (e.g., using a Stomacher ho-

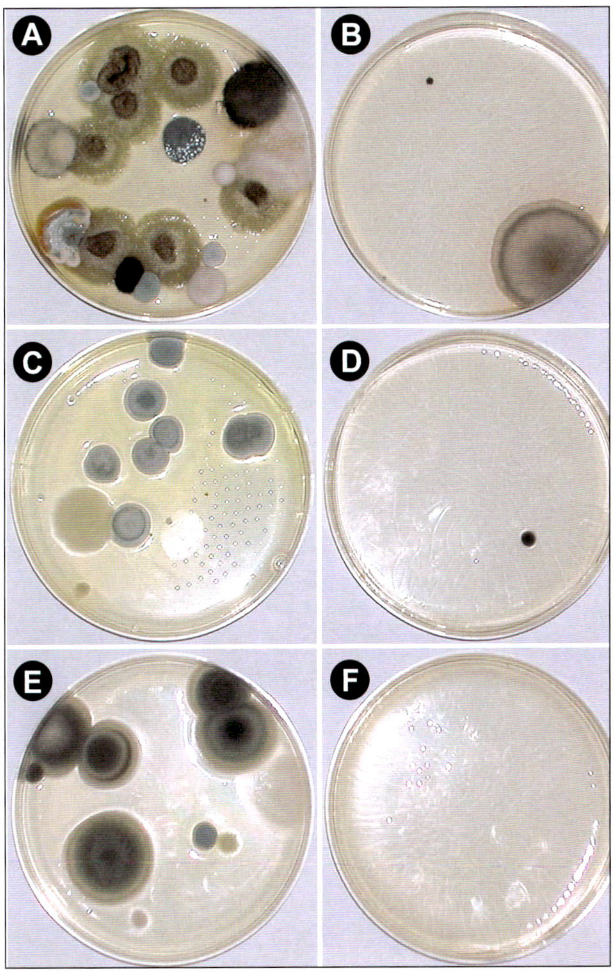

Fig. 31. Fungal propagules in the air of an insect rearing room possessing point-source filtered air. (A, C, E) Samples obtained with a single-stage Andersen air sampler run for 5, 2 and 1 min, respectively. (B, D, F) Samples obtained using the sedimentation plate method with exposures for 120, 60, and 5 min, respectively. Note the gross imprecision and insensitivity of the sedimentation plate method for estimating fungal propagules present in air.

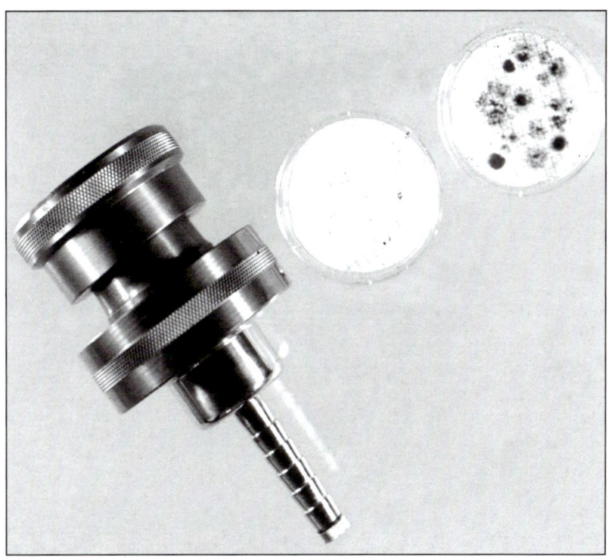

Fig. 32. Air-filtration apparatus using a membrane filter to sample airborne propagules. The two filters show environments with different levels of contamination. [Based on Sikorowski (1975).]

mogenizer; Sec. V. C. 2) and quantify the number of CFU using the spread-plate method (Sec. V. C. 2. a). The advantages of the filtration system are its simplicity, low cost, and the fact that small propagules are detected. This method does not allow enumeration of individual colonies, particularly when microorganisms are present in large numbers, and is thus semiquantitative. The semiquantitative nature of the method and the relatively high labor required are the primary disadvantages of the filtration method.

2. Particle Sampling. Perhaps the most advanced samplers of airborne particles—whether microbial propagules or not—are the *laser air-particle counters* (Fig. 33). These samplers draw air through a laser that measures particles by size class. In most models, the samplers are able to measure particles 0.3–5.0 μm in size in defined, total volumes of air. Many models are able to enumerate very high densities of particles even up to 3 x 10^7 particles per m^3. The model shown in Fig. 33 is an older one (Met One Instruments, Inc.) routinely used at Mississippi State University. This model is not particularly mobile and is reliant on an AC electrical source; but several hand-held, battery-powered models are now available. Laser counters provide very reproducible results and are used to assess clean room standards. Their salient disadvantage is that they do not distinguish between microbial propagules and other particles.

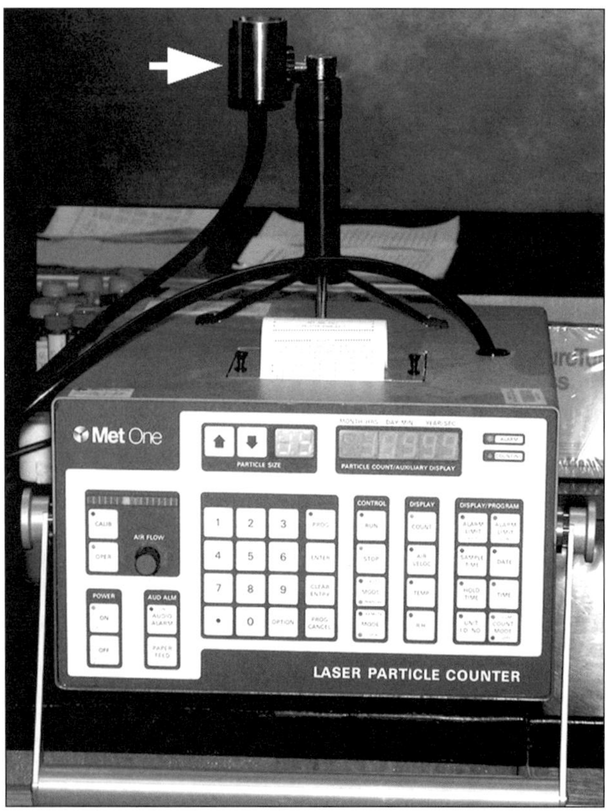

Fig. 33. A Met One laser air-particle sampler. This sampler draws air through a hose (arrow) and into the machine where a laser counts and measures particles by size class.

B. Personnel

Although personnel are the major source of bacterial contamination in insectaries, rarely are routine sampling procedures used to assess hygiene in workers. However, the application of basic microbiological methods in training exercises can be exceptionally beneficial in reinforcing the importance of proper hygiene (Sec. VI. D. 1).

C. Diets and Diet Ingredients

Artificial diets are usually complex media composed of numerous organic and inorganic components, and unsterilized or incompletely sterilized diets are readily spoiled by microorganisms. These microorganisms are often associated with diet ingredients, but they also may originate from other sources through the air or contact with equipment. In order to assess the potential impact of microorganisms and strategies for their management, it is imperative to monitor the microbiological properties of the diet (Chap. 9 *Entomopathogens and Insect Rearing*; Cohen 2003). Monitoring microorganisms in artificial diets utilizes basic qualitative and quantitative microbiological techniques developed for foods. There are numerous variations of the basic methods, and modifications to these methods will be developed with experience.

1. Qualitative Methods. Qualitative methods for monitoring microorganisms in artificial diets often involve the placement of substrates in contact with an appropriate microbiological medium. Small pieces of solid diet can be placed on or in a microbiological medium, or semisolid diets can be streaked onto a solid culture medium using a bacteria loop (Fig. 34). The media are then scored for microbial growth. Although this streak method is qualitative, it does provide some information on the degree of contamination. Qualitative methods are relatively rapid and labor efficient, and they are often the method of choice to ensure sterility of diets. An example of a qualitative method involves the testing of corn cob grits for sterility. Grits are used as a carrier for lepidopteran eggs, but they are almost always contaminated with saprophytic fungi such as *Aspergillus* and *Penicillium*. Appropriate sterilization methods must be used (Davis

1987). *Sterilization* means the total destruction of life—there are no degrees of sterilization. To assess the efficacy of the sterilization procedure, a subsample of grits can be aseptically sprinkled onto an appropriate medium (Fig. 28). The medium is then incubated at an appropriate temperature for fungi and bacteria. The temperature selected will influence differentially the growth of particular microorganisms. In many instances, the best incubation temperature is that at which the insects are reared. However, for incubation of medically important bacteria, a temperature of 37°C is used. If high incubation temperatures are chosen, it is important to incubate the Petri dishes upside down to avoid condensation falling onto the agar surface. After an appropriate period of time, the presence of microbial growth is recorded. Many bacterial colonies will appear within 24–48 hours, but fungi are typically slower to develop and are typically first examined at 48–72 hours.

2. Quantitative Methods. While qualitative methods are adequate to confirm sterility and to isolate

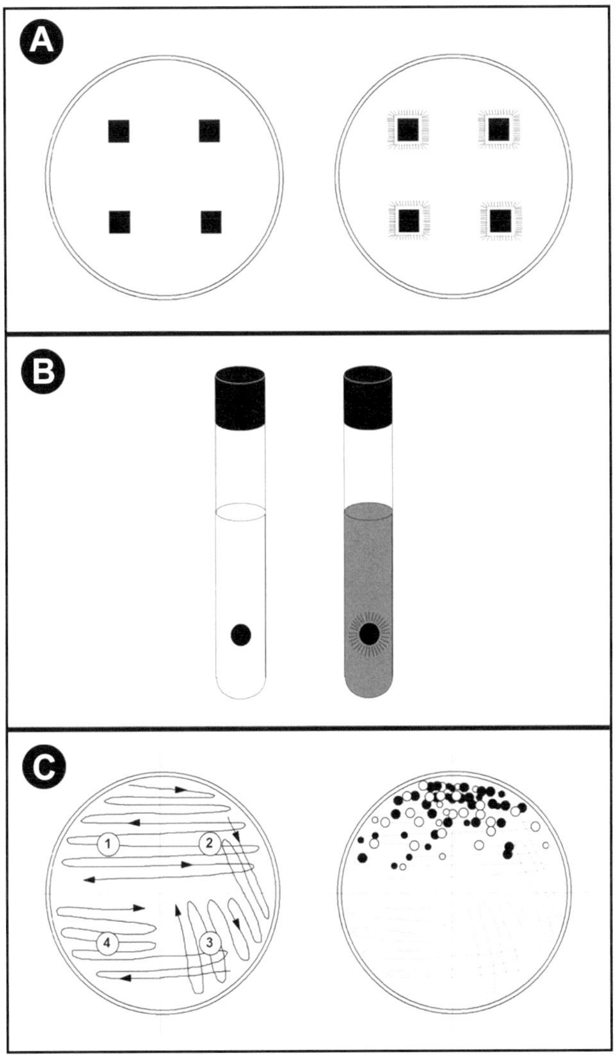

Fig. 34. Qualitative assessment of microbial contamination. (A) Diet is placed directly on a solid microbiological medium, incubated at an appropriate temperature for 3–7 d, and microbial growth assessed. On the left, no microbial development is observed. On the right, fungal mycelium is observed. (B) Diet is placed in liquid microbiological medium, incubated at an appropriate temperature for 3–7 d, and microbial growth assessed. On the left, no microbial development is observed. On the right, the medium turned turbid, and fungal mycelium emanating from the diet pieces is observed. (C) Diet is streaked directly onto a solid microbiological medium using a bacteria loop as indicated on the left, incubated at an appropriate temperature for 3–7 d, and microbial growth assessed. Numbers within circles indicate quadrants of the dish. On the left, no microbial development is observed. On the right, bacterial and/or yeast colonies are observed on the medium.

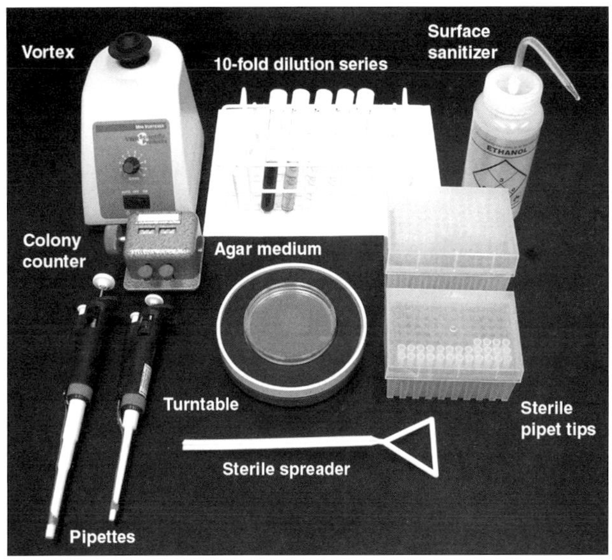

Fig. 35. Basic equipment used to assess microbial contamination quantitatively.

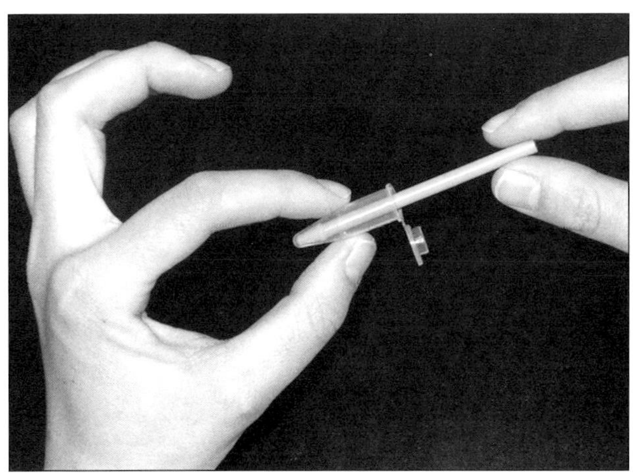

Fig. 36. A mortar and pestle homogenizer consisting of a polypropylene pestle and a 1.5-ml microcentrifuge tube.

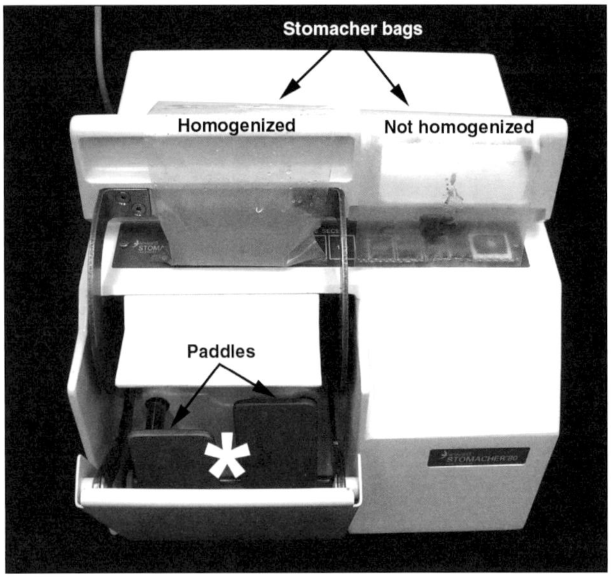

Fig. 37. A Stomacher homogenizer. A sample is placed in a sterile plastic bag and inserted into the machine at the point indicated with an "*". The action of the paddles provides excellent release of microbial propagules from a wide range of substrates typically encountered in insectaries.

and characterize many microorganisms, quantitative methods are usually required to investigate the causes of particular instances of diet spoilage. Since a multitude of microorganisms are usually present in spoiled diet, one must determine the relative abundances of the various taxa involved to begin to assign relative levels of culpability. The most commonly used methods are the spread-plate and most probable number techniques. A minimal investment in equipment is required for either technique. The basic pieces of equipment are pictured in Fig. 35. Defined quantities of diet are homogenized in sterile water or buffer using any of a variety of homogenizers: pestles (Fig. 36), homogenization tubes, tissue tearers, or more specialized equipment such as a *Stomacher homogenizer* (Fig. 37). Stomacher homogenizers were developed for the food industry and provide excellent distribution of microbial propagules in the homogenate. A sample is placed in a sterile plastic bag with an appropriate liquid (i.e., water or buffer), and the bag is placed be-

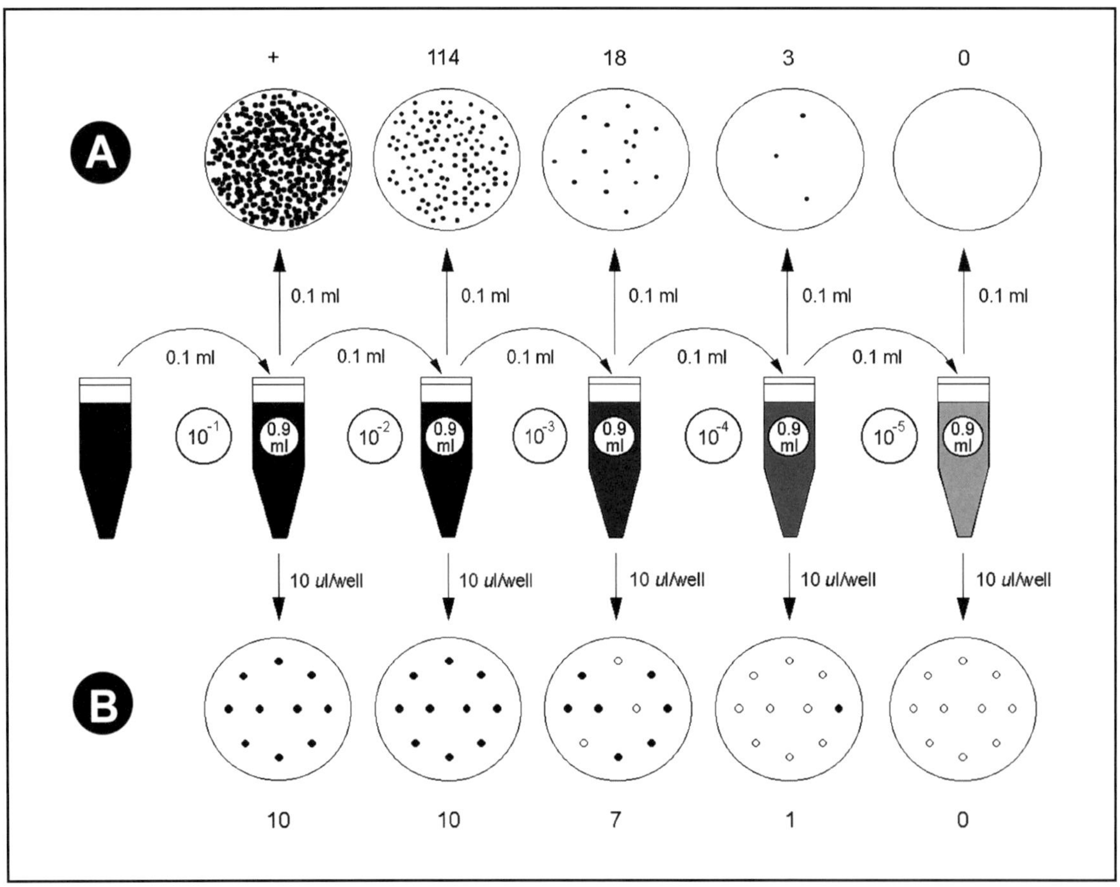

Fig. 38. Methods for enumeration of propagules. (A) Spread-plate. (B) Most-probable number (MPN). See the text for further discussion of each of the two methods. [From Goettel and Inglis (1997) with permission from Elsevier.]

tween two paddles and the outer wall of the homogenizer. The homogenization results within ca. 1 min from the action of the two paddles moving back and forth against the sample. Following homogenization, the homogenate is diluted in a dilution series: in most instances, a 10-fold series is used (i.e., 1 part sample into 9 parts dilutant) with vortexing for 10 sec following each transfer, but other dilution rates (e.g., 4-fold) can be used (Fig. 38).

a. Spread-Plate Method. For the *spread-plate* method (Fig. 38A), aliquots of 100 to 200 ml are aseptically removed from each dilution and spread onto an appropriate medium using a sterile applicator (Fig. 39) or glass beads. A nonselective or selective medium can be used depending on the target microorganism to be isolated (Chap. 9 *Entomopathogens and Insect Rearing*). To prevent spreading of motile bacteria, the solution should be allowed to absorb into the medium before the cultures are transferred to an appropriate incubation temperature. To estimate the level of uncertainty in the resulting propagule density estimates, it is recommended that two or more Petri dishes should be prepared at each dilution. Cultures are incubated 3–7 days depending on the medium or organism, and colonies are counted at the dilution yielding 30 to 300 CFU per Petri dish (Fig. 40). To increase the accuracy of the counts, especially for smaller colonies, a 2 or 3X stereomicroscope can be used. In the example illustrated in Fig. 38A, the culture resulting from the 10^{-1} dilution [*TD* (tube dilution) = $1/10^{-1} = 10$] and a sample of 100 µl [10^{-1} ml, *PD* (plate dilution) = $1/(10^{-1}$ ml) = 10 ml^{-1}] contained too many colonies to obtain an accurate count: colonies were confluent and crowded each other. The 10^{-2} dilution culture with $TD = 1/10^{-2} = 10^2$ and the same value for PD contained 114 well-distributed and easily counted colonies [*CC* (colony count) = 114 propagules]. The culture from the 10^{-3} dilution re-

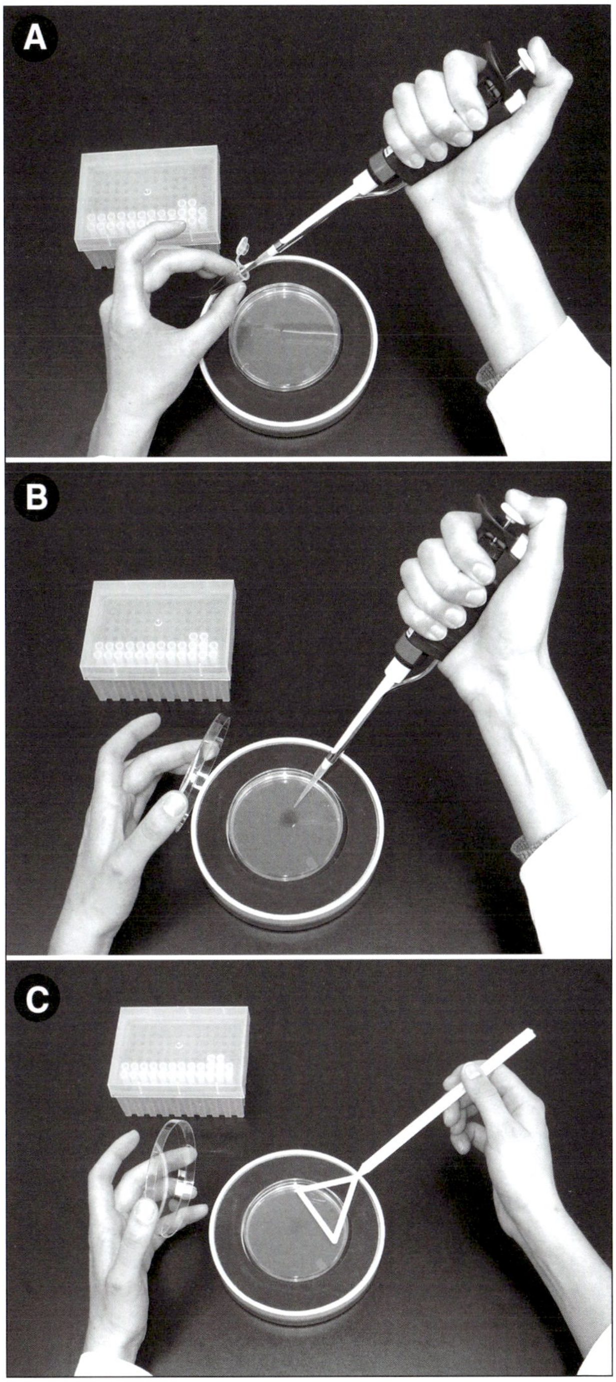

Fig. 39. The spread-plate method. (A) 100 µl of the homogenate is removed from a 1.5-ml microcentrifuge tube using a pipettor fitted with a sterile plastic tip. (B) The homogenate is aseptically dispensed onto a solid medium. (C) The homogenate is uniformly spread over the surface of the solid medium using a sterile spreader. A turntable is used to facilitate uniform spreading of the homogenate. The culture is incubated at an appropriate temperature for a specified period of time.

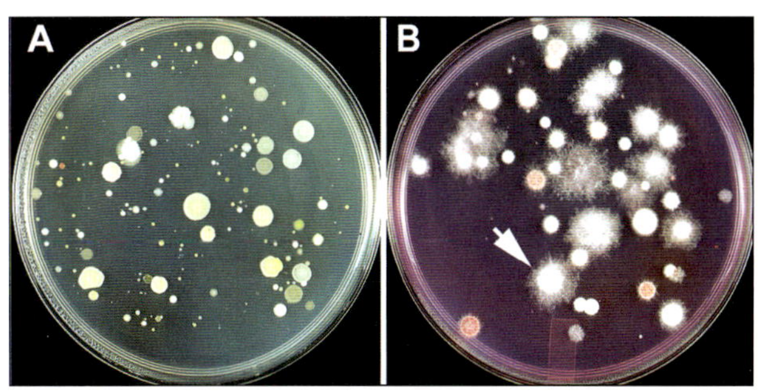

Fig. 40. Results of the spread-plate method. (A) Bacteria growing on trypticase soy agar medium containing antifungal agents after 72 h of growth. (B) Fungi growing on a mycological solid medium containing rose bengal and antibacterial agents. The function of rose bengal is to slow down rapidly growing fungal colonies such as the one indicated by the arrow.

sulted in *CC* = 18 propagules, seven more than would be predicted from the 10^{-2} culture. The percentage error due to variation in *CC* for a high dilution makes such a count unreliable, but duplicate or triplicate cultures may be used to reduce this uncertainty. *DH* (the density of propagules per ml homogenate) is given by application of the following equation as indicated:

$$DH = CC \cdot TD \cdot PD \qquad \text{Eq. 2}$$
$$DH = (114 \text{ propagules}) \cdot 10^2 \cdot 10 \text{ ml}^{-1}$$
$$= 1.14 \times 10^5 \text{ propagules per ml.}$$

Densities are usually presented as CFU per g diet on a fresh or dry weight basis. To determine the moisture content of the diet, a weighed subsample is dried at approximately 80°C for a minimum of 24–48 h, the sample reweighed, *WF* (the fraction of water in the sample) is calculated, and the densities of CFU per unit weight adjusted accordingly. Given *MD* [(mass dilution during sample preparation) = final volume homogenate/fresh mass of diet used], one can calculate *DDF* (the density of propagules in the diet on a fresh weight basis) or *DDD* (the density of propagules in the diet on a dry weight basis) as follows:

$$DDF = DH \cdot MD \qquad \text{Eq. 3}$$
$$DDD = DDF/(1 - WF) \qquad \text{Eq. 4}$$

This method was used to enumerate the taxa of bacteria and fungi decomposing the meat-based, entomophage diet presented in Fig. 21.

In an attempt to reduce error that may occur during dilution or plating and to expedite the procedure, mechanized plating systems have been developed. Spiral platers, initially shown to be effective for bacteria (Gilchrist et al. 1973), have also been used for the enumeration of molds (Fargues et al. 1996). Image analysis systems can expedite the enumeration of colonies and provide a permanent record of cultures. However, as a result of rapid hyphal growth and other characteristics exhibited by fungi, it is often necessary to make somewhat subjective decisions when fungal colonies grow together or are found in close proximity to one another.

b. Most Probable Number (MPN) Method. An alternative to the spread-plate method is the MPN technique (Woomer 1994; Fig. 38B). The MPN method was developed for the enumeration of bacteria in water and has been adapted for use with bacteria and filamentous fungi from a variety of substrates including insect diets. As with the spread-plate method, it is first necessary to homogenize the microorganisms from the diet. Once in suspension, the sample is diluted to the point of propagule extinction. Each dilution step is usually 4- or 10-fold, and three to ten replicate samples per dilution are required. Although a liquid medium may be used, a solid substrate is generally used for filamentous fungi and many bacteria. As with the spread-plate method, it may be desirable to use a selective medium. A physical barrier is usually required to prevent spreading of the suspension across the agar surface, and this is accomplished by placing a glass or polypropylene ring in the medium or by making a well in the agar surface with a cork borer. Volumes of diluted suspension placed within the wells should vary according to the size of the well and the absorptive properties of the medium. Once inoculated, the Petri dishes are incubated; and after an appropriate amount of time, the wells are scored as positive or negative for growth of the target microorganism.

In the example illustrated in Fig. 40B, 10 replicate samples of 10 µl [10^{-2} ml, *PD* (plate dilution) = $1/(10^{-2}$ ml) = 10^2 ml^{-1}] were placed on the solid medium from each dilution tube. Three consecutive dilutions should be selected at or immediately before propagule extinction and the numbers of positive wells recorded. In this case, the ranges of dilutions that may be used and the corresponding numbers of positive wells observed are 10^{-2}–10^{-4} (10, 7, 1) and 10^{-3}–10^{-5} (7, 1, 0). *MPNP* (most probable number of propagules per well sample at the minimum end of the chosen dilution range) is acquired from an MPN table based on the Poisson statistical distribution (e.g., Meynell and Meynell 1970). If 10^{-2} is the minimum dilution, *MPNP* = 11.6 propagules. If 10^{-3} is the minimum, *MPNP* = 1.16 propagules. For 10^{-2} as the minimum dilution [*TD* (tube dilution) = 10^2], *DH* (density of propagules) is given by application of the following equation as indicated:

$$DH = MPNP \cdot TD \cdot PD \qquad \text{Eq. 5}$$
$$DH = 11.6 \text{ propagules} \cdot 10^2 \cdot 10^2 \text{ ml}^{-1}$$
$$= 1.16 \times 10^5 \text{ propagules per ml.}$$

In this example, the dilution plating and MPN methods provide similar estimates of propagule density. Although it is considered less accurate than the spread-plate method, the MPN method may provide comparable results (Harris and Sommers 1968). The salient advantages of the MPN technique are its greater overall flexibility and scope of application. A major disadvantage of the MPN technique relative to the spread-plate method is the difficulty in recovering specific microbial taxa.

c. Plants. The surfaces of plants contain considerable numbers of microorganisms. Although microbial propagules may be strongly attached to the cuticle of plants, they can be dislodged by vigorous washing. To facilitate removal of microorganisms from the surfaces of plants, the wash solution may be amended with a surfactant (e.g., 0.01% Tween 80™). However, caution must be exercised in choosing a surfactant because they can be toxic to microorganisms. For washing, both reciprocal and rotary shakers are used, but the former is preferred since it provides more vigorous agitation. If a rotary shaker is employed, it may be necessary to add coarse, sterile sand or small glass beads to the wash solution to aid in the mechanical detachment of propagules. Such a strategy can cause mechanical damage to microorganisms, and the speed at which the shakers are operated should be carefully chosen. The duration of the washing also varies, but periods of at least 2 h are generally used. The volume of the wash solution and the type of vessel used also influence the efficacy of the wash procedure. Wash volumes in excess of 5 ml in relatively wide containers result in the greatest turbulence of the wash solution. Because large wash volumes are required, it may be necessary to wash several leaves together in order to increase the density of propagules released into the wash solution. The recovery of propagules from the external surfaces only and relatively low labor requirements are the primary advantages of the wash method. The requirement for large wash volumes, incomplete detachment of propagules, and clumping of propagules are disadvantages. The latter problem is the primary disadvantage because it results in a conservative estimate of microbial densities.

As with artificial diets, a number of homogenization techniques can be used to release microorganisms within plant tissues; but the Stomacher homogenizer is generally the preferred method. It should be realized that plant tissues, particularly moribund tissues, fruits, processed tissues (e.g., wheat bran), and tissues in contact with soil can be heavily colonized/contaminated by bacteria and fungi (e.g. > 10^6 CFU/g. Antimicrobial compounds (e.g., phenolics) released from homogenized tissues can be toxic to microbial cells, and care should be taken when using homogenization devices (e.g., blenders) that disrupt plant cells. The use of chilled buffers may limit the detrimental impacts of antimicrobial compounds released from plants.

D. Insects

It is often necessary to measure microorganisms associated with various developmental stages of insects and their frass before and after exposure to sanitation methods. The sterility of eggs treated with surface sanitizers can be confirmed by aseptically placing them on agar media or incubating them in a liquid culture medium. Alternatively, microorganisms on or in eggs, larvae, nymphs, adults, and frass can be quantitatively assessed by homogenization followed by the enumeration of CFU on appropriate media using spread-plate or MPN methods. One of the major limitations of the spread-plate method is that motile bacteria can rapidly spread over the surface of the medium and obscure nonmotile bacteria forming dis-

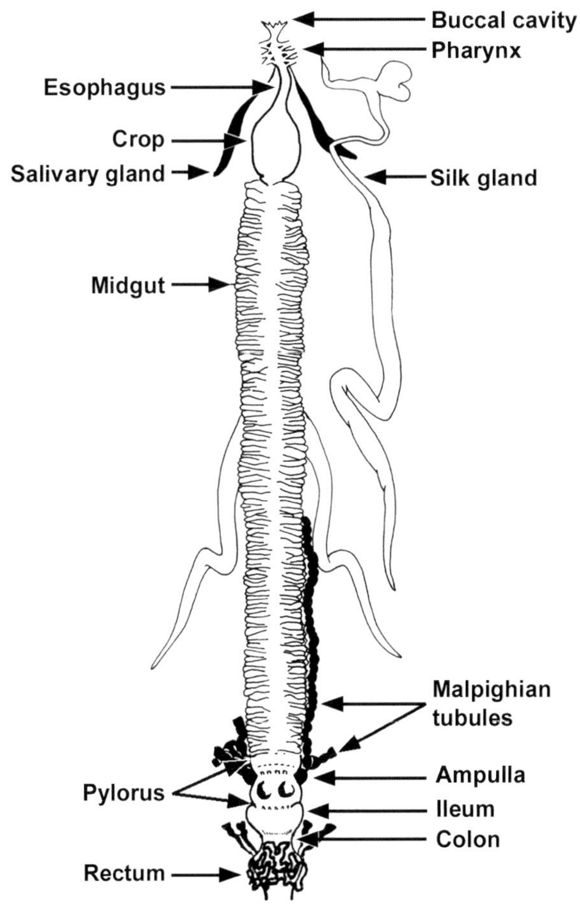

Fig. 41. Alimentary canal of a corn earworm, *Helicoverpa zea* (Lepidoptera: Noctuidae), larva. [Drawing provided by Matt MacGown, Mississippi State University, MS.]

crete colonies. To overcome this problem, the *pour plate* method is often used. Instead of spreading the homogenate over the surface of the medium, the homogenate is uniformly mixed in molten agar (45–50°C) before it is dispensed into the Petri dish. The temperature of the medium does not kill most microorganisms, and once suspended in the medium, motile bacteria are able to grow within the medium but are unable to spread. The disadvantage of the pour plate method is that it is relatively labor intensive, and it can be difficult to enumerate and recover bacteria that are relatively rare. Lowering the incubation temperature may also limit spreading of bacteria.

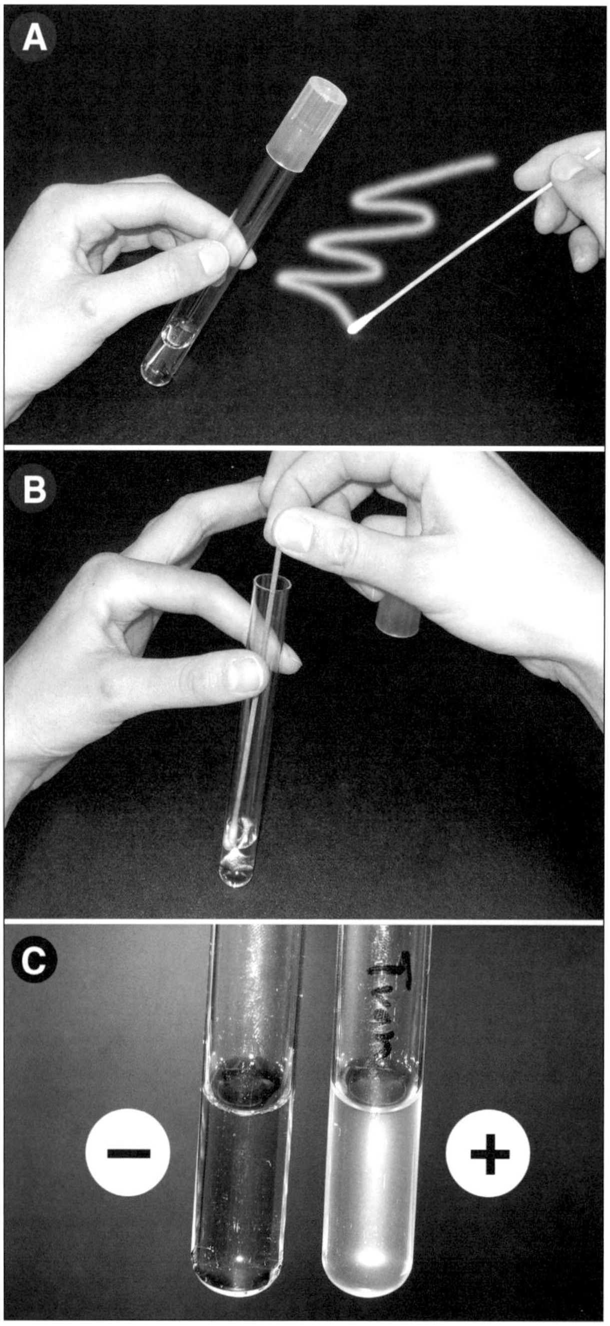

Fig. 42. Swab-rinse method for qualitatively assessing contamination of a variety of surfaces. (A) A sterile cotton swab is rubbed over the surface. (B) The swab is placed in a tube containing liquid culture medium and vortexed to dislodge microbial cells. (C) The liquid culture is incubated at an appropriate temperature and scored for turbidity. A turbid culture is indicative of a positive sample (+), whereas a clear sample indicates a negative sample (-).

The above methods do not discriminate between microorganisms on the surface versus within the insect (e.g., in the alimentary canal), and to do so requires additional steps. One method to recover microorganisms from the alimentary canal is to surface-sanitize the insect, aseptically remove the alimentary canal, and then homogenize the gut or specific portions of the gut (Fig. 41). Another method is to surface-sanitize the insect prior to homogenization. The assumption behind this strategy is that all microbial cells on the surface of the integument will be killed by the sanitizer and that the hemolymph and tissues are devoid of microorganisms. In reality, surface-sanitization procedures are rarely 100% efficacious, and it is good practice initially to determine the effectiveness of a sanitation procedure. To differentially recover microorganisms from the insect integument, it may be necessary to utilize wash methods. Because microorganisms can be released in regurgitated foregut contents and in frass, it may be necessary to coat the mouthparts and anus with a material such as molten paraffin prior to the washing procedure.

In some instances, quality control of insect rearing operations has included an analysis of microbial contamination (e.g., Sikorowski 1984). Generally, there is a positive relationship between the number of contaminating microorganisms and the quality of reared insects. However, simply stating that insect quality is inversely proportional to the number of microorganisms harbored can lead to erroneous results. The composition of the microflora also is important, and in order to fully understand the relationship between bacteria and insect quality, additional observations are required. Furthermore, no single method or medium is capable of detecting all bacteria or fungi that may contaminate an insect (Atlas 1993), and combinations of methods are necessary.

E. Processing Equipment and Workrooms

Various methods for detection of microorganisms on workroom surfaces are reviewed by Patterson (1971) and Block (1991). We summarize two basic methods that are commonly used to detect microbial contamination of equipment and workrooms.

1. Swab-Rinse Method. The swab-rinse method is commonly used to qualitatively assess contamination of a variety of surfaces. The simplest method involves rubbing a sterile, dry, cotton swab over the surface or object to be sampled (Fig. 42). The tip of the swab is broken into a tube of liquid culture medium and vortexed to dislodge microbial cells. The tubes are then incubated at an appropriate temperature for 24–72 h, and the tubes are scored for turbidity due to proliferation of microorganisms within the medium. The primary disadvantages of this method are the following: 1) it is strictly qualitative, 2) there is often poor recovery of bacteria from the surfaces sampled, 3) the cotton can retain microorganisms, and 4) the results are often not reproducible between different workers and laboratories.

To obtain quantitative measures of surface contamination, a number of variations on the swab method have been applied. A common strategy is to dislodge microbial cells from the swab into a sterile buffer using defined conditions and to spread the rinse fluid onto an appropriate medium as described previously for the spread-plate method. Some workers use the sponge-swipe method. For this method, a sterile sponge moistened in buffer but with excess buffer squeezed out is used to sample the substrate of interest by wiping the entire area. The sponge is placed in a Stomacher bag, an appropriate buffer is added to the bag, and the sponge is homogenized using a Stomacher homogenizer or by hand for 30–60 sec. The buffer is then processed using the spread-plate or MPN

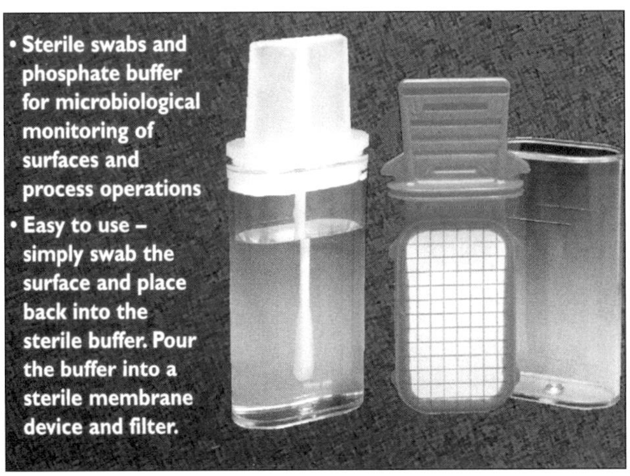

Fig. 43. Example of a commercial swab-rinse kit.

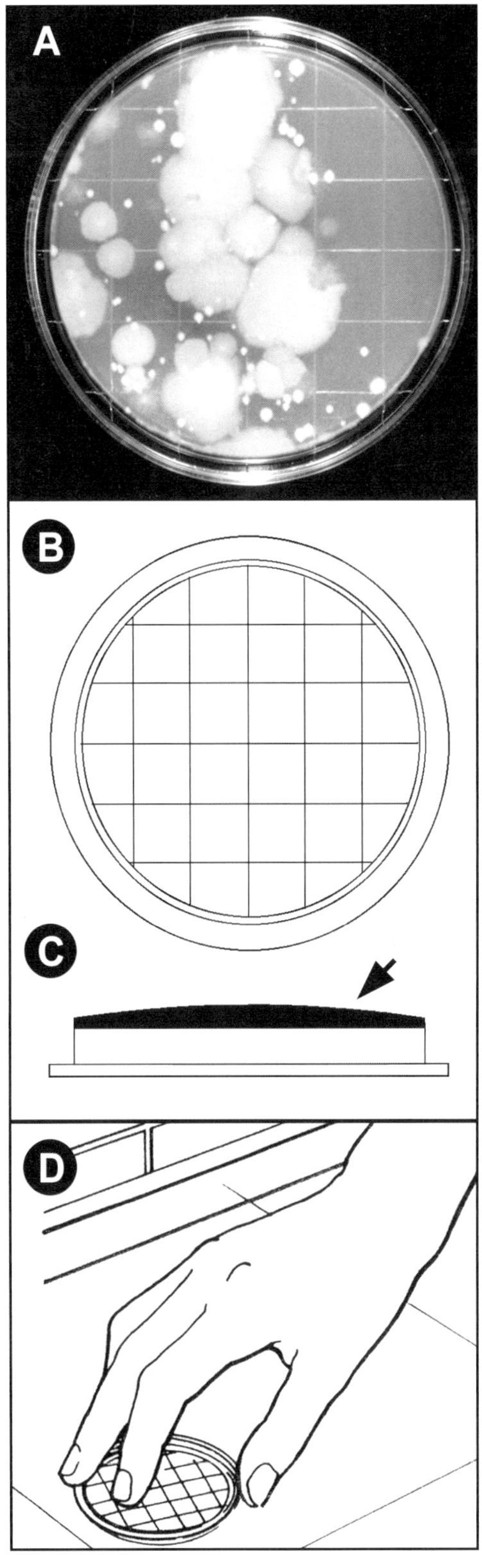

Fig. 44. The contact plate method using Rodac® plates. (A) A Rodac® plate containing a bacteriological medium that was incubated for 72 h at 37°C after exposure. Note the large number of bacterial colonies. (B) A top view of a Rodac® plate showing the grid pattern to facilitate counting. (C) A side view showing how the agar bed (arrow) is higher than the plastic side of the dish. (D) The agar bed of a Rodac® plate being pressed against the surface to be tested by exertion of moderate vertical pressure.

method. A number of sponges are commercially available. Although the swab method is generally considered to be a qualitative sampling method, a variety of commercially marketed swab kits touted to provide quantitative assessments of surface contamination are currently available (Fig. 43).

2. Agar Contact Method. The agar contact method is often used to monitor the degree of contamination of flat or nearly flat surfaces such as floors, bench tops, and some equipment surfaces. This method often uses *Rodac® plates*: modified, polyethylene Petri dishes in which the agar medium is raised above the plastic rim (i.e., a convex surface of agar results from surface tension of the plate edge with the molten agar during plate preparation; Fig. 44). Rodac plates can be purchased already prepared, but they are typically purchased without medium. Within 72 h before use, molten agar medium (16.5–17.5 ml) is dispensed into the Rodac plate taking care to prevent bubble formation or overflowing. General-purpose media such as trypticase soy agar for bacteria and potato dextrose agar for fungi are typically used (Chap. 9 *Entomopathogens and Insect Rearing*). The plates also possess a grid to facilitate counting of colonies, and the method can be categorized as semiquantitative. To sample a surface with a Rodac plate, the lid is removed, and the agar bed is placed in contact with the surface to be tested while exerting moderate pressure. Microbial cells present on the test surface are transferred to the agar surface. On a curved surface, the Rodac plate is rocked to achieve contact over the entire surface of the plate, but caution should be exercised to avoid rubbing the plate over the surface to be tested. The plate is then removed from the test surface, the lid is replaced, and the plate is incubated at a temperature appropriate for the medium being used.

Rodac plates are often used in hospitals to assess cleanliness. The following guidelines have been established for hospital floors immediately after cleaning: Good, ≤ 25 colonies; Fair, 26–50 colonies; and Poor, ≥ 50 colonies. With careful sanitation, it is possible to achieve counts of < 12 colonies per plate.

The major advantage of the Rodac plate method is simplicity. The major disadvantages of the Rodac plate method are the following: 1) the plates cannot fit into the confined areas common in various types of equipment, 2) accuracy is dependent on efficient transfer of microbial cells from the test surface to the agar medium, 3) residual agar left after sampling can promote growth of microorganisms and must be removed, and 4) a relatively large number of plates are required to sample a large area.

VI. Managing Microbial Contamination

Efficacious management of microbial contaminants requires a multifaceted approach relying on knowledge of what the sources of contamination are, how contamination can be detected, and application of the appropriate controls. Here we discuss a number of control categories including air filtration, physical and chemical methods of diet preservation, sanitation of insects, education of personnel, and insectary design and management.

A. Air Filtration

1. Whole Room. As we have discussed previously, a primary source of contamination in insectaries is air. Consequently, air filtration is one of the most important tools that can be used to control microbial contamination. In addition to insectaries, air filtration is used in numerous industries including aeronautics, hospitals, and electronics. Clean room technology standards have been established to categorize the degree to which air in a facility is contaminated with particulate matter, including microbial propagules. The *American Federal Clean Room Standards* were established first and specify six classes (*1*; *10*; *100*; *1,000*; *10,000*; *100,000*; Fig. 45A). These classes were named based on the maximum number of particles ≥ 0.5 μm in size per ft^3 of air allowed. For example, class *100* rooms allow no more than 100 particles ≥ 0.5 μm per ft^3 of air. The distribution of larger particle sizes is also constrained; and, for example, class *100* rooms allow no more than 1 particle ≥ 4.0 μm per ft^3 of air. The American standards have been largely supplanted by the *International Standard Organization Clean Room Standards* (ISO) in which particle density is measured per m^3 of air and the stricter classes are named based on the density of particles ≥ 0.1 μm (Fig. 45B). ISO classes *3* through *8* correspond approximately to American Federal classes *1* through *100,000*, respectively. In the ISO system, two cleaner classes (ISO *1* and *2*) and one dirtier class (ISO

9) were established. In establishing clean room standards, laser particle counters are typically used, and samples are required from various sample locations, volumes, particle numbers, and size limits. For more information on clean room standards, the reader is referred to iProcesSmart (http://www.iprocessmart.com/techsmart/tech_assist_1.htm).

Achieving ISO class *5* or better standards requires a special air filtering and air handling system to achieve *laminar* (nonturbulent) airflow. In industries possessing very strict requirements for clean air, laminar airflow rooms are used. A *laminar airflow room* is designed so that laminar flow of air occurs vertically or horizontally through the entire volume of the room (Fig. 46). No insectaries to our knowledge use laminar airflow rooms.

Insect rearing facilities typically utilize *conventional, point-source filtration*: air enters the clean room through a ceiling port or wall port fitted with a filter; and the filtered air is subsequently exhausted through ports or grills located in the ceiling, floor, or walls as a function of differential air pressure (Fig. 47). Airflow is typically adjusted so that 15 to 20 changes of room air capacity occur per hour. The industry standard filter used in clean rooms is the *HEPA* (high efficiency particulate air) filter. Although variations in construction exist, the basic design of HEPA filters consists of a continuous web of fiberglass pleated over corrugated separators (Fig. 48). The separators are made from paper, aluminum, or plastic. They add strength to the filter core and form air passages between the folds. The filter is sealed to the casing. Gasket material on the face of the filter frame is used to attach the whole filter unit securely to the mounting frame. This arrangement provides a very high surface area through which air must pass. HEPA filters remove ≥ 99.97% of particles ≥ 0.3 μm in size. Since the pores of HEPA filters are exceptionally small, it is imperative that prefilters be used to remove larger particles. Many HEPA filter installations possess a flow rate monitor to indicate how much flow resistance is being encountered. A routine maintenance program for prefilters should be considered a standard operating procedure.

Some insectaries have opted to use filters other than HEPA filters as a cost saving measure, and a number of options are available. For example, some so-called *pocket filters* (e.g., the Vildedon MF90™) remove ≥ 99% of particles ≥ 1 μm in size. The reduced filtering efficacy imparted by such filters relative to HEPA filters would not be expected to have a significant impact on the air quality of point-source filtration rooms.

Two additional devices that are sometimes used to remove viable microorganisms from air in insectaries include stand-alone, particle filter units and ultraviolet irradiators. Stand-alone filter units work in a manner similar to biosafety cabinets (Sec. VI. A. 2) in that air is drawn from the insectary room through a prefilter and then a secondary filter (e.g., HEPA filter) where particles are trapped (Fig. 49A, B). Many of these units are marketed for home use, but they may be useful as a secondary measure in clean rooms. However,

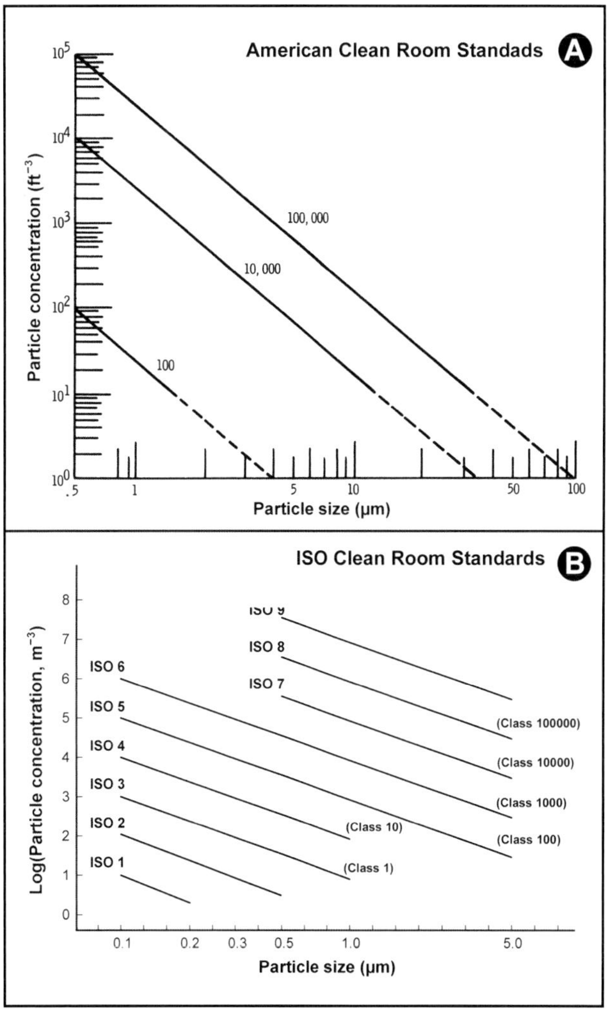

Fig. 45. Clean room standards based on numbers and sizes of particles in air (maximum density of particles ≥ a given size). (A) American Clean Room Standards. (B) International Standard Organization (ISO) Clean Room Standards. The American standards have largely been supplanted by ISO standards in which air volume measurements are based on m³ rather than ft³ of air. However, there is an approximate correspondence between the standards of the two systems as indicated in (B). [(A) from NASA (1969).]

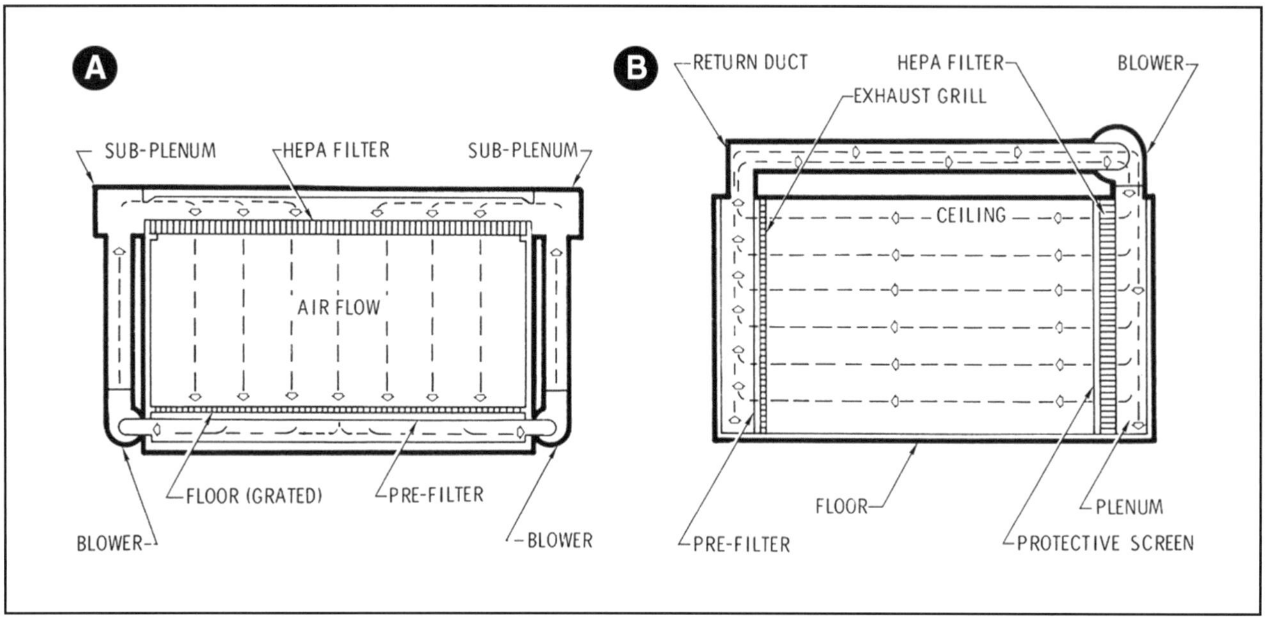

Fig. 46. Two types of laminar flow rooms. (A) Vertical. (B) Horizontal. [NASA (1967).]

their capacity is limited; and they cannot eliminate microorganisms that have been permitted to enter the facility due to other failures. *Ultraviolet radiation* (UV, electromagnetic radiation of 240–280 nm wavelength) is sometimes used to kill microorganisms in air through nonselective damage to nucleic acids. A number of commercial systems for killing microorganisms in air are currently available. The basic principle behind these systems is that microorganisms in air are drawn through a chamber within which they are exposed to a UV source. Exposure to UV irradiation has been shown to reduce, but not eliminate, bacteria and fungi in some situations (Levetin et al. 2001, Cundith et al. 2002, Xu et al. 2003). However, it is doubtful that UV irradiation alone is capable of sterilizing air entering an insect rearing facility.

Fig. 47. Airflow patterns in conventional, point-source filtration clean rooms. In this illustration, filtered air enters the room through ceiling ports and is exhausted through wall ports. The airflow is nonlaminar; and numerous air eddies, indicated with an "*", occur in the room. Eddies occur according to the shape of the room, objects within the room, and activity of personnel. [Based on NASA (1967).]

Conventional clean rooms possess very limited self-cleaning capability relative to laminar flow rooms. The air entering conventional clean rooms through well-maintained filters will meet or exceed ISO class *5* standards. However, once the air enters the room, its pattern of flow is affected by a variety of factors including people, movement of people, equipment, walls, and position of the exhaust vents (Fig. 47). Within conventional clean rooms, air eddies are a common occurrence. Any particles liberated from people (e.g., skin flakes) or from the insectary environment (e.g., diet ingredients) can persist within the insectary for prolonged periods of time in air eddies. Furthermore, many aerosols and fungal spores are electrostatically charged, and the airflow patterns in conventional clean rooms cause them to come in contact with objects to which they adhere. In conventional clean rooms, it may be possible to achieve ISO *7* standards with rigorous attention to the sanitation of personnel and continuous janitorial cleaning, but this is not cost effective. Typically, achieving ISO class *8* or *9* standards is more realistic in insectaries. For example, ISO class *8* rooms can contain up to ca. 35,000 particles $\geq$ 5.0 μm per m^3. The inability of point-

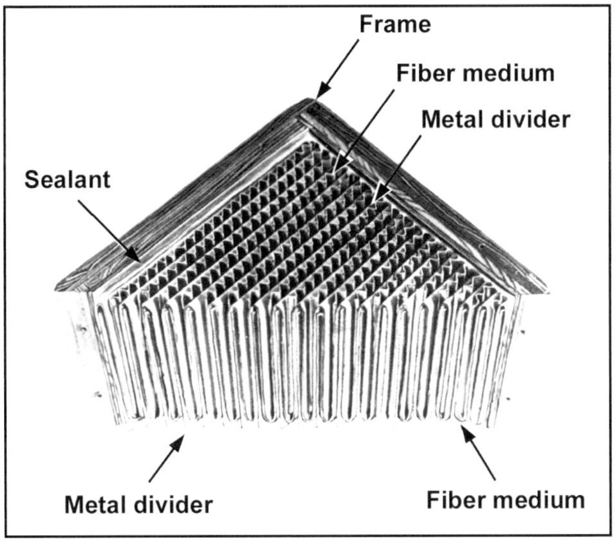

Fig. 48. Example of a HEPA (high efficiency particulate air) filter. Note the continuous web of fiberglass pleated over corrugated separators to increase the surface area of the filter.

source filtration systems to remove fungal propagules was illustrated in Fig. 31: > 190 fungal CFU/m^3 were detected in the air of a rearing room in which incoming air was filtered. Obviously, point-source filtration of air in insect rearing facilities does not preclude the need for additional, good management practices to deal with airborne contamination.

2. Work Stations. Although it is not possible to achieve ISO class 5 standards within insectary rooms, it is possible to establish ISO class 5 work stations within an insectary. Clean work stations are exceptionally valuable for operations that are sensitive to or at high risk of contamination. There are two major types of clean work stations: 1) laminar flow hoods and 2) biosafety cabinets. Many sizes and designs are commercially available and can be fitted for almost any size facility. If these devices are used properly, they will achieve ISO Class 5 standards or better. Both types of these work stations are equipped with HEPA filters and prefilters. The primary difference between the two is whether the filtration step takes place before or after the air has passed through the work area. This difference in operation dictates which application is appropriate. In a *laminar flow hood*, air drawn through a prefilter and then a HEPA filter by a blower fan is then directed in a laminar flow manner over the work surface (Fig. 50A). If the surfaces inside the hood (i.e., after the HEPA filter) are properly maintained and if the laminar flow of air is not disturbed,

Fig. 49. Example of a stand-alone, air filter unit. (A) Exterior of the unit showing the grate through which air is drawn. (B) Interior of the unit showing the three filters. Note the dirty prefilter, clean prefilter, and exposed HEPA (high efficiency particulate air) filter. Airflow is monitored as an indication of proper HEPA filter operation; however, the life of the HEPA filter will be extended by proper maintenance of the prefilters.

then air within the hood will meet ISO Class 5 standards.

It is a common practice of many untrained workers to load various pieces of equipment into a laminar flow hood. This will seriously affect airflow patterns and create air eddies (Fig. 51). Eddies can allow particles from the room to enter the work station where they can contaminate a substrate. Another aspect that is sometimes overlooked by insectary workers is that conventional laminar flow hoods are only designed for working with materials that are free of contaminants or materials in which contaminants can be contained. If a contaminated material is opened within a laminar flow hood, the contaminant will be blown directly into the face of the worker and into the insectary. Examples of operations for which laminar flow hoods are designed include dispensing of autoclaved diets, filtering of diet ingredients, and egg placement on diet.

For practices in which release of contaminants is expected, *biosafety cabinets* (also known as *biohazard cabinets*) should be utilized to preclude exposure of staff to these contaminants and of liberation of particles into the insectary. With biosafety cabinets, air is drawn into the unit over the work area (Fig. 52). The air, with any biological contaminants, is then drawn through a prefilter followed by a HEPA filter where particles liberated from the work surface are trapped before the air returns to the room. Additional examples of chamber designs can be viewed at AirClean® Systems (http://www.aircleansystems.com).

The proper operation of biosafety cabinets requires periodic decontamination of the filters—usually by a trained technician. To determine percentage leakage, the passage of smoke particles (0.3 μm average diameter) through the HEPA filter is measured with an aerosol photometer. In most jurisdictions, a leak limit of 0.01% is prescribed. To obtain ISO class 5 standards, biosafety cabinets typically provide a vent at the front

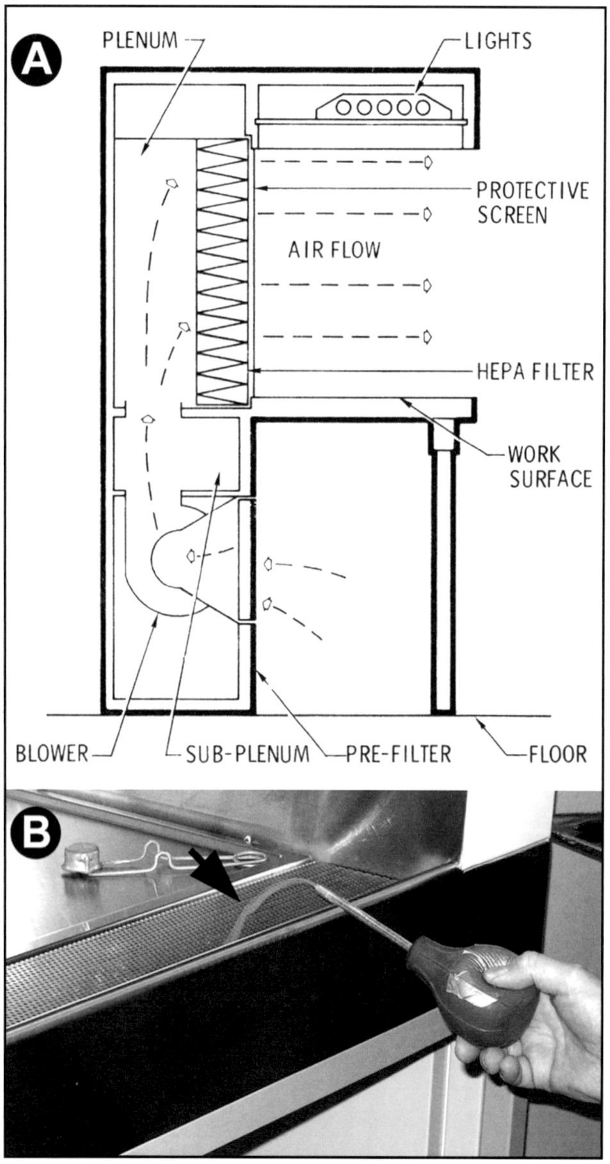

Fig. 50. Clean work stations. (A) A laminar flow hood showing unidirectional movement of air over the work surface. (B) A smoke test conducted in a biosafety cabinet. The arrow indicates the test stream of smoke. [(A) from NASA (1967).].

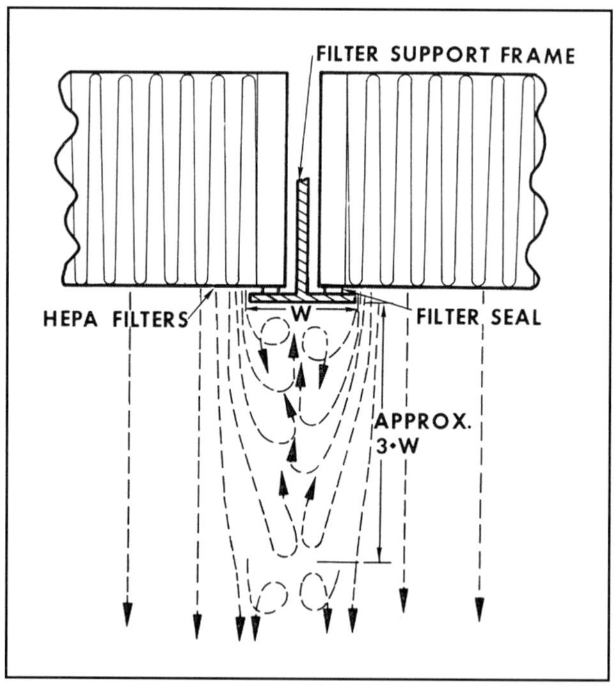

Fig. 51. Influence of a HEPA (high efficiency particulate air) filter support on air currents. Note the presence of eddies.

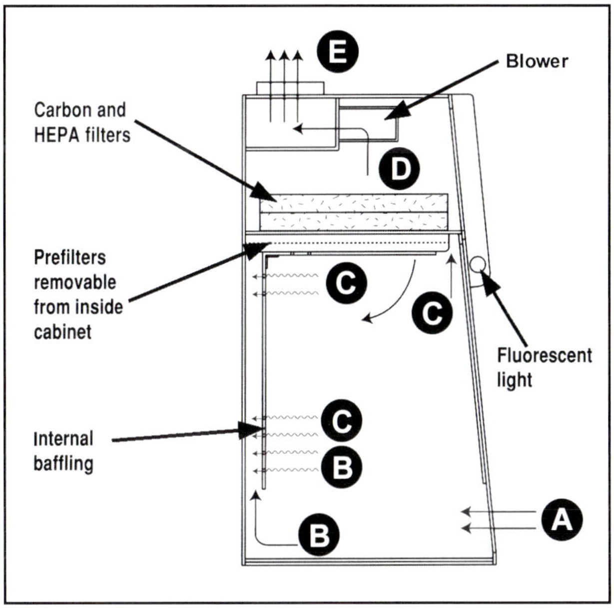

Fig. 52. Example of a ductless biosafety cabinet. Room air flowing away from the operator enters the enclosure in a horizontal direction at (A). Contaminated air (B) containing heavier particles and (C) containing lighter particles is drawn behind the baffle via a low pressure area established at the base of the baffle and behind the horizontal baffle exhaust slots. Contaminated air is directed through the prefilter and HEPA (high efficiency particulate air) filter (D) where the particles are trapped. The cleaned air (E) is recirculated into the room. [Based on an AirClean® Systems (URL) design.]

of the work area into which air from the work surface is drawn, thus preventing it from leaving the cabinet (Fig. 50B). Additional vents at the back of the work surface also draw biological contaminants away from the work surface. UV bulbs are often used in biosafety cabinets to decontaminate them after use. However, their efficacy in such applications is questionable due to inadequate UV flux at the distances of the surfaces from the bulb. Furthermore, they represent a health concern to workers who may be exposed to the radiation through UV-transmitting glass.

We have previously discussed the inability of conventional clean rooms to remove contaminants after they are liberated into the air. For this reason, it is highly desirable to separate the facility into cleaner and less clean areas and to utilize the air handling system to separate these areas through differential air pressures. The insectary should be designed so that cleaner areas possess positive air pressure. This is achieved by engineering the air handling system so that more filtered air enters the room than is removed by the outlet vents. With this arrangement, contaminated air in less clean areas within the insectary cannot enter the cleaner room when doors are opened because air movement is outward. Airflow in insectaries should follow the following pattern: filtered air enters the cleanest area of the insectary, moves to the less clean areas, and leaves the insectary from the least clean area. Insect harvesting is a particularly high-risk activity, and even a single contaminated diet container can severely contaminate an insectary (Figs. 8 and 53). It is important to emphasize that this system will only be effective if the clean areas within a facility are never contaminated. Therefore, strict adherence to personal hygiene and operational practices (e.g., use of clean work stations) must also be implemented.

In addition to monitoring microorganisms in air, airflow patterns should be monitored routinely for leaks in barriers and to ensure that positive air pressure is maintained and that excessive eddies are not occurring. To accomplish this, smoke or fog is typically used to visualize airflow patterns within rooms, through doors and other openings, and in hoods (Fig. 50B). Commercial smoke generators are often used although there are concerns with toxicity of the smoke, residues, and cost. As an alternative, dry ice fog or smoke from incense sticks can be used. The advantages of the dry ice fog method are that it is relatively inexpensive, nontoxic, and leaves no residue. The major disadvantage of the dry ice fog method is that the fog is very dense and low-lying so that it may not be appropriate for large areas or areas with minimal airflow. Dry ice can be bought cheaply from frozen food distributors or obtained from university physics departments, and it can

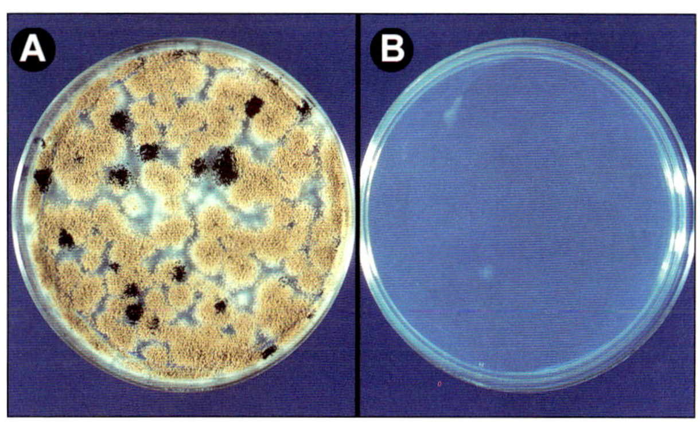

Fig. 53. Results of air samples taken using an Andersen air sampler in an insectary possessing point-source filtration. (A) Air sampled in a pupal harvesting room. Note the black (*Aspergillus niger*) and lighter (*Aspergillus ustus*) conidia. The origin of the contamination was traced to the harvesting of pupae from heavily contaminated rearing cells (Fig. 8). (B) Air sampled in an adjacent room under positive air pressure relative to the pupal harvesting room.

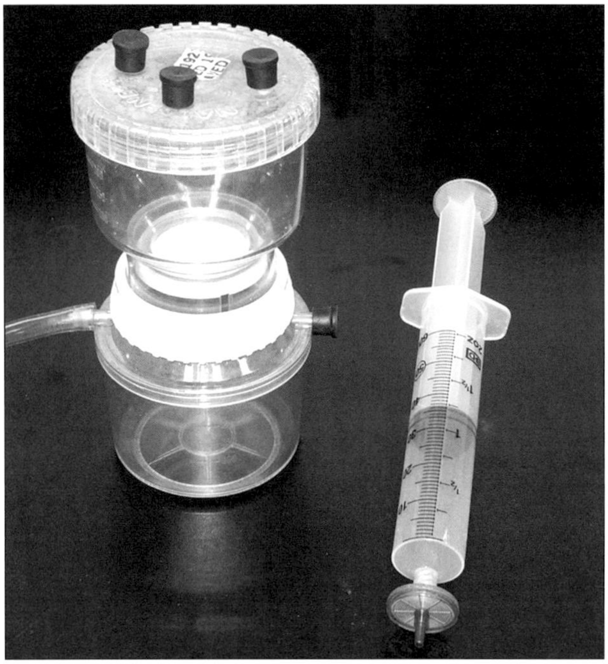

Fig. 54. Two examples of liquid filtration equipment. On the right is a disposable syringe filter equipped with a 0.22-μm filter. The syringe filter can be fitted to a number of different size syringes—a 50-ml syringe is illustrated. This apparatus is ideally suited for the filtration of small volumes. On the left is a side-arm filtration system, also equipped with a 0.22-μm filter. This system requires an externally supplied vacuum to facilitate movement of the liquid through the filter. The filtration system shown in this photograph is a reusable model and must be autoclaved prior to use. Other similar sterile disposable filtration systems are available.

be stored in a polystyrene box or cooler for 2–4 days. However, dry ice should not be stored in a small room with improper ventilation because gaseous CO_2 causes asphyxiation. Furthermore, the storage container used must not be air tight or else excessive pressures can develop. Dry ice pellets can be added to hot water contained within a small ShopVac style container and the hose utilized to direct the resultant fog. In addition to the use of smokes and fogs, differential pressure transmitters (e.g., Dwyer Instruments Inc., Michigan City, IN) work well for monitoring pressure differences between rooms.

B. Insect Diets

Most diets for terrestrial insects are solid or semisolid and are excellent substrates for microorganisms. Cohen (2003) reviews the variables that make artificial insect diets suitable substrates for insects. With respect to microbial contamination, he emphasizes that insect diet specialists must balance the goal of minimizing the effects of microorganisms with the goal of providing nourishing diets and hospitable conditions for insects utilizing these diets. All strategies that are implemented to manage microorganisms can have adverse effects on insects, and this will be a recurring theme through the following discussion. Cohen (2003) also emphasizes that insect diet specialists should utilize the vast array of information and technologies developed for managing microorganisms in human foods. We totally concur; however, managing microorganisms in insect diets is in some respects unique. For example, the diversity of nutritional requirements among insect species requires a diversity of diets and, consequently, a diversity of tactics to manage microorganisms in those diets. Furthermore, insect diets are typically maintained at relatively high temperatures and humidities for prolonged periods of time—ideal conditions for microbial proliferation. Many of the common strategies used to control microorganisms in human foods (e.g., drying, high osmotic pressure, low pH, low water activity, and modification of the atmosphere by packaging) can have serious negative effects on insects. We believe that no single method can be used to control microorganisms in insect diets. The goal of this chapter is to present a number of strategies that insect rearing specialists can apply in combination to manage microbial contaminants in their particular situation. Throughout this chapter we stress management rather than the control of microorganisms, i.e., reduction of contamination to acceptable levels—not its eradication. The four major categories of methods used to manage microorganisms in insect diets are as follows: 1) preventing introduction of microorganisms into diets; 2) killing microorganisms in diets, usually during preparation; 3) preventing recontamination of diets; and 4) inhibiting microorganisms in diets.

1. Preventing the Introduction of Microorganisms into Diets. *a. Storage of Diet Ingredients.* Diet ingredients often contain a variety of microorganisms, and they may proliferate to unacceptably high levels during storage in the insect rearing facility before use. While completed diets are generally better substrates for growth of microorganisms, diet ingredients also can support microbial development. In general, the higher the water activity of a substrate (Sec. VI. B. 4. a), the greater the chance of microbial development. Manipulation of the storage environment is frequently used to prevent microbial development. The most common storage condition to limit microbial spoilage of diet ingredients is low temperature. Low temperatures generally do not kill microorganisms, but they do slow down their metabolism and thereby reduce the rapidity of decomposition. However, a number of *psychrophilic* (low-temperature-loving) or *psychrotolerant* microorganisms can cause substantial spoilage at low temperatures. It is important to realize that storage conditions, often as a function of time, can seriously affect the nutritional quality of diets and diet ingredients. For ingredients whose water activity (Sec. VI. B. 4. a) is < 0.5 and with moisture contents $< 10\%$, storage at low temperatures often preserves their nutritional value and palatability for months (Cohen 2003). In contrast, ingredients with higher moisture content are subject to a number of physical and chemical changes. For information on the influence of low temperature storage on artificial diets and diet ingredients, the reader is referred to Cohen (2003).

b. Liquid Filtration. Physical removal of bacteria, fungi, larger multicellular microorganisms (e.g., protozoa), and many viruses in liquids can be achieved using filtration. Filtration is commonly used for heat-labile components (e.g., vitamins and antimicrobial agents). The most commonly used pore sizes for filters used to remove bacteria are 0.45 µm and 0.22 µm. Although some bacteria will pass through 0.45-µm filters, the vast majority will be removed; however, for more critical applications, a 0.22-µm filter is recommended. The disadvantage of 0.22-µm filters is the higher rate at which the pores become plugged if there is any particulate matter present in the solution. Virions of all entomopathogenic viruses are considerably smaller than 0.22 µm; and, with the exception of those forming proteinaceous occlusion bodies (Chap. 9 *Entomopathogens and Insect Rearing*), virions will pass through filters used for bacteria. Fortunately, entomopathogenic viruses are rarely associated with non-insect-derived diet components. There are a variety of methods used for aseptically filtering liquids, some of which require an external vacuum source. For small volume filtrations, syringes fitted with a syringe filter are used. For larger volume applications, sterile, disposable or autoclavable, side-arm filtration systems are typically used (Fig. 54). Careful attention to aseptic technique is required.

2. Killing Microorganisms in Diets. A number of methods have been used to kill microorganisms in insect diets. The objective of some methods is sterilization, whereas the objective of others is *pasteurization* (reducing populations of microorganisms). Pasteurization was developed by Louis Pasteur, who found that heating wine to a temperature well below the boiling point destroyed the majority of spoilage-causing bacteria without affecting flavor characteristics of the wine. Pasteurization is commonly applied to a variety of human foods such as dairy products. Every method used to kill microorganisms in insect diets possesses advantages and disadvantages. While some methods are less destructive, it should be realized that all methods cause varying degrees of chemical alteration to the diet. On the other hand, as discussed below, some of these chemical alterations can be beneficial.

A valuable measure of the efficacy of sterilization and pasteurization methods is one used in human food microbiology: *D-value* (*D*, time required for the microbial population to be reduced 1 $\log_{10}$ unit, i.e., to 10% of its initial value) (Fig. 55). *S* (survival rate of the microbial population) is inversely proportional to *D*. For example, if the time required to reduce a viable population of bacteria from 10^8 to 10^7 bacteria per g is 1.0 h, then $D = 1.0$ h and $S = 0.1/D = 0.1$ h^{-1} for the control method used. It is possible for *D* to vary as a function of population level, i.e., to be density dependent. In this case, the straight-line relationship illustrated in Fig. 55 is replaced by a curvilinear survivorship curve. *D* is most often used for heat deactivation of microorganisms, but it is applicable to other methods as well.

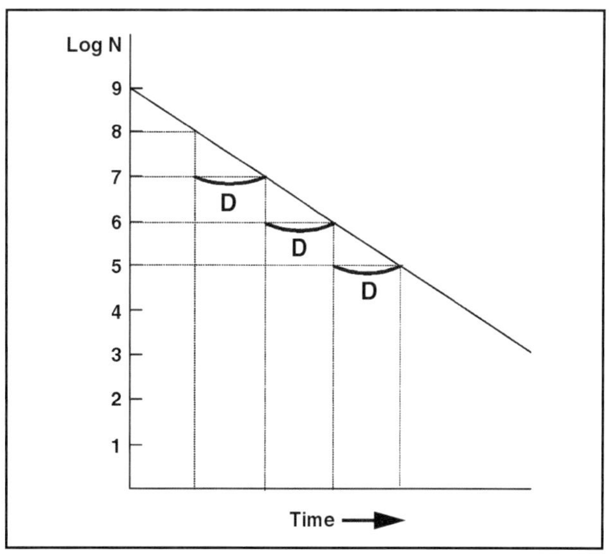

Fig. 55. $\log_{10}$-linearized survival curve for a population of microorganisms subject to a constant rate of mortality. The D-value is the length of time for the population to be reduced 1 $\log_{10}$ unit (i.e., the time for 90% of the population to die). [Based on Farkas (2001).]

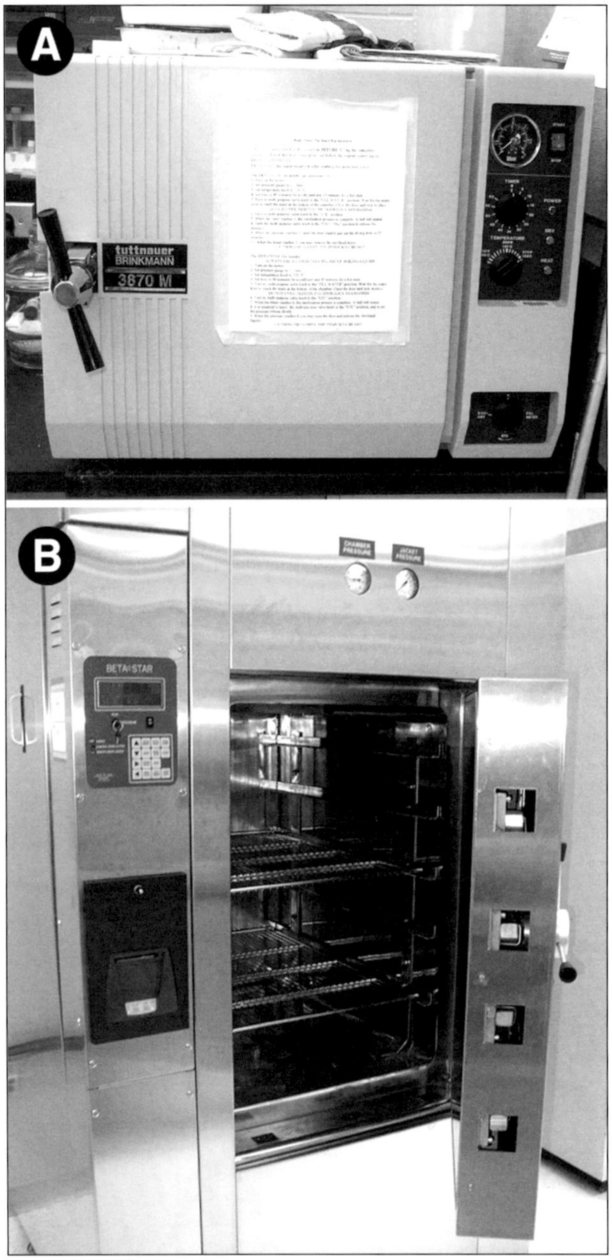

Fig. 56. Two types of autoclaves used to sterilize materials. (A) A bench-top autoclave that generates its own steam and is used to sterilize small volumes of diet and other materials. (B) A large, floor-model autoclave that requires an external source of steam and is used for large-scale sterilization processes.

a. Heat Sterilization/Pasteurization. By far the most common method of killing microorganisms in insect diets involves the use of heat (with or without pressure) with the goal of sterilizing or pasteurizing the diet. *TDT* (thermal death time) is the minimum length of time required to sterilize a substrate at a specified temperature. *TDP* (thermal death point) is the lowest temperature at which a substrate is sterilized within 10 min. It is important to realize that a variety of characteristics of the substrate to be treated influence the efficacy of heating. For example, water (including pH), fat, salt, carbohydrate, and protein content; number, age, and type of microorganism (e.g., endospores versus vegetative cells); temperature; time; and pressure have all been demonstrated to affect thermal deactivation (Jay 1996). For this reason, it is essential to monitor the efficacy of the sterilization method using microbiological methods and to determine *TDT* and *TDP* for each substrate of interest. The general rule of thumb is to use the minimum combination of temperature and time to render a substrate sufficiently free of viable microorganisms to make it acceptable for its intended use. A number of devices are commonly used in insectaries to kill microorganisms in diets by heating. These include autoclaves, kettles, flash sterilizers, and twin-screw extruders. In addition to the impact of heat on killing microorganisms, it is important to consider other effects, both positive and negative. Positive impacts of heat on diets include activation of gelling agents, increase in protein digestibility, destruction of digestive inhibitors and enzymes, enhanced flavor, and acceleration of desirable chemical reactions, to name a few (Cohen 2003).

One of the most popular devices used to sterilize diets is an autoclave. An *autoclave* is a chamber in which steam sterilization is carried out under pressure. At 2.0 atm pressure (= 203 kPa = 29.4 psi), water boils at 121°C. The steam at this temperature provides > 7X as much available heat as air at the same temperature (Ernst 1977). Although all autoclaves work on the same principle, a variety of different models are available (Fig. 56). In most applications, all microorganisms and their endospores are killed in about 15 min. The primary cause of death in microorganisms exposed to high temperatures is the coagulation of proteins, denaturation of nucleic acids, destruction of lipids, and disruption of lipids' normal associations with components of cell membranes (Wolin 1979). Autoclaves are not generally used to sterilize large quantities of complete diet. This is due to the relatively poor penetration of heat into the diet and the fact that heat-labile components of the diet may be substantially degraded during the prolonged heating period. It is important to realize that the size and shape of the container within which the diet is placed, the thermal conductivity of the container, and the physical characteristics of the diet itself can have profound effects on the efficacy of autoclaving. Some rearing specialists have advocated adding excessive amounts of heat-sensitive components to compensate for heat-related losses, but extreme care must be exercised in using this strategy because changing the relative concentrations of diet components can severely affect insect nutrition (Chap. 7 *Insect Nutrition, Feeding, and Artificial Diets*). A superior strategy is to sterilize heat-labile and nonlabile nutrients separately (Vanderzant 1966, Reinecke and Sikorowski 1989): nonlabile nutrients can be autoclaved, and other sterilization methods not involving heating, such as liquid filtration, can be employed for heat-labile nutrients. For a detailed review of heat sterilization, the reader is referred to Joslyn (1991).

If autoclaving is used to sterilize substrates, we recommend that sterility of the treated substrates be routinely monitored. In addition, the efficacy of the autoclave should be periodically assessed using endospores of *Geobacillus stearothermophilus* (formerly, *Bacillus stearothermophilus*), one of the most heat-resistant cells known to science, using one of a variety of commercially available kits. Physical and chemical indicators also may be used to assess the efficacy of sterilization (Joslyn 1991).

A *flash sterilizer* operates by heating the diet rapidly to a high temperature (130–160°C) as it moves through thin coils in a heat exchanger (Fig. 57A). Because the coils constitute a closed system, sufficiently high pressure is maintained to prevent the diet from boiling. Once the diet passes through the heating coils, it is cooled to 40°C. Because of the high temperatures encountered, *TDT* is typically only 2 to 3 min. Flash sterilization is a reliable but relatively expensive method of diet sterilization due to the large capital expenditure required to purchase the equipment. In some situations, the flash sterilizer can be coupled with a form-fill-seal machine (Cohen 2003). The major advantages of this method are the high rate at which diet can be processed and the relatively minimal adverse effects on heat-labile diet components such as vitamins, amino acids, and lipids (Fennema 1996).

Of particular concern to managing microbial contaminants is the impact of heat on heat-labile antimicrobial compounds. Hedin et al. (1974) did not detect any degradation of potassium sorbate or methyl-p-hydroxybenzoate, two commonly used microbial inhibitors. Reinecke and Sikorowski (1989) showed that both flash sterilization and autoclaving reduced the activity of penicillin, chloramphenicol, kanamycin monosulfate, and erythromycin but that flash sterilization was substantially less detrimental than autoclaving to all of the antibiotics tested.

In addition to cost, an additional disadvantage of flash sterilization is the difficulty in sanitizing the coils after use. Interior surfaces are cleaned with an acid wash and fresh water, requiring ½ h per 4 h of run time (Cohen 2003). For additional information on flash sterilization of insect diets, the reader is referred to Griffin et al. (1974) and Cohen (2003).

Steam kettles are the most popular method of heat-treating artificial diets, particularly in small to medium-sized insectaries (Fig. 57B). With this method, the diet is heated at atmospheric pressure under conditions of vigorous stirring to facilitate transfer of heat. The name

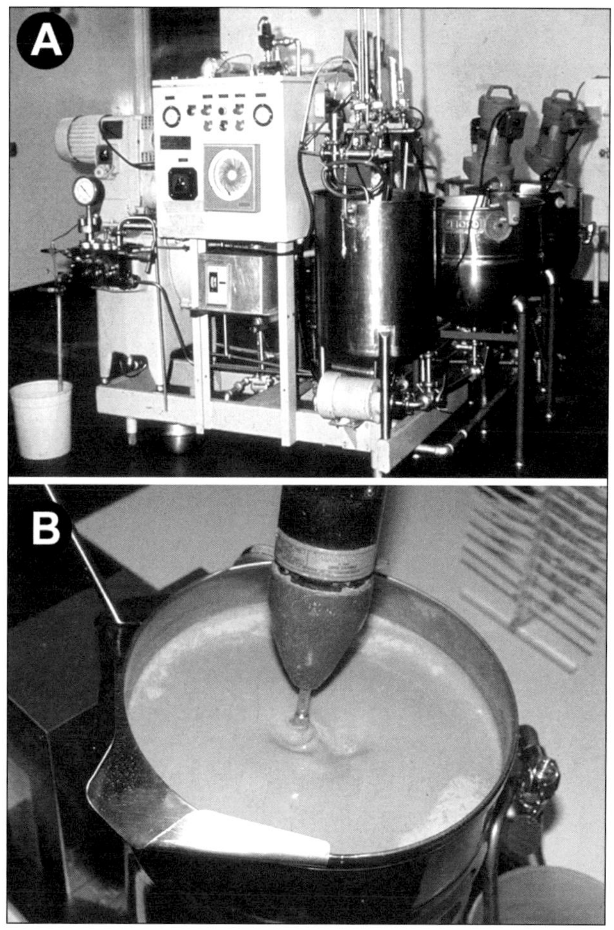

Fig. 57. Two types of equipment used to heat-treat insect diets. (A) A flash sterilizer. Sterilization of diets is achieved by exposing a thin stream of diet to high temperatures under moderate levels of pressure for a relatively short period of time. (B) A steam kettle. The jacket of the kettle is supplied with steam under pressure, and the diet is heated at atmospheric pressure with vigorous stirring to facilitate transfer of heat. This method of heat treatment will not kill endospores and should be considered pasteurization rather than sterilization.

steam kettle derives from the design: a jacket constituting the wall of the kettle containing the diet is supplied with steam under pressure. Typically the temperature of the jacket rises higher than the boiling point of water. The major disadvantages of steam kettles are inefficient transfer of heat from the jacket wall to the bulk of the diet and maximum diet temperatures insufficient to kill heat-resistant microbial cells. Most of the non-sporeforming bacteria and vegetative cells of sporeformers will be killed by exposures to 80ºC for 5 to 10 min; however, this treatment will not kill all bacterial endospores and other heat-resistant microorganisms. Consequently, steam kettles should be regarded as providing pasteurization rather than sterilization. It is imperative that the microbial status of diets heat-treated in steam kettles be assessed for microorganisms. The major advantages of steam kettles include low purchase cost, the high rate at which diet can be processed, and ease of sanitation following use.

Pasteurization of diets is accomplished by exposing the diet to relatively low temperatures (60–75°C) for short periods of time (15 sec to 30 min). Although pasteurization enjoys the advantage of causing less severe effects on dietary components, it generally does not kill *thermophilic* (high temperature–tolerant) microorganisms, bacterial endospores, or cells in a desiccated state. Blanching (short-term heat treatment of foodstuffs by boiling or steaming in order to destroy natural enzymes, soften tissue, and/or remove raw flavoring) is commonly used to heat-treat diet components (Cohen and Smith 1998). Although blanching is not primarily intended as an antimicrobial measure, it may be considered a form of pasteurization. When bacteria producing endospores (e.g., *Bacillus*, *Paenibacillus*, and *Clostridium*) are present in diets or diet components, a process called *tyndallization* may be used as an alternative to high temperature sterilization. Tyndallization involves the repeated (2 or 3 times), short-term exposure of diet or diet components to high temperatures (e.g., 100ºC) at intervals of usually 24 hours. The rationale for tyndallization is that exposure to a moderately high temperature stimulates endospore germination and that the resulting vegetative cells are vulnerable to subsequent exposures to those same, moderately high temperatures.

Other technologies have been developed for the sterilization or pasteurization of foods, and a number of these technologies have been used to reduce microbial spoilage of insect diets. One technology that is frequently used in the food industry and in some insectaries is the *twin-screw extruder*. The two "screws" of this device may be corotated or counterrotated, intermeshed or nonintermeshed. In addition, the configurations of the screws themselves may be varied using forward conveying elements, reverse conveying elements, kneading blocks, and other designs in order to achieve particular mixing characteristics; and they can produce great force to propel the food through the extruder while heating the matrix. With respect to the management of microorganisms, conditions of pressure and temperature can be controlled along the path of the food. A high processing temperature (157ºC) in combination with a low water activity (Sec. VI. B. 4. a) can result in undetectable microbial populations in the extrudate (Murphy et al. 2003). Bulut et al. (1999) concluded that shear forces, rather than heat, were primarily responsible for death of even heat-resistant bacterial cells.

Additional technologies based on heating that are used to kill microorganisms in foods include microwave heating, ohmic heating, and ultrasound treatment. *Microwave* (electromagnetic radiation of 0.3–30 cm wavelength) heating of food has gained widespread acceptance, but this technology is not particularly effective for deactivating microorganisms because spatial heterogeneity in heating means that microorganisms within some portions of the food will not be exposed to lethal temperatures (Farkas 2001). *Ohmic* (direct electrical) heating of foods is still in the development stage but shows considerable promise for sterilizing substrates. *Ultrasound* can kill microorganisms but used on its own is not particularly practical (Farkas 2001). However, ultrasound has been shown to sensitize microorganisms to heat, and its application in combination with heat may prove useful.

b. Radiation. Two technologies that have demonstrated considerable promise for the sterilization of foods are irradiation and *e-beam* (accelerated electron beam) technology. Two forms of radiation that can be used to sterilize substrates include UV and ionizing radiation. One can categorize spoilage microorganisms in order of decreasing sensitivity to UV radiation as follows: Gram-negative bacteria; Gram-positive bacteria; many viruses, bacterial endospores, and molds; and molds that produce melanin, a dark pigment (Farkas 2001). Besides the insensitivity of some microorganisms to UV, the major limitations of using UV radiation for treating insect diets are poor penetration into bulk diet (Lee et al. 1989) and difficulty in obtaining uniform exposure over all points of the food surface (Farkas 2001).

Ionizing radiation includes high-energy electromagnetic radiation, such as gamma rays and X-rays, and high-energy particles, such as *alpha* and *beta* particles (helium nuclei and electrons, respectively). *X-rays* are high-energy, penetrating radiation with wavelengths of 0.01–10 nm. *Gamma rays* are even higher energy and have wavelengths < 0.01 nm. Except for X-rays, ionizing radiation is typically generated by radioactive sources. For example, gamma rays are usually obtained from Co-60 or Cs-137 sources. Ionizing radiation causes damage to nucleic acids and by preventing accurate DNA synthesis resulting in death of the microbial cells. As a function of dose (5–10 MeV), this type of radiation can kill microorganisms in food without causing serious changes to the food itself (Farkas 2001). A major advantage of ionizing radiation is its excellent penetration characteristics. For example, gamma rays can penetrate water-based foods up to a depth of 4.0 cm (Urbain 1986). The primary disadvantages of gamma irradiation of foods are that specialized, expensive equipment is required, the isotope emits gamma rays continuously in all directions, and the technology is highly regulated for safety reasons. As a result, irradiation of foods is restricted to specialized facilities. However, at least one insect diet component, corn cob grits, is currently being sterilized by gamma rays (Bio-Serv, Frenchtown, NJ; http://www.insectrearing.com/index.html). As with heat treatment, the characteristics of the substrate influence the ability of gamma radiation to sterilize food. Some salient factors that decrease the efficacy of irradiation include the following: high numbers of microorganisms, high levels of protein, absence of oxygen, low water content of microorganisms, and bacteria that are not actively growing (i.e., in lag versus exponential phase of growth) (Jay 1996). While ionizing electromagnetic radiation can provide excellent sterilization of insect diets with reduced chemical alterations relative to heat, it must be stressed that it still alters the chemical composition of the diet.

In the case of ionizing radiation consisting of high-energy particles, alpha particles are not suitable for killing microorganisms in insect diets because of their low penetrating power. Electrons in the form of beta particles from radioactive decay are generally not suitable because they are of insufficiently high energy. E-beams are produced by acceleration of electrons to sufficiently high energies to be an effective sterilizing agent, and this technology offers certain advantages over other forms of radiation. E-beam technology is very versatile, and high-energy e-beams are able to sterilize much thicker substrates than gamma rays (Koch and Eisenhower 1965, Jay 1996). Although e-beams are being used to sterilize foods, we are not aware of instances where this technology is currently being used to sterilize insect diets.

c. Chemical Fumigants. A variety of chemical fumigants may be used to kill microorganisms in insect diets. The most common chemicals used are ethylene oxide, propylene oxide, formaldehyde, and paraformaldehyde. Poor penetration into diets, human toxicity, and residual toxicity to insects have largely limited their use. Ethylene oxide is a gas that kills microorganisms by alkylation of cellular proteins and nucleic acids. It is commonly used to sterilize items that would be damaged by heat, such as materials made of plastic (e.g., Petri dishes and insect rearing containers). It is toxic to humans at concentrations of 100 ppm (Dunham 1977) and is not generally used to sterilize foods. Mortality of southwestern corn borer placed on diets sterilized with ethylene oxide is nearly 100% (Frank Davis, pers. com.).

Propylene oxide is slower acting than ethylene oxide but typically presents fewer toxicity problems. Natural diets of the alfalfa leafcutter bee, *Megachile rotundata* (Hymenoptera: Megachilidae) have been effectively sterilized with propylene oxide (Inglis et al. 1992). Fumigation had no effect on the protein content or distribution of pollen within the diet but increased the pH and reduced total sugars and quantities of fructose and glucose. However, developing larvae were very sensitive to residual propylene oxide in the diet, and exposure of the treated diet to ca. 120 h of vacuum was required to remove the residual propylene oxide. Unfortunately, this step reduced the water content of the provisions by 40%.

Historically, formaldehyde has been used as a vapor-phase decontaminant in hospitals (Hoffman 1971). A major limitation of its use is that formaldehyde as a vapor, as a solid (*paraformaldehyde*), and as an aqueous solution (*formalin*) are all very toxic and carcinogenic to humans (Sun 1981). The vapor phase can produce paraformaldehyde on dry surfaces and formalin on damp surfaces. Removal of paraformaldehyde from conventional rooms can be a lengthy process, often requiring days of aeration. Formaldehyde vapor is now used to decontaminate rooms and equipment in containment level 3 and 4 pathogen compliant facilities. Formaldehyde vapor also is used extensively to decontaminate HEPA filters from biosafety cabinets. It is imperative that a certified technician be contracted to decontaminate equipment in an insect rearing facility using these materials. In contrast to formaldehyde, paraformaldehyde is available as a powder. It is readily converted to formaldehyde gas with heat. It has been used to kill spores of the fungal pathogen *Ascosphaera aggregata* in diet of *M. rotundata* with minimal toxicity towards eggs or larvae (Goettel and Duke 1996).

d. Surface Sanitation of Natural Diets. In many instances, plant parts are used to rear herbivorous insects. In some instances, it is prudent to kill microorganisms on the surfaces of such tissues to prevent extensive development of microorganisms in the moribund tissues. A variety of different surface sanitation methods can be applied. For example, Sikorowski and Lawrence (1994) washed cotton leaves in 0.5% sodium hypochlorite for 10 min, followed by a 1- to 2-h rinse in running tap water. Chlorinated tap water is adequate for rinsing because it typically contains very few viable microorganisms. Sanitizing cotton leaves in such a manner had no significant effect on either foliage consumption or vigor of 3-d-old *H. virescens* larvae. In some instances, residual sodium hypochlorite can be detrimental, and additional measures should be taken (Sec. VI. B. 3. b). As mentioned previously, surface sanitation will not be efficacious against microorganisms within plant tissues.

3. Preventing Contamination of Diets. An often overlooked aspect of managing microorganisms is preventing contamination of diets following sterilization and pasteurization. This involves the implementation of aseptic techniques including the use of clean work stations (Sec. VI. A. 2).

a. Aseptic Packaging. Following sterilization of artificial diet, it must be aseptically dispensed into rearing containers. Typically, plastic containers, even those that are specified as microorganism free by the manufacturer, should be treated with ethylene oxide or some other type of gaseous fumigant. Periodic testing of diet containers for microorganisms should be conducted. In some diet sterilization procedures (e.g., flash sterilizers), diet is dispensed immediately after the sterilization procedure. With other methods (e.g., autoclaves and steam kettles), it is necessary to remove the molten diet from the device and allow it to cool before it can be dispensed into rearing containers. This process can be a high-risk period for microbial contamination from air or directly from humans. Judicious use of a clean work station and careful attention to worker hygiene is important during this period. Whether automated, semiautomated, or manual dispensing devices are used, contamination of these devices will in turn contaminate the diet. Therefore, dispensing devices must be adequately sanitized prior to and after use, and the efficacy of the sanitization method should be monitored (Sec. VI. C. 3). Insects within rearing chambers require adequate air exchange, and pores in the lids of rearing containers are a potential portal by which microorganisms can gain access to the diet. The number and size of pores should be minimized as much as possible given other constraints.

b. Surface Sanitation of Insects. Insects represent an important source of microbial contaminants for diets. Reducing the introduction of specified microorganisms contaminating the insects themselves can be an important step in the management of contamination in insectaries. By far the most common and efficacious procedures used for surface sanitation of insects are those directed toward the egg stage. Eggs are typically contaminated by various microorganisms present in the rearing environment including those carried by both the insects and humans. The primary reasons for targeting eggs are the following: 1) eggs of many species possess a thick chorion, which protects the embryo from the toxic effects of sanitizers; 2) eggs are relatively easy to handle, so the process is amenable to mechanization; and 3) few microorganisms occur within egg cytoplasm and embryo tissues—the exceptions being vertically transmitted entomopathogens and mutualistic symbionts. It is important to note that surface sanitation will not be effective against microorganisms within the egg without having a detrimental impact on the egg or embryo. The major objective of egg sanitation is to kill microorganisms on the chorionic surface. The primary factor determining the impact of the sanitation procedure on the egg itself is the thickness of the chorion: the thicker the chorion, the lower the impact. The chorion is a complex structure that varies in thickness from < 1 μm in eggs of parasitic Hymenoptera to over 50 μm in some orthopteran and lepidopteran eggs (Chapman 1998). Other factors, such as the composition of the chorion, also are likely important. In many medium-

and large-scale insectaries, sanitation of eggs is mechanized; however, extreme care must be taken to ensure that eggs are not mechanically damaged during this process (Leppla et al. 1974).

When the chorion is thin or the eggs are very small, it is not possible to surface-sanitize them. In these cases, it may be useful to surface-sanitize other stages such as larvae, nymphs, pupae, or adults. However, the detrimental impacts of this procedure on these developmental stages tend to be high, and sanitizers have minimal efficacy against microorganisms within the alimentary canal. Microorganisms present within the alimentary canal will be released in frass and contaminate the diet. Another problem with surface sanitation of insects is that the procedure is rarely 100% effective. Insect cuticles are typically hydrophobic, and this, along with complex surface morphology, may limit contact between microorganisms and the active ingredient of the sanitizer through formation of microbubbles. Therefore, surface sanitation of nonegg stages of insects should be considered a secondary option.

There is an extensive array of commercial sanitizers available. However, most of the antimicrobial chemicals they contain belong to a small number of chemical classes. Differences in sanitizers are largely due to formulation (e.g., type of surfactants). The most common chemical classes used to surface-sanitize insects include sodium hypochlorite, formaldehyde, and quaternary ammonium compounds. The ideal sanitizer should be fast acting, effective against a wide range of microorganisms, exhibit good penetration characteristics without damaging or discoloring substrates, be easy to prepare and stable, be readily available and inexpensive, and exhibit low toxicity to humans and insects. The primary modes of action of sanitizers are protein denaturation, membrane disruption, nucleic acid damage, and inhibition of metabolic processes.

Because of its potent, broad-spectrum microbicidal activity, sodium hypochlorite is used as a general-purpose sanitizer. However, the effectiveness of sodium hypochlorite depends on its reaction with water to form hypochlorous acid, which is the actual killing agent. Mole for mole, hypochlorous acid is one of the most powerful antimicrobial sanitizers known (Wolin 1979) and is effective against both viruses and bacterial endospores. How hypochlorous acid kills is not completely understood (Tortora et al. 1989); however, hypochlorous acid liberated when water is added to the sodium hypochlorite in its nondissociated form is highly microbicidal, and the acid pH favors its maintenance in this form. Thus, its killing action decreases sharply as pH of the solution increases above neutrality. Because hypochlorous acid oxidizes organic matter indiscriminately, a sufficient quantity must be added to satisfy the demand of the treated substrate and leave sufficient hypochlorous acid to react with the targeted microorganisms. Sodium hypochlorite is unstable (i.e., it decomposes to sodium chloride and sodium chlorate), which limits its ability to be stored; and it can represent a health concern to workers in confined spaces because it releases chlorine gas. Because it is such a strong oxidizing agent, it is very corrosive, and it can severely damage metal work surfaces including those constructed of stainless steel.

Various concentrations of sodium hypochlorite have been used to surface-sanitize insect eggs. For example, 0.05% sodium hypochlorite for 45 min was used to surface-sanitize eggs of gypsy moth, *Lymantria dispar* (Lepidoptera: Lymantriidae), (Leonard and Doane 1966); and 1.0% sodium hypochlorite solution for 10 min at 30°C was used to sanitize eggs of pink bollworm, *Pectinophora gossypiella* (Lepidoptera: Gelechiidae), (Stewart et al. 1976). Sodium hypochlorite is sporacidal, and treatment with 0.05 and 0.5% sodium hypochlorite for 1 to 10 min reduces both the viability and infectivity of *Bac. thuringiensis* endospores (Ignoffo and Dutky 1963). Given the potential for damage of insect eggs due to exposure to sodium hypochlorite, it is imperative that the optimum concentration and exposure time be determined experimentally for the eggs of each newly colonized insect species. Residual sodium hypochlorite can affect insects, and a detoxification step may be required. This could involve rinses with water or the use of a detoxicant such as *antichlor* (1% sodium thiosulfate). However, it is important to remember that the detoxification step may result in its own negative side effects such as additional mechanical damage.

When formalin is used in a sufficiently high concentration (generally available at a maximum concentration of 37%), all microorganisms are destroyed, including bacterial endospores and viruses, through its methylating action (i.e., addition of methyl groups to proteins, DNA, etc.). It is used to kill microorganisms in vaccines.

Formalin is normally used at concentrations ranging from 5–10% for 5–90 min to sanitize insect eggs, as in the following examples: 1) 15 min with 5% formalin to sanitize eggs of Diptera (White 1931); 2) 90 min with 10% formalin to sanitize eggs of the alfalfa caterpillar, *Colias eurytheme* Boisduval (Lepidoptera: Pieridae) (Thompson and Steinhaus 1950); and 3) 20 min with 10% formalin to sanitize eggs of *A. grandis* (Sikorowski et al. 1977). Even at low concentrations of formalin, germination of *Bac. subtilis* endospores is inhibited (Trujillo and David 1972). It is essential that the insect eggs are rinsed for at least 30 min with sterile water following exposure to formaldehyde because eggs with even a trace of residual formaldehyde typically will not hatch. Although formaldehyde is an excellent sanitizer with respect to its microbicidal properties, it is volatile; and it is imperative that all work be conducted in an externally exhausted fume hood. Because of its toxicity, the use of formaldehyde has been discontinued in many insectaries.

Quaternary ammonium salts are effective against Gram-positive bacteria but less effective against Gram-negative bacteria. Certain bacteria, such as some species of *Pseudomonas*, not only survive in quaternary ammonium salts but actively grow in them. Quaternary ammonium salts also possess good activity against some viruses and fungi but are not effective against bacterial endospores. The mode of action of quaternary ammonium salts is unknown but most likely involves the plasma membrane (Tortora et al. 1989). In general, the quaternary ammonium salts are not as effective as the hypochlorites in killing bacterial endospores or Gram-negative bacteria. They are inactivated by many soaps, anionic detergents, and by calcium and magnesium. They also are inhibited by cotton and other fibrous or porous substances. Quaternary ammonium salts are typically formulated with other agents to overcome some of the above problems and to increase their effectiveness. Examples of commonly used formulations of quaternary ammonium salts include Zephiran™ and Cepacol™.

Quaternary ammonium salts have been used in concentrations of 0.1–0.4% for 5–25 min to sanitize insect eggs and other life stages, as in the following examples: 1) 25 min with 0.1% aqueous solution of Zephiran to sanitize eggs of blowflies (*Calliphora* spp., Diptera: Calliphoridae) (Brust and Fraenkel 1955); (2) Zephiran™ (unspecified concentration and exposure) to sanitize eggs of the southern corn rootworm, *Diabrotica undecimpunctata* (Coleoptera: Chrysomelidae), (reduced egg viability and larval vigor) (Sutter et al. 1971); and 3) 5 min with 0.2–0.4% Hyamine 10-X™, an obsolete formulation, to sanitize larvae and pupae of various Lepidoptera (Martignoni and Milstead 1960). Water rinses are typically required to remove residues.

The use of alcohols for surface sanitation of insects is limited because they possess relatively poor penetration characteristics and they do not kill bacterial endospores or resistant cells. Considering methyl (C-3) through amyl (C-5) alcohol, microbicidal activity increases with increasing chain length. In practice, however, butyl and amyl alcohol are not used because of their odor and poor solubility in water. Ethyl and isopropyl alcohol are widely used as skin antiseptics, and both are excellent for this purpose. Alcohols disrupt proteins and membranes and dissolve lipids. The ease with which proteins are denatured depends on the extent to which the cell is hydrated. Because absolute alcohols are dehydrating agents, they are much less effective than are alcohol-water mixtures. Alcohol-water mixtures of 50–70% alcohol provide the greatest antimicrobial activity, with isopropyl being somewhat better than ethyl. Alcohol-water mixtures are also more penetrating than pure alcohols. Because alcohols are lipid solvents, they possess some activity against viruses, which have lipid components in their envelopes. The lipid-solvent action of alcohols will cause drying of the skin if used repeatedly and excessively. A major advantage of the alcohols is that they evaporate rapidly and leave a minimal residue. Sometimes alcohols are used in conjunction with other surface sanitizers to improve the wetting of hydrophobic surfaces such as most insect cuticles.

Other types of sanitizers include iodine, iodophores, copper sulfate, phenol (including halogenated phenolic agents), and hydrogen peroxide compounds. However, these sanitizers are infrequently used to kill microorganisms on the surfaces of insects. Peroxigard®, a relatively new product, contains hydrogen peroxide formulated with a number of biodegradable adjuvants. It is touted as being nontoxic and more highly microbicidal than quaternary ammonium salts, phenol, and sodium hypochlorite. To our knowledge, this product has not been tested as a surface sanitizer of insects. For more information on these sanitizers, see Sec. VI. C. 2.

4. Inhibition of Microorganisms in Diets. Even if the appropriate steps are taken to kill microorganisms in the diet and the diet is aseptically packaged, this does not preclude growth of residual cells or subsequent contamination. Therefore, methods to inhibit the growth of microorganisms in insect diets are often required. The choice of a method depends on the microorganisms of concern, the physical and chemical properties of the diet, the impact of the agent on the insects, and a variety of additional factors such as human toxicity. In recent years, many insectary managers have become dependent on the use of a few antimicrobial agents and antibiotics used in human foods. Despite the recent development and application of a variety of new strategies for managing microorganisms in human foods, the management of microbial contaminants in insect diets has not advanced correspondingly. The first step in remedying this deficiency is for rearing specialists to realize that the diet-microorganism-insect interaction is complex and that a single method to suppress microorganisms in all situations does not and will likely never exist. If efficacious management of microbial contaminants of insect diets is to be realized, it is imperative that the microbial taxa causing the spoilage be determined and the degree to which various management strategies affect not only the microorganisms but also the insects utilizing the diet be determined as well. Unfortunately, this information has rarely been obtained. Simply reporting the presence of total bacteria or fungi is inadequate. Furthermore, information concerning the physical and chemical alterations to diets caused by microorganisms and how these changes affect insects is limited. This information will be obtained only once strong working collaborations are formed among microbiologists, diet specialists, and rearing specialists.

a. Diet Characteristics. The physical and chemical properties of human foods are often considered important elements in the suppression of microorganisms (Jay 1996), but they are often overlooked in the development of insect diets. Some of the more important, intrinsic variables that affect microorganisms include: 1) pH, 2) water activity and moisture content, and 3) oxidation-reduction potential. These variables often do not act independently, and this interaction also influences the efficacy of other management tactics. Unfortunately, most insects are relatively intolerant to substantive manipulation of these variables, and the challenge is to find a combination that inhibits microorganisms while minimizing adverse effects on insects (Cohen 2003). While intrinsic characteristics of diets may provide varying levels of suppression of spoilage microorganisms, it is unlikely that they will provide adequate control on their own. Therefore, it will be necessary to utilize these characteristics as components of an integrated management strategy.

Table 3. Minimum pH values allowing growth of some food-borne bacteria (Jay 1996)

Bacterium	pH
Aeromonas hydrophila	6.0
Alicyclobacillus acidocaldarius	2.0
Bacillus cereus	4.9
Clostridium botulinum (Group I)	4.6
Clostridium botulinum (Group II)	5.0
Clostridium perfringens	5.0
Escherichia coli O157:H7	4.5
Gluconobacter spp.	3.6
Lactobacillus brevis	3.2
Lactobacillus plantarum	3.3
Lactococcus lactis	4.3
Listeria monocytogenes	4.1
Plesiomonas shigelloides	5.0
Salmonella spp.	4.1
Shewanella putrefaciens	5.0
Shigella flexneri	4.0
Shigella sonnei	5.0
Staphylococcus aureus	4.0
Vibrio parahaemolyticus	4.8
Yersinia enterocolitica	4.2

The pH of foods is influenced by a variety of abiotic factors including temperature and concentration of salts (Jay 1996). It has long been established that most bacteria grow best at pH values near neutrality (i.e., pH ≈ 7), and although yeasts and molds are less fastidious than bacteria, they are generally favored by slightly acidic (i.e., pH < 7) conditions. Exceptions to these generalizations include lactic acid bacteria, which are common spoilage microorganisms and are used to ferment foods. They produce organic acids as a by-product of fermentation and can grow at pH levels as low as 3. These low pH levels (Table 3), along with the antimicrobial proteins they produce, exclude many other bacterial competitors. However, some molds and yeasts grow at pH levels as low as 2. The pH of the diet also influences many aspects of its nutritional properties including texture and flavor, activity of enzymes, and overall palatability (Cohen 2003). Another important aspect of diet pH is the impact it has on the activity of many antimicrobial compounds. For example, the activity of sorbic acid is higher at lower pH levels (Sec. VI. B. 4. c).

The impact of water in diets on insects is exceptionally complex, affecting diet palatability, texture, nutritional value, and stability (Cohen 2003). Water is an essential nutrient for all organisms, and the sole source of water in most insect rearing settings is the diet. An important concept is *water activity* (a_w), the ratio of the vapor pressure of water in equilibrium with the diet substrate to the vapor pressure of water in equilibrium with pure, liquid water at the same temperature. a_w is closely related to relative humidity (RH) (Christian 1963). In essence, water activity of an insect diet is the tendency of water to leave the diet relative to the tendency of water to evaporate from pure water (Cohen 2003). Water activities have a range of 0–1. By definition, pure water has $a_w = 1$. A solution of 22% NaCl has $a_w = 0.86$, so water is less easily taken up by a microorganism from the salt solution than from pure water. A variety of factors influence the water activity of insect diets. Increasing concentrations of a variety of compounds (e.g., salts, sugars, amino acids, glycerol, urea) will lower the a_w of insect diets. In general, smaller molecules affect a_w more than large molecules (e.g., starches and proteins) (Cohen 2003).

Another related concept is *osmotic pressure* (the pressure required to counterbalance the tendency of water to move from the side of a membrane with a lower solute concentration to the other side) (Abercrombie et al. 1980). For example, when a solution of sugar in water is separated from pure water by a membrane permeable to water but not to the sugar, water will pass across the membrane into the sugar solution. This movement of water is *osmosis*. If external pressure is applied to the sugar solution, the movement of water into it will be opposed; and the pressure required to stop the movement completely is the osmotic pressure of the sugar solution. All cells possess a semipermeable plasma membrane that separates their cellular contents from the environment. Thus, in diets possessing low a_w, the concentration of solutes in the diet may be higher than the concentration of solutes in the cytoplasm of microorganisms utilizing this substrate, so the osmotic pressure would be positive, causing a tendency for the microorganism to lose water. However, some microorganisms are capable of growing under conditions of high osmotic pressure, and they are commonly referred to as *osmophilic* or *halophilic* (salt-loving).

The a_w of most insect diets is > 0.99. An obvious exception would be diets containing high concentrations of sugar. For honey, $a_w < 0.6$ (Cohen 2003). To compensate for the low a_w of honey, bees largely obtain water from other sources (e.g., pollen and water in the environment). Cohen (2003) stresses that water availability is essential for insect nutrition, and even when diets contain high water content, their water activity may be inappropriate. Water activity also has a profound effect on microorganisms, and reducing a_w of human foods by drying is one of the oldest methods of preventing microbial spoilage. In general, bacteria require higher a_w values than fungi. The majority of spoilage bacteria required a_w values ≥ 0.90 to grow, whereas most spoilage yeasts and molds require a_w values ≥ 0.88 and 0.80, respectively (Jay 1996). However, some halophilic bacteria can grow at a_w values of 0.75, and some osmophilic yeasts and molds can grow at a_w = 0.60. In general, Gram-negative bacteria require higher water activities than Gram-positive taxa, but considerable variability exists within these groups. Despite widespread use of methods to adjust the moisture content of human foods to suppress spoilage microorganisms, water activity has not been extensively studied as a method of managing microorganisms in insect diets. Cohen (2003) points to the following four reasons for this difference: 1) difficulty in meeting the insects' water and textural requirements if water concentration is reduced, 2) dilution of the bulk concentration of nutrients if nonnutritional solutes are added, 3) reduced acceptance of the diet, and 4) general increase in physiological stress. It is clear that reducing a_w to levels where all microbial activity will be terminated (i.e., < 0.60) is not a tenable approach in most instances. However, by lowering a_w to some degree, it may be possible to delay spoilage by increasing the lag phase (i.e., a state of apparent inactivity preceding a response) and thus decreasing the growth rate of microbial populations. It is important to reiterate that the interaction between microorganisms and water activities in diets is complex and influenced by other variables in a taxon-specific manner. Whether manipulating a_w in diets will prove effective in managing microbial contamination remains to be determined.

Another diet characteristic for which various microorganisms have different requirements is *Eh* (oxidation-reduction potential). This is a measure of the tendency of the substrate to gain electrons (reducing environment, $Eh < 0$) or to give up electrons (oxidizing environment, $Eh > 0$) relative to a standard electrode. Obligate anaerobes such as *Clostridium* require reducing conditions for growth ($Eh \approx -200$ mV), microaerophilic bacteria (e.g., lactobacilli) grow best under slightly reducing conditions, and obligate aerobes grow best under oxidizing conditions ($Eh > +200$ mV). Most plant products tend to have *Eh* values of +300 to 400 mV (Jay 1996), so it is not surprising that aerobic bacteria and fungi are the primary spoilage microorganisms encountered in such diets. A number of successful entomophage diets are based on meats (Cohen 2003); solid meats possess *Eh* values of −100 mV, whereas in minced meats $Eh \approx +200$ mV. However, minced-meat entomophage diets are often presented to insects in parafilm packets, which are impervious to gas exchange. It is well documented that microorganisms can modify the *Eh* of their environments during growth just as they do pH (Jay 1996; Fig. 21). For example, the decline in *Eh* of a *Chyrsoperla* diet (Cohen and Smith 1998) presented in parafilm packets may result from the initial activity of aerobic bacteria that results in the more reducing environment favoring the microaerophilic taxa (e.g., *Lactobacillus* and *Carnobacterium* species) ultimately responsible for spoilage of the diet (Inglis and Cohen 2004).

b. Environmental Factors. Abiotic environmental factors that affect both diets and spoilage microorganisms include temperature, relative humidity, and the composition of atmospheric gases. In most instances, it is not possible to adjust environmental conditions sufficiently to interfere appreciably with microbial development because of the similarity of the requirements of most insects and microorganisms; however, examples do exist. In the case of rearing the house cricket, *Acheta domesticus* (Orthoptera: Gryllidae), the growth of microorganisms in the diet and in the frass and insect debris that accumulate on the bottom of the rearing containers can be eliminated by manipulation of the temperature, humidity, and availability of water (Chap. 6 *Environmental Biology of Insect Rearing*).

c. Antimicrobial Agents. There are a vast number of antimicrobial agents available. However, only a relatively small number have been used in insect diets. To illustrate the limited diversity of antimicrobial agents that are employed in insect rearing, a compilation of 58 synthetic diets used for phytophagous insects showed that chlortetracycline (aureomycin) was used in 19 diets (32%), methyl-p-hydroxybenzoate in 58 diets (100%), sodium benzoate in six diets (10%), sorbic acid in 42 diets (72%), and formaldehyde in six diets (10%) (Singh and House 1970). Unfortunately, a reliance on a restricted number of antimicrobial agents in insect diets continues to this day. In this chapter, we divide antimicrobial agents into seven broad categories: 1) antibacterial antibiotics, 2) antifungal antibiotics, 3) synthetic fungicides, 4) human food antimicrobial agents, 5) antimicrobial proteins, 6) plant natural products, and 7) microbial inoculants.

Most antimicrobial chemicals are known under a variety of names. Each chemical possesses a *chemical name*, a *generic name* (common name; trivial name; simple, chemical name), and most possess one or more *trade names* (marketing names). In this chapter we primarily use generic names for antimicrobial agents. The chemical name is based on strict rules in chemical nomenclature and is often associated with a Chemical Abstracts Registry Number. The chemical name is rarely used by insectary managers, and most insect rearing specialists know a chemical by its generic or trade names. In the case of drugs, the generic name possesses a drug code number. Trade names are usually capitalized unless the trade name is the same as the generic name. It is important to realize that trade names often refer to specific formulations of a given material; whereas, the generic name refers to the specific chemical (i.e., the active ingredient). For example, consider the antibacterial antibiotic known by the generic name of "chlortetracycline." Its chemical name is 7-chloro-4-dimethylamino-1,4,4a,5,5a,6-11,12a-octahydro-3,6,10,12,12a-pentahydroxy-6-methyl-1,-11-dioxo-2-n aphthacenecarboxamide. It is marketed under the trade names Acronize, Aureocina, Aureomycin, Biomitsin, Biomycin, and Chrysomykine. A useful reference for insect rearing specialists is the Merck Index (O'Neil et al. 2001). In addition to names of chemicals, the Merck Index also contains information on physical data (e.g., solubility and stability), common uses, and therapeutic applications. The Merck Index does not provide adequate information on doses, so the readers must consult additional references. For additional information on antimicrobial compounds, the reader is referred to Block (1991).

The term *antibiotic* (literally "anti-life") is a broadly-used term but is typically used to describe a class of secondary metabolites produced by microorganisms to reduce the competitive threat posed by other microorganisms. Some use the more specific term *antimicrobic* for this class of chemicals, but the traditional term will be used here. Some antibiotics kill the target microorganism, and others inhibit microbial development. Antibiotics that inhibit development are referred to as *-static* agents (e.g., bacteristatic antibiotics). In contrast, those that kill the target microorganism are referred to as *-cidal* agents (e.g., bactericidal antibiotics).

Most antibacterial antibiotics are derived from fungi or other bacteria (primarily actinomycetes), the exception being the sulfonamide antibacterial agents. In 1929, penicillin, the first antibiotic, was discovered by Alexander Fleming. Fleming discovered that 6-aminopenicillamic acid (Fig. 58A), a secondary metabolite produced by the fungus *Penicillium chrysogenum*, inhibits cell wall synthesis of Gram-positive bacteria (Table 4). Penicillin was subsequently used to treat various diseases including syphilis (*Treponema pallidum*), gonorrhea (*Neisseria gonorrhoeae*), diphtheria (*Corynebacterium diphtheriae*), scarlet fever (*Streptococcus* spp.), tuberculosis (*Mycobacterium tuberculosis*), pneumonia (e.g., *Streptococcus pneumoniae*), and various bacteria causing gangrene. The discovery of this secondary metabolite revolutionized the treatment of infectious pathogens and saved millions of lives. Furthermore, it spawned the multi-billion dollar pharmaceutical industry. Many derivatives of the original 6-aminopenicillamic acid molecule have been developed that vary in their activities. For example, Penicillin G possesses increased stability at decreased pH levels, and ampicillin is active against Gram-positive and some Gram-negative bacteria. Additional classes and examples of antibacterial antibiotics, along with their basic properties, are presented in Table 4.

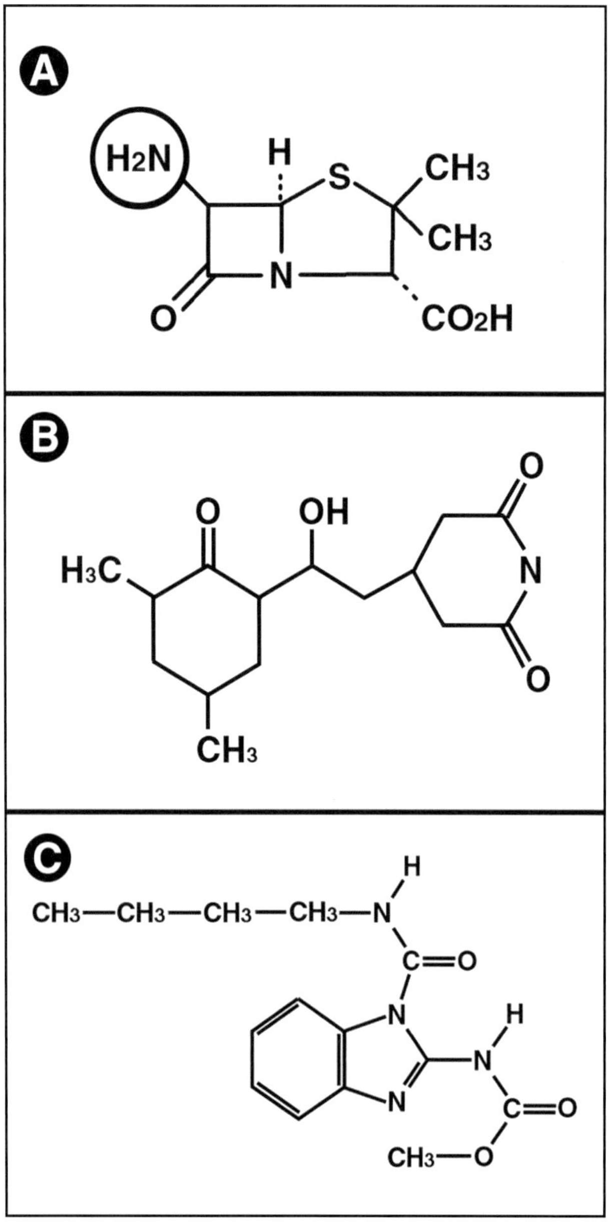

Fig. 58. Antimicrobial agents. (A) Aminopenicillamic acid, the parent molecule of all penicillins. The circled functional group determines the type of penicillin. (B) Cycloheximide. (C) Benomyl.

Although antibacterial antibiotics were initially predicted to eliminate bacteria-caused disease in humans, this has not occurred. Even as early as 1946, Fleming realized the Achilles' heel of penicillin (and other antibiotics). He wrote, "The administration of too small doses leads to the production of resistant strains of bacteria" (Amyes 2001). Three examples of antibacterial agents possessing a wide spectrum of activity that are commonly used in insect diets are chlortetracycline, streptomycin, and chloramphenicol. Bacteria of medical importance have demonstrated the ability to develop resistance to these and other antibiotics. Although it has largely been ignored in insect rearing, it is almost certain that the selection pressure resulting from the continuous inclusion of antibiotics in insect diets will result in the development of antibiotic resistance in bacteria associated with diets and with insects reared on the diets. Early on, Sikorowski et al. (1980) indicated that antibiotic use in insect diets should only be a temporary remedy. Unfortunately, this recommendation has largely been overlooked; and to this day, many rearing specialists still rely on the repeated, prophylactic incorporation of antibiotics in insect diets.

Often one is preoccupied with the effects of antibacterial antibiotics on spoilage bacteria, and the impact on mutualistic symbionts is overlooked. For example, incorporation of tetracycline antibiotics into insect diets kills *Wolbachia* and *Rickettsia* symbionts (Tanada and Kaya 1993).

Although fungi are a major problem as contaminants in insectaries, antifungal antibiotics are often overlooked. There are hundreds of antifungal antibiotics, most of them produced by a few species of actinomycetes (e.g., *Streptomyces*) and by fungi of the genera *Penicillum*, *Aspergillus*, and their relatives (Griffin 1981). Well-known examples of such agents include cycloheximide (Naramycin, Actidione) (Fig. 58B), griseofulvin (Fulcin, Fulvicin, Grifulvin, Grisactin, Grisovin, Gris-PEG, Grysio, Lamoryl, Likuden, Neo-Fulcin, Poncyl, Spirofulvin, Sporostatin), amphotericin B (Fugizone, Fungilin, Ampho-Moronal), and nystatin (Fungicidan, Diastatin, Mycostatin, Moronal, Nilstat, Nystatan, Nystavescent). Cycloheximide, amphotericin B, and nystatin are produced by *Streptomyces* sp., while griseofulvin is produced by *Penicillium griseofulvum*. Cycloheximide tends to be quite toxic to eukaryotic organisms, and it is useful as an antifungal agent only when the host is more resistant than the fungus (Griffin 1981). Griseofulvin has been used as a systemic fungicide against plant pathogenic fungi and is taken orally in doses of up to 1 g per day to combat recalcitrant dermatophyte infections of *Tinea* in humans. Griseofulvin affects fungal membranes and inhibits mitosis, but it is generally more fungistatic than fungicidal. Griseofulvin is considered

Table 4. Common antibacterial antibiotics

Class	Antibiotic	Mechanism	Target/Comments
Aminoglycosides	Gentamicin	Bactericidal; inhibit protein synthesis	Broad spectrum including antibiotic-resistant bacteria
	Neomycin		Broad spectrum
	Streptomycin		Broad spectrum
Cephalosporins	Cephalothin (1st generation) Cephoperazone (3rd generation)	Bactericidal; inhibit cell wall synthesis (similar to penicillins)	Activity of 1st–4th generations: most Gram (+) & some Gram (–), most Gram (+) & many Gram (–), many Gram (–) rods, some Gram (+) & many Gram (–) bacteria, respectively
Macrolides	Erythromycin	Bacteristatic; inhibit protein synthesis	Mainly Gram + bacteria
Penicillins	Penicillin	Bactericidal; inhibit cell wall synthesis	Gram (+) bacteria
	Pen G		Gram (+) & some Gram (–) bacteria
	Ampicillin		Gram (+) & Gram (–) bacteria
	Amoxicillin		Acid stable
	Carbenicillin		
Polypeptides	Bacitracin	Bactericidal; increase membrane permeability	Gram (–) bacteria
	Polymyxin B		Gram (–) bacteria
Quinolones	Ciprofloxacin	Bactericidal; inhibit gyrase activity	Gram (–) bacteria including *P. aeruginosa* & some Gram (+) cocci
	Norfloxacin		Some Gram (–) bacteria
Tetracyclines	Tetracycline	Bacteristatic; inhibit protein synthesis	Broad spectrum; low water solubility
	Chlortetracycline		Water soluble
	Oxytetracycline		
Sulfonamides	Sulfamethoxazole	Bacteristatic; metabolic inhibitors	Broad spectrum; resistance common; not isolated originally from microorganisms
	Sulfisoxazole		
	Sulfacetamide		
Miscellaneous	Chloramphenicol	Bacteristatic; inhibits protein synthesis	Broad spectrum; high mammalian toxicity
	Vancomycin	Bactericidal; inhibits cell wall synthesis	Mainly Gram (+) bacteria; resistance in most Gram (–) and some Gram (+) bacteria

to be carcinogenic, but it is not readily absorbed in the digestive tract of humans. Nystatin and amphotericin B belong to the polyene macrolide antibiotics, which damage the sterol components of fungal cell membranes. While they have high fungicidal activity, insolubility in water has limited their use (Griffin 1981). However, nystatin has been shown to suppress fungi in some rearing systems with apparently minimal effects (Chamberlain 1989, Jarosz 1989, Temeyer 1998, Buyukguzel and Sevki 2002). Other, lesser known antifungal antibiotics include flucytosine, ketoconazole, and pentamidine (Atlas 1997). Although the antifungal antibiotics are often incorporated into selective media to prevent fungal contamination in the isolation of bacteria, they have not been extensively tested for managing microbial contamination in insectaries. However, Temeyer (1998) found that nystatin was compatible with use in horn fly, *Haematobia irritans* (Diptera: Muscidae), larval bioassays; and Buyukguzel and Sevki (2002) found that nystatin had positive effects on food consumption by larvae of a parasitic wasp *Pimpla turionellae* (Hymenoptera: Ichneumonidae). Others have also reported the successful use of nystatin in insect diets (Chamberlain 1989, Jarosz 1989).

Synthetic fungicides (fungicides not originally derived from microorganisms) have been used to suppress fungi since the 1800s (Griffin 1981). They can be either fungicidal or fungistatic. As a group they are heterogeneous and possess a wide spectrum of activity. Many fungicides have been reported, and a detailed examination of all of them is beyond the scope of this chapter. Two of the most common synthetic fungicides used to manage fungal contaminants in insect rearing are Benlate (active ingredient benomyl) (Fig. 58C) and Captam (active ingredient captan).

Benomyl was first reported as a fungicide in the 1960s. It interferes with mitosis and intracellular transportation of fungi through its deleterious effects on microtubules of the mitotic spindle and cytoskeleton, respectively. Ascomycota fungi are more sensitive to benomyl than are other types of fungi. Benomyl has a very low acute toxicity in mammals (WHO 1994), but the US Environmental Protection Agency (USEPA) has classified benomyl as a possible human carcinogen (USEPA 1996).

Benlate has been used in a variety of insect rearing systems, and a variety of effects on insects have been observed—both beneficial and deleterious. For example, treatment of western corn rootworm, *Diabrotica virgifera virgifera* (Coleoptera: Chrysomelidae), eggs with benomyl decreases colonization by the fungus *Cylindrocarpon destructans* and increases egg eclosion (Oloumi-Sadeghi and Levine 1989). Treatment of vermiculite with benomyl causes minimal detrimental effects on lesser cornstalk borer, *Elasmopalpus lignosellus* (Lepidoptera: Pyralidae), larvae and is highly effective in controlling *Asp. niger* (Lynch and Reed 1985). We have observed that a sand-vermiculite matrix containing grasshopper egg pods treated with Benlate minimized fungal contamination with no apparent effect on egg hatching. A few studies have addressed the effects of benomyl incorporated into artificial diets on insects. Teague et al. (1985) reared larvae of four species of noctuids [corn earworm, *Helicoverpa zea* (Lepidoptera: Noctuidae); beet armyworm, *Spodoptera exigua* (Lepidoptera: Noctuidae); soybean looper, *Pseudoplusia includens* (Lepidoptera: Noctuidae); and yellowstriped armyworm, *Spodoptera ornithogalli* (Lepidoptera: Noctuidae)] on a semisynthetic diet containing various concentrations of benomyl. Benomyl did not affect the noctuid larvae, but parasitism of the larvae by the parasitic wasp *Cotesia marginiventris* (Hymenoptera: Braconidae) was reduced. In a subsequent study, Horton et al. (1986) showed that benomyl did not affect the survival of unparasitized *H. zea* larvae but did reduce emergence from its host by *M. croceipes*. Finally, benomyl applied to plants at concentrations ≥ 50 ppm adversely affects symbionts within mycetocytes in the bird cherry-oat aphid, *Rhopalosiphum padi* (Homoptera: Aphididae) (Akhtar and Van Emden 1996).

A number of other synthetic fungicides have been used in rearing systems to manage microbial contaminants. At Mississippi State University, corn cob grits have been routinely treated with Captam to control fungal contamination of semisynthetic diets for Lepidoptera. While synthetic fungicides may prove useful in managing fungal contaminants, appropriate quality control assessments must be made before these chemicals are incorporated into a rearing system.

The *food antimicrobial agents* are a group of chemicals used in the food industry to suppress microbial contaminants. Many were designated as "generally recognized as safe" (GRAS) for human consumption by the United States Congress in 1958. As a result, this group of chemicals is excluded from regulations concerning food additives, and they can be added to foods by manufacturers without rigorously establishing their safety. It is important to stress that, even though GRAS chemicals apparently possess low toxicity towards humans, they could have toxic effects on insects.

An array of food antimicrobial agents is currently available, and each possesses its own combination of weaknesses and strengths (Table 5). The majority of food antimicrobial agents are relatively simple molecules. As a group they should be considered bacteristatic or fungistatic and not bactericidal or fungicidal. The selection of the appropriate agent or combination of antimicrobial agents depends on the insect and the composition of the array of microorganisms present. Furthermore, characteristics of the diet (e.g., pH), the degree of and manner of manipulation of the diet by insectary workers (e.g., introduction of microbial contaminants), and the environment in which the diet is maintained all influence the effectiveness of food antimicrobial agents. Food antimicrobial agents in the following five categories will be considered here: 1) organic acids and esters, 2) parabens, 3) nitrites, 4) phosphates, and 5) sulfites (Davidson 2001). We also introduce a number of nonprotein antimicrobial chemicals (e.g., plant-derived essential oils) that have not been extensively tested as inhibitors of microbial spoilage of insect diets.

The organic acids and their salts are one of the most common classes of food antimicrobial agents used in insect diets. The most active of these acids are acetic, benzoic, lactic, propionic, and sorbic acids (Fig. 59). The antimicrobial activity of these acids depends strongly on pH. The undissociated form of these molecules is required for activity, so activity is significantly reduced at pH levels greater than their pKa's—generally 3.0 to 5.0—at which the concentration of the dissociated form exceeds that of the undissociated form. Therefore, at neutral or basic pH levels, the organic acids are not effective. For example, benzoic acid is 100 times as effective in acid solutions as in neutral solutions (Rahn and Conn 1944). Unfortunately, the pH of the diet is often not considered by insect rearing specialists attempting to use this class of food antimicrobial agents. The mechanism of action of organic acids is not definitively understood but probably includes a variety of effects involving cytoplasmic membranes and metabolite transport enzymes (Davidson 2001).

Acetic acid and its salts are some of the oldest food antimicrobial agents (Table 5). Bacteria within the genera *Bacillus*, *Pseudomonas*, and *Staphylococcus* are

Table 5. Common antimicrobial agents used in foods.

Class	Agent	Target				Mechanism	Comments	
		Mold	Yeast	Bacterium				
				Gram (+)	Gram (–)			
Organic acids	Acetic acid	+	++	++	++	Interference with membrane function, metabolite transport	pKa = 4.75; sodium acetate & diacetate salts	
	Benzoic acid	++	++	+	+		pKa = 4.19	
	Proprionic acid	++	+	+	+		pKa = 4.87	
	Sorbic acid	++	++	+	+		pKa = 4.75; potassium sorbate highly soluble	
	Lactic acid	–	–	+	+	Low pH; reduced water activity	Typically 25,000-50,000 ppm	
Parabens	Methyl-	+++	++	+	+	Not well understood; possible disruption of membrane transport, inhibition of nucleic acid synthesis or of key enzymes	2.5 g/L	Increased activity methyl to heptyl
	Propyl-						0.5 g/L	
	Heptyl-paraben						> 0.5 g/L	
Nitrites		–	–	+	+	Multifactorial		
Phosphates	Sodium tripoly-phosphate	+	+	++	+	Multifactorial; chelation of metal ions	Typically 1,000–100,000 ppm; enhances lysozyme, nisin activity	
	Trisodium-phosphate							
Sulfites	Sulfur dioxide	–	–	+	+	Impairment of metabolism (e.g., glycolysis)	Antioxidants	

sensitive to acetic acid, as are fungi of the genera *Aspergillus*, *Penicillium*, *Rhizopus*, and *Mucor* (Davidson 2001). Acetic acid is not effective against bacterial endospores. Benzoic acid and sodium benzoate were the first antimicrobial agents permitted in foods by the U.S. Food and Drug Administration. Sodium benzoate is highly soluble in water compared to benzoic acid. Both these agents are primarily antifungal, but at high concentrations, they also exhibit some antibacterial activity. *Aspergillus*, *Penicillium*, and *Mucor* are all inhibited by the sorbates. However, at high initial densities, molds are capable of degrading sorbic acid (Davidson 2001). Sorbic acid primarily inhibits catalase-producing bacteria (Chap. 9 *Entomopathogens and Insect Rearing*). Generally, the sorbates have somewhat higher activity against molds than the proprionates. Although less commonly used as a food antimicrobial agent, lactic acid and the lactates (e.g., sodium lactate) do possess antimicrobial properties and have been used to inhibit spoilage bacteria.

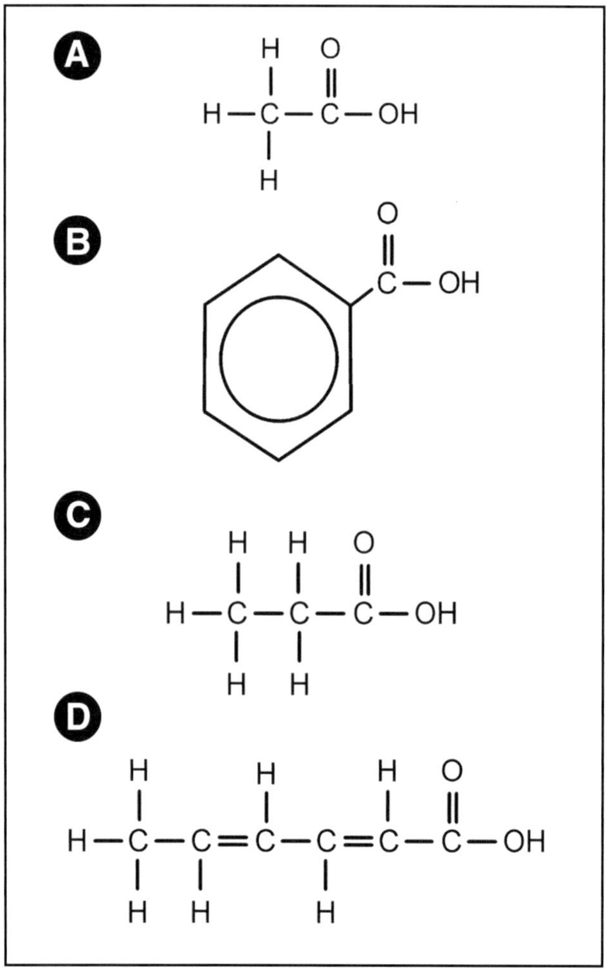

Fig. 59. Molecular structures of some major classes of food antimicrobial agents. (A) Acetic acid. (B) Benzoic acid. (C) Propionic acid. (D) Sorbic acid.

Many additional organic acids and esters have demonstrated some activity against spoilage microorganisms, but relative to the agents described above, their activity is minimal or they are active against a limited number of microorganisms. For example, citric acid retards the growth of *Aspergillus parasiticus* and *Aspergillus versicolor* and their production of mycotoxins but has neither effect on *Penicillium expansum* (Reiss 1976).

The antimicrobial properties of the *parabens* were first realized in the 1920s. They are derived from benzoic acid by esterification with alcohols to add alkyl chains of various lengths to the base structure (Fig. 60). This esterification allows the molecule to remain undissociated at pH levels up to 8.5 (Davidson 2001). Therefore, in diets that are neutral to somewhat basic, the parabens are a potentially superior alternative to the organic acids and their salts. A number of different parabens are available, but only methyl-, propyl-, and heptylparaben (Fig. 60) are used in foods. Among the bacteria, Gram-positive bacteria are generally more sensitive than Gram-negative bacteria. Interestingly, methylparaben, the most commonly used paraben in insect diets, is the least active form. Its extensive use in insect diets may be due to its relatively high water solubility. In foods, methyl- and propylparaben are often used in combination at a ratio of 2:1 to 3:1. Given the increased activities of the longer alkyl chain parabens, diet specialists should test their efficacy and toxicity towards insects. Low solubilities in water (e.g., heptylparaben) should not necessarily preclude their use, because they might be effective at low concentrations or they might be incorporated into diets at higher concentrations by first dissolving them in ethanol, propylene glycol, or some other appropriate solvent. However, it should be remembered that the toxicity of the solvent towards the insect must be tested.

Nitrites, phosphates, and sulfites are not commonly used in insect diets, but they will be treated here because their use in insect diets has not been adequately explored. The nitrites are commonly used in cured meats to inhibit spore-forming bacteria (e.g., *Clostridium botulinum*). While the nitrites are more active under anaerobic conditions, they do have some activity in aerobic environments. The mechanism of action of nitrates against spore-forming and non-spore-forming bacteria appears to differ. The polyphosphates (e.g., sodium tripolyphosphate, sodium tetrapolyphosphate) appear to act by chelating metal ions (Sofos 1986). Trisodium phosphate acts as a potentiating agent for both lysozyme and nisin (Carneiro de Melo et al. 1998). The sulfites are primarily used in fruits, vegetables, and wine to control spoilage yeasts and certain types of bacteria (Ough 1993). Like the organic acids, sulfites are most active at low pH (< 4.0). Some sulfites (e.g., sulfur dioxide) are active against bacteria even in low concentrations (Davidson 2001). If used correctly, food antimicrobial agents can retard microbial growth and prevent spoilage for several days, which might even constitute the entire feeding stage of the insect. However, they can also be totally ineffective and may severely affect the health of insects consuming them (Sec. VI. B. 4. d).

In this chapter we limit our discussion of antimicrobial proteins to the bacteriocins and lysozymes. *Bacteriocins* are a heterogeneous group of small proteins (3–10 kDa) that normally act against closely related bacteria but not against the producer (Montville et al. 2001). Many bacteriocins are produced by the lactic acid bacteria. Nisin is the first and best characterized bacteriocin. It is produced by the bacterium *Lactococcus lactis*, and it is commercially available under the trade names Nisaplin and Novasin. Nisin is a GRAS compound that is primarily used in foods to inhibit pathogenic, Gram-positive bacteria such as *Clo. botulinum* and *Listeria monocytogenes* but has activity against other Gram-positive bacteria as well. It is both heat and acid stable, but stability and solubility decrease with increasing pH. Furthermore, nisin is inactivated by certain proteases (e.g., chymotrypsin). To our knowledge, nisin has not been used in insect diets. It is important to stress that nisin and other bacteriocins have very restricted ranges of activity. Their efficacious use in insect diets will require an understanding of the primary bacterial taxa causing spoilage of the diets. It is expected that other bacteriocins will become commercially available in the near future, and these may have utility to suppress spoilage bacteria in insect diets.

Lysozyme is a 14.6 kDa enzyme present in avian eggs, mammalian milk, tears and other secretions. It is stable to moderate heating (80°C for 2 min) at low pH. The optimum temperature for activity of lysozyme is 55–60°C, but it retains 50% of its activity at 10–25°C (Inovatech 2000). It is most active against Gram-positive bacteria, but susceptibility among these bacteria is variable. Lysozyme degrades peptidoglycan in the cell wall of Gram-positive bacteria. Gram-negative bacteria are less susceptible to lysozyme owing to the presence of an outer membrane comprised of lipopolysaccharide and lipoprotein and also because the cell wall contains less peptidoglycan (Wilkins and Board 1989). However, the susceptibility of Gram-negative bacteria to lysozyme can be increased by disrupting the outer membrane using chelators such as ethylene diaminetetraacetic acid (EDTA), aminoglycosides, or by the antibacterial antibiotic polymyxin B (Tranter 1994). In addition, Gram-negative bacteria can be sensitized by subjecting the cells to pH shock, heat shock, osmotic shock, drying-wetting cycles, freeze-thaw cycles, or trisodium phosphate (Wilkins and Board 1989, Tranter 1994, Carneiro de Melo et al. 1998). These potentia-

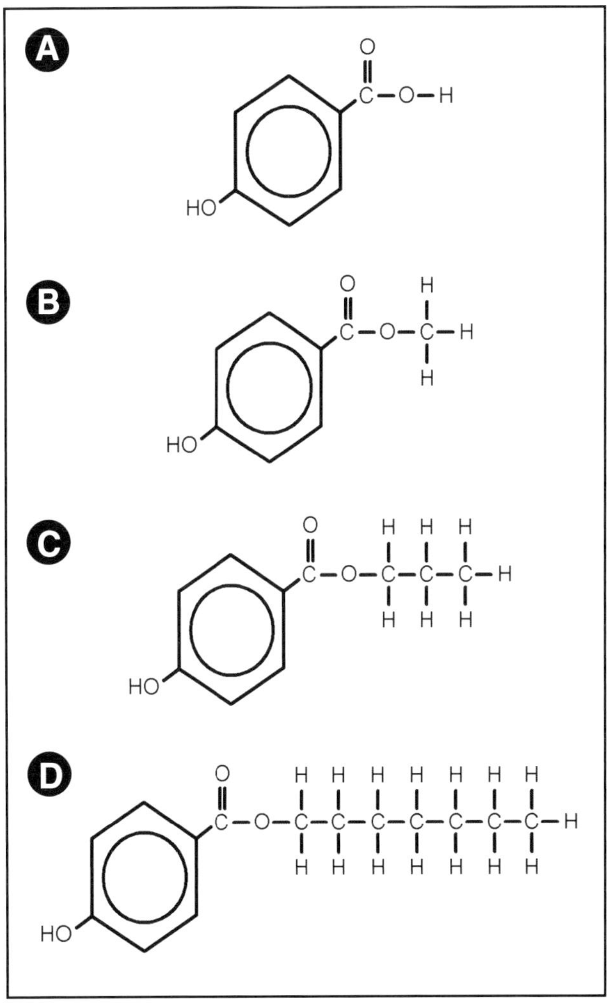

Fig. 60. Molecular structures of the alkyl esters of benzoic acid (i.e., the parabens). (A) Benzoic acid parent molecule. (B) Methylparaben. (C) Propylparaben. (D) Heptylparaben. In general, as the length of the alkyl chain increases, the antimicrobial activity of the molecule increases but aqueous solubility decreases.

tion interactions further illustrate the integrated approach that is required to adequately manage spoilage microorganisms. Lysozyme is commercially available and is one of the few naturally occurring antimicrobial agents approved by regulatory agencies for use in foods.

A number of other proteins have been reported to possess antimicrobial properties and include lactoperoxidases, lactoferrin, and avidin to name a few. These proteins occur naturally in milk, eggs, colostrum, and saliva. For more information on these and other antimicrobial proteins, the reader is referred to Davidson (2001).

A number of nonprotein, antimicrobial chemicals are found in various plant species including the spices, members of the genus *Allium*, members of the family *Brassicaceae*, and others. Essential oils from spices such as cloves, cinnamon, oregano, thyme, and—to a lesser extent—sage and rosemary are antimicrobial. The primary antimicrobial components of cloves and cinnamon are eugenol and cinnamic aldehyde, respectively, which exhibit various levels of activity against bacteria and fungi. Oregano and thyme contain the essential oils carvacrol and thymol, respectively. These compounds exhibit activity against a number of bacteria, as well as various molds and yeasts (Davidson and Naidu 2000). Minimum inhibitory concentrations of the essential oils from these spices were 0.02 to 0.1% (Smith-Palmer et al. 1998). The terpenes from sage (primarily thujone) and rosemary (primarily boneol) are active against Gram-positive and, to a lesser extent, Gram-negative bacteria. These spices are bacteristatic at a concentration of 0.3% and bactericidal at 0.5% (Shelef et al. 1980). Oils from sweet basil are active against certain fungi (e.g., *Mucor* and *Penicillium*) but have very limited activity against bacteria (Lachowicz et al. 1998). Vanillin from vanilla beans is most active against molds and non-lactic acid-producing, Gram-positive bacteria (Jay and Rivers 1984). A number of antimicrobial agents are found in the juices and vapors of onions and garlic, including allicin, protocatechuic acid, and catechol (Cavallito and Bailey 1944). Activity of onion extracts have been reported against Gram-negative bacteria and various yeasts. Another group of antimicrobial compounds is the isothiocyanates, commonly found in mustard seeds. The isothiocyanates are inhibitory to fungi, yeasts, and various bacteria at 0.01–0.6% in liquid media (Mari et al. 1993). Other examples of antimicrobial compounds include various phenolic compounds, hydroxycinnamic acid, fatty acids, and various flavonoids (Davidson 2001). To our knowledge, the effects of these substances on spoilage microorganisms of insect diets and on the insects themselves have not been determined. It is doubtful that they will prove to be adequate on their own, but they could be useful in combination with other agents to increase efficacy and to broaden the spectrum of activity. Of course, it is possible that, at the concentrations required to inhibit spoilage microorganisms, these compounds may impart undesirable sensory effects. For more information on the antimicrobial properties of these compounds, the reader is referred to Davidson (2001).

Chitosan is another nonprotein, antimicrobial chemical that has received significant attention. Chitosan is a deacetylated from of chitin, a polysaccharide; and both are found in the exoskeletons of arthropods. Chitosan possesses antimicrobial activity against a range of molds, yeasts, and bacteria; and it is designated as a GRAS compound in the USA. Recently, chitosan (0.005%) was shown to potentiate the activity of sodium benzoate against spoilage yeasts *in vitro* (Sagoo et al. 2002). The impact of chitosan on suppressing microbial spoilage of insect diets, alone or in combination with other agents, has not been determined.

Another class of antimicrobial agent, which may prove to be effective, is the microbial inoculant—benign microorganisms that are antagonistic to spoilage microorganisms. This type of preservation is often classified as *biopreservation*. Microbial inoculants have been utilized in a variety of foods but not in artificial, insect diets. Fermentation is one such example. Lactic acid bacteria such as *Lactobacillus*, *Leuconostoc*, *Lactococcus*, *Carnobacterium*, and *Streptococcus* have been used to preserve a variety of foods (e.g., sauerkraut) and agricultural products (e.g., silage). For inoculants to be useful in suppressing spoilage microorganisms in insect diets, they must not have adverse effects on the insects to be reared. In the case of fermentation, lactic acid bacteria produce organic acids (e.g., lactic acid) as a by-product; and, as discussed previously, the resulting low pH can directly inhibit the growth of many spoilage microorganisms. Factors determining the effectiveness of *in situ* acidification include the initial pH of the diet and its buffering capacity, the type and level of the spoilage microorganisms, the nature and concentration of fermentable carbohydrate in the diet, other components of the diet that influence the viability and growth rate of the lactic acid bacteria relative to the spoilage microorganisms, and abiotic environmental factors that influence relative growth rates (Montville et al. 2001). The production of bacteriocins, diacetyl, and hydrogen peroxide may also contribute to the overall inhibition obtained (Montville et al. 2001). In addition to lactobacilli, a variety of other microorganisms may prove useful in biopreservation of insect diets, and increased research in this area is warranted.

d. Developing an Antimicrobial Management Plan. The following steps should be followed in developing an antimicrobial management plan for insect diets: 1) determine the microorganisms causing contamination, 2) assess the effects of the candidate antimicrobial agents and factors on the microorganisms, 3) assess the effects of the candidate antimicrobial agents and factors on the insects, and 4) assess the compatibility of the chosen antimicrobial agent(s) and factor(s) with one another in an integrated management system.

First, it is essential to know what microorganisms are causing contamination of the diet. Molds are easily visualized once they produce spores; however, yeasts and bacteria are not always so easily recognized. Basic microbiological techniques should be applied to determine what groups of microorganisms are present and, if possible, what taxa are responsible for the spoilage. In the case of antimicrobial agents, the taxa present determine the best agent(s) to use and the minimum effective concentration(s) required. For example, many antimicrobial chemicals are restricted to Gram-positive or Gram-negative bacteria. In the case of diet characteristics (e.g., pH, a_w, and *Eh*), the microbial taxa present determine the degree to which conditions acceptable to the insect may be suboptimal for the spoilage microorganisms. For example, bacteria generally require $a_w > 0.90$, while some insects thrive on considerably drier diets. Microorganisms should be isolated and stored for testing of candidate antimicrobial agents and factors.

Before an antimicrobial agent or factor can be incorporated into a rearing program, it is important to assess the susceptibility of microorganisms to that agent or factor. For assessment of diet characteristics and environmental factors, one must quantify microbial population growth rates in the presence and absence of the candidate control factor. In general, this will require measuring the sizes of microbial populations over time. For a discussion of methods to accomplish these observations, see Sec. V. C. 2 and the discussion of methods to assess antimicrobial agents, below.

For assessment of antimicrobial agents, a number of methods can be used. Two most commonly used methods are the Kirby-Bauer method and the minimum inhibitory concentration method. The *Kirby-Bauer method*, also known as the *diffusion assay method* (Fig. 61), works best for water-soluble agents; and it is designed for rapidly growing bacteria. In its standard form, it is not well suited for filamentous fungi or slow-growing bacteria or yeasts, although modifications to the method may provide suitable results for these microorganisms as well. A more quantitative method for assessing the activity of an antimicrobial agent *in vitro* is the *minimum inhibitory concentration* (MIC) method, also known as the agar dilution method (Fig. 62). This method involves testing various dilutions of the antimicrobial agent to determine the lowest concentration that is effective in preventing growth of the test microorganism. The method can be conducted using liquid or solid media. For a further description

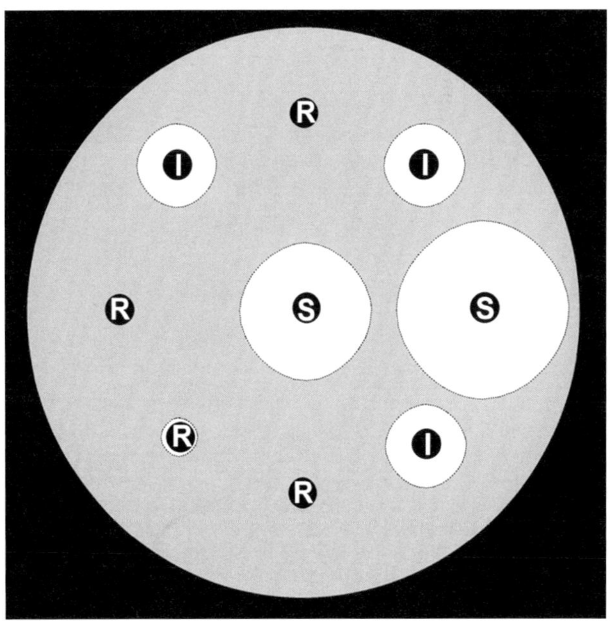

Fig. 61. Kirby-Bauer method for determining the sensitivity of fast-growing microorganisms to antimicrobial antibiotics. The diameter of the zone of inhibition around the disk impregnated with the test antibiotic indicates sensitivity of the microorganism to that antibiotic. Three classes of susceptibility have been defined under well-specified conditions: S (> 12 to > 29 mm, depending on the agent), I (8–12 to 20–29 mm, depending on the agent), and R (< 8 to < 20 mm, depending on the agent).

of these two methods, the reader should consult any of a number of microbiological references and develop standard operating procedures.

To directly measure proliferation of microorganisms in diets containing the antimicrobial agent typically involves the application of quantitative microbial enumeration methods (Sec. V. C. 2). The dilution spread-plate method typically works well for bacteria. While this method can also be used for fungi, it favors the isolation of yeasts and prolific sporulators such as *Aspergillus*, *Penicillium*, and *Rhizopus*. Thus, it typically provides a gross underestimation of fungal biomass (e.g., hyphae present in the diet) and diversity. This is particularly important given that deleterious secondary metabolites and proteins including catabolic enzymes are produced by hyphae. Moreover, we have observed instances where fungistatic agents inhibit sporulation but not hyphal development. In this case, if one were to rely on the spread-plate method or on a visual assessment of contamination, an erroneous conclusion concerning the usefulness of the agent would be made. The best strategy is to measure fungal mass rather than quantity of spores produced. However, even measuring total mass—much less taxon-specific masses—of filamentous fungi is not an easy task. Total fungal mass has been quantified using ATP, chitin, ergosterol, fatty acids, and antigens to hyphal proteins; but this approach does not separate viable from nonviable cells. Real-time, quantitative polymerase chain reaction may prove useful for determination of taxon-specific fungal masses; however, this technology has yet to be extensively tested for this purpose. Another strategy that may provc useful to determine the efficacy of antifungal agents is the quantification of specific secondary metabolites. For example, *Aspergillus* and *Pencillium* spp. produce mycotoxins, which can be quantified.

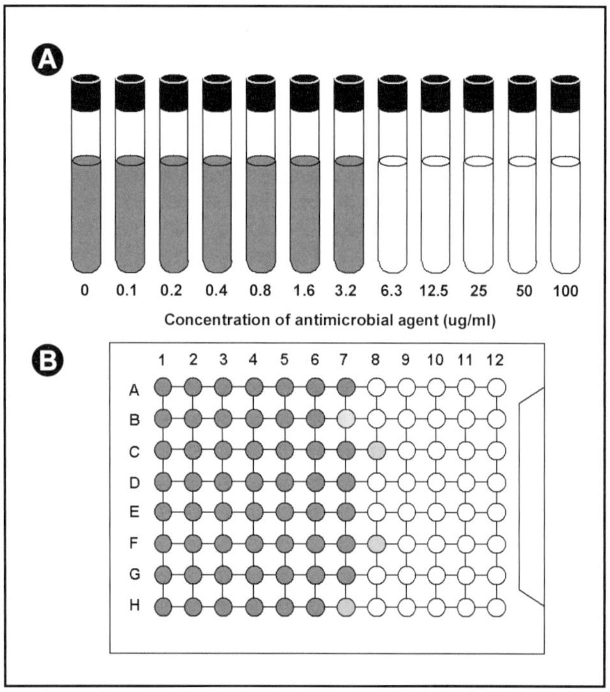

Fig. 62. MIC (minimum inhibitory concentration) methods for quantitatively determining the lowest concentration of an antimicrobial that prevents growth. (A) Tube MIC method. Positive growth is indicated by the proliferation of cells, as indicated by turbidity, within the liquid culture medium containing varying concentrations of an antimicrobic. In this example, the MIC is 6.3µg/ml: the lowest concentration that precludes growth. Not illustrated, replicate tubes could be used for each concentration. (B) Microtiter plate MIC method. This method is analogous to the tube method except that the liquid culture medium is housed in the wells of a microtiter plate. In this example, replicate samples are represented in rows A–H with columns 1–12 corresponding to the same concentrations used for the tube MIC method. The advantage of this method is the reduced volumes required (100 µl). In some instances, pH dyes are used to facilitate visualization of positive samples.

The next step in developing an antimicrobial management plan for insect diets is to elucidate the effect of the antimicrobial agent or factor on the insect. Antimicrobial agents generally are not benign to insects, and both diet characteristics and environmental factors can have profound effects on insect survival, growth, and development. A high dose of antibacterial antibiotics reduces *H. virescens* adult emergence and egg hatch (Sikorowski et al. 1984). Although the tetracyclines possess remarkably low toxicity to animals, chlortetracycline is very detrimental to *D. grandiosella* (Frank Davis, unpublished data). Antibacterial antibiotics ingested by *H. virescens* and *H. zea* larvae enter the hemolymph and persist for periods of time depending on concentration (Sikorowski and Thompson 1985, Sikorowski et al. 1988). As with antibacterial antibiotics, food antimicrobial agents can have detrimental effects on insects; and care must be exercised in selecting both the agent and the concentration used. Singh and House (1970) studied the effects of 21 antimicrobial agents on development of the parasitoid *Agria housei* (= *affinis*) (Diptera: Sarcophagidae) when incorporated into an artificial diet. As a function of concentration, all of the agents caused effects such as inhibition of larval development and increased mortality in both larval and pupal stages. Alverson and Cohen (2003) observed that methylparaben, one of the antimicrobial agents most commonly used in insect diets, adversely affected western tarnished plant bugs, *Lygus hesperus* (Heteroptera: Miridae). Many other researchers have similarly noted inhibition of insects reared on diets containing antimicrobial compounds, and Dunkel and Read (1991) reviewed the literature on the effects of sorbic acid and other antimicrobials on survival of various insects. Most researchers who have studied antimicrobial agents in insect diets now believe that the minimum effective concentration should be used to minimize detrimental effects on the quality of insects. This requires knowledge of the microorganisms causing spoilage and how candidate antimicrobial agents interact with spoilage microorganisms, insects, and environmental variables. Examples of deleterious effects of environmental conditions on insects can be found in Chap. 6 *Environmental Biology of Insect Rearing*.

In addition to the direct, adverse impacts of antimicrobial agents on insects, other considerations include the possible occurrence of iatrogenic diseases and the development of antibiotic resistance in spoilage bacteria. Disease caused by the destruction of beneficial bacteria through the use of wide-spectrum antibacterial antibiotics would be considered an iatrogenic disease. As indicated previously, resistance has been reported to almost all antibacterial antibiotics by medically important bacteria. As a result of overuse and misuse of antibiotics, unnecessarily high selection pressure on the microorganisms accelerates the evolution and proliferation of antibiotic-resistant strains. A number of strategies have been devised by the medical profession to prevent or delay the development of resistance to antibiotics. It is now recommended to perform culture and sensitivity studies prior to selection of an antibiotic and to alternate antibiotics with different modes of action across treatment episodes. To date, this strategy has not been adopted in insect rearing, and most programs utilize wide-spectrum antibacterial antibiotics over long periods of time regardless of the microbial taxa present.

In addition to their adverse effects on insects, antimicrobial agents also can be toxic to insectary personnel depending on dose. It is a prerequisite that appropriate precautions be taken in handling and disposing of these materials and that the appropriate regulations for discarding these compounds be strictly enforced (Chap. 4 *Safety and Health Issues in the Rearing of Arthropods*).

The final step in developing an antimicrobial management plan for insect diets is to integrate the components of the plan. The "hurdle" concept in food preservation involves the integration of multiple factors to effect control of microorganisms (Jay 1996). While combinations of antimicrobial compounds have been incorporated into insect diets, systematic research concerning the integration of strategies to manage spoilage in insect diets largely has been ignored to date. Research on the microbial ecology of diet spoilage and the impact of all control methods, alone and in combination, on reducing spoilage and on the insects reared is necessary if one is to achieve efficacious management of spoilage microorganisms. Given the difficulty in many cases of manipulating intrinsic characteristics of diets (e.g., pH and a_w) to manage microbial contamination without severely reducing the ability of insects to utilize those diets, the goal generally should be to use a combination of antimicrobial agents that exert additive or synergistic affects against microorganisms but are tolerated or even preferred by the insects.

C. Workrooms and Equipment

1. Designation of Clean Areas. An essential component of any microbial management program is proper insectary design (Chap. 2 *Insectary Design and Construction*). In particular, it is essential that clean areas be designated and their integrity enforced. Airflow patterns should be manipulated using differential pressures so that no contaminants move from less clean areas. Examples of insectary activities conducted in clean areas include dispensing sterilized diets, egg sanitation, insect placement on diets, and aseptic packaging procedures. The utility of air filtration systems and of differential air pressure in maintaining clean areas within an insectary have been discussed (Sec. VI. A). Given that people are a major source of bacterial contamination, insectary workers play a critical role in maintaining the integrity of clean areas. Access to clean areas should be restricted to trained personnel who utilize appropriate cleanliness standards. Furthermore, appropriate sanitation measures should be strictly followed. Visitors typically should not be allowed into clean areas. Activities that should not be permitted in clean areas include immature insect maintenance, pupal harvesting, adult maintenance, and egg harvesting. In addition, care must be taken to ensure that insects are prevented from entering clean areas—whether they originate within the insectary or without. Obviously, every effort should also be made to prevent rodents and other animals from gaining access to any areas within the facility.

2. Sanitation of Workrooms. Floors in work areas, particularly areas designated as clean areas, should be mopped to reduce the liberation of airborne particles. The frequency of sanitation will be dependent on the type of activity (e.g., after a particular procedure). In other instances, predetermined sanitation schedules may work best (e.g., in clean rooms). Dry/wet mopping with the Swiffer Sweeper™ Mop System (Procter and Gamble, Cincinnati, OH) works well. Sweeping of workroom floors should never be conducted; this method of cleaning results in the liberation of considerable numbers of microorganisms into the air. Another strategy involves the use of an exterior vacuum system. It is desirable that the vacuum machine be housed outside of the insectary to prevent liberation of microorganisms into the air. Some commercial vacuum cleaners possess HEPA filters, but a breach in this barrier would result in serious contamination if the base unit is located within the insectary.

Walls and working surfaces should be thoroughly wiped with an appropriate sanitizing solution. From the numerous chemicals that possess antimicrobial properties, the most popular sanitizers are chlorine compounds, iodophores, quaternary ammonium, and hydrogen peroxide compounds (Sec. VI. B. 3. *b* and Sec. VI. C. 4).

3. Sanitation of Equipment. In choosing or designing insectary equipment, it is important to consider ease of sanitation. The equipment should be designed such that nutrients supporting microbial growth do not accumulate in inaccessible points (e.g., joints and valves) to serve as sources of continual contamination. Stainless steel is most frequently used for surfaces that come into contact with diet because of its durability and resistance to most corrosive sanitizing agents. Heat under pressure can be used to sterilize small, heat-resistant equipment and utensils. Large equipment is usually cleaned and sanitized simultaneously using combined detergent/sanitizers. This single operation of cleaning and sanitizing saves time and labor and is used in food industries (Banwart 1979). The effectiveness of sanitizers depends on concentration, temperature, pH, contact efficiency, time of exposure, and the number and species of contaminants. The use of high concentrations of a sanitizing agent as a substitute for thorough cleaning is inefficient, expensive, and in some cases may cause corrosion of the metal (Banwart 1979). A large equipment system, such as an integrated, media-processing apparatus, should be chemically cleaned and sanitized according to the procedures recommended by the manufacturer.

4. Sanitizers. The most widely used sanitizing agents in food-processing industries are the hypochlorites, which are effective at relatively low concentrations against a wide range of bacteria and bacterial spores, fungi, bacteriophages, and some viruses. However, because of their high corrosion potential, they should not be used at high temperatures (Banwart 1989). For more information on sanitizers see Sec. VI. B. 3. *b*.

Aqueous solutions or alcoholic tinctures of iodine are commonly used as antiseptics but are infrequently used to sanitize equipment. As a substitute, iodophors are now commonly used for that purpose because they generally have a broader spectrum of activity, stain less permanently, and are less irritating to tissues. The *iodophors* are solutions in which the iodine is combined with a solubilizing carrier. Such solutions are sold as concentrates, and when they are diluted in water, they provide a slow, continuous release of free iodine. Their activity is based on this release of iodine, which is well recognized for its microbicidal activity against a wide variety of microorganisms including spore-forming bacteria. Although iodophores possess sporocidal properties, in general they are less effective than sodium hypochlorite. According to Banwart (1979), iodophores cannot be used at temperatures above 49°C because sublimation of the iodine decreases its effectiveness. Povidone-iodine (PVP-iodine), which uses polyvinylpyrrolidine as the solubilizing carrier, is a commonly used iodophore.

The quaternary ammonium salts may be bacteristatic at low concentrations and bactericidal at high concentrations. They are frequently used to sanitize equipment in the food and dairy industries. The most commonly used quaternary ammonium salts include: Zephiran™, a mixture of alkyldimethyl benzylammonium chlorides; benzethonium chloride, a substituted benzalkonium chloride; cetrimide, a mixture of hexadecyl-, tetradecyl-, and dodecyltrimethylammonium bromides; 1-hexadecylpyridinium chloride; and domiphen bromide, dodecyldimethyl-2-phenoxyethylammonium bromide.

Hydrogen peroxide is an oxidizing agent that can be an effective microbial sanitizer at low concentrations (≤ 0.1%) and a sterilant at high concentrations (10–25%). It kills by several mechanisms but primarily by the generation of hydroxyl radicals that cause DNA damage (Imlay and Linn 1998). Hydrogen peroxide is a common, household antiseptic available as 3–10% solutions. The antimicrobial activity of such solutions may be largely mechanically: catalase present in tissues causes a rapid evolution of oxygen that can dislodge contaminating foreign matter, including bacterial cells. Furthermore, the highly oxygenated conditions that are created by this foaming action can inhibit the development of anaerobic taxa. The bacterial sporocidal activity at high concentrations of hydrogen peroxide is enhanced by ultrasonic energy and by certain metal ions, particularly Cu^{2+}. Vaporized hydrogen peroxide (VHP) is created by a machine typically termed a "VHP generator" (e.g., Steris VHP® 1000ED Biodecontamination System; Steris Corporation, Mentor, OH) and is often used as an antimicrobial pesticide for decontaminating sealed enclosures such as scientific workstations, isolators, rooms, medical and diagnostic devices, and for other biological safety applications. VHP is the standard decontamination technology used in the pharmaceutical industry, and the by-products are safe (water vapor and oxygen). VHP possesses efficacy against a wide variety of microorganisms, but some bacteria (e.g., *Esc. coli* and *Salmonella typhimurium*) have developed resistance to hydrogen peroxide. The activity of hydrogen peroxide is adversely affected by heat, the enzymes catalase and peroxidase, and pH. Hydrogen peroxide solutions are generally more stable at low pH. In addition, hydrogen peroxide does not possess good penetration ability, and it has a relatively limited shelf life once the container is opened. The hydrogen peroxide–based product Peroxigard® is now commonly used as a sanitizer in a variety of industries. It contains hydrogen peroxide formulated with a number of biodegradable adjuvants, including surfactants. It is reported to be highly microbicidal yet nontoxic to humans and to leave no active ingredient residues on surfaces once dry (Sattar and Springthrope 1998, Rochon and Sullivan 1999).

Phenol (also know as carbolic acid) was used by Joseph Lister in his attempts to reduce infections resulting from surgical procedures. It proved to be very effective, but it is no longer of any practical value as a sanitizer because its chemical derivatives have proven to be considerably more effective and less toxic. At low concentrations, the phenolics cause cell membrane damage and cell lysis; and at higher concentrations, they can denature and precipitate cellular proteins. Many of the phenolics reduce surface tension and, when used with other surfactants, possess excellent wetting properties. This thorough wetting increases their residual activity. The extensive variety of possible chemical substitutions to the phenol molecule has given rise to the numerous phenolic sanitizers now available. By adding a halogen (chlorine being the most important), antimicrobial activity is increased. If aliphatic or aromatic groups are added to halogenated phenols, the antimicrobial activity can be increased even further. Cresols (i.e., methylphenols) are a good example of how a simple substitution markedly improves the antimicrobial activity of phenol. The cresols (ortho-, meta-, and para-) are normally used in a mixture known as *tricresol*. Tricresol combined with soap as an emulsifier constitutes the active ingredient of such commercial sanitizers as Lysol™.

D. Personnel

Humans are important causes of bacterial contamination of the insectary. They act as both direct (e.g., from hair and skin) and indirect (e.g., inadequate management practices) sources of microorganisms. Once microorganisms are liberated into the air of an insectary, even one possessing air filtration, they are difficult to manage, so the key is to prevent the liberation of contaminants.

1. Education. Education of insectary personnel is one of the fundamental requirements of any successful management program for microbial contamination. This will rely primarily on intramural training. Such training should instruct workers concerning the following: 1) characteristics of major microbial taxa; 2) from where microbial contaminants originate in insectary settings; 3) that not all microbial contaminants are detrimental and that some are actually beneficial; 4) how microbial contaminants affect insect survival, growth, development, and reproduction; 5) the basic methods used to monitor microbial contamination; 6) strategies to manage microbial contaminants in insectary settings; and 7) how microbial contaminants can affect human health. Of particular importance is that workers recognize the criteria used to assess insect quality. Producing high quality insects is not an easy task; and, to paraphrase Allen Cohen (Cohen 2003),

"Producing high quality insects is not rocket science—it is much more difficult." A feeling of pride in the production of high quality insects should be instilled in workers, and they should be rewarded for achieving this goal.

A few examples of simple training exercises to reinforce the importance of proper hygiene include the following: 1) fingers, before and after washing, are pressed onto a general-purpose culture medium (Fig. 63); 2) skin flakes or hairs are placed onto a culture medium (Fig. 64A); 3) individuals are allowed to cough onto a solid culture medium (Fig. 64B); and 4) Rodac® impressions of clothing and skin are taken. The media are incubated, and insectary personnel observe the degree of contamination from their own samples. The microbiological methods presented in this chapter are not difficult to master, and workers should be encouraged to utilize them.

2. Worker Sanitation. Minimizing the liberation of bacteria from personnel and visitors is essential. Hand washing after restroom visits and wearing clean, protective clothing while in clean areas of the insec-

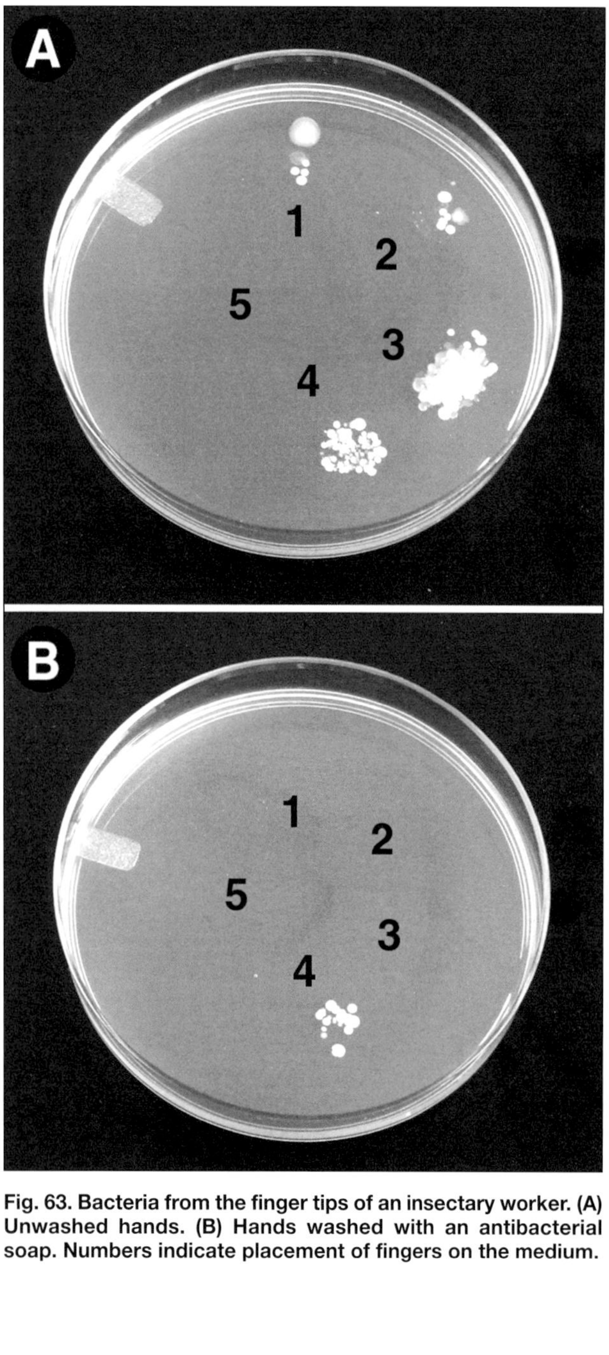

Fig. 63. Bacteria from the finger tips of an insectary worker. (A) Unwashed hands. (B) Hands washed with an antibacterial soap. Numbers indicate placement of fingers on the medium.

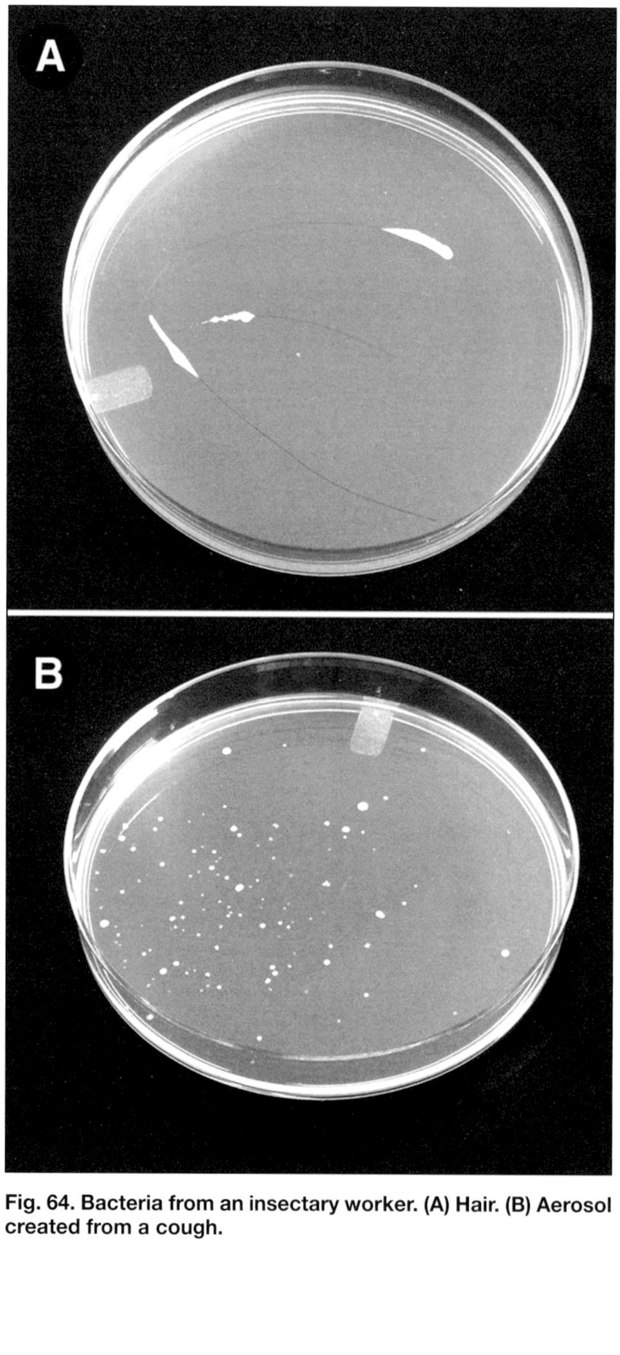

Fig. 64. Bacteria from an insectary worker. (A) Hair. (B) Aerosol created from a cough.

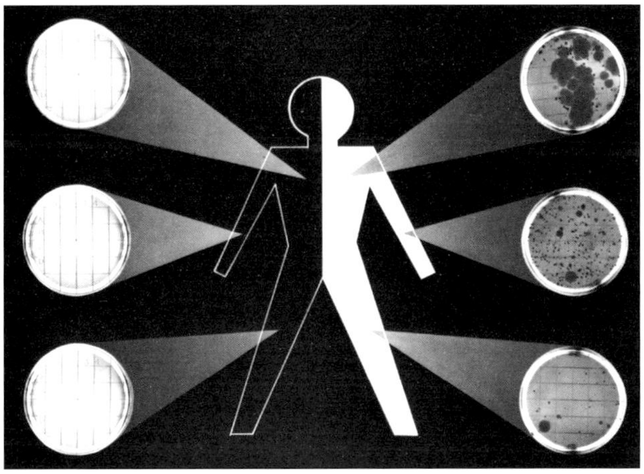

Fig. 65. Influence of laundering on microbial contamination of clothing. Detection of microorganisms on garments was by the agar contact method using Rodac® plates. On the left, samples were taken from a freshly laundered garment, which was determined to be sterile. On the right, samples were taken after a worker had worn the garment for 4 h. Note the high degree of contamination observed. [Based on Sikorowski (1975).]

tary are crucial. Multiple studies have shown that 80% of men and 40% of women do not wash their hands after visiting the restroom. It is well documented that hand washing reduces the number of viable microorganisms present on the skin. For example, schools with regular hand washing programs have measurably fewer bacterial infections within their student population than do schools with sporadic hand washing policies. Furthermore, schools without hand washing programs have high levels of fecal bacteria detectable on classroom floors and carpets. To achieve adequate decontamination during hand washing, it is essential to wash with water and a cleansing agent, not just water alone. The best type of cleansing agent is open to debate. Soaps and detergents have very little inherent antimicrobial activity; however, their value lies in facilitating the mechanical removal of microorganisms during washing. Although the term *soap* (a mixture of sodium and potassium salts of fatty acids derived through saponification) is used throughout this section, in most instances the cleansing agents commercially available are actually *detergents* (synthetic cleansing agents containing compounds with both hydrophilic and hydrophobic regions to facilitate solubilization of fats and oils in water). Soaps and detergents share many properties, but detergents are synthetic products of the modern chemical industry.

Whether an antibacterial soap should be used over conventional soaps is debatable. The scientific studies comparing antibacterial soaps to regular soaps give apparently contradictory results; however, it now appears that antibacterial soaps are effective at reducing some microorganisms but less so against others. There are additional considerations. The antibacterial components of soaps (usually triclosan or, less commonly, triclocarbon) need to be left on the skin surface for not less than about 2 min in order to work; but most people are not this patient. Some scientists have expressed concern over the potential for development of resistance to the active ingredients used in antibacterial soaps, thus decreasing their effectiveness for other uses. Finally, regular soaps are available that decrease the incidence of dry or cracked skin relative to available antibacterial soaps. Far more important perhaps than the choice of soap is consistent, thorough hand washing.

In high-risk areas, it is important that workers wear clothing covering the legs, arms, and neck to minimize the liberation of skin flakes. Also, regular clothing is contaminated with large numbers of microorganisms, so insectary-only garments, worn over or as a replacement for regular clothing, are strongly

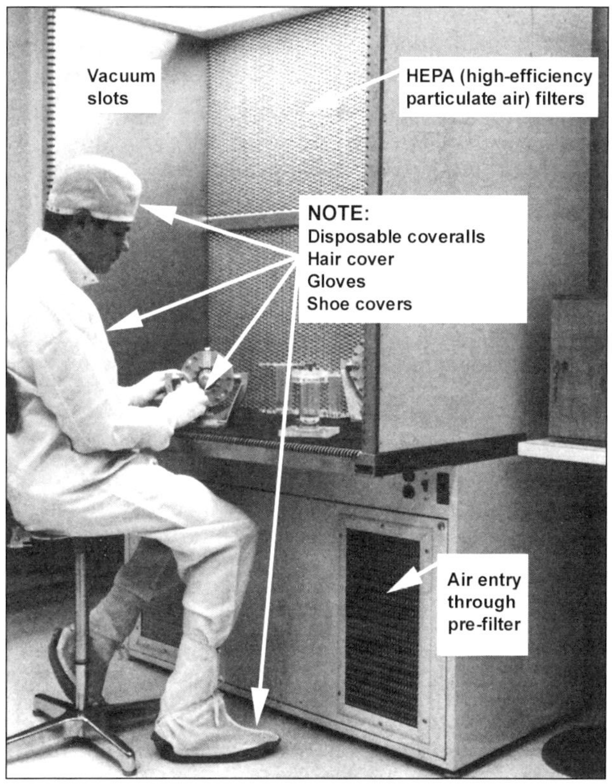

Fig. 66. Clean room technician wearing attire appropriate for working in a laminar flow, ISO class 5 work station. In insectary settings, such attire would be used in high-risk areas. [NASA (1969).]

advised. Freshly laundered garments are sterile; but even after a short time outside a clean environment, they become contaminated (Fig. 65). It is advisable that garments for use in the insectary either be laundered frequently or that disposable garments (e.g., laboratory coveralls) be used (Fig. 66). It is important that such garments be restricted to the insectary. Bacteria on hair and bacteria in aerosols generated from sneezes or coughs are both potential sources of bacterial contaminants. Therefore, in clean areas hats or hair nets and masks (Fig. 67) are recommended. Street shoes also frequently introduce soil and organic debris into an insectary, so footwear dedicated for use only in the insectary is advisable. In some instances, it may prove useful to use clean booties that fit over street shoes. A changing room with showers should be provided within the insectary. It is essential that managers demonstrate the importance of protective clothing and that they ensure garments fit well and are comfortable.

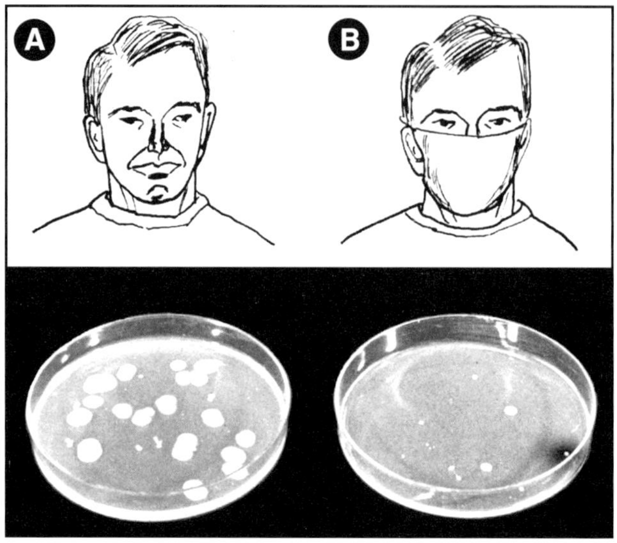

Fig. 67. Effects of a face mask on the liberation of bacterial aerosols by coughing. (A) Bacteria liberated when a worker was not wearing a mask and coughed on a culture plate. (B) Bacteria liberated when a worker was wearing a mask and coughed on a culture plate. [From Sikorowski (1975) with permission from the Mississippi Agricultural and Forestry Experiment Station.]

VII. Conclusions

The efficacious management of microbial contaminants in insectaries is a prerequisite for rearing high quality insects (Fig. 68). Perhaps the greatest obstacle that microbiologists face is convincing insectary managers and workers that the recommended procedures actually are necessary to achieve efficacious management of microbial contaminants. Insectary personnel and managers typically have little or no training in microbiology and tend to rely on the prophylactic use of antimicrobial agents, such as antibacterial antibiotics in diets. Sikorowski et al. (1980) stated that "… use of antibiotics should be the method of choice only under certain unusual circumstances when sanitary measures are temporarily interrupted. Prolonged use of an antibiotic insures that only resistant cells survive; and, in time, the progeny of these cells become the dominant bacterial population of the local environment." Unfortunately, this statement and similar statements concerning the use of antibiotics in more traditional types of animal husbandry have been ignored, and the real possibility exists that governmental regulation may restrict the use of these agents for nonmedical purposes. A panacea for preventing microbial contamination does not exist. Successful management of microorganisms requires an integrated approach. Of paramount importance is that insect rearing specialists embrace microbial management strategies. These strategies are used in analogous industries (e.g., the food industry), yet they are largely ignored by insectary managers.

Fig. 68. Healthy fall armyworm, *Spodoptera frugiperda* (Lepidoptera: Noctuidae), pupae produced at the USDA-ARS Corn Host Plant Resistance Research Unit, Mississippi State University.

Prevention should be the goal of all management strategies. However, despite the implementation of preventative management strategies, it is inevitable that microbial contamination will be encountered from time to time. When such problems occur, it is critical that the appropriate infrastructure be in place so that effective measures can be taken to correct the problem.

Acknowledgments

We are very grateful to the following persons who supplied images or specimens for this chapter: Janet Alverson (USDA-ARS, Washington State University, Pullman, WA), Hilma Busz (Agriculture & Agri-Food Canada, Lethbridge, AB), Erin Cadeau (Agriculture & Agri-Food Canada, Lethbridge, AB), Kathaleen House (Agriculture & Agri-Food Canada, Lethbridge, AB), Amanda Lawrence (Department of Entomology & Plant Pathology, Mississippi State University, MS), Joe MacGown (Department of Entomology & Plant Pathology, Mississippi State University, MS), Matt MacGown (Department of Entomology & Plant Pathology, Mississippi State University, MS), and Jay Yanke (Agriculture & Agri-Food Canada, Lethbridge, AB). We also are thankful to Jay Yanke and Amanda Lawrence for providing critical reviews of this chapter. Some definitions were drawn from the following references: Atlas (1993), Hawksworth et al. (1995), Schrezenmeir and de Vrese (2001), Singleton and Sainsbury (2001). This is Lethbridge Research Centre contribution 03105.

References

Abercrombie, M., C. J. Hickman, and M. L. Johnson.1980. The penguin dictionary of biology. Penguin Books, Harmondsworth, UK.

Akhtar, S., and H. F. Van Emden. 1996. Effect of the synthetic fungicide benomyl on the symbionts and mycetocytes of the bird cherry-oat aphid (*Rhopalosiphum padi*) (Homoptera: Aphididae) reared on wheat plants. Bull. Entomol. Res. 86: 319–328.

Alverson, J., and A. C. Cohen. 2003. Effect of antifungal agents on biological fitness of *Lygus hesperus* (Heteroptera: Miridae). J. Econ. Entomol. 95: 256–260.

Amyes, S. G. B. 2001. Magic bullets, lost horizons: the rise and fall of antibiotics. Taylor & Francis Inc., London, UK.

Andrews, H. H. 1992. Fungal life-history strategies, pp. 119–145. *In* C. G. Carroll and D.T. Wicklow (eds.), The fungal community, 2nd ed. Marcel Dekker, New York, NY.

Atlas, R. M. 1993. Handbook of microbiological media. CRC Press, Boca Raton, FL.

Atlas, R. M. 1997. Principles of microbiology. Wm. C. Brown Publishers, Dubuque, IA.

Banwart, G. J. 1979. Basic food microbiology. AVI, Westport, CT.

Banwart, G. J. 1989. Basic food microbiology. Van Nostrand Reinhold, New York, NY.

Barnett, J. A., R. W. Payne, and D. Yarrow. 1990. Yeasts: characteristics and identification. Cambridge University Press, Cambridge, UK.

Barron, G. L. 1968. The genera of Hyphomycetes from soil. Williams and Wilkins Company, Baltimore, MD.

Beneke, E. S., J. W. Rippon, and A. L. Rogers. 1984. Human mycoses. Upjohn Company, Kalamazoo, MI.

Block, S. S. (ed.). 1991. Disinfection, sterilization, and preservation. Lea and Febiger, Philadelphia, PA.

Bong, C. F. J., and P. P. Sikorowski. 1991. Effects of cytoplasmic polyhedrosis virus and bacterial contamination on growth and development of the corn earworm, *Helicoverpa zea* (Lepidoptera: Noctuidae). J. Invertebr. Pathol. 57: 406–412.

Brooks, M. A. 1963. The microorganisms of healthy insects, pp. 215–243. *In* E. A. Steinhaus (ed.), Insect pathology, vol. 1. Academic Press, New York, NY.

Brust, M., and G. Fraenkel. 1955. The nutritional requirements of the larvae of a blowfly, *Phormia regina* (Meig.). Physiol. Zool. 28: 186–204.

Bucher, G. E., and R. Williams. 1967. The microbial flora of laboratory cultures of the greater wax moth and its effect on rearing parasites. J. Invertebr. Pathol. 9: 467–473.

Bulut, S., W. M. Waites, and J. R. Mitchell. 1999. Effects of combined shear and thermal forces on destruction of *Microbacterium lacticum*. Appl. Environ. Microbiol. 65: 4464–4469.

Buyukguzel, K., and Y. Sevki. 2002. Effects of antimicrobial agents on the survival and development of larvae of *Pimpla turionellae* L. (Hymenoptera: Ichneumonidae) reared on an artificial diet. Turkish J. Zool. 26: 111–119.

Carneiro de Melo, A. M. S., C. A. Cassar, and R. J. Miles. 1998. Trisodium phosphate increases sensitivity of gram-negative bacteria to lysozyme and nisin. J. Food Prot. 61: 839–844.

Cavallito, C. J., and J. H. Bailey. 1944. Allicin, the antibacterial principle of *Allium sativum*. I. Isolation, physical properties and antibacterial action. J. Am. Chem. Soc. 16: 1950–1951.

Chamberlain, W. F. 1989. Survival of second and third instars of the cattle grub *Hypoderma lineatum* Villers and *Hypoderma bovis* L. in artificial media. Southwest. Entomol. 14: 233–240.

Chapman, R. R. 1998. The insects: structure and function. Cambridge University Press, Cambridge, UK.

Christian, J. H. B. 1963. Water activity and growth of microorganisms, pp. 248–255. *In* J. M. Leitch and D. N. Rhodes (eds.), Recent advances in food science. Butterworths, London, UK.

Cohen, A. C. 2003. Insect diets: science and technology. CRC Press, Boca Raton, FL.

Cohen, A. C., and L. K. Smith. 1998. A new concept in artificial diets for *Chrysoperla carnea*: the efficacy of solid diets. Biol. Contr. 13: 49–54.

Cundith, C. J., C. R. Kerth, W. R. Jones, T. A. McCaskey, and D. L. Kuhlers. 2002. Microbial reduction efficiencies of filtration, electrostatic polarization, and UV components of a germicidal air cleaning system. J. Food Sci. 67: 2278–2281.

Curtis, T. P., W. T. Sloan, and J. W. Scannell. 2002. Estimating prokaryote diversity and its limit. Proc. Nat. Acad. Sci. 99: 10494–10499.

Dark, R. A., and H. J. Harper. 1972. Aerosol sampling, pp. 37–46. *In* D. A. Shapton and R. G. Board (eds.), Safety in microbiology. Academic Press, New York, NY.

Davidson, P. M. 2001. Chemical preservatives and natural antimicrobial compounds, pp. 593–627. *In* M. P. Doyle, L. R. Beauchat, and T. J. Montville (eds.), Food microbiology: fundamentals and frontiers, 2nd ed. ASM Press, Washington, DC.

Davidson, P. M., and A. S. Naidu. 2000. Phyto-phenols, pp. 266–294. *In* A. S. Naideu (ed.), Natural food antimicrobials systems. CRC Press, Boca Raton, FL.

Davis F. M. 1989. Rearing the southwestern corn borer and fall armyworms at Mississippi State, pp. 27–36. *In* J. Keyser and N. Russell (eds.), Toward insect resistant maize for the third world. Proceedings of the International Symposium on Methodologies for Developing Host Plant Resistance to Maize Insects, 9–14 March 1987, El Baton, Mexico. International Maize and Wheat Improvement Center (CIMMYT), Mexico City, Mexico.

Day, M. F., and I. I. Oster. 1963. Physical injuries, pp. 29–63. *In* E. A. Steinhaus (ed.), Insect pathology, vol. 1. Academic Press, New York, NY.

Domsch, K. H., W. Gams, and T. H. Anderson. 1980. Compendium of soil fungi. Academic Press, London, UK.

Dunham, W. B.1977. Virucidal agents, pp. 426–441. *In* S. S. Block (ed.), Disinfection, sterilization, and preservation. Lea and Febiger, Philadelphia, PA.

Dunkel, F. V., and N. R. Read. 1991. Review of the effect of sorbic acid on insect survival in rearing diets with reference to other antimicrobials. Am. Entomol. 37: 172–178.

Ellis M. B. 1971. Dematiaceous Hyphomycetes. Commonwealth Mycological Institute, Surrey, UK.

Ernst, R. R. 1977. Sterilization by heat, pp. 481–521. *In* S. S. Block (ed.), Disinfection, sterilization, and preservation. Lea and Febiger, Philadelphia, PA.

Fargues, J., M.S. Goettel, N. Smits, A. Ouedraogo, C. Vidal, L.A. Lacey, and M. Rougier. 1996. Variability in susceptibility to simulated sunlight of conidia among isolates of entomopathogenic Hyphomycetes. Mycopathol. 135: 171–181.

Farkas, J. 2001. Physical methods of food preservation, pp. 567–591. *In.* M. P. Doyle, L. R. Beuchat, and T. J. Montville (eds.), Food microbiology: fundamentals and frontiers. ASM Press, Washington, DC.

Favero, M. S., J. R. Puleo, J. H. Marshall, and G. S. Oxborrow. 1966. Comparative levels and types of microbial contamination detected in industrial clean rooms. Appl. Microbiol. 14: 539–551.

Feller, W. 1950. An introduction to the probability theory and its application. John Wiley and Sons, Inc., New York, NY.

Fennema, O. R. 1996. Water and ice, pp. 18–94. *In* O. R. Fennema (ed.), Food chemistry. Marcel Dekker, New York, NY.

Freeman, B. A. 1979. The enteric bacilli: classification and properties; coliform bacteria and *Proteus*, pp. 500–517. *In* B. A. Freeman (ed.), Textbook of microbiology. Saunders, Philadelphia, PA.

Gifawesen, C., B. R. Funke, and F. I. Proshold. 1975. Control of antifungal-resistant strains of *Aspergillus niger* mold contaminants in insect rearing media. J. Econ. Entomol. 68: 441–444.

Gilchrist, J. E., J. E. Campbell, C. B. Donnely, J. T. Peeler, and J. M. Delanay. 1973. Spiral plate method for bacterial determination. Appl. Microbiol. 43: 149–157.

Goettel, M. S., and G. M. Duke. 1996. Decontamination of *Ascosphaera aggregata* spores from alfalfa leafcutting bee cadavers and bee cells by fumigation with paraformaldehyde. Bee Sci. 4: 26–29.

Goettel, M. S., and G. D. Inglis. 1997. Fungi: Hyphomycetes, pp. 213–249. *In* L. A. Lacey (ed.), Manual of techniques in insect pathology. Academic Press, San Diego, CA.

Goodwin, R. H. 1968. Nonsporeforming bacteria in the armyworm, *Pseudaletia unipuncta*, under gnotobiotic conditions. J. Invertebr. Pathol. 11: 358–370.

Greany, P. D., G. E. Allen, J. C. Webb, J. L. Sharp, and D. L. Chambers. 1977. Stress-induced septicemia as an impediment to laboratory rearing of the fruit fly parasitoid *Biosteres* (*Opius*) *longicaudatus* (Hymenoptera: Braconidae) and the Caribbean fruit fly *Anastrepha suspensa* (Diptera: Tephritidae). J. Invertebr. Pathol. 29: 153–161.

Greenberg, B. 1962. Host-contaminant biology of muscoid flies. II. Bacterial survival in the stable fly, false stable fly, and the little house fly. J. Insect. Pathol. 4: 216–223.

Greenberg, B., and M. Klowden. 1972. Enteric bacterial interactions in insects. Am. J. Clin. Nutr. 25: 1459–1466.

Griffin, D. H. 1981. Fungal physiology. John Wiley & Sons, New York, NY.

Griffin, J. G., O. H. Lindig, and R. E. McLaughlin. 1974. Flash sterilizers: sterilizing artificial diets for insects. J. Econ. Entomol. 67: 689.

Gueldner, R. C., P. P. Sikorowski, and J. M. Wyatt. 1977. Bacterial load and pheromone production in the boll weevil *Anthonomus grandis*. J. Invertebr. Pathol. 29: 397–398.

Habib, M. E. 1978. A bacterial disease of the American cotton leaf worm, *Alabama argillacea* (Hubner) (Lep., Noctuidae), with notes on its histopathological effects. Z. Angew. Entomol. 85: 76–81.

Harris, R. F., and L. E. Sommers. 1968. Plate-dilution frequency technique for assay of microbial ecology. Appl. Microbiol. 16: 330–334.

Hawksworth, D. L. 2001. The magnitude of fungal diversity: the 1.5 million species estimate revisited. Mycol. Res. 12: 1422–1432.

Hawksworth, D. L., P. M. Kirk, B. C. Sutton, and D. N. Pegler. 1995. Ainsworth and Bisby's dictionary of the Fungi. CAB International, Wallingford, UK.

Hedin, P. A., A. C. Thompson, R. C. Gueldner, and R. D. Henson. 1974. Analysis of the antimicrobial agents, potassium sorbate and methyl-p-hydroxybenzoate, in boll weevil diets. J. Econ. Entomol. 61: 147–149.

Hensley, S. D., and A. M. Hammond, Jr. 1968. Laboratory techniques for rearing the sugarcane borer on an artificial diet. J. Econ. Entomol. 61: 1742–1743.

Hintikka, E. L. 2004. The role of *Stachybotrys* in the phenomenon known as sick building syndrome. Adv. Appl. Microbiol. 55: 155–173.

Hocking, A. D. 2001. Toxigenic *Aspergillus* species, pp. 451–465. *In* M. P. Doyle, L. R. Beauchat, and T. J. Montville (eds.), Food microbiology: fundamentals and frontiers, 2nd ed. ASM Press, Washington, DC.

Hoffman, R. K. 1971. Toxic gases, pp. 225–258. *In* W. B. Hugo (ed.), Inhibition and destruction of the microbial cell. Academic Press, New York, NY.

Holt, J. G., N. R. Krieg, P. H. A. Sneath, and D. Bergy. 1994. Bergey's manual of determinative bacteriology. Williams & Wilkins, Baltimore, MD.

Horton, D. L., T. G. Teague, J. R. Phillips, and W. C. Yearian. 1986. Fungicide interference with parasitization of *Heliothis zea* by *Microplitis croceipes*. J. Agric. Entomol. 3: 186–191.

Howell, J. F. 1971. Problems involved in rearing the codling moth on diet in trays. J. Econ. Entomol. 64: 631–636.

Hurej, M., P. P. Sikorowski, and H. W. Chambers. 1982. Effects of bacterial contamination on insecticide-treated boll weevils (Coleoptera: Curculionidae). J. Econ. Entomol. 75: 651–654.

Ignoffo, C. M. 1965. The nuclear-polyhedrosis virus of *Heliothis zea* (Boddie) and *Heliothis virescens* (F.). II. Biology and propagation of diet-reared *Heliothis*. J. Insect Pathol. 7: 217–226.

Ignoffo, C. M., and S. R. Dutky. 1963. The effect of sodium hypochlorite on the viability and infectivity of *Bacillus* and *Beauveria* spores and cabbage looper nuclear-polyhedrosis virus. J. Insect Pathol. 5: 422–426.

Imlay, J. A., and S. Linn. 1998. DNA damage and oxygen radical toxicity. Science 240: 1302–1309.

Inglis, G. D., and A. C. Cohen. 2004. Influence of antimicrobial agents on the spoilage of a meat-based entomophage diet. J. Econ. Entomol. 97: 235–250.

Inglis, G. D., M. S. Goettel, and L. Sigler. 1992. Analysis of alfalfa leafcutter bee (*Megachile rotundata*) provisions pre- and post-sterilization with propylene oxide. Apidologie 23: 119–132.

Inovatech. 2000. Inovapure product description. Canadian Inovatech, Inc., Abbotsford, Canada.

Jarosz, J. 1979. Gut flora of *Galleria mellonella* suppressing ingested bacteria. J. Invertebr. Pathol. 34: 192–198.

Jarosz, J. 1989. Simplified technique for preparing germ-free specimens of greater wax moth Lepidoptera Pyralidae larvae. J. Econ. Entomol. 82: 1478–1481.

Jawetz, E., J. L. Melnick, and E. A. Adelberg. 1978. Review of medical microbiology. Lange Medical, Los Altos, CA.

Jay, J. M. 1996. Modern food microbiology. Chapman and Hall, New York, NY.

Jay, J. M., and G. M. Rivers. 1984. Antimicrobial activity of some food flavoring compounds. J. Food Safety 6: 129–139.

Jennison, M. W. 1942. Atomizing of mouth and nose secretions into the air as revealed by high-speed photography. Society for Medical Science Aerobiology, Publication 17. American Association for the Advancement of Science, Washington, DC.

Joslyn, L. J. 1991. Sterilization by heat, pp. 495–526. *In* S. S. Block (ed.), Disinfection, Sterilization, and preservation. Lea and Febiger, Philadelphia, PA.

Kendrick, B. 2000. The fifth kingdom, 3rd ed. Mycologue Publications, Sidney, Canada.

Koch, H. W., and E. H. Eisenhower. 1965. Electron accelerators for food processing, pp. 149–180. *In* Radiation preservation of foods. National Research Council, National Academy of Science, Washington, DC.

Krieg, A. 1987. Diseases caused by bacteria and other prokaryotes, pp. 323–354. *In* J. R. Fuxa and Y. Tanada (eds.). Epizootiology of insect diseases. Wiley, New York, NY.

Kurtzman, C. P., and J. W. Fell. 2000. The yeasts, a taxonomic study, 4th ed. Elsevier, Amsterdam, The Netherlands.

Lachowicz, K. J., G. P. Jones, D. R. Briggs, F. E. Bienvenu, J. Wan, A. Wilcock, and M. J. Coventry. 1998. The synergistic preservative effects of the essential oils of sweet basil (*Ocimum basilicum* L.) against acid-tolerant food microflora. Lett. Appl. Microbiol. 26: 209–214.

Lee, B. H., S. Kermasha, and B. E. Baker. 1989. Thermal, ultrasonic and ultraviolet inactivation of *Salmonella* in thin films of aqueous media and chocolate. Food Microbiol. 6: 143–152.

Leonard, D. E., and C. C. Doane. 1966. An artificial diet for the gypsy moth *Porthetria dispar* (Lepidoptera: Lymantriidae). Ann. Entomol. Soc. Am. 59: 462–464.

Leppla, N. C., S. L. Carlyle, and T. C. Carlysle. 1974. Effects of surface sterilization and automatic collection on cabbage looper eggs. J. Econ. Entomol. 67: 33–36.

Levetin, E., R. Shaughnessy, C. Rogers, A. Christine, and R. Scheir. 2001. Effectiveness of germicidal UV radiation for reducing fungal contamination within air-handling units. Appl. Environ. Microbiol. 67: 3712–3715.

Lindberg, R. B. 1974. Culture and identification of commonly encountered gram-negative bacilli: *Pseudomonas, Klebsiella-Enterobacter, Serratia, Proteus, and Providencia*, pp. 19–35. *In* J. E. Prier and H. Friedman (eds.), Opportunistic pathogens. University Park Press, Baltimore, MD.

Lynch, R. E., and T. Reed. 1985. Rearing the lesser cornstalk borer: fungicides for control of *Aspergillus niger*. J. Entomol. Sci. 20: 26–33.

MacGown, M. W., and P. P. Sikorowski. 1980. Histopathology of midgut of mass reared, irradiated boll weevils contaminated with *Staphylococcus aureus* and *Streptococcus* sp. J. Econ. Entomol. 73: 81–87.

Mari, M., R. Iori, O. Leoni, and A. Marchi. 1993. *In vitro* activity of glucosinolate derived isothiocyanates against postharvest fruit pathogens. Ann. Appl. Biol. 123: 155–164.

Martignoni, M. E., and J. E. Milstead. 1960. Quaternary ammonium compounds for the surface sterilization of insects. J. Insect Pathol. 2: 124–133.

McLaughlin, R. E. 1962. The effect of temperature upon larval mortality of the armyworm, *Pseudaletia unipuncta* (Haworth). J. Insect. Pathol. 4: 279–284.

McLaughlin, R. E., and P. P. Sikorowski. 1978. Observations of boll weevil midgut when fed natural food or on bacterially contaminated artificial diet. J. Invertebr. Pathol. 32: 64–70.

Meynell G. G., and E. Meynell. 1970. Theory and practice in experimental bacteriology, 2nd ed. Cambridge University Press, Cambridge, UK.

Montville, T. J., K. Winkowski, and M. L. Chikindas. 2001. Biologically based preservation systems, pp. 629–647. *In* M. P. Doyle, L. R. Beauchat, and T. J. Montville (eds.), Food microbiology: fundamentals and frontiers, 2nd ed. ASM Press, Washington, DC.

Murphy, M. G., D. I. Skonberg, M. E. Camire, M. P. Dougherty, R. C. Bayer, and J. L. Briggs. 2003. Chemical composition and physical properties of extruded snacks containing crab-processing by product. J. Sci. Food Agric. 83: 1163–1167.

(NASA) National Aeronautics and Space Administration. 1967. Contamination control principles. NASA SP-5045. US Government Printing Office, Washington, DC.

(NASA) National Aeronautics and Space Administration. 1969. Clean room technology. NASA SP-5074. US Government Printing Office, Washington, DC.

Oloumi-Sadeghi, H., and E. Levine. 1989. Controlling fungi that colonize eggs of the western corn rootworm in the laboratory. Entomol. Exp. Appl. 50: 271–280.

O'Neil, M. J., A. Smith, P. E. Heckelman, and J. R. Obenchain (eds.). 2001. Merck index: an encyclopedia of chemicals, drugs, and biologicals. Merck and Co., White House Station, NJ.

Ough, C. S. 1993. Sulfur dioxide and sulfites, pp. 137–190. *In* P. M Davidson and A. L. Branen (eds.), Antimicrobials in foods. Marcel Dekker, New York, NY.

Patterson, J. T. 1971. Microbiological assessment of surfaces. J. Food. Technol. 6: 63–72.

Phillips, G. B., and R. S. Runkle. 1973. Biomedical applications of laminar air flow. CRC, Cleveland, OH.

Pitt, J. I. 1988. A laboratory guide to common *Penicillium* species. Commonwealth Scientific and Industrial Research Organization, North Ryde, Australia.

Pitt, J. I. 2001. Toxigenic *Penicillium* species, pp. 467–480. *In* M. P. Doyle, L. R. Beauchat, and T. J. Montville (eds.), Food microbiology: fundamentals and frontiers, 2nd ed. ASM Press, Washington, DC.

Rahn, O., and J. E. Conn. 1944. Effect of increase in acidity on antiseptic efficiency. Ind. Eng. Chem. 36: 185.

Raper, K. B., and D. I. Fennell. 1965. The genus *Aspergillus*. Williams and Wilkins Company, Baltimore, MD.

Raper, K. B., C. Thom, and D. I. Fennell. 1968. A manual of the Penicillia. Hafner Publishing Company, New York, NY.

Reinecke, J. P., and P. P. Sikorowski. 1989. Residual activity of four antibiotics in an artificial insect diet after three heat sterilization treatments. Southwest. Entomol. 14: 339–343.

Reiss, J. 1976. Prevention of the formation of mycotoxins in whole wheat bread by citric acid and lactic acid. Experientia 32: 168.

Rochon, M., and N. Sullivan. 1999. Products based on accelerated and stabilized hydrogen peroxide: evidence for cleaning and sanitizing efficiency, environmental and human safety and non-corrosiveness. Can. J. Infect. Contr. 14: 123–130.

Runkle, R. S., and G. B. Phillips. 1969. Microbial contamination control facilities. Van Nostrand Reinhold, New York, NY.

Sagoo, S. K., R. Board, and S. Roller. 2002. Chitosan potentiates the antimicrobial action of sodium benzoate on spoilage yeasts. Lett. Appl. Microbiol. 34: 168–172.

Saikkonen, K., S. H. Faeth, M. Helander, and T. J. Sullivan. 1998. Fungal endophytes: a continuum of interactions with host plants. Annu. Rev. Ecol. Syst. 29: 319–343.

Sattar, S. A., and V. S. Springthrope. 1998. A product based on accelerated and stabilized hydrogen peroxide: evidence for broad-spectrum germicidal activity. Can. J. Infect. Contr. 13: 51–54.

Schrezenmeir, J., and M. de Vrese. 2001. Probiotics, prebiotics, and synbiotics—approaching a definition. Am. J. Clin. Nutr. 73(suppl): 361S–364S.

Shapiro, M. 1984. Microorganisms as contaminants and pathogens in insect rearing, pp. 130–142. *In* E. G. King and N. C. Leppla (eds.), Advances and challenges in insect rearing. USDA-ARS, U.S. Government Printing Office, Washington, DC.

Shelef, L. A., O. A. Naglik, and D. W. Bogen. 1980. Sensitivity of some common foodborne bacteria to the spices sage, rosemary, and allspice. J. Food Sci. 45: 1042–1044.

Sigler, L., S. P. Abbott, and H. Gauvreau. 1996. Assessment of worker exposure to airborne molds in honeybee overwintering facilities. Am. Ind. Hyg. Assoc. J. 57: 484–490.

Sikorowski, P. P. 1975. Microbiological monitoring in the boll weevil rearing facility. Miss. Agric. For. Exper. Stn. Tech. Bull. 71. Mississippi State University, Mississippi State, MS.

Sikorowski, P.P. 1984. Pathogens and microbiological contaminants: their occurrence and control, pp. 115–169. *In* P. P. Sikorowski, J. G. Griffin, J. Roberson, and O. H. Lindig (eds.), Boll weevil mass rearing technology. University Press of Mississippi, Jackson, MS.

Sikorowski, P. P. 1985. Pecan weevil pathology, pp. 87–101. In W. W. Neel (ed.), Pecan weevil: research perspective. Quail Ridge Press, Jackson, MS.

Sikorowski, P. P., and A. M. Lawrence. 1991. Effects of bacterial contamination on gamma-irradiated and nonirradiated corn earworm (Lepidoptera: Noctuidae). J. Econ. Entomol. 84: 1441–1447.

Sikorowski, P. P., and A. M. Lawrence. 1994. Microbial contamination and insect rearing. Am. Entomol. 40: 240–253.

Sikorowski, P. P., and A. C. Thompson. 1979. Effects of bacterial load on mass reared boll weevils *Anthonomus grandis*, pp. 195–196. *In* Proceedings, Symposium: Progress in Invertebrate Pathology. International Colloquium on Invertebrate Pathology, Prague, Czechoslovakia, 11–17 September 1978.

Sikorowski, P. P., and A. C. Thompson. 1980. Effects of internal bacterial contaminants on mass reared boll weevils (*Anthonomus grandis* Boheman). Miss. Agric. For. Exper. Stn. Tech. Bull. 103. Mississippi State University, Mississippi State, MS.

Sikorowski, P. P., and A. C. Thompson. 1984. Effects of bacterial contamination on development and blood chemistry of *Heliothis virescens*. Comp. Biochem. Physiol. 77(A): 283–285.

Sikorowski, P. P., and A. C. Thompson. 1985. Residual quantities of streptomycin sulfate in haemolymph and tissue of *Heliothis virescens* reared on treated diet. Pharmacol. Biochem. Behav. 23: 525–528.

Sikorowski, P. P., J. M. Wyatt, and O. H. Lindig. 1977. Method for surface-sterilization of boll weevil eggs and the determination of microbial contamination of adults. Southwest. Entomol. 2: 32–36.

Sikorowski, P. P., A. D. Kent, O. H. Lindig, G. Wiygul, and J. Roberson. 1980. Laboratory and insectary studies on the use of antibiotics and antimicrobial agents in mass-rearing of boll weevils. J. Econ Entomol. 73: 106–110.

Sikorowski, P. P., G. R. Burkett, A. C. Thompson, and A. D. Kent. 1984. Effects of antimicrobials on development of *Heliothis virescens* (F.) (Lepidoptera: Noctuidae). J. Georgia Entomol. Soc. 19: 327–332.

Sikorowski, P. P., A. D. Kent, and J. L. Willers. 1988. Distribution of tetracycline in hemolymph and tissue of all stages of corn earworm *Heliothis zea* Boddie. Comp. Biochem. Physiol. Toxicol. Endocrinol. 91: 437–442.

Sikorowski, P. P., J. E. Powell, and A. M. Lawrence. 1992. Effects of bacterial contamination on development of *Microplitis croceipes* [Hym: Braconidae]. Entomophaga 37: 475–481.

Singh, P., and H. L. House. 1970. Antimicrobials: "safe" levels in a synthetic diet of an insect., *Agria affinis*. J. Insect Physiol. 16: 1769–1782.

Singleton, P., and D. Sainsbury. 2001. Dictionary of microbiology and molecular biology, 3rd ed. John Wiley and Sons, Chichester, UK.

Smalley, D. L., and D. D. Ourth. 1979. Bacterial genera isolated from field-collected (diapausing) and laboratory-reared cotton boll weevils, *Anthonomus grandis* (Coleoptera: Curculionidae). J. Invertebr. Pathol. 34: 158–163.

Smid, T., E. Schokkin, J. S. Boleij, and D. Heederik. 1989. Enumeration of viable fungi in occupational environments: a comparison of samples and media. Am. Ind. Hyg. Assoc. J. 50: 235–239.

Smith, F. W., and M. Bruch. 1969. Reduction of microbiological shedding in clean rooms. Dev. Ind. Microbiol. 10: 290.

Smith-Palmer, A., J. Stewart, and L. Fyfe. 1998. Antimicrobial properties of plant essential oils and essences against five important food-borne pathogens. Lett. Appl. Microbiol. 26: 118–122.

Sofos, J. N. 1986. Use of phosphates in low-sodium meat products. Food Technol. 40: 52–68.

Steinhaus, E. A. 1941. A study of the bacteria associated with thirty species of insects. J. Bacteriol. 42: 757–790.

Steinhaus, E. A. 1947. Insect microbiology. Comstock Publishing Company, Ithaca, NY.

Steinhaus, E. A. 1949. Principles of insect pathology. McGraw-Hill, New York, NY.

Steinhaus, E. A. 1953. Diseases of insects reared in the laboratory or insectary. Calif. Agric. Exp. Stn. Ext. Serv. Leafl. No. 9. University of California, Berkeley, CA.

Steinhaus, E. A. 1975. Disease in a minor chord. Ohio State University Press, Columbus, OH.

Steinhaus, E. A., and M. E. Martignoni. 1970. An abridged glossary of terms used in invertebrate pathology. Pacific Northwest Forest and Range Experiment Station, U.S. Forest Service, Corvallis, OR.

Stephens, J. M. 1962. A strain of *Streptococcus faecalis* Andrews and Horder producing mortality in larvae of *Galleria mellonella* (Linnaeus). J. Insect Pathol. 4: 267–268.

Stewart, F. D., M. R. Bell, A. J. Martinez, J. L. Roberson, and A. M. Lowe. 1976. The surface sterilization of pink bollworm eggs and spread of a cytoplasmic polyhedrosis virus in rearing containers. USDA-APHIS-81-27. Government Printing Office, Washington, DC.

Sun, M. 1981. Study shows formaldehyde is carcinogenic. Science 213: 1232.

Sutter, G. R., J. L. Krysan, and P. L. Guss. 1971. Rearing the southern corn rootworm on artificial diet. J. Econ. Entomol. 64: 65–67.

Tanada, Y., and H. K. Kaya. 1993. Insect pathology. Academic Press, New York, NY.

Teague, T. G., D. L. Horton, W. C. Yearian, and J. R. Phillips. 1985. Benomyl inhibition of *Cotesia marginiventris* survival in four lepidopterous hosts. J. Entomol. Sci. 20:76–81.

Temeyer, K. B. 1998. Antifungal agents and solvents for use in horn fly larval rearing or bioassays. Southwest. Entomol. 23: 237–245.

Thompson, A. C., and P. P. Sikorowski. 1978. The effect of bacterial load on amino acids in the boll weevil *Anthonomus grandis*. J. Invertebr. Pathol. 32: 388–389.

Thompson, A. C., and P. P. Sikorowski. 1982. Haemolymph analysis of irradiated and Dimilin-treated boll weevils, *Anthonomus grandis*. J. Invertebr. Pathol. 39: 158–163.

Thompson, C. G., and E. A. Steinhaus. 1950. Further tests using a polyhedrosis virus to control the alfalfa caterpillar. Hilgardia 19: 411–415.

Thompson, A. C., P. P. Sikorowski, and J. M. Wyatt. 1977. The effect of bacterial load on fatty acids in the boll weevil *Anthonomus grandis* (Coleoptera: Curculionidae). J. Invertebr. Pathol. 30: 274–275.

Tortora, G. J., B. R. Funke, and C. L. Case. 1989. Microbiology, an introduction. Benjamin-Cummings, Redwood City, CA.

Tranter, H. S. 1994. Lysozyme, ovotransferrin and avidin, pp 65–97. *In* V. M. Dillon and R. G. Board (eds.), Natural antimicrobial systems and food preservation. CAB International, Wallingford, UK.

Trujillo, R., and T. J. David. 1972. Sporostatic and sporocidal properties of aqueous formaldehyde. Appl. Microbiol. 23: 618–622.

Urbain, W. M. 1986. Food irradiation. Academic Press, Orlando, FL.

(USEPA) U.S. Environmental Protection Agency. 1996. List of chemicals evaluated for carcinogenic potential. U.S. EPA Office of Pesticide Programs, Washington, DC.

Vago, C. 1963. Predispositions and interrelations in insect diseases, pp 339–370. *In* E. A. Steinhaus (ed.), Insect pathology, an advanced treatise. Academic Press, New York, NY.

Vanderzant, E. S. 1966. Defined diets for phytophagous insects, pp. 274–303. *In* C. N. Smith (ed.), Insect colonization and mass production. Academic Press, New York, NY.

Von Arx, J. A. 1981. The genera of Fungi sporulating in pure culture. Gunter Verlag, Vaduz, Germany.

Webster, J. 1980. Introduction to Fungi, 2nd ed. Cambridge University Press, Cambridge, UK.

White, G. F. 1931. Production of sterile maggots for surgical use. J. Parasitol. 18: 133.

Whitfield, W. J. 1963. The design of a dust-controlled vented hood utilizing laminar air flow. Research Report No. SC-4905 (RR). Sandia Laboratories, Albuquerque, NM.

(WHO) World Health Organization. 1990. Biological contaminants in indoor air. EURO Reports and Studies 113. World Health Organization Regional Publications, Copenhagen, Denmark.

Wilkins, K. M., and R. G. Board. 1989. Natural antimicrobial systems, pp. 285–362. *In* G. W. Gould (ed.), Mechanisms of action of food preservation procedures. Elsevier Applied Science, London, UK.

Wiygul, G., and P. P. Sikorowski. 1981. The effect of various levels of contamination by two species of bacteria (*Staphylococcus aureus* and *Streptococcus* sp.) on oxygen uptake of the boll weevil. Comp. Biochem. Physiol. 68(A): 527–530.

Wiygul, G., and P. P. Sikorowski. 1986. The effect of staphylococcal enterotoxin B on pheromone production in fat bodies isolated from male boll weevils. J. Invertebr. Pathol. 47: 116–119.

Wolin, M. J. 1979. Physical agents, bactericidal substances (disinfectants), and chemotherapeutic drugs, pp. 121–156. *In* B. A. Freeman (ed.), Textbook of microbiology. Saunders, Philadelphia, PA.

Woomer, P.L. 1994. Most probable number counts, pp. 329–350. *In* S.H. Mickelson (ed.), Methods of soil analysis, part 2: microbiological and biochemical properties. Soil Science Society of America, Madison, WI.

Xu, P., J. Peccia, P. Fabian, J. W. Martyny, K. P. Fennelly, M. Hernandez, and S. L. Shelly. 2003. Efficacy of ultraviolet germicidal irradiation of upper-room air in inactivating airborne bacterial spores and mycobacteria in full-scale studies. Atmos. Environ. 37: 405–419.

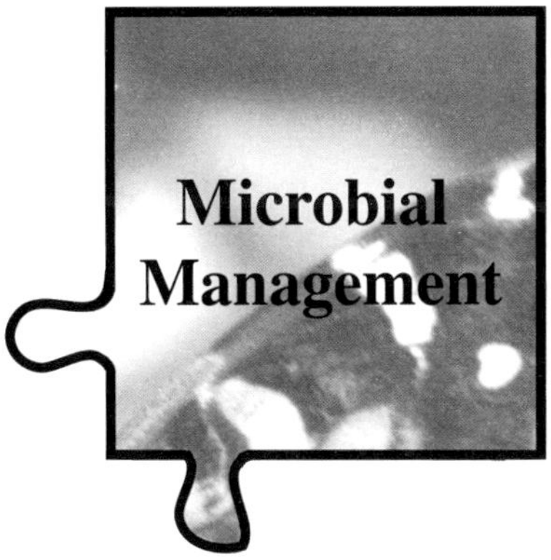

Chapter 9

Entomopathogens and Insect Rearing

G. DOUGLAS INGLIS
Research Scientist
Zoonotic Bacteriology
Agriculture and Agri-Food Canada
Lethbridge, AB Canada

PETER P. SIKOROWSKI, Deceased
Professor of Entomology
Department of Entomology and Plant Pathology
Mississippi State University
Mississippi State, MS

I. Introduction

Reared insects are essential to many aspects of entomology, and the effects of microorganisms on insects are of paramount importance in the success of rearing programs. Microorganisms can affect insects directly as *entomopathogens*[1] (i.e., disease-causing microorganisms) or as mutualistic (i.e., mutually beneficial) or commensalistic (i.e., no effect on the host) symbionts. Equally important are the adverse, indirect effects of microorganisms on insects (e.g., as microbial contaminants). Neglecting the insidious nature of microorganisms on insects often has dire consequences, resulting in the production of poor-quality individuals, which in turn compromises their utility. To avoid costly failures and wasted effort, it is imperative that insectary workers practice appropriate monitoring and management strategies for microorganisms. This chapter focuses on diseases caused by entomopathogens and is designed to provide an overview of the impact of entomopathogenic organisms on reared insects with emphasis on the principles of detection and management. While microorganisms can be highly detrimental to a rearing program, in some instances they are intrinsically essential. Some examples include insects reared to propagate entomopathogens used as microbial control agents (e.g., viruses, protozoa, and nematodes), to produce specialized entomopathogens (e.g., baculoviruses) used as vectors of genes encoding heterologous proteins (i.e., proteins encoded by a gene from a different species), and to serve as expression vectors of these proteins (e.g., vaccines). In any insect rearing facility, problems with microorganisms will eventually occur; and it is imperative that an insectary has a suitable infrastructure in place to identify and deal with the problem. The success of the rearing program depends on it.

II. Disease Agents

The word *pathology* is derived from the Greek words *pathos* meaning suffering and *logos* meaning reason; so *insect pathology* is the application of reason to understand the origin, nature, and course of suffering (i.e., disease) in insects. Before proceeding, it is important to first understand the meaning of disease. *Disease* can be defined as a dynamic interaction between an organism and its environment that results in deleterious, abnormal, physiological and often morphological or neurological changes in the organism (Merrill 1980). Thus disease is simply a deleterious, abnormal state. Specifying the causal agent is not part of the definition. Unfortunately, the term *disease* is often misused. One often encounters statements such as "the insect caught the disease" or the "mosquito transmitted malaria." An insect can neither catch nor transmit an interaction. Rather, an insect is infected by and transmits a pathogen, respectively, not the disease. When describing the effects of disease on an insect, two terms are used. The term *symptom* refers to a functional or behavioral disturbance (e.g., abnormal movement, digestive disturbance), and the term *sign* refers to visual evidence of the pathogen (e.g., changes in color, malformation of body parts). In many cases, the symptoms and signs of disease can be used to determine the category of causal agent. However, unless a symptom or sign of disease is both conspicuous and unique [e.g., pebrine disease of the silkworm, *Bombyx mori* (Lepidoptera: Bombycidae), caused by *Nosema bombycis* and producing pepper-like spots on the integument], it cannot usually be used alone to make a definitive diagnosis.

Disease is almost always more common and severe in insectary than in field settings. A number of factors can play a role in the relatively high frequency of disease in insectaries, but the most common factor appears to be physiological stress. Stress occurs when an agent disturbs or interferes with the normal physiological equilibrium of an insect, and stress can cause disease directly or indirectly by predisposing the insect to other factors such as entomopathogens. Insects in artificial rearing systems are almost certainly exposed to a variety of suboptimal conditions at some point during their development; and if entomopathogens are not managed, disease outbreaks can occur and devastate a colony.

[1]Definitions of technical terms are given at their points of first occurrence. Definitions of related terms can be found in the following references: Atlas (1993), Chapman (1998), Hawksworth et al. (1995), Steinhaus and Martignoni (1970), and Tanada and Kaya (1993).

A *pathogen* is any agent that incites disease. In most instances, the term *pathogen* is used to describe biotic agents such as viruses or bacteria that incite infectious disease. Biotic pathogens can be further divided into: 1) obligate pathogens, i.e., those that develop only within a living host (all viruses, some bacteria, some fungi, all protozoa, and some nematodes) and 2) facultative pathogens, i.e., those that may or may not develop in a living host (primarily bacteria and fungi). When a facultative pathogen utilizes a non-living host or substrate, it is termed a *saprotroph*. Pathogens can be abiotic as well. Abiotic pathogens can be mechanical (e.g., physical damage by improper handling); environmental (e.g., temperature, humidity, or radiation); chemical (e.g., toxins); genetic (e.g., inbreeding); or nutritional (e.g., inadequate nutrition). Often abiotic pathogens act in concert, and the result is severe physiological stress that may lead to disease. Therefore, an essential goal of insect rearing is the utilization of husbandry practices that minimize physiological stress.

Biotic pathogens of insects are comprised of five major groups including: 1) viruses, 2) bacteria, 3) fungi, 4) protozoa, and 5) nematodes. In this chapter, each of these pathogens is considered in turn. For more detailed information on microorganisms associated with insects, the reader should consult: Steinhaus (1947, 1949), Weiser (1977), Poinar and Thomas (1984), Chukhrii (1987), Samson et al. (1988), Tanada and Kaya (1993), Tzean et al. (1997), and Boucias and Pendland (1998). A recent review by Björnson and Schütte (2003) provides a list of entomopathogens that infect a number of mass-produced natural enemies and pollinators.

A. Viruses

Insect pathogenic viruses have been placed into 12 viral families, but many remain unclassified (Tanada and Kaya 1993). Many insect viruses are assigned to families of vertebrate viruses, but some are in families specific for insects and related invertebrates (e.g., Ascoviridae, Baculoviridae, and Polydnaviridae). Representatives of all virus families have been observed to infect insects in rearing settings. Some can destroy a colony, while others cause chronic or sublethal infections. Some examples of viruses frequently encountered in insectaries include the nuclear polyhedrosis virus (synonym: nucleopolyhedrosis virus), granulosis virus (synonym: granulovirus), nonoccluded baculovirus, gonad-specific virus, entomopox virus, cytoplasmic polyhedrosis virus, iridescent virus, and densovirus.

Viruses are a diverse group of biological macromolecules that have the ability to multiply within living cells as obligate parasites. All insect viruses are comprised of nucleic acid (dsRNA, ssRNA, dsDNA, or ssDNA) within a proteinaceous capsule termed a *capsid*. Collectively, the nucleic acid and capsule are termed a *nucleocapsid*. Depending on the virus, some nucleocapsids are enclosed within a lipid envelope. A *virion* is the infectious unit of a virus. In an *enveloped virus*, the virion consists of the nucleocapsid and the envelope. In a *nonenveloped virus*, the virion is comprised only of the nucleocapsid. Viruses do not possess the ability to replicate themselves independent of a living host and thus cannot be cultured on microbiological media. In essence, they highjack the metabolic machinery of the host cell and trick it into producing progeny viruses. Furthermore, virions are very small and below the resolution of light microscopes (< 0.1 μm). Therefore, diagnosing many viral infections relies on *symptomatology* (i.e., the science of the symptoms and signs of disease), electron microscopy, and/or molecular detection methods. However, a number of entomopathogenic viruses (Baculoviridae, Poxviridae, and Reoviridae) produce occlusion bodies (OB) in which the virions are embedded within a paracrystalline protein matrix, the aggregate size of which is much greater than that of an individual virion. The OB protects the virions (e.g., from ultraviolet light) and increases their persistence outside of the host; consequently, their formation may dramatically increase the virion's disease-producing potential. They also may serve an important function in the infection process. Because OBs are relatively large and can be seen with a light microscope, their formation greatly facilitates disease diagnoses. Depending on the virus, the location in which the OBs form (tissues and organelles) provides important clues for identification.

1. Baculoviridae. *a. Occluded.* All baculoviruses produce rod-shaped, enveloped virions that contain dsDNA. The family Baculoviridae contains two economically important genera that produce virions embedded within OBs: 1) nuclear polyhedrosis viruses (NPV) and 2) granulosis viruses (GV) (Murphy et al. 1995; Fig. 1). Nuclear polyhedrosis viruses produce polyhedral (i.e., faceted) OBs composed of the protein polyhedrin. In contrast, GVs have OBs made up of a protein called *granulin*. The OBs produced by NPVs are relatively large and can be easily visualized with the microscope (0.5 to 15 μm in size) (Tanada and Kaya 1993). In contrast to NPVs, OBs of GVs are smaller (0.3 to 0.5 μm). They can be detected with a light microscope by an experienced diagnostician, but care must be exercised. The nucleocapsid is similar in all members of Baculoviridae and has a DNA core surrounded by a protein capsid. Virions typically range in size from 40–60 x 200–400 nm (note: 1 nm = 10^{-6} mm = 10^{-3} μm); and they are encased by an outer lipoprotein envelope. Granulosis virus OBs always contain a single virion (Figs. 1, 2). In contrast, NPV OBs contain multiple virions. Another distinguishing characteristic between the two families is that GV envelopes always contain a single nucleocapsid. In contrast, NPV nucleocapsids can be packaged singly (designated by an *S*) or multiply (designated by an *M*) within the lipid envelope (Fig. 2). These letters are sometimes used in naming types of NPV as MNPV or SNPV.

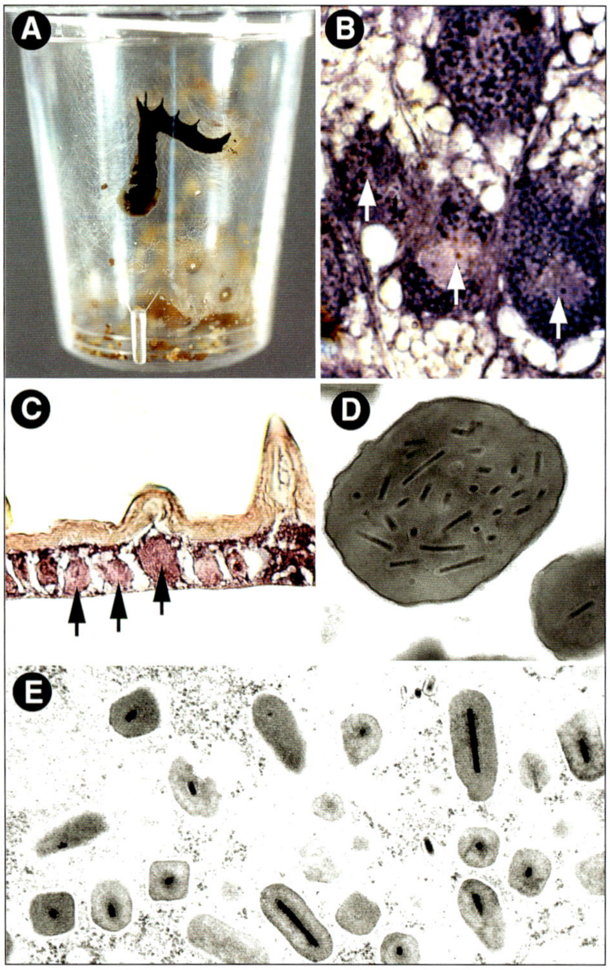

Fig. 1. Occluded baculoviruses. (A–D) tobacco budworm, *Heliothis virescens* (Lepidoptera: Noctuidae), infected with nuclear polyhedrosis virus (NPV), and (E) clover cutworm, *Dicestra trifolii* (Lepidoptera: Noctuidae), infected with granulosis virus (GV). (A) Grossly disintegrated larva killed by NPV. (B) Fat body cells showing enlarged nuclei packed with polyhedra (arrows). (C) Epidermis showing enlarged nuclei (arrows). (D) Transmission electron microscope (TEM) micrograph of a cross-section of an occlusion body showing virions randomly scattered throughout. (E) TEM micrograph showing occlusion bodies of GV-infected fat bodies (note the single virion within each elongated occlusion body). [Images A–D from Sikorowski and Lawrence (1997) with permission from the Mississippi Agricultural and Forestry Experiment Station, Mississippi State, MS; image E provided by Brian A. Federici (University of California, Riverside, CA).]

More than 500 NPVs have been identified from over 400 insect species, with susceptible hosts primarily in the Lepidoptera and to a lesser extent in the Hymenoptera, Diptera, Coleoptera, and Neuroptera (Martignoni and Iwai 1986, Tanada and Kaya 1993). The SNPVs have been reported from all of these orders, but MNPVs and GVs are restricted to the Lepidoptera. The MNPVs typically affect a number of lepidopteran insects, but certain types are restricted to one or a few species within the order.

The most common mode of entry of baculoviruses into the host is by ingestion. Other less common ports of entry and infection are through eggs and *spiracles* (i.e., external respiratory openings) and by parasitism. Once ingested, OBs dissolve in the alkaline conditions of the midgut, and the liberated virions breach the *per-*

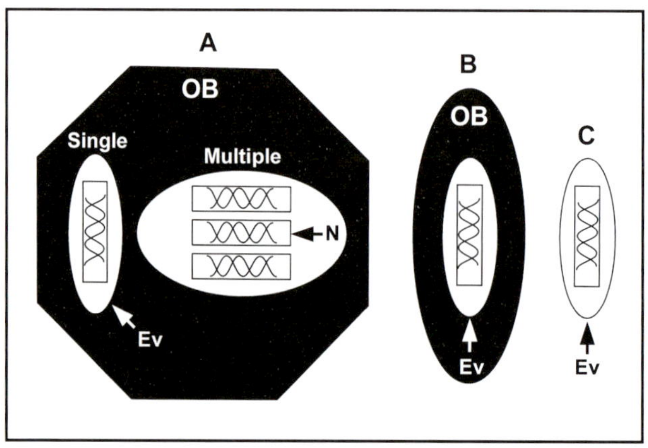

Fig. 2. Schematic diagram of three baculovirus families. The two helical lines within each individual rectangle represent double-stranded DNA, the rectangle represents the capsid within an envelope (EV), and the dark material represents protein of an occlusion body (OB). (A) Nuclear polyhedrosis virus (NPV) in which nucleocapsids (N) are packaged singly or in multiples within an envelope within a polyhedral occlusion body. (B) Granulosis virus (GV), which always contains a single, enveloped nucleocapsid within an ellipsoidal occlusion body. (C) Nonoccluded baculovirus (NOBV) in which a single nucleocapsid within an envelope is produced without an occlusion body.

itrophic matrix (PM, a complex structure found within the midguts of many insects). Penetration of the chitinous PM may be facilitated by specialized proteins termed *enhancins* that are present in the OBs (Derkson and Granados 1988, Wood and Granados 1991). Enhancins produce holes in the PM and allow the virions access to the midgut cells. Once the virions have passed through the PM, antireceptor proteins associated with the virion envelope bind to receptor proteins in/on midgut microvilli. The nucleocapsid then enters the cell through an opening formed at the site of attachment, with the envelope remaining at the fusion site. The nucleocapsids travel to the nucleus of the host cell where viral replication occurs. Progeny nucleocapsids formed in the nucleus are transported to the exterior of the cell and, upon exit from the cell, obtain a new lipid envelope containing antireceptor proteins from the host-cell membrane. The released virions then infect other cells within the insect. This process is repeated, resulting in a systemic infection.

In NPV, these progeny virions, termed *budded virion phenotype*, are specialized to reinfect cells within the body of the insect and are distinct from the *occluded virion phenotype* that initially infect the insect through the midgut cells. In most Lepidoptera, NPVs cause systemic infections and multiply in the nuclei of a variety of different tissue types including the epidermis, trachea (i.e., system of branching tubes constituting the respiratory system), fat bodies, and blood cells (Tanada and Kaya 1993). In contrast, most GVs are confined to one or a few tissues, with the fat bodies being the most commonly affected cells (Tanada and Kaya 1993). At some point, the occluded virion phenotypes are produced, and OBs form around these virions within the nuclei of infected cells (Fig. 1). This causes *hypertrophy* (i.e., conspicuous swelling) of the nuclei (Figs. 1, 3). In NPV, the cells eventually burst due to the large numbers of OBs formed. In systemically infected insects, widespread cell disintegration is responsible for the characteristic liquefaction sign of disease.

In individuals systemically infected with NPV, disease is almost always fatal. In lepidopterous insects infected with NPV, early instar larvae may die within 1 to 2 d of infection (Ignoffo 1966). However, after inoculation of older (5- to 6-d-old) larvae, signs and/or symptoms of disease may not become apparent for several days. Affected larvae usually exhibit loss of appetite and a color change may be evident. The integument becomes opaque and assumes a shiny and

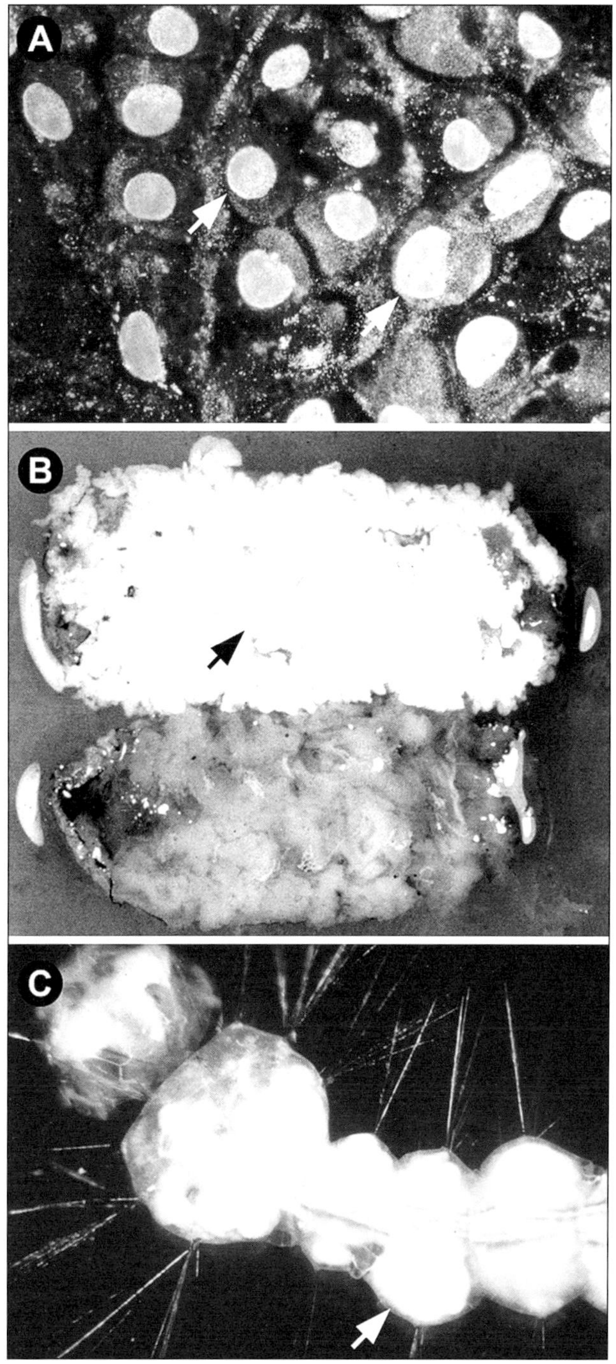

Fig. 3. Signs of disease in three insects. (A) Midgut epithelial cells of the European pine sawfly, *Neodiprion sertifer* (Hymenoptera: Diprionidae), infected with nuclear polyhedrosis virus show hypertrophy of the nuclei (arrows). [Image provided by John Cunningham (Forestry Canada, Sault St. Marie, ON).] (B) Proliferation of granulosis virus in the fat body of cabbage looper, *Trichoplusia ni* (Lepidoptera: Noctuidae). The fat body of the infected larva (top, arrow) is opaque and bright white due to the accumulation of millions of GV occlusion bodies in this tissue. This is in contrast to the greyer fat body of a healthy larva (bottom). [Image provided by Brian A. Federici (University of California, Riverside, CA).] (C) Hypertrophy of fat body (arrow) of the yellowfever mosquito, *Aedes aegypti* (Diptera: Culicidae), infected by the microsporidian *Edhazardia aedis* (Microsporidia: Culicosporidiae). [Image provided by James. J. Becnel (USDA-ARS, Gainesville, FL).]

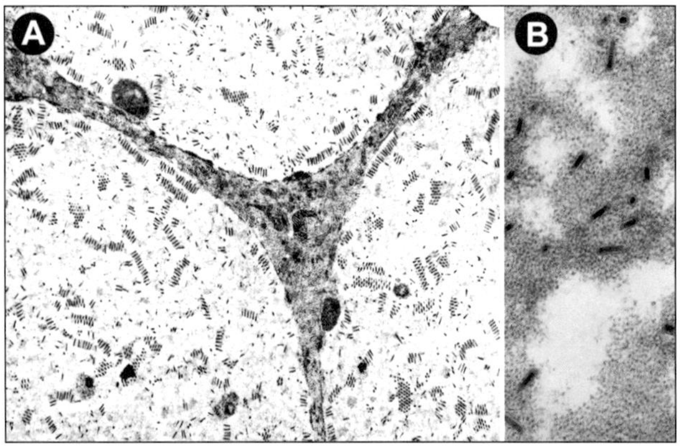

Fig. 4. Electron micrographs showing nonoccluded baculovirus (NOBV) within the nuclei of the parasitoid *Microplitis croceipes* (Hymenoptera: Braconidae). (A) Parts of three hypertrophic nuclei fat body cells. (B) Enlarged, enveloped nucleocapsids. [From Sikorowski and Lawrence (1997) with permission from the Mississippi Agricultural and Forestry Experiment Station, Mississippi State, MS.]

swollen appearance, and the internal tissues enter a state of disintegration. Such larvae are difficult to remove from their habitats without breaking the skin and liberating the liquefied contents (Fig. 1). In pupae, there is typically no external change observed during the greater part of the incubation period; but at the end of the disease cycle, the integument is very easily ruptured on handling. Even though death usually occurs in the larval stage, individuals infected in the late instars may survive to the adult stage. Such infections are followed by breakdown of the skin and subsequent liberation of body contents into the environment. The hemolymph (i.e., blood) of an infected larva is filled with OBs. Smears of hemolymph prepared from diseased larvae show the presence of large numbers of OBs in a light microscope. The virions may be transferred from individual to individual and from parents to their progeny, although the latter method of transmission is less common (Rothman and Myers 1996).

In GV infections, hypertrophy of the fat body is often observed (Fig. 3). The hemolymph of infected insects may turn milky due to the release of large numbers of occlusion bodies. In individuals infected with systemic GVs, death is often rapid, and infected individuals are typically smaller than normal (Tanada and Kaya 1993). Similar to insects infected with systemic NPV, larvae possess a fragile integument, and the dead larvae are typically "wilted" in appearance. The small size of OBs makes their detection difficult with the light microscope; they are often described as "pepper grains" in the hemolymph.

b. Nonoccluded. The nonoccluded baculoviruses (NOBV) contain dsDNA and remain unoccluded at all stages in their replication cycle (Fig. 2). This makes diagnosis of diseases incited by these viruses much more difficult because the size of the virions (75 x 270 nm) is below the resolution of the light microscope. However, they are easily visualized by electron microscopy (Fig. 4). Similarly to NPVs and GVs, NOBVs replicate within nuclei, particularly within fat bodies. Infections can become systemic, and hemolymph may increase in turbidity in such circumstances.

Infections by NOBVs have been reported in a number of insects, and infection of the parasitoid *Microplitis croceipes* (Hymenoptera: Braconidae) by NOBV (Mc-NOBV) is representative. Mc-NOBV was first reported to be pathogenic in *M. croceipes* by Hamm et al. (1988), but Gothama et al. (1998) found that Mc-NOBV also replicated in the tobacco budworm, *Heliothis virescens* (Lepidoptera: Noctuidae). The principal sites of infection in *M. croceipes* are the fat body and midgut (Hamm et al. 1992), but the circu-

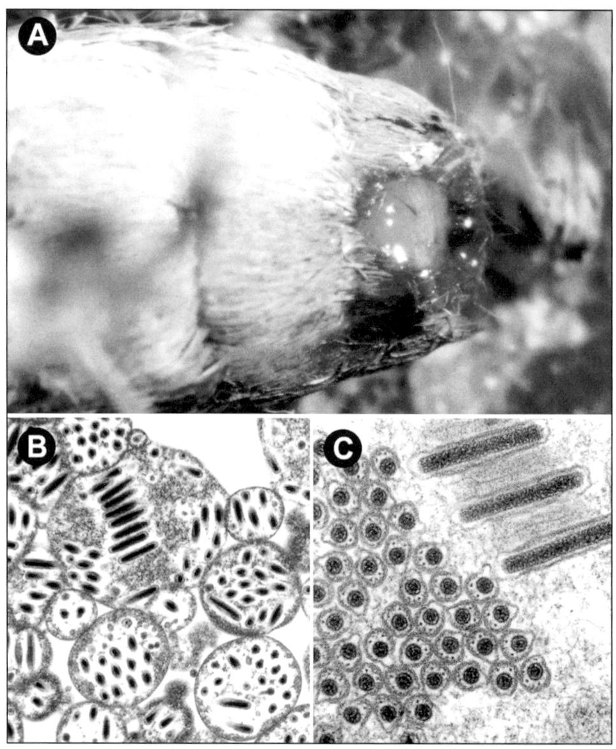

Fig. 5. Bollworm, *Helicoverpa zea* (Lepidoptera: Noctuidae), infected with gonad-specific virus (GSV). (A) Abdomen of a GSV-infected female showing the waxy plug replacing the reproductive tissues. (B) Section of the waxy plug showing vesicles containing GSV virions. (C) Intranuclear virions of GSV showing longitudinal and cross-sections. [Images provided by John J. Hamm (USDA-ARS, Tifton, GA) from Hamm et al. (1996) with permission from Elsevier.]

latory system, respiratory system, and integument can also be affected (Chittihunsa and Sikorowski 1995a). In *H. virescens*, the virus is present in the midgut, tracheal matrix, fat body, and hemolymph (Gothama et al. 1998). Infection of *H. virescens*, as well as in other hosts of Mc-NOBV, occurs *per os* (by mouth) and during parasitization (Sikorowski and Lawrence 1997). No noticeable evidence of disease is observed during the early stages of *H. virescens* larval infection. However, some larvae fail to pupate, and some pupae fail to eclose to adults. In general, infected larvae weigh less than healthy larvae (Gothama et al. 1998). Furthermore, the feces from Mc-NOBV infected *H. virescens* and *M. croceipes* contain virulent virions capable of producing infections. Thus, virus-contaminated foliage may provide inoculum for *H. virescens* larvae feeding on it. During parasitization of infected *H. virescens* larvae, healthy *M. croceipes* females may contaminate their ovipositors with Mc-NOBV. These females may in turn transmit the virus to *H. virescens* larvae during oviposition by direct insertion of a virus-contaminated ovipositor into the body of a healthy larva. Infected males may also infect females venereally.

Another example of NOBV is the gonad-specific virus (GSV) that was originally described in laboratory-reared bollworm, *Helicoverpa zea* (Lepidoptera: Noctuidae) (Raina and Adams 1995, Hamm et al. 1996). The virus affects the ovarian and testicular organs of *H. zea* and the old world bollworm, *Helicoverpa armigera* (Lepidoptera: Noctuidae), and was reported for the first time by Herzog and Phillips (1982). The disease primarily manifests itself in adults and is detected by dissection. Infected females have grossly deformed common and lateral oviducts. Nucleocapsids of GSV measure 382 x 77 nm (Raina and Adams 1995; Fig. 5B, C). However, most agonadal females could be identified externally by a white, waxy plug protruding from the vulva (Hamm et al. 1996; Fig. 5A). Fecundity of infected females is greatly reduced or nonexistent because of the degradation or elimination of reproductive organs (Herzog and Phillips 1982, Raina and Adams 1995, Hamm et al. 1996).

2. Poxviridae. The Poxviridae is a large family containing members that infect both vertebrates and invertebrates. For example, diseases such as smallpox, cowpox, and fowlpox are caused by poxviruses. There are currently two subfamilies within the Poxviridae: the Chorodopoxvirinae and the Entomopoxvirinae. Viruses within the Entomopoxvirinae only infect insects (Boucias and Pendland 1998). This subfamily contains three genera (Murphy et al. 1995): 1) Entomopoxvirus A includes the poxviruses of Coleoptera; 2) Entomopoxvirus B includes the poxviruses of Lepidoptera and Orthoptera; and 3) Entomopoxvirus C includes the poxviruses of Diptera. The virions of entomopox viruses (EPVs) are brick-shaped or oval, from 150–470 nm long and from 165–300 nm wide, with a folded outer membrane giving the appearance of a mulberry-like surface (Tanada and Kaya 1993; Fig. 6). Two types of OBs may occur in EPVs, spheroid- or spindle-shaped. The spheroid-shaped OBs contain virions, whereas the spindle-shaped OBs do not. Both types of OBs are found in Lepidoptera and Coleoptera infected with EPV. However, only spheroid OBs are produced in infected Orthoptera. In most insects, the primary tissues affected are fat bodies. The host range of EPVs is relatively restricted with no

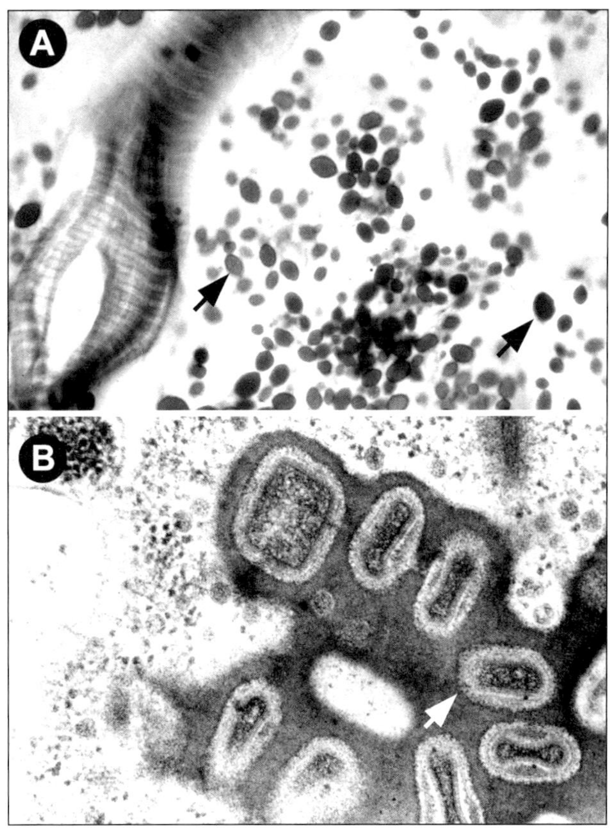

Fig. 6. Entomopox virus (EPV). (A) Light micrograph showing spindle-shaped occlusion bodies infecting a *Chironomus* species. (B) Electron micrograph illustrating virions within an occlusion body in a larva of the midge *Goeldichironomus holoprasinus* (Diptera: Chironomidae) infected with EPV. [Images provided by Brian A. Federici (University of California, Riverside, CA)]

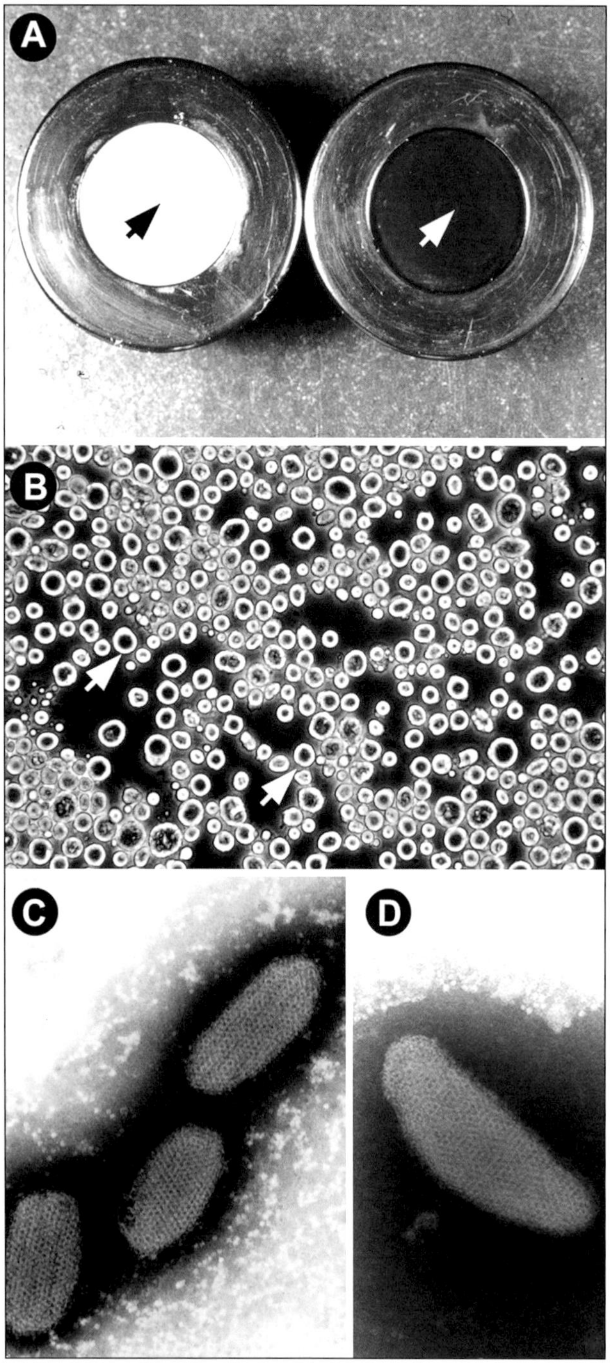

Fig. 7. Ascovirus. (A) Hemolymph (arrows) from a diseased (left) and healthy (right) larva of the cabbage looper, *Trichoplusia ni* (Lepidoptera: Noctuidae), infected with the *Trichoplusia ni* ascovirus (TAV). The opaque white color of the hemolymph from the diseased larva is due to the presence of virion-containing vesicles. (B) Light micrograph of hemolymph obtained from a fall armyworm, *Spodoptera frugiperda* (Lepidoptera: Noctuidae), infected with *Spodoptera* ascovirus (SAV). The refractile spheres (arrows) are virion-containing vesicles. (C–D) Electron micrographs of negatively stained virions of Ascovirus infecting the tobacco budworm, *Heliothis virescens* (Lepidoptera: Noctuidae). Note the reticulate surface of the capsid. [Images A and B provided by Brian A. Federici (University of California, Riverside, CA). Images C and D provided by John J. Hamm (USDA-ARS, Tifton, GA) and Eloise Styer (Veterinary Diagnostic Laboratory, University of Georgia, Athens, GA)].]

cross-infections occurring between different families (Kurstak and Garzon 1977). Infection occurs *per os* and once ingested, the occlusion bodies disintegrate in the alkaline gut environment of the digestive tract. The liberated virions attach and fuse to the plasma membrane of the microvilli, and the viral core from each enters the cell (Granados 1973). How EPVs penetrate the PM is not well understood. Some EPVs demonstrate considerable tissue specificity (e.g., restricted to fat bodies), whereas, others affect a multitude of tissues and cause systemic infections. Signs and symptoms of disease vary by host. In most insects, development is slowed. In lepidopteran larvae, an external whitening and softening often occurs, and larvae may grow to more than double their normal size and weight due to swelling of the fat bodies (Bird et al. 1971).

3. Ascoviridae. Members of the family Ascoviridae infect only noctuid larvae, in which they cause both chronic and acute diseases (Federici et al. 1991). These virions contain dsDNA, and they are enveloped. The envelope exhibits a distinctive hexagonal pattern in negatively stained preparations observed with a transmission electron microscope (TEM) (Fig. 7). Virions are approximately 130 x 400 nm, and enveloped virions are allantoid (i.e., sausage-shaped) or bacilliform. The virions often accumulate in vesicles (Fig. 7); this is unique among the entomopathogenic viruses.

Infection can occur *per os*, but ascoviruses generally possess poor infectivity after ingestion (Federici 1993). Parasitoids can also transmit the virus to lepidopteran larvae (Hamm et al. 1985). In infected cells, the nucleus is greatly hypertrophied. Fat bodies are the primary tissue affected by some strains, but a broad range of tissues are affected by others. After fragmentation of the nuclear membrane, sheets of cytoplasmic membrane assemble throughout the cell and coalesce, partitioning the cell into a cluster of vesicles. Once vesicle formation is complete, the vesicles dissociate from each other and accumulate within the tissue in which they are formed. As the disease progresses, the basement membrane of infected tissues is disrupted, and vesicles are released into the hemolymph causing the hemolymph to appear opaque (Federici et al. 1991; Fig. 7). Ascoviruses often cause chronic infections, and infected *T. ni* and *H. virescens* larvae turn an opaque yellowish-white, have difficulty in molting, and show reduced feeding activity, retarded growth rates, and possibly death (Tanada and Kaya 1993). However for many species, infected individuals differ little if any in appearance from healthy individuals (Federici et al. 1991).

4. Iridoviridae. More than 30 types of iridoviruses have been described from insects in the orders Diptera, Lepidoptera, Coleoptera, Hymenoptera, Orthoptera, and Hemiptera (Boucias and Pendland 1998). Two genera infect insects: 1) *Iridovirus* and 2) *Chloroviridovirus* (Hall 1985, Williams 1996, Boucias and Pendland 1998). Iridescent viruses (IVs) are dsDNA viruses, and virions measure 120 to 140 nm in diameter. Virions are neither enveloped nor occluded.

Most iridoviruses are not highly infections when ingested (Tanada and Kaya 1993). For example, the infection rate of *H. zea* larvae ingesting virions ranged from only 11 to 47% (Sikorowski and Tyson 1984). In contrast, virions are highly infectious if they are injected directly into the *hemocoel* (i.e., the body cavity in which hemolymph circulates). Carey et al. (1978) observed IV infection of an insectary stock of *H. armigera*. Although virions do not typically incite disease *per os*, in many insects the virus can be transmitted via cannibalism (e.g., Linley and Nielson 1968, Carter 1973, Boucias et al. 1987). Once virions reach the hemocoel, infections often become systemic, affecting most tissues (Tanada and Kaya 1993). Fat bodies are particularly affected. Systemic infections are typically lethal, but low dosages and exposure in older larvae and pupae may allow individuals to reach adulthood although such adults often are sterile (Day 1965). The first sign of disease is the development of iridescent patches (Fig. 8), although some infected individuals may not exhibit discoloration (Houston 1991). Typically the iridescent patches initially form in the extremities of the body such as the mouthparts and prolegs. The iridescence may spread to other parts of the body and increase in intensity. Sometimes, infected larvae develop a widespread lavender-blue, blue, or blue-green coloration. Affected insects may often appear deranged (Tanada and Kaya 1993). Examination of affected organs (e.g., fat bodies, silk glands, muscles) with the electron microscope typically reveals the presence of numerous intracytoplasmic virions that occur in paracrystalline arrays (Fig. 8). The iridescence results from the Bragg reflection of visible light through the microcrystals of virions in the arrays (Tanada and Kaya 1993). Iridescent viruses do not infect nuclei.

5. Parvoviridae. The family Parvoviridae contains six genera but only three infect invertebrates. These include: 1) *Contravirus*, 2) *Iteravirus*, and 3) *Densovirus* (Berns 1990). *Densovirus* (DNV; synonym: densonucleosis virus) has been isolated from Diptera, Lepidoptera, Blattodea, Orthoptera, and Odonata. The DNVs are a heterogeneous group of viruses, but they are all nonoccluded and contain ssDNA. They are among the smallest viruses known, typically ranging in size from 19–24 nm (Fig. 9). The first DNV to be described was from greater wax moth, *Galleria mellonella* (Lepidoptera: Pyralidae), and has since been observed in other Lepidoptera, Diptera, Orthoptera, Blattodea, and Odonata (Kawase 1985). Perhaps the best studied insect parvoviruses are the *Bombyx* DNVs, which cause the disease densonucleosis in *B. mori*. Two types of DNVs have been described from silkworms (DNV-1 and DNV-2), and both replicate in the nuclei of the midgut columnar cells (Bando et al. 1995). Both forms

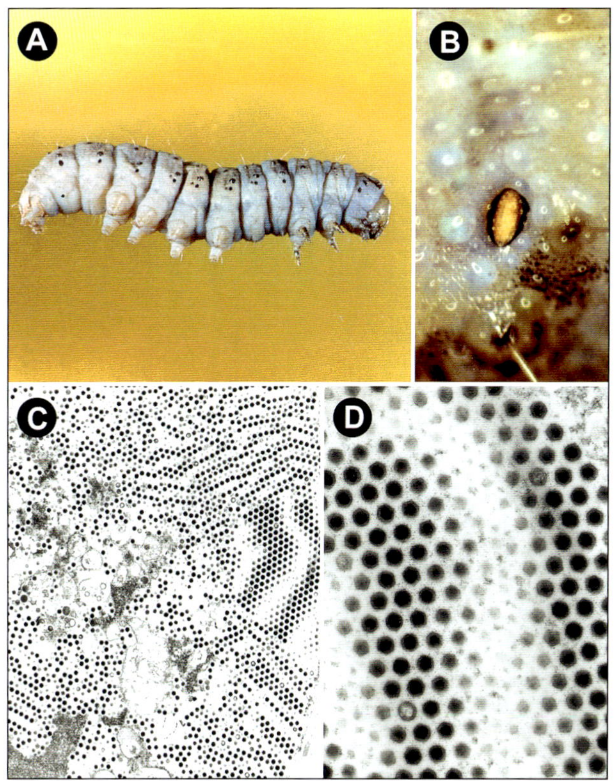

Fig. 8. Bollworm, *Helicoverpa zea* (Lepidoptera: Noctuidae), infected with iridescent virus (IV). (A) A larva infected with IV showing iridescent lavender-blue coloration over the entire body. (B) Small iridescent patches that frequently are the only signs of infection. (C) Electron micrograph of an infected fat body. (D) High magnification showing the characteristic array formed by IV virions. [From Sikorowski and Lawrence (1997) with permission from the Mississippi Agricultural and Forestry Experiment Station, Mississippi State, MS.]

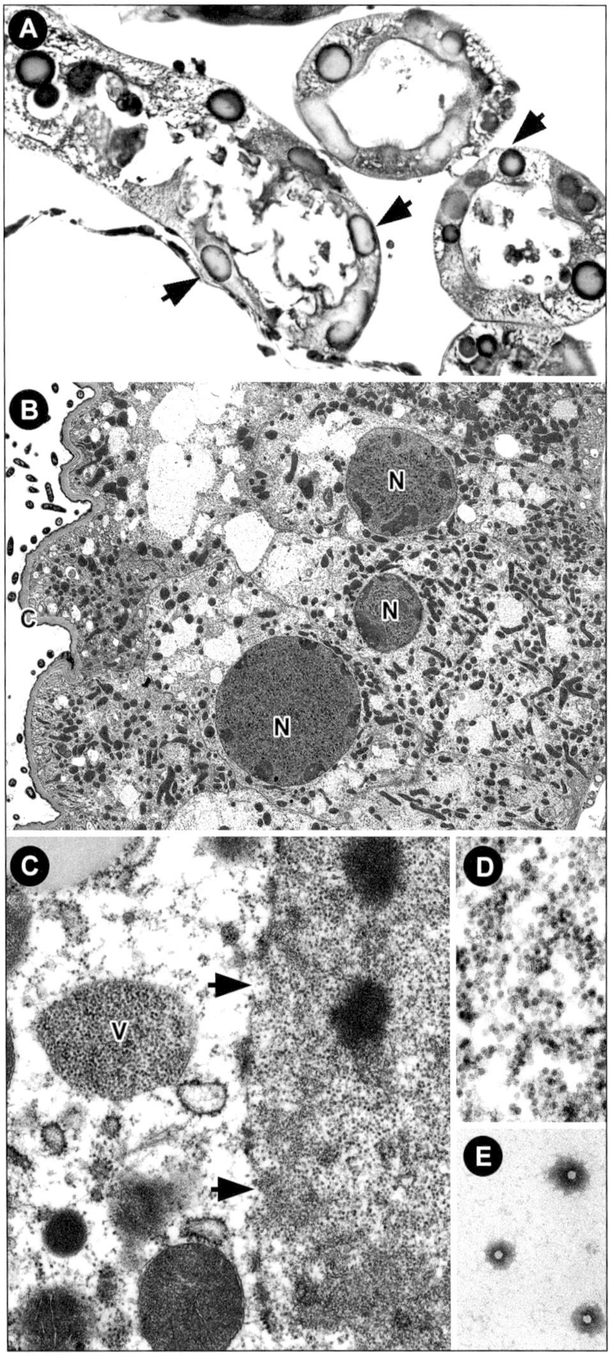

Fig. 9. Densovirus-infected house crickets, *Acheta domesticus* (Orthoptera: Gryllidae). (A) Malpighian tubules showing hypertrophied nuclei with intranuclear inclusions (arrows). (B) Epidermis with hypertrophied nuclei (N) infected with densovirus (C, cuticle). (C) Fat body showing a cytoplasmic vesicle containing virions (V). The arrows point to the disrupted nuclear membrane containing large numbers of virions. (D) Electron-dense virions and empty capsids from a cytoplasmic vesicle. (E) Negatively stained virions from homogenized dead crickets. [Images provided by Eloise Styer (Veterinary Diagnostic Laboratory, University of Georgia, Athens, GA) and from Styer and Hamm (1991) with permission from Elsevier.]

of the virus cause chronic infections in the mulberry pyralid, *Glyphodes pyloalis* (Lepidoptera: Pyralidae), and putatively the virus overwinters in *G. pyloalis* and is transmitted to *B. mori* via contamination of mulberry foliage (Boucias and Pendland 1998). The virus causes conspicuous hypertrophy and rupture of the host nuclei (Fig. 9) thereby resulting in general debilitation. From 1988 to 1989, a cricket producer in southern Georgia, USA, noted increased sluggishness and early mortality in house cricket, *Acheta domesticus* (Orthoptera: Gryllidae) adults, followed by gradually increasing mortality rates in progressively younger crickets (Styer and Hamm 1991). The affected crickets possessed enlarged nuclei, and electron microscopy showed large numbers of nonenveloped, electron-dense, icosahedral (twenty-faceted) virions typical of *Denosvirus* (Fig. 9).

6. Reoviridae. The reoviruses infect animals, plants, and fungi. All reoviruses contain dsRNA encapsulated within an icosahedral OB (Fig. 10). The cytoplasmic polyhedrosis viruses (CPV) are similar to other reoviruses but can be distinguished by host range, morphology of the viral particle (lack of a double capsid), phylogeny, and the production of viral-coded proteins (polyhedrin) from which occlusion bodies are comprised (Tanada and Kaya 1993). Only CPVs infecting insects form OBs. Cytoplasmic polyhedrosis viruses have a very wide host range (about 250 species), affecting mainly lepidopterous insects (80%) (Tanada and Kaya 1993). However, infections of Diptera, Hymenoptera, Coleoptera, and Neuroptera by CPVs have been reported. Infections by CPVs are very common in reared lepidopteran insects. Some CPVs possess a broad host range and infect insects of multiple species, genera, or families (Tanaka 1971). The formation of OBs is a major diagnostic characteristic of CPV infections. Occlusion bodies are normally located in the cytoplasm of virus-infected cells in the midgut epithelium (Fig. 10). They typically vary widely in size, ranging from 0.1 to 10 μm in diameter. A few to several hundred virions may be found within the OB.

Cytoplasmic polyhedrosis virus infections can be a serious problem in insectary-reared insects. Infections typically produce a chronic disease (Sec. III. B) as a result of structural damage to the midgut epithelium (Fig. 11). The virus infection causes cessation of feeding and reduction of larval size. As a result, CPV-infected larvae produce smaller pupae and adults, and require a longer time to complete the larval stage. The infection process has been investigated in a variety of hosts. In many insects, ingestion of OBs on contaminated food is the most common mechanism of entry of the virus into the host. Once in the gut, the OBs dissolve in the alkaline gut fluids of the larva. The liberated virions are able to breach the PM and then infect susceptible cells in the epithelial cells of the midgut. Virions are icosahedral in shape and are typically 50–80 nm in size (Tanada and Kaya 1993). Capsids usually possess 12 protein projections (Fig. 10), which are involved in the infection process. Small OBs are often observed in the columnar cells of the midgut 1 or 2 d after infection, but the microvilli of infected columnar cells are not affected until immediately prior to rupture of the cell. At this time, the microvilli are partially or almost completely absent from infected cells (Fig. 11). Infected columnar cells may rupture to release OBs into the gut lumen 2 or 3 d after infection. OBs are also released into the lumen by extrusion of heavily infected columnar cells. In the advanced stage of disease development, OBs are excreted with the feces (Sikorowski et al. 1973, Bong and Sikorowski 1991). Occlusion bodies are highly stable and serve as the vehicle for transmission of infectious virions from one susceptible host to another. In addition to occurring in feces, OBs are present in the *meconium* [i.e., the substance excreted after adult eclosion from the pupal stage and believed to contain the waste products of pupal metabolism such as uric acid (Chapman 1998)]. In addition, CPV may be acquired as a result of injury (e.g., by parasitism) and by cannibalism. Sikorowski et al. (1973) reported that a major means of transmission of CPV to the next generation is via surface contamination of eggs. Larvae are infected during emergence from the egg when they consume a portion of the CPV-contaminated chorion (egg shell).

7. Picornaviridae. The insect picornaviruses are extremely small viruses that are morphologically similar to the vertebrate picornaviruses (e.g., polio virus, common cold rhinoviruses, and foot-and-mouth disease viruses). The genomic material is ssRNA. To date, more than 20 insect viruses have been described as picornavirus-like. The best characterized are the cricket paralysis virus, *Drosophila* C virus, Queensland fruit fly virus, flacherie virus of the silkworm, and various honey bee viruses (e.g., the sacbrood virus, acute bee paralysis virus, chronic bee paralysis virus). The virus primarily infects healthy insects *per os*, but it can be transmitted from parents to offspring

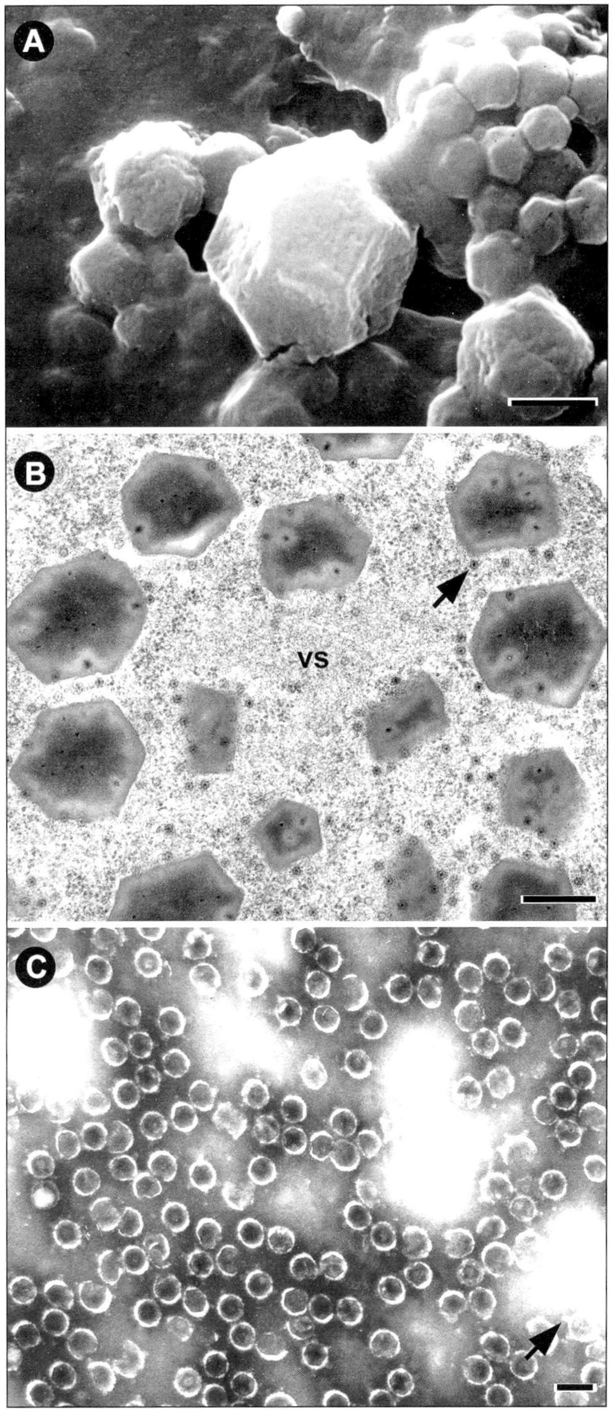

Fig. 10. The tobacco budworm, *Heliothis virescens* (Lepidoptera: Noctuidae), infected with cytoplasmic polyhedrosis virus (CPV). (A) Scanning electron micrograph of polyhedral occlusion bodies (bar = 2 µm). (B) Electron micrograph of occlusion bodies showing virions (arrow) and virogenic stroma (vs) (bar = 2 µm). (C) Negatively stained virions released from occlusion bodies by treatment with dilute alkali (note the protein spikes (arrow) projecting from the capsid). [From Sikorowski and Lawrence (1997) with permission from the Mississippi Agricultural and Forestry Experiment Station, Mississippi State, MS.]

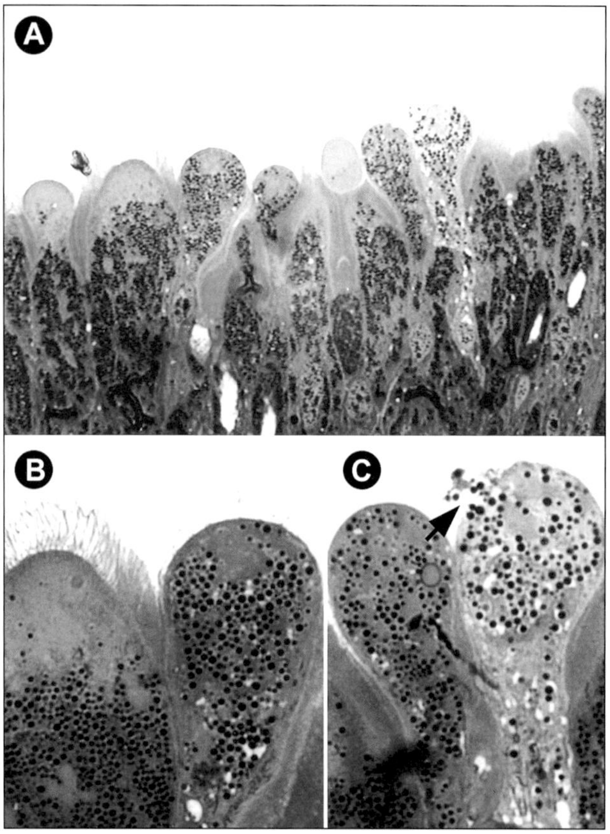

Fig. 11. Midgut epithelium of tobacco budworm, *Heliothis virescens* (Lepidoptera: Noctuidae), larvae infected with cytoplasmic polyhedrosis virus (CPV). (A–B) Light micrographs of infected epithelium showing occlusion bodies (stained spheres) within epithelial cells. (C) Degenerated epithelium immediately prior to rupture (note loss of microvilli) and following cellular rupture (arrow). [From Sikorowski and Lawrence (1997) with permission from the Mississippi Agricultural and Forestry Experiment Station, Mississippi State, MS.]

as well. Since the picornaviridae are a heterogeneous assemblage of viruses, they produce a variety of pathologies in their insect hosts. However, most cause dysentery following infection of gut tissues, they often produce copious anal discharges, and the virions are commonly spread through the contamination of food.

8. Other. A number of additional viruses have been reported from laboratory-reared insects including members of the Polydnaviridae, Birnaviridae, Tetraviridae, Nodaviridae, and Rhabdoviridae. For the most part, these viruses have not been extensively studied as entomopathogens of reared insects, and they will not be discussed in detail. Members of the Polydnaviridae are dsDNA viruses, and they are restricted to braconid and ichneumonid wasps. They are considered mutualistic rather than pathogenic to parasitoid hosts: they help protect the parasitoid egg from the immune response of the host by altering its physiology in favor of the parasitoid (Hamm et al. 1992). The Birnaviridae are dsRNA viruses and are known from *Drosophila* (*Drosophila* X virus), and *D. melanogaster* infected with the virus exhibit anoxia sensitivity (i.e., they do not recover after being anesthetized with CO_2 or N_2) (Boucias and Pendland 1998). Tetraviridae, Nodaviridae, and Rhabdoviridae are all ssRNA viruses. The Tetraviridae are associated exclusively with lepidopteran insects. Infections are often chronic, leading to retarded growth and abnormally small pupae and adults (Boucias and Pendland 1998). Nodaviridae can establish persistent infections in mosquitoes, hymenopterans, various lepidopterans, and ticks. In some insects, no symptoms occur, in others, the virus can be lethal (e.g., wax moths, honey bees and certain mosquitoes). The Rhabdoviridae are a diverse group of viruses that can cause disease in both plants and animals. The only well-characterized insect rhabdovirus is the Sigma virus of *D. Melanogaster*, which causes anoxia sensitivity (Boucias and Pendland 1998).

B. Bacteria

Bacteria are a large group of *prokaryotes* (i.e., lacking a nucleus) that are usually unicellular and obtain nutrition as saprotrophs, facultative parasites, or obligate parasites. Although they are the smallest cells known to science, they can be observed with the compound light microscope (≥ 0.5 µm). Bacterial cells vary in shape including: 1) spherical or coccus; 2) rod-shaped or bacillus; 3) curved; 4) helical or spiral, hyphal, and pleomorphic (i.e., indefinite) forms (Fig. 12A). Bacteria are the most common group of microorganisms associated with insects, and most insect pathogenic taxa occur in the families Bacillaceae, Pseudomonaceae, Enterobacteriaceae, Streptococcaceae, and Micrococcaceae (Tanada and Kaya 1993). Some taxa produce internal spores termed *endospores* (Fig. 12B). Endospores are heat-resistant cells, and the formation of endospores is important in the management of taxa that produce them (e.g., *Bacillus*, *Clostridium*, and *Paenibacillus*). Many entomopathogenic bacteria are saprotrophic and can be cultured on appropriate microbiological media (e.g., Lacey 1998 and references therein). Once in pure culture, they are typically identified based on morphological (e.g., cell shape and cell wall structure), physiological (e.g., assimilation), and/or molecular (e.g., 16S rDNA sequence) characters (e.g., Murray et al. 2003, Garrity 2005). Bacteria can be divided into two major groups based on the morphology of their cell walls (CW). One group (gram positive) produces a thick CW comprised of peptidoglycan, whereas the other group (gram negative) possesses a relatively thin CW and an outer membrane consisting of lipopolysaccharide and protein (Fig. 13). The two groups exhibit different characteristics (e.g., sensitivities to antibacterial agents), and distinguishing them is not only an important step in identification of a pathogenic bacterium but is often required to determine its most efficacious management strategy. The most common method is the Gram staining method in which Gram positive bacteria appear purple in color and Gram negative bacteria appear pink in color. Most bacteria are not readily culturable, and a variety of such bacteria are associated with insects, including *Rickettsiella* and *Wolbachia*. Detection of these bacteria requires the use of molecular and microscopic methods.

Bacteria infect insects mostly through the mouth and digestive tract and less commonly through the egg, integument, and tracheae. They may also enter an insect by means of parasites and predators. Most pathogenic bacteria are extracellular pathogens except for the pathogenic Rickettsiaceae and Mollicutes (mycoplasma and spiroplasma). Generalized symptoms of insects associated with bacterial infections are cessation of feeding, discharge of watery feces, and regurgitation (Tanada and Kaya 1993). In the alimentary canal, the bacteria produce enzymes (e.g., lecithinase, proteinase, and chitinase) that damage the midgut epithelium and may enable the bacteria to enter the hemocoel. The invasion of bacteria into the hemocoel usually results in *septicemia* (i.e., a morbid condition caused by the multiplication of a microorganism in the

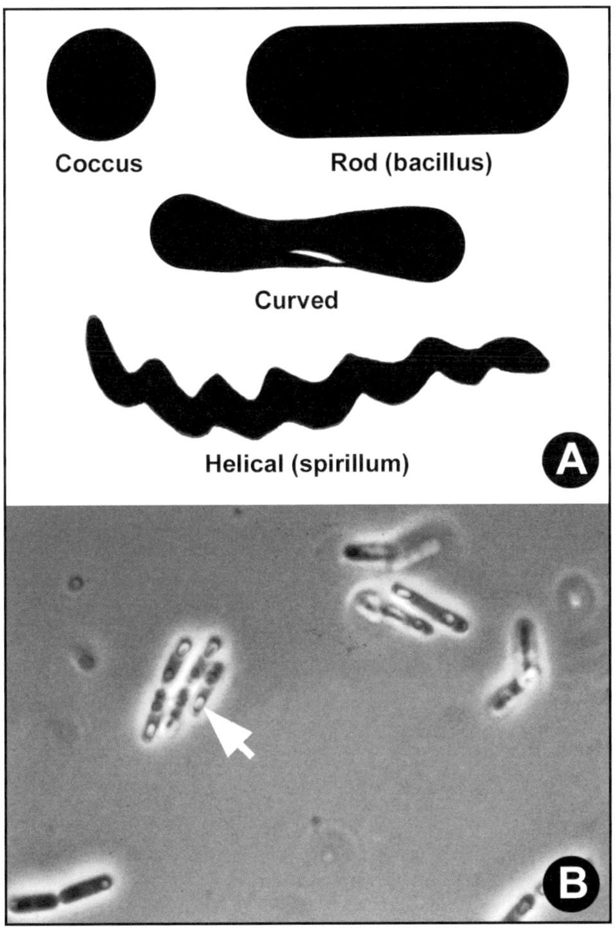

Fig. 12. Morphology of bacteria. (A) Common bacterial shapes. (B) A light micrograph showing rod-shaped cells (*Bacillus*) containing refractive *endospores* (i.e., the vegetative cells serve as sporangia for the endospores).

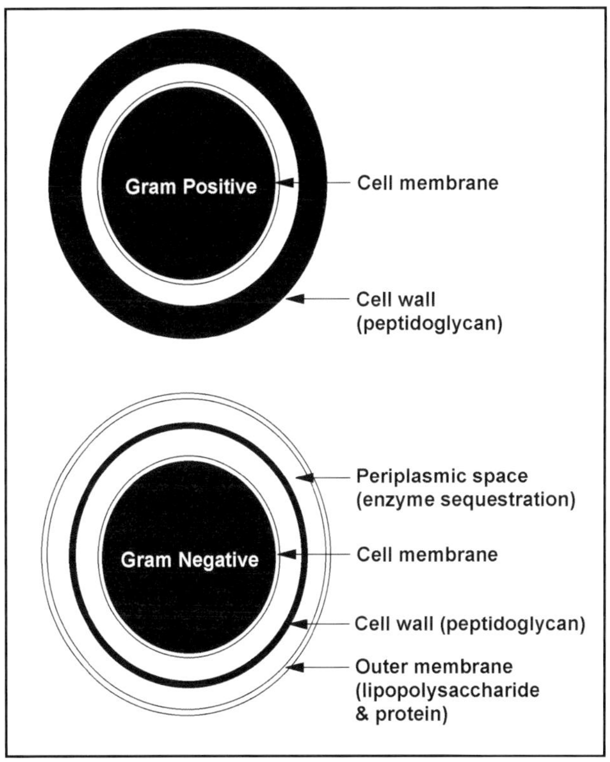

Fig. 13. Differential cell wall morphology of Gram positive and Gram negative bacteria. Note the thick peptidoglycan layer produced by Gram positive bacteria in comparison to the thin peptidoglycan layer produced by Gram negative bacteria. Also note the presence of an outer lipopolysaccharide membrane in Gram negative bacteria.

hemolymph) and death of the insect. Insects killed by bacteria usually darken in color and soften, but the integument remains intact.

1. Gram Positive. A number of Gram positive bacteria are important pathogens of insects. Two of the most important genera are *Bacillus* and *Clostridium*. Both taxa produce endospores. The traditional genus, *Bacillus*, has been substantially changed in recent years, and it has been divided into a number of new genera (e.g., *Brevibacillus*, *Aneurinibacillus*, *and Paenibacillus*). Members of the genus *Bacillus* and its relatives are aerobes (require oxygen) or facultative anaerobes (can live in the presence or absence of oxygen); whereas, all *Clostridium* spp. are obligate anaerobes (require the absence of oxygen). Another characteristic difference between the two taxa is that *Bacillus* species usually produce catalase, an enzyme catalyzing the oxidation of hydrogen peroxide to oxygen and water as well as the oxidation of various toxins, while species of *Clostridium* do not. Also, all *Bacillus* and related genera (except *Paenibacillus*) do not produce gas from glucose, while *Clostridium* generally produces copious amounts of gas.

a. Bacillus thuringiensis. This bacterium is soil borne and characterized by its ability to produce crystalline, proteinaceous parasporal bodies (Fig. 14). Many of these proteins are highly toxic to specific orders of insects. Susceptible insects are mainly in the orders Lepidoptera, Diptera, and Coleoptera; but with the discovery of new strains, the range of insects susceptible to *Bac. thuringiensis* (*Bt*) has also increased. At present, *Bt* is the most widely used bacterium for insect control. Despite the disease-producing potential of *Bt*, the bacterium does not commonly cause serious disease outbreaks in insectary settings. However, *Bt* can be a problem in some instances, especially when the bacterium is present in or near the insectary (e.g., in laboratories conducting research with *Bt*) or when foliage is provided as a diet. For example, a high *prevalence* (i.e., number of hosts expressing a disease at a given time) of *Bt* has been found in silkworm insectaries. Researchers discovered that from 3 to 75% of leaves sampled from 96% of mulberry trees were positive for *Bt* (Ohba 1996). In insectaries in which artificial diets are used, *Bt* is less commonly encountered. However, the *kurstaki* strain of *Bt* was frequently isolated from mass-reared tobacco budworm in Mississippi, albeit at relatively low levels (Sikorowski unpublished). Although the bacterium did not cause significant levels of acute disease, the sublethal impacts of infection are unknown.

The *parasporal body* is a crystalline particle associated with the spores of *Bacillus* spp. It consists of a proteinaceous toxin generally bipyramidal in shape

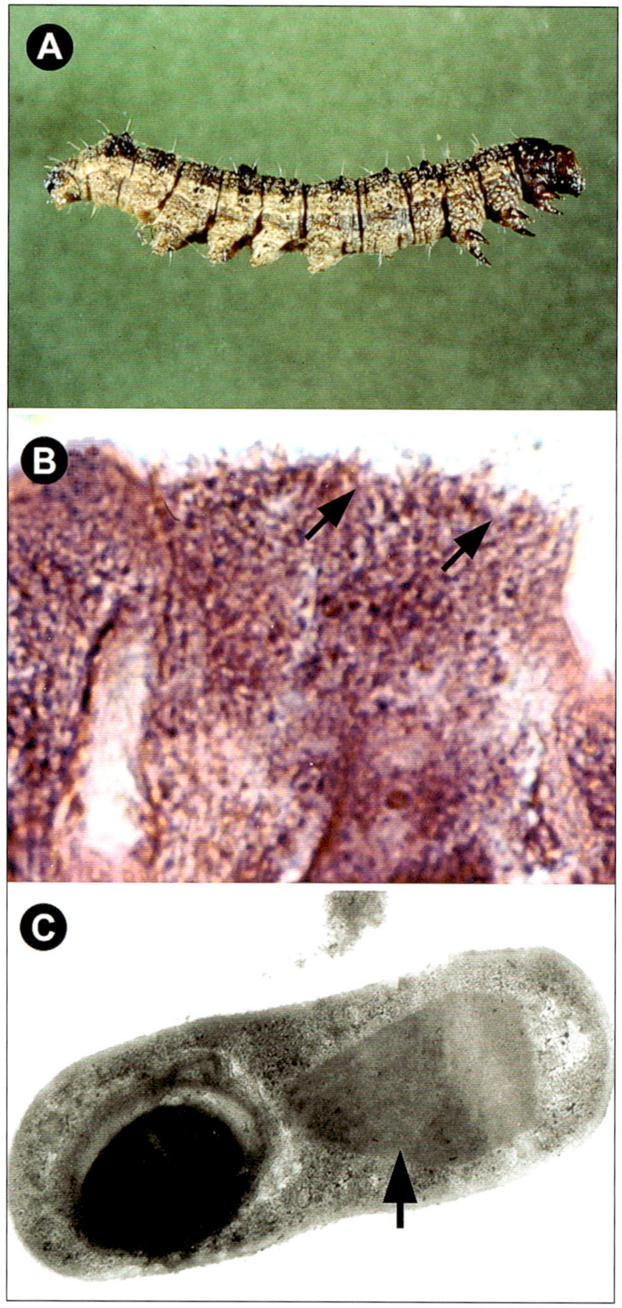

Fig. 14. Tobacco budworm, *Heliothis virescens* (Lepidoptera: Noctuidae), infected with *Bacillus thuringiensis* (*Bt*). (A) A larva killed by *Bt*. (B) Light micrograph of infected *H. virescens* midgut showing cellular degeneration (arrow). (C) An electron micrograph of a *Bt* sporangium showing the parasporal body (arrow) and an electron dense endospore. [From Sikorowski and Lawrence (1997) with permission from the Mississippi Agricultural and Forestry Experiment Station, Mississippi State, MS.]

in strains of *Bt* that are mainly active against lepidopterans (Fig. 14) but may be triangular, cubical, or flat in other subspecies of *Bt*. The parasporal body can be seen at the light microscope level, but differential staining is often required. The proteinaceous crystal must be activated before it is insecticidal, and activation results through the action of proteolytic enzymes at alkaline pH within the alimentary tract of the insect host. The toxic peptide is termed the *endotoxin* (i.e., a toxin retained within the cell). Hofte and Whitely (1989) proposed a nomenclature and classification scheme based on the insect specificity (toxicity) and the primary structure of the proteins. In their system, cryI toxins ("cry" is an abbreviation for crystal) were active against lepidopterans, cryII proteins against lepidopterans and dipterans, cryIII proteins against coleopterans, and cryIV proteins against dipterans alone. While this system has provided a useful framework for classifying cry toxins, inconsistencies exist in the scheme (e.g., related cry genes that did not encode toxins with similar insecticidal spectra). As a result, a new nomenclature scheme based on gene sequence data has been proposed. See *Bacillus thuringiensis* delta endotoxin nomenclature (http://www.biol.susx.ac.uk/Home/Neil_Crickmore/Bt/index.html).

Following ingestion of *Bt*, the first pathological reaction is paralysis of the gut leading to cessation of feeding. According to Cooksey (1971), there is ballooning, exfoliation, and breakdown of the insect's gut epithelium. Cells of the midgut of affected larvae have enlarged nuclei, alteration of the endoplasmic reticula to vacuole-like configurations, disintegration of the microvilli (Fig. 14), and deformation of the basal infoldings of the epithelium. Along with the intracellular damage, the apical microvilli of midgut cells also swell. Often there are no conspicuous external signs of disease (Fig. 14). In addition to endotoxins, a number of *exotoxins* may also be produced during bacterial growth. As the name indicates, exotoxins are typically excreted during growth of the bacterium. A number of exotoxins are produced (e.g., α-exotoxin and β-exotoxin or thuringiensin), and they are toxic to insects as well as other animals, including some vertebrates.

b. Paenibacillus larvae. *Paenibacillus larvae* (previously known as *Bacillus larvae*) is the causal agent of American foulbrood, a serious disease of honey bees that despite its name occurs worldwide. American foulbrood is a persistent disease, and it can severely weaken a colony and may eventually destroy it depending on conditions such as nutrition (Tanada and Kaya 1993). The bacterium only infects larvae. After ingestion, endospores germinate in the gut lumen, bacterial cells enter the midgut cells, and eventually enter the hemocoel causing septicemia. Sporulation occurs within the decomposing larva. A dead larva or pupa dries to a "scale" that adheres tenaciously to the brood cell wall. The mouthparts of a dead pupa may protrude as a fine thread, commonly referred to as the "stuck up tongue" of the pupa. Dead bees often give off a characteristic odor of burnt glue. The removal of cadavers from the hive by adult bees is a major factor in limiting transmission of the pathogen from cadavers to healthy individuals.

c. Other Bacillus *and* Paenibacillus *spp.* The genera *Bacillus* and *Paenibacillus* are very diverse. Some members are entomopathogens while others are saprotrophs and important contaminants. A number of additional species are entomopathogens of economically important insects in the orders Coleoptera, Diptera, and Hymenoptera. *Paenibacillus popillae* is the primary causal agent of milky disease of scarab larvae, and *Bacillus sphaericus* consists of a heterogeneous group of hundreds of nonpathogenic and pathogenic strains. Some genotypes produce insecticidal toxins that are active against various mosquito larvae (Boucias and Pendland 1998). While these entomopathogens are important from a microbial control perspective, they are not widely encountered in rearing settings.

d. Clostridium. Members of the genus *Clostridium* are Gram positive, motile or nonmotile obligate anaerobes. A number of species are pathogens of insects. For example, Bucher (1957, 1961) described two species of *Clostridium* (*Clo. brevifaciens* and *Clo. malacosoma*), which are the causal agents of the disease called *brachytosis* in tent caterpillars, *Malacosoma* spp. (Lepidoptera: Lasiocampidae).

e. Other. Members of the genus *Corynebacterium, Melissococcus, Micrococcus, Staphylococcus,* and *Streptococcus* have all been reported as facultative pathogens of insects (Tanada and Kaya 1993, Boucias and Pendland 1998). In most cases, bacteria belonging to these genera rarely incite disease in insectary-reared insects. However, they may cause mortality in feral insects transferred to insectaries, and care should be taken when bringing field-collected insects into insectaries. *Melissococcus pluton* (= *Streptococcus pluton*) is the causative agent of European foulbrood affecting honey bees.

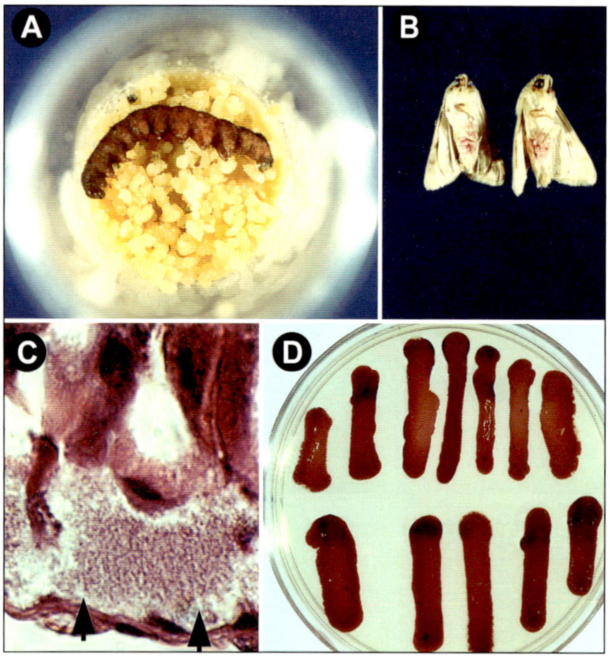

Fig. 15. Tobacco budworm, *Heliothis virescens* (Lepidoptera: Noctuidae), infected with *Serratia marcescens*. (A) An infected larva exhibiting red coloration caused by the production of the water-insoluble pigment prodigiosin. (B) Infected adults also showing red coloration. (C) Light micrograph of an infected larval midgut showing accumulation of the bacterium above the basement membrane, which underlies the epithelium (arrows). (D) Characteristic coloration of *Ser. marcescens* colonies isolated from infected larvae. [From Sikorowski and Lawrence (1997) with permission from the Mississippi Agricultural and Forestry Experiment Station, Mississippi State, MS.]

2. Gram Negative. The Gram negative bacterial pathogens of insects are all nonsporeforming. Bucher (1963) classified these into three groups: 1) obligate pathogens, 2) potential pathogens, and 3) facultative pathogens. The obligate pathogenic bacteria are very difficult or impossible to culture *in vitro*. In nature, they multiply only within the bodies of host insects where they cause specific diseases. The potential pathogens are bacteria that multiply extracelluarly in the hemocoel of insects and may subsequently produce a lethal septicemia. They grow readily in culture and attack a wide range of insects. An example of a potential pathogen is *Pseudomonas aeruginosa*. The facultative pathogens possess some mechanism for infecting susceptible body tissues, for damaging host tissue by growing in the gut, and are normally present in the environment. The most common facultative bacterial pathogen of insects in insectaries is *Serratia marcescens*. While there are numerous publications describing the ravaging effects of *S. marcescens* on insect colonies (e.g., Steinhaus 1959, Bell et al. 1981, Krieg 1987, Sikorowski and Lawrence 1998), most nonsporeforming bacteria do not cause conspicuous signs and/or symptoms of disease unless the insects are also exposed to stress factors. For example, *Ser. marcescens* and *Pse. aeruginosa* were able to overwhelm Caribbean fruit fly, *Anastrepha suspensa* (Diptera: Tephritidae), larvae and pupae only when they were exposed to thermal stress (rearing temperatures > 30°C) (Greany et al. 1977).

a. Pseudomonas. The traditional genus *Pseudomonas* has undergone considerable taxonomic revision in recent years. The genus *Pseudomonas* still encompasses a large number of species, many of which are not of importance in insect rearing. The two most commonly isolated species from insectaries are *Pse. aeruginosa* and to a lesser extent *Pseudomonas fluorescens*. Pecan weevil, *Curculio caryae* (Coleoptera: Curculionidae), larvae infected with *Pse. aeruginosa* exhibit a decline in vigor and a gradual color change; at the time of death, the body turns black (Sikorowski unpublished). Larvae infected with *Pse. fluorescens* become gray and remain gray for several days. The pigment fluorescin often diffuses into the agar medium, and it is light green in appearance and fluoresces under long-wave ultraviolet light. *Pseudomonas* spp. may be recovered from tissues and fluids of human patients, and *Pse. aeruginosa* is a serious pathogen of immunocompromised individuals (Lindberg 1974).

b. Serratia marcescens. The genus *Serratia* is composed of several species, some of which produce prodigiosin, a nondiffusible, water-insoluble red pigment (Fig. 15). The majority of strains isolated from man are nonpigmented, but those isolated from insects are red pigmented. However, similar red pigments are produced by many species of bacteria (e.g., *Vibrio* spp. and *Micrococcus* spp.); and therefore, the production of red pigment should not be used exclusively as an indicator of *Serratia* infection. *Serratia* species frequently have been recovered from healthy, diseased, or dead insects by numerous workers (Steinhaus 1959, Bucher 1963, Bell 1969, Sri-Arunotai et al. 1975, Sikorowski 1985). *Serratia marcescens* has been isolated from a number of insect orders including Coleoptera, Hymenoptera, Lepidoptera, and Diptera (Thomas and Poinar 1973, Krieg 1987). Many insect species are susceptible to infection upon ingestion of high doses of red-pigmented strains. Other pathogenic but nonpigmented strains of *Serratia* have also been isolated from a variety of insects (Krieg

1987). *Serratia marcescens* has been isolated from the inside of eggs of insectary-reared *H. zea* by Bell (1969) and *H. virescens* by Sikorowski and Lawrence (1998) (Fig. 16) and from field-collected egg masses of the European corn borer, *Ostrinia nubilalis* (Lepidoptera: Crambidae), by Lynch et al. (1976). However, *Ser. marcescens* is most frequently reported as a pathogen of reared insects. Human safety may also be affected by nonpigmented biotypes of *Ser. marcescens* that are encountered with increasing frequency in hospital patients (Neter 1974), but the broader significance of pigmented biotypes in human infections is still uncertain (Grimont and Grimont 1978, Tanada and Kaya 1993).

In lepidopterans, *Ser. marcescens* is able to infect eggs, larvae, pupae, and adults (Sikorowski and Lawrence 1998). Neonate larvae are particularly sensitive, succumbing to *Ser. marcescens* infection generally within 24 h after exposure to the inoculum. Older larvae infected with the bacterium often exhibit a decreasing responsiveness to external stimuli (Sikorowski and Lawrence 1997). In general, *Ser. marcescens* is not pathogenic to insects when present in the digestive track in small numbers and is able to multiply and cause disease only when insect vigor is greatly reduced (Sikorowski 1985). The mode of infection is not well understood, but under some conditions, the bacterium is able to breach the PM and penetrate the epithelium to accumulate above the basement membrane (Fig. 15). Once the bacterium enters the hemocoel, it typically multiplies rapidly and causes a lethal septicemia in 1 to 3 days (Sri-Arunotai et al. 1975, Tanada and Kaya 1993). Infected larvae often turn red after death (Fig. 15), and only occasionally will a red larva still be living. Some larvae infected in the late larval stage pupate and produce red pupae postmortem. Adults may become infected after feeding on contaminated food, and they too turn red postmortem (Fig. 15). Eventually, the bacterium invades all insect tissues.

Serratia marcescens is very infectious and has the capacity to very quickly bring an insect rearing program to a virtual standstill. Furthermore, it can cause sublethal infections in which insect vigor is greatly compromised (Inglis and Lawrence 2001, Sikorowski et al. 2001). *Serratia marcescens* is also an important pathogen in insectaries involved with rearing of parasitoids (Sikorowski et al. 1992).

c. Xenorhabdus *and* Photorhabdus. Bacteria within the genera *Xenorhabdus* and *Photorhabdus* form symbiotic relationships with entomopathogenic nematodes. They are not encountered as pathogens of insects independent of nematodes. *Xenorhabdus* spp. are associated with *Steinernema* species, whereas *Photorhabdus* spp. are associated with *Heterorhabditis* species. Both types of nematodes carry their specific, mutualistic bacteria in their alimentary canal. After infective juveniles enter their insect host, bacterial cells are released into the host's hemocoel. This typically occurs within hours of the nematode entering the hemocoel. The bacteria multiply rapidly resulting in septicemia, typically within 48 h. *Xenorhabdus* and *Photorhabdus* are not capable of infecting insects independent of the nematodes, but some *Steinernema* species are pathogenic in the absence of their *Xenorhabdus* symbiont. The bacteria produce numerous toxins and provide a source of nutrients for the nematodes. *Photorhabdus* spp. may produce a red

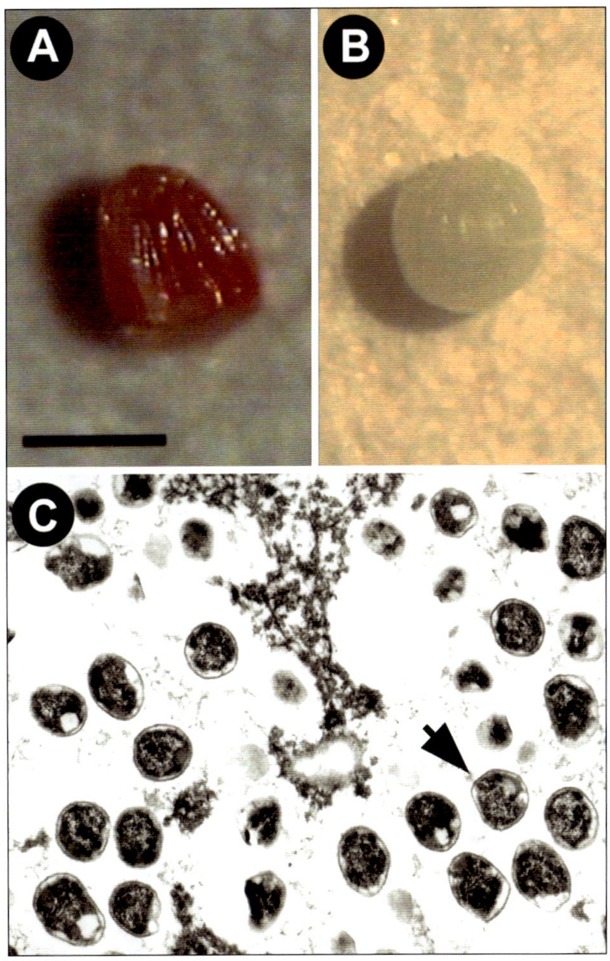

Fig. 16. Tobacco budworm, *Heliothis virescens* (Lepidoptera: Noctuidae), infected with *Serratia marcescens*. (A) An infected egg. (B) A healthy egg. (C) An electron micrograph of *Ser. marcescens* cells within the infected egg (arrow).

pigment in infected insects resulting in the reddish appearance of cadavers (Fig. 17). For more information on *Xenorhabdus* and *Photorhabdus*, the reader is referred to Kaya and Stock (1997).

d. Other. A variety of other genera of bacteria including *Aeromonas, Escherichia, Proteus, Salmonella*, and *Vibrio* have been reported as facultative or opportunistic pathogens of a variety of insects (Tanada and Kaya 1993, Boucias and Pendland 1998).

3. Nonculturable. Many species of rickettsia and rickettsia-like microorganisms cause diseases in insects. The majority of rickettsias associated with insects are found within the genera *Rickettsia*, *Rickettsiella*, and *Wolbachia*; but for convenience, we include the *Chlamydia* in this section as well. These bacteria are obligate parasites, they possess Gram negative cell walls, and they are typically rod shaped but may be either ovoid or pleomorphic (Boucias and Pendland 1998). Many species grow intracellularly, whereas other species may develop epicellularly or extracellularly. *Rickettsiella* comprises a heterogeneous group of intracellular bacteria that are all pathogens of a variety of different arthropods. Well-documented pathogenic *Rickettsiella* in insects include *Rickettsiella popilliae*, the causal agent of blue disease in larvae of the Japanese beetle; *Popillia japonica* (Coleoptera: Scarabaeidae); *Rickettsiella grylli* in *A. domesticus* nymphs; and *Rickettsiella chironomi* in larval chironomids. *Rickettsiella* infections are often lethal. In *C. caryae* infected by *Rickettsiella*, blood cells, the midgut, muscles around the midgut, the trachea, silk glands, fat, the heart, *Malpighian tubules* (closed tubules originating near the junction of the midgut and hindgut and involved in the excretion of nitrogenous wastes), and muscular sheaths around the Malpighian tubules are infected. In latter stages of disease, the fat tissues are completely broken down (Sikorowski unpublished). At advanced stages of infection, *C. caryae* larvae often become swollen (frequently to twice the size of normal larvae), turn whitish-yellow and are rubber-like in consistency in contrast to the light yellow color and soft consistency of uninfected larvae. Diseased larvae show little response to external stimuli and remain in this quiescent state for several days. Symptomatic larvae usually die before pupation.

The genus *Rickettsia* is not considered a serious problem in most rearing settings. However, members of *Rickettsia* cause a number of human diseases (e.g., typhus fever), and they are often transferred to humans from invertebrate vectors (e.g., insects, ticks, and mites).

Wolbachia is a heterogeneous group of small rods and *cocci* (i.e., spheroids), and they are morphologically similar to the *Rickettsia*. Many *Wolbachia* are considered to be obligate symbionts of arthropods and are not directly pathogenic to either insects or vertebrates (Boucias and Pendland 1998). Some species are parasitic and cause *cytoplasmic incompatibility* (CI). Cytoplasmic incompatibility is a form of male sterility in which males carrying the *Wolbachia* symbiont suffer a complete or partial loss of fertility when mated with uninfected females. However, fertility is restored if the female harbors the same *Wolbachia* variant. Uninfected females thus suffer a fertility deficit relative to infected ones, the intensity of which depends on the frequency of infected males. It is called cytoplasmic incompatibility because the bacterium is transmitted by females via the egg cytoplasm. CI has been reported in various Diptera, Hymenoptera, Coleoptera, and Lepidoptera. Fortunately, *Wolbachia* are highly sensitive to tetracycline, and CI caused by *Wolbachia* can be prevented in rearing settings by antibacterial treatments.

Chlamydia spp. are most well known from vertebrate hosts. They are also frequently associated with a vast number of insects but usually only manifest themselves as stunting agents and do not normally cause conspicuous disease outbreaks (Goodwin 1984). Although this group of bacteria possesses a developmen-

Fig. 17. Pupa of a southwestern corn borer, *Diatraea grandiosella* (Lepidoptera: Crambidae), overwintering in a corn crown and infected with the nematode *Heterorhabditis bacteriophora*. Note the characteristic brick-red color produced by symbiotic bacteria associated with the nematode.

tal cycle similar to the *Rickettsiella*, DNA hybridization studies have demonstrated that members of the two genera are not closely related (Frutos et al. 1994). *Chlamydia* spp. have been isolated from epidermal tissue (sometimes causing black spotting of the integument) and occasionally from the midgut and nerve tissue. Examples of insects from which *Chlamydia* spp. have been isolated include the German cockroach, *Blattella germanica* (Blattodea: Blattidae); various cutworms *Agrotis* spp. (Lepidoptera: Noctuidae); *D. grandiosella*; *H. virescens*; *H. zea*; *T. ni*; *O. nubilalis*; pink bollworm, *Pectinophora gossypiella* (Lepidoptera: Gelichiidae); tobacco hornworm, *Manduca sexta* (Lepidoptera: Sphingidae); cereal leaf beetle, *Oulema melanopus* (Coleoptera: Chrysomelidae); sugarbeet root maggot, *Tetanops myopaeformis* (Diptera: Otitidae); melon aphid, *Aphis gossypii* (Homoptera: Aphididae); and the alfalfa weevil parasitoids *Bathyplectes spp.* (Hymenoptera: Ichneumonidae).

C. Fungi

Fungi represent a diverse assemblage of nonphylogenetically related microorganisms representing at least three Kingdoms. They are grouped together since they are all eukaryotes, they usually produce *hyphae* (tubular filaments that grow from the tip and excrete enzymes and various other metabolites; sing. *hypha*) and possess rigid cell walls, and they are all *heterotrophs* (i.e., utilize organic matter as a source of energy). Reproduction is by spores and may be sexual or asexual. Some types of fungi produce both sexual and asexual spores, and others produce one or the other. Asexual spores of many entomopathogenic taxa are borne naked on specialized cells termed *sporogenous cells*. In most cases, these nonmotile asexual spores are called *conidia* (sing. *conidium*), and they are borne on specialized sporogenous cells termed *phialides* or more generally as *conidiogenous cells*. Often conidiogenous cells are formed from a stalk termed a *conidiophore* (pl. *conidiophores*). Examples of fungi that produce asexual spores that infect insects include *Aspergillus*, *Beauveria*, *Entomophaga*, *Entomophthora*, *Hirsutella*, *Metarhizium*, *Nomuraea*, *Paecilomyces*, and *Penicillium*. Characteristics of fungi are primarily discerned using a compound light microscope. They are primarily identified based on their taxonomically characteristic asexual or sexual spores, and the way spores are formed (i.e., conidiogenesis). There are four divisions of true fungi including the Chytridiomycota, Zygomy-

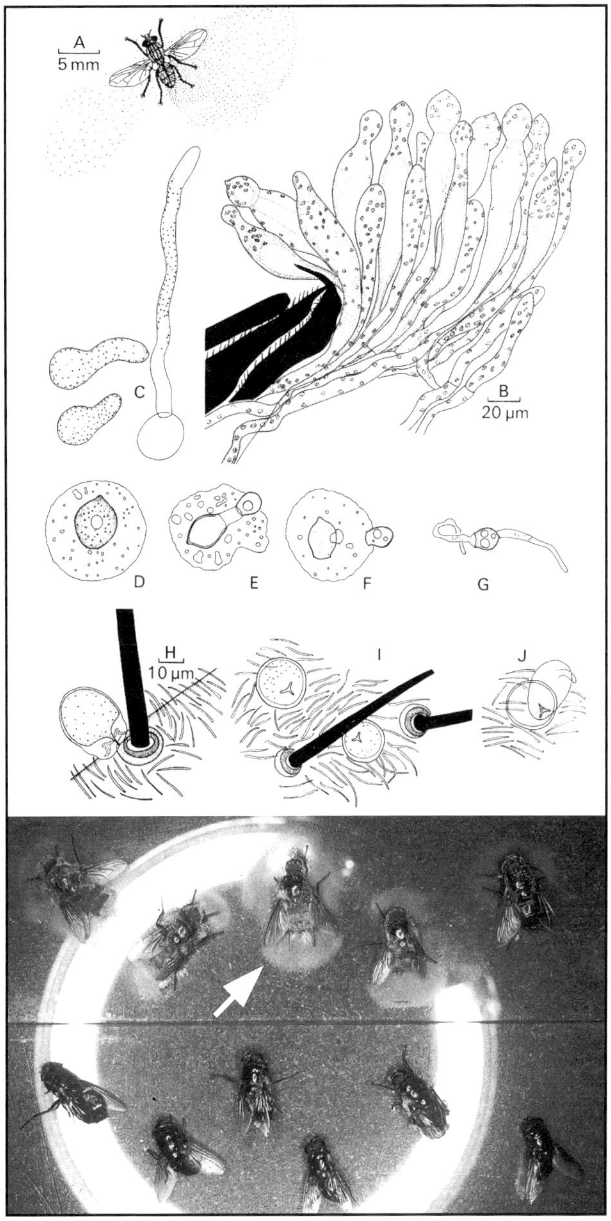

Fig. 18. The top plate shows the stages of infection of the house fly, *Musca domestica* (Diptera: Muscidae), by the entomopathogenic fungus *Entomophthora muscae*. (A) A cadaver adhering to a window pane, surrounded by a halo of conidia. (B) A cross-section of a cadaver showing the unbranched conidiophores projecting between the segments of the exoskeleton. (C) Hyphal bodies from the body of a recently dead fly. (D) A conidium immediately after discharge surrounded by cytoplasm from the conidiophore. (E–F) Germination of a primary conidium to form a secondary conidium within 12 h of discharge. Note the septum cutting off the secondary conidium. (G) Germination of the secondary conidium to produce two germ tubes. (H) Attachment of the primary conidium to the integument of a fly. Note the thickened appressorium and the narrow point of penetration. (I) Two primary conidia attached to the integument and penetrating it by a tri-radiate fissure. (J) Penetration from within the integument. Note the bladder-like expansion inside the tri-radiate fissure. [From Webster (1980) with permission from Cambridge University Press, Cambridge, MA.] The bottom plate shows a photograph of dead flies that were infected by *E. muscae* (top) surrounded by a shower of primary conidia, compared to fly cadavers not infected with the fungus (bottom).

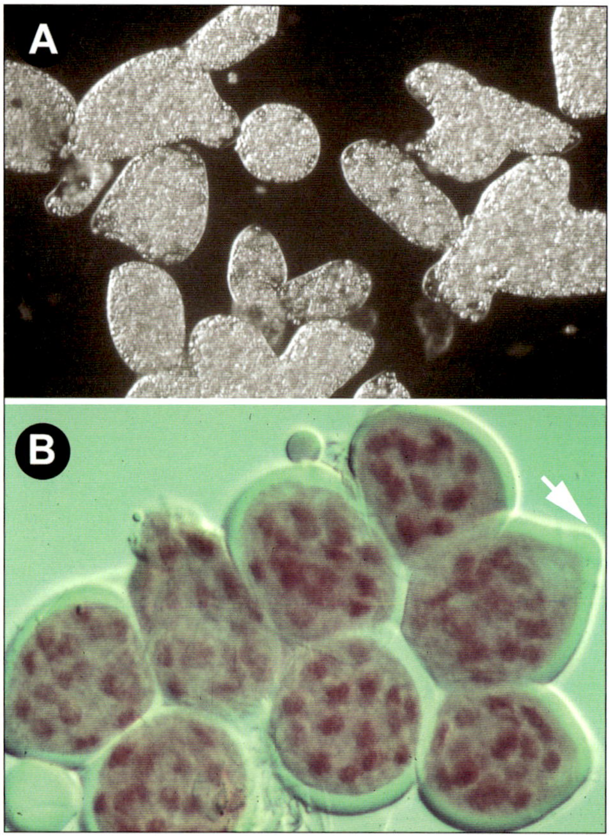

Fig. 19. The entomopathogenic fungus *Entomophaga grylli* from infected grasshoppers. (A) Protoplasts within the hemolymph. (B) Conidia stained with aceto-orcein showing multinucleate status of the spores. Note the papilla (arrow). [Images provided by Dan L. Johnson, Agriculture and Agri-Food Canada, Lethbridge, AB, Canada.]

cota, Ascomycota, and Basidiomycota. Fungi are unique among the insect pathogens in that they primarily infect their hosts through the cuticle (outer layers of the integument), although a few taxa may invade through their host's alimentary canal.

1. Zygomycota. Entomogenous zygomycetes are located principally in the order Entomophthorales. The genera *Conidiobolus, Entomophaga, Entomophthora*, *Neozygites*, and *Pandora* all have species recognized for being able to cause *epizootics* (i.e., outbreaks of disease in which there is an unusually large number of cases) that decimate arthropod populations. The genus name *Entomophthora* means "insect destroyer," aptly named because of the mass die-offs this group of fungi can cause. The insect hosts of these fungi occur in more than 32 families in the orders Hemiptera, Homoptera, Diptera, Lepidoptera, Coleoptera, Orthoptera, and Hymenoptera (MacLeod and Muller-Kogler 1973). Species of *Entomophthora* have a worldwide distribution and play an effective role in destroying many insects of economic importance (MacLeod 1963).

The conidia of entomophthoralean fungi are forcibly discharged from the conidiophore, and the adherence of conidia to the cuticle of a susceptible host is facilitated by a mucilaginous sticky substance (Eilenberg et al. 1986). It is common to see a powdery film of conidia *en masse* on glass surrounding fly cadavers (Fig. 18). In some taxa, the conidia are multinucleate, a character which is diagnostic (Fig. 19). When the conidium lands on a moist substrate other than the host, it may produce a secondary conidiophore and secondary conidium (Fig. 18), which may in turn produce tertiary conidiophores and conidia. This process may continue until energy reserves are depleted or a suitable host is found (Tanada and Kaya 1993). Secondary and tertiary conidia may become resting spores, or *capilliconidia* (secondary conidia) may be borne on narrow capillary conidiophores. The surface of the capilliconidia has sticky substances that cause the spores to adhere to objects with which they come in contact. Upon germination, the conidium produces a hyphal germ tube that penetrates the insect cuticle and enters the hemocoel. Inside the hemocoel, the fungus forms *protoplasts* (i.e., cells lacking a CW), which eventually produce hyphae. The protoplasts appear early in infection and exhibit amoeboid movement (Fig. 19A). Since they do not possess β-glucan-containing cell walls, they are not readily detected by the insect and are not subject to a defense response. The time required for entomophthoralean species to kill an insect is dependent upon many factors such as the host species and developmental stage and environmental conditions such as temperature and moisture. The period from infection to death of the insect may be from 3 to 12 days, with most deaths occurring between 5 and 8 days (MacLeod 1963). The fungus invades most of the organs within the insect's body before host death. Once the host is colonized, conidiophores emerge through the intersegmental membranes. This typically requires high humidity and occurs shortly after the death of the insect. Conidia are produced on the conidiophores and serve as a source of infectious propagules. Conidia are not typically long-lived. However, entomophthoralean fungi also produce various types of resistant or resting structures, primarily *chlamydospores* (asexual, thick-walled resting spores formed within hyphae) and *zygospores* (thick-walled, sexual spores). In particular, zygospores are important for survival in temperate climatic zones. Germinating resting zygospores produce a germ tube, which may produce infectious primary-discharge conidia or secondary adhesive spores.

2. Ascomycota. The Ascomycota are a large group of fungi that if they produce *ascospores* (i.e. sexual spores,) do so in *asci* (i.e. sacs, sing. *ascus*). Quite often, asci are formed in fruiting structures referred to generally as *ascocarps*. However, many of

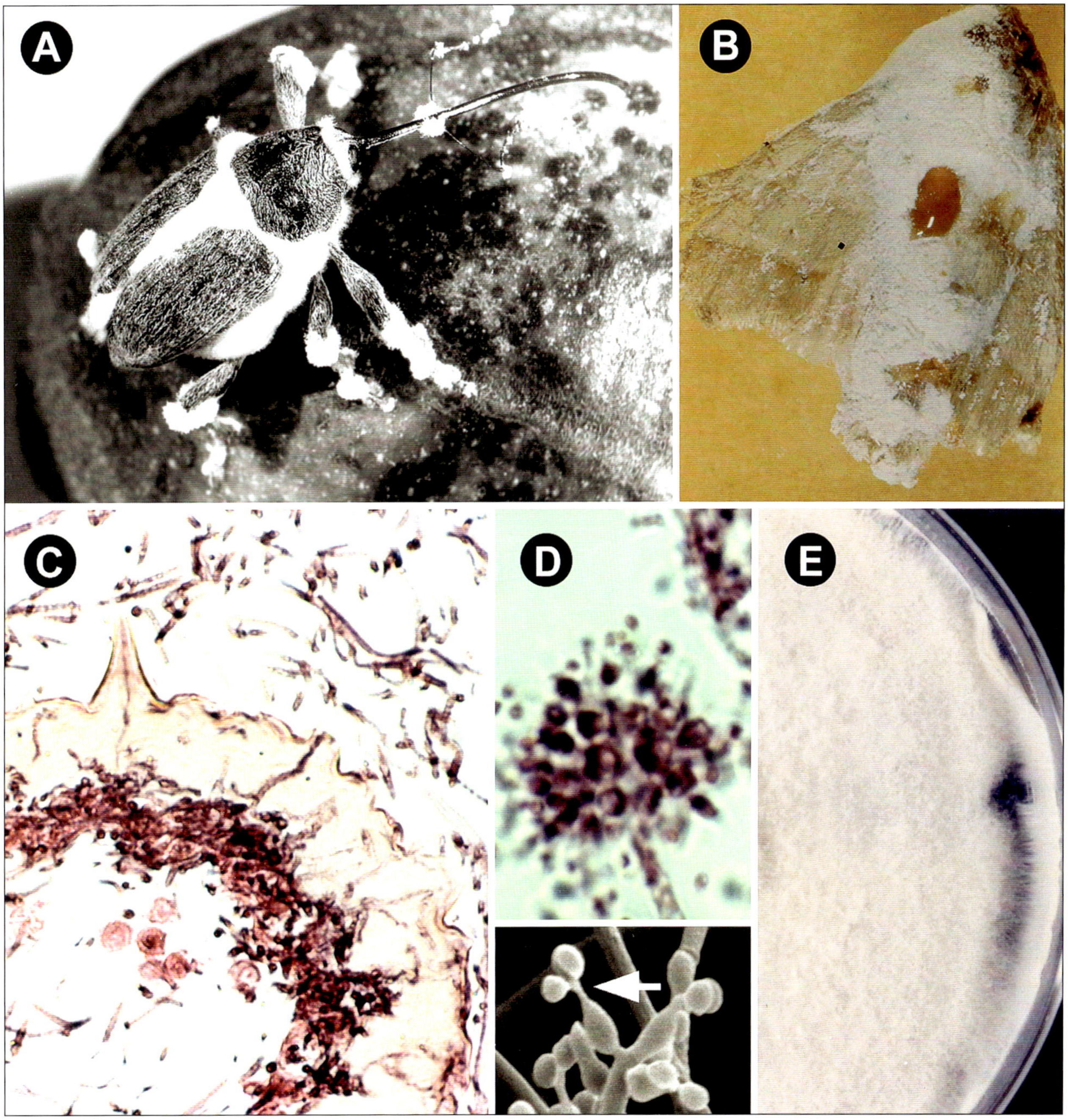

Fig. 20. The entomopathogenic fungus *Beauveria bassiana*. (A) An infected pecan weevil, *Curculio caryae* (Coleoptera: Curculionidae), exhibiting the characteristic tufts of white hyphae erupting through the intersegmental membranes of the integument. (B) An adult tobacco budworm, *Heliothis virescens* (Lepidoptera: Noctuidae), colonized by the fungus. (C) Invasion of the cuticle and hemocoel of an *H. virescens* larva by hyphae. (D) A light micrograph (bright field) and scanning electron micrograph showing conidiogenesis. Note the conidiophore producing ampulliform conidiogenous cells from which the round unicellular conidia are borne on narrow stalks (arrow). (E) Characteristic white floccose (i.e., cottony) appearance of a colony originating from an infected insect. [Images B–E, with the exception of the scanning electron micrograph, are from Sikorowski and Lawrence (1997) with permission from the Mississippi Agricultural and Forestry Experiment Station, Mississippi State, MS.]

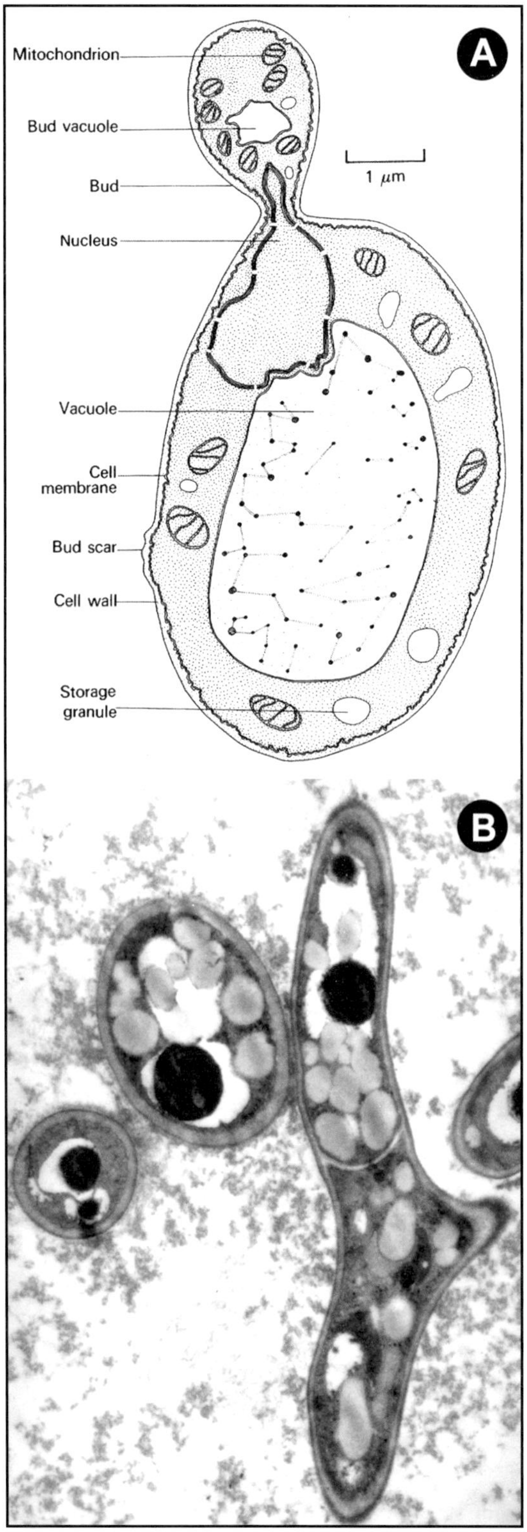

Fig. 21. (A) Diagrammatic representation of a section of a budding brewer's yeast cell (*Saccharomyces cerevisiae*) as seen under an electron microscope. [From Webster (1980) with permission from Cambridge University Press.] (B) Fungal hyphal bodies of *Beauveria bassiana* growing in the hemolymph of a tobacco budworm, *Heliothis virescens* (Lepidoptera: Noctuidae), larva. Note the conspicuous cell walls and the presence of 1- and 2-celled bodies. Also, note that these cells are sometimes incorrectly referred to as "blastospores."

the entomopathogenic taxa of ascomycetous fungi have lost the ability to or rarely produce sexual spores. These fungi traditionally have been placed in the formal division *Deuteromycota* (i.e., asexual or mitosporic fungi) within the class *Hyphomycetes* (i.e., producing naked or exposed asexual spores). Most mycologists no longer accept Deuteromycota as a formal taxonomic category, and the majority of these fungi are now handled within the Ascomycota. Although few taxa of entomopathogenic fungi possess a sexual state, they still exhibit other characteristics common to fungi in the Ascomycota including: 1) septate hyphae possessing relatively simple septal pores; 2) lamellate hyphal walls with a thin electron-dense outer layer and a relatively thick electron-transparent inner layer; and 3) sensitivity to the fungicide benomyl, which interferes with spindle formation during nuclear division. The placement of many hyphomycetous fungi within the Ascomycota has been substantiated on the basis of homology of 18S rDNA. Nevertheless, entomopathogenic fungi that produce no sexual state and reproduce by asexual spores borne exposed on sporogenous cells are still referred to as "hyphomycetous fungi" largely due to tradition and convenience. Here we focus on *Beauveria bassiana*, *Metarhizium anisopliae*, *Nomuraea rileyi*, *Paecilomyces* spp., *Aspergillus* and *Penicillium*. However, a number of other hyphomycetous taxa cause disease in insects (Humber 1997).

a. Beauveria bassiana. This fungus is soil borne, occurs worldwide, and has one of the largest host lists among the entomopathogenic fungi (Tanada and Kaya 1993). Diseases caused by this fungus are often referred to as *white muscardine* resulting from the characteristic sign of disease incited by this fungus. Following death of the insect, white hyphae erupt through the intersegmental folds of the integument; and conidia, which are also white *en masse*, are produced on conidiophores that develop from hyphae, producing a thick covering about the animal (Fig. 20). These conidia are passively discharged, and once they come in contact with a host, they strongly adhere to the cuticle, facilitated by nonspecific adhesion mechanisms mediated by the hydrophobicity of the CW (Boucias et al. 1991). Under appropriate conditions, conidia germinate and produce germ tubes that penetrate the host cuticle using a combination of enzymatic and mechanical mechanisms. Once the fungus reaches the hemocoel, it grows as hyphal bodies, which are single- or multi-celled yeast-like structures that possess a well-developed CW (Fig. 21). However, before the fungus can proliferate in the hemocoel, it often must overcome the insect's defense response. Host death may result from a combination of actions, including depletion of nutrients, physical obstruction or invasion of organs, and toxinosis. Following death, the fungus often grows saprotrophically as hyphae within the host (Fig. 20C). Under favorable conditions (e.g., high ambient humidity), hyphae then emerge from the cadaver producing the characteristic tufts of white hyphae (Fig. 20A, B). Conidia are borne on characteristically shaped *ampulliform* (i.e., flask-shaped) conidiogenous cells (Fig. 20D). As mentioned previously, conidia are passively dispersed, relying principally on wind; but other factors such as rain can play a role in their dissemination. In reared field-collected grasshoppers, we have observed mortality caused by *B. bassiana* that has exceeded 90% of individuals.

b. Metarhizium anisopliae. Commonly referred to as the *green-muscardine fungus*, this fungus possesses a relatively wide host range affecting many economically important insects (McCoy et al. 1988), particularly insects maintained in unpasteurized soils. The hyphae of this fungus are white *en masse* but conidia appear green (Fig. 22). Conidia are the infectious propagule, and infection generally takes place through the integument. However, the exact site of infection is dependent on the stage of the insect, environmental conditions, and opportunity. Under appropriate conditions, conidia germinate to produce a germ tube that produces a swollen structure termed an *appressorium* (pl. *appressoria*) that attaches to the cuticle. From the appressorium, a specialized hypha termed an *infection peg* penetrates the cuticle, facilitated by enzymes secreted at the apex of the peg. The penetrant hypha gives rise to hyphal bodies. Hyphal bodies become distributed throughout the body cavity and give rise to secondary hyphae. Once in the hemocoel, the fungus produces a variety of toxins termed *destruxins* that contribute to insect death (Suzuki et al. 1966, 1970, 1971). In moist environments, *Met. anisopliae* forms a white hyphal mat on the surface of the cadaver from which green conidia are produced. In contrast to *Bea. bassiana*, conidia are formed in long *basipetal* chains (i.e., chain extension is from the base).

c. Nomuraea rileyi. The fungus *Nomuraea rileyi* (= *Spicaria prasina* = *Spicaria rileyi*) is pathogenic to a number of economically important lepidopterous pests in the Noctuidae (Ignoffo et al. 1976) and frequently causes natural epizootics in noctuid populations. Following death in a high humidity environment, infected hosts are covered with a dense white mat of hyphae that, upon conidial formation, turn pale green. Often conidia form over the larva's entire body. These conidia are important in the transmission of infective conidia to healthy individuals in the population. The infection process is similar among hosts, and here we provide a description of infection of *H. zea* (Mohamed et al. 1978). Under conditions of high humidity, conidia *germinate* (i.e., produce germ tubes) within 2 days after they are deposited on the cuticle of a susceptible larva. Germ tubes enter the cuticle, and points of entry are often darkened. Branched hyphae develop from the germ tubes, and by 5–6 days, hyphae gain access to the

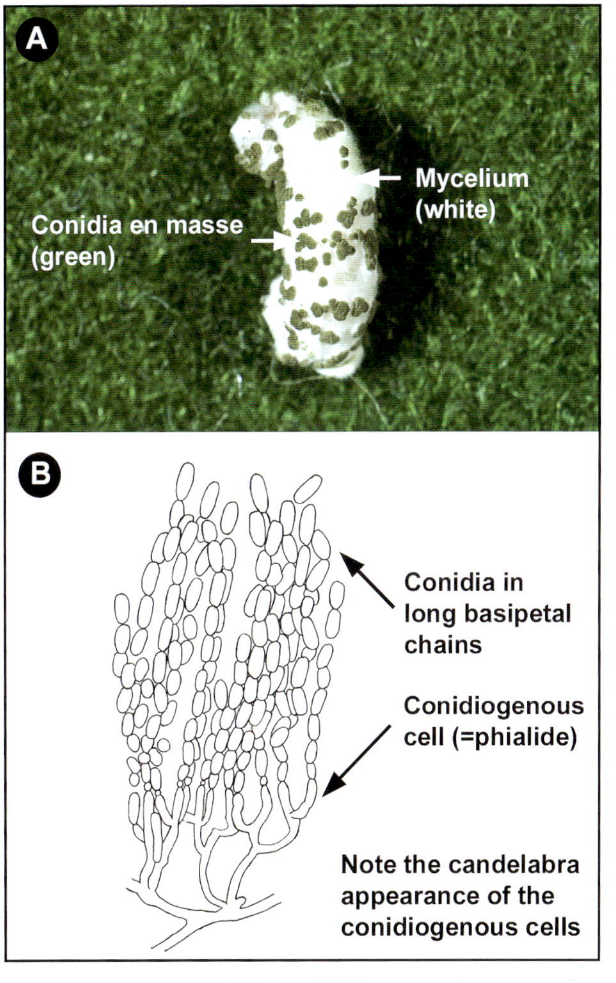

Fig. 22. ***Metarhizium anisopliae*. (A) White mycelium producing green-colored conidia on the integument of an infected southern pine beetle, *Dendroctonus frontalis* (Coleoptera: Curculionidae), larva. (B) A drawing showing the candelabra appearance of the conidiogenous cells producing long, basipetal chains of green-colored conidia.**

hemocoel. Hyphal bodies form from the penetrant hyphae and proliferate within the hemocoel. Hyphal bodies are typically short, thick, mostly 1-, 2-, or 3-celled filaments and distinctly nucleated. Hemolymph cells are the first to be invaded, followed by fat bodies, Malpighian tubules, and muscles. At death of the host, hyphae emerge from the host, and conidiogenesis occurs on the external integument. Conidia are green *en masse* and typically cover the entire body of the larva under conditions of high humidity.

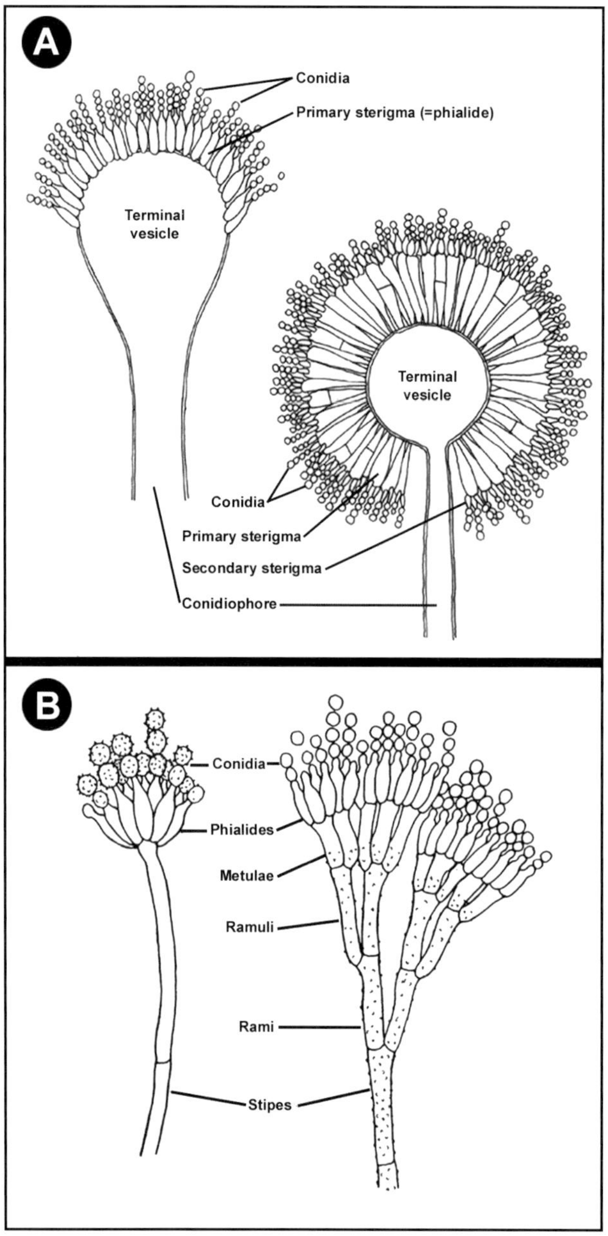

Fig. 23. (A) Two types of conidial heads formed by *Aspergillus*. (A) Head with a single series of conidiogenous cells (i.e., *sterigma*) (left) and a head with two series of conidiogenous cells (i.e., primary and secondary sterigmata) (right). Note the conspicuous vesicle at the apex of the conidiophore. (B) Penicillate heads formed by *Penicillium*. A single-series species (left) and a four-series species (right). Note the brush-like appearance of the heads formed at the apex of a long stipe (i.e., *conidiophore*). The conidiophores can be smooth-walled or rough-walled. Primary, secondary, tertiary, and quaternary conidiogenous cells are referred to as *phialides*, *metulae*, *ramuli*, and *rami*, respectively. [From Pitt (2000) with permission from Food Science Australia, North Ryde, NSW Australia.]

d. Paecilomyces *spp.* The genus *Paecilomyces* contains a number of plant-, nematode-, and insect-pathogenic species. Of the 15 entomopathogenic species, *Paecilomyces farinosus* and *Paecilomyces fumosoroseus* are the most common species encountered in insectaries. This genus is characterized by the production of *penicillate* (i.e., brush-like) heads with ampulliform phialides; this type of conidiogeneous cell is similar to that observed with *Penicillium* (Fig. 23). In contrast to *Penicillium*, conidia of *Paecilomyces* are oval to spindle-like in shape, and the conidiogenous cell is usually swollen towards the base and tapers gradually towards the apex. Similar to other entomopathogenic fungi, conidia possess hydrophobic CWs to facilitate attachment to insect cuticles. Both species have a wide host range, attacking a number of lepidopteran, coleopteran, hymenopteran, and homopteran pests.

e. Aspergillus *and* Penicillium *spp.* These fungi are mainly saprophytic, but they may be facultative pathogens when conditions are favorable (e.g., when insects are physiologically stressed). There is a tremendous diversity of both *Aspergillus* and *Penicillium* species with over 250 recognized species. For more information, the reader is referred to Chap. 8 *Microbial Contamination and Insect Rearing*.

A number of species (e.g., *Aspergillus flavus* and *Aspergillus niger*) can be pathogenic to insects, particularly young larvae in insectary settings (e.g., Sikorowski and Lawrence 1994; Fig. 24). *Aspergillus* and *Penicillium* are two of the groups of fungi most easily identified to genus. *Aspergillus* is characterized by the formation of ampulliform conidiogenous cells on a vesicle that forms at the apex of a conspicuous conidiophore (Fig. 23A). In contrast, *Penicillium* is characterized by the formation of ampulliform conidiogenous cells that form directly from a conidiophore (i.e., singly) or from dichotomously branched conidiogenous cells in penicilliate heads (Fig. 23B). The occurrence of *Aspergillus* spp. in commercially available insect diets is unavoidable because key diet ingredients are almost always contaminated with various species of *Aspergillus* and *Penicillium*. Insect rearing containers contaminated with *Aspergillus* or *Penicillium* spp. should be autoclaved before they are opened or discarded for the following reasons: 1) to avoid contamination of the laboratory with airborne conidia; and 2) *A. flavus* and *A. niger* are among species of this genus that can cause aspergillosis, a human disease (see Chap. 8 *Microbial Contamination and Insect Rearing*). In addition, *Asp. flavus* and *Asp. parasiticus* (a closely related species) are producers of aflatoxin B_1, one of the most potent carcinogens known.

g. Laboulbeniales. Another group of often overlooked ascomycetous fungi commonly associated with reared insects is the Laboulbeniales. These fungi are minute and often inconspicuous, but they can often be detected at minimal magnification using a hand lens (Fig. 25). The Laboulbeniales are associated with a variety of insect species, and the majority of species only infect adults. Some species are host-specific, but others have a broad host range (Alexopoulos et al. 1996). Like the nonculturable bacteria, the Laboulbeniales are obligate parasites and cannot be cultured on agar media. They are generally considered to have a limited detrimental impact on their insect hosts, but a number of taxa have been documented to cause disease (Tanada and Kaya 1993).

The Laboulbeniales are typically nonhyphal and produce *haustoria* (sing. *haustorium*). Haustoria are absorbing organs originating on a hypha that penetrates the host exoskeleton and that invaginate the host cell plasma membrane. Haustoria are often produced by obligate parasites. In the Laboulbeniales, haustoria typically consist of one to several peg-like or extensively branched structures. This group of fungi is different from those groups discussed previously in that they produce sexual spores (i.e., ascospores). Asci typically contain ascospores that are two-celled. Ascospores in asci are produced in a specialized ascocarp termed a *perithecium* (pl. *perithecia*) (Fig. 25). A perithecium is a closed ascocarp with only a pore at the top through which ascospores are released, usually by force. Perithecia often form on the exterior of the host and may possess visible appendages that probably function in the release of ascospores (Fig. 25). Released ascospores attach to an insect exoskeleton facilitated by the basal cell of the ascospore which is sticky. Soon thereafter, a hypha penetrates the exoskeleton and a haustorium extends into the hemocoel.

f. Other. A number of other ascomycetous fungi including *Aschersonia*, *Ascosphaera*, *Culicinomyces*, *Tolypocladium*, and *Lecanicillium lecanii* (previously known as *Verticillium lecanii*) can cause sporadic infections in insectary settings. With the exception of *Ascosphaera*, these occur only in the anamorphic (asexual) state. Ascomycetous yeasts (Fig. 21) are common in the alimentary canal of insects, and a few taxa have demonstrated pathogenicity (e.g., *Metschnikowia* and *Saccharomyces*). Some yeasts are exclusively single-celled microorganisms whereas others are dimorphic producing both a single-celled yeast phase and a hyphal phase. In the yeast phase, yeasts

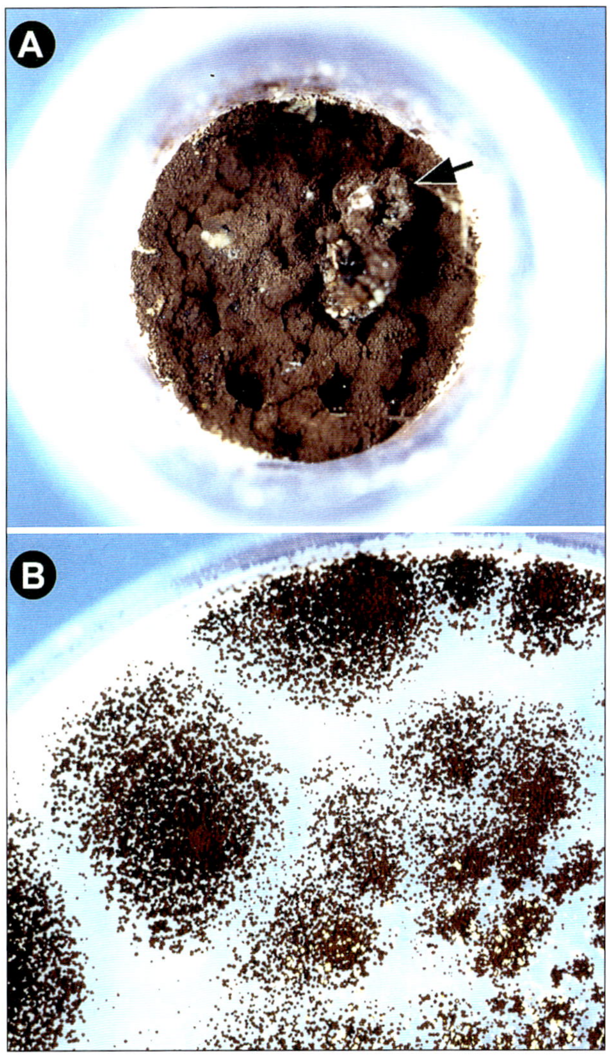

Fig. 24. Tobacco budworm, *Heliothis virescens* (Lepidoptera: Noctuidae), infected with *Aspergillus niger*. (A) Infected larva (arrow) on a grossly contaminated artificial diet. (B) A colony originating from an infected larva showing the characteristic black-colored conidia. [From Sikorowski and Lawrence (1997) with permission from the Mississippi Agricultural and Forestry Experiment Station, Mississippi State, MS.]

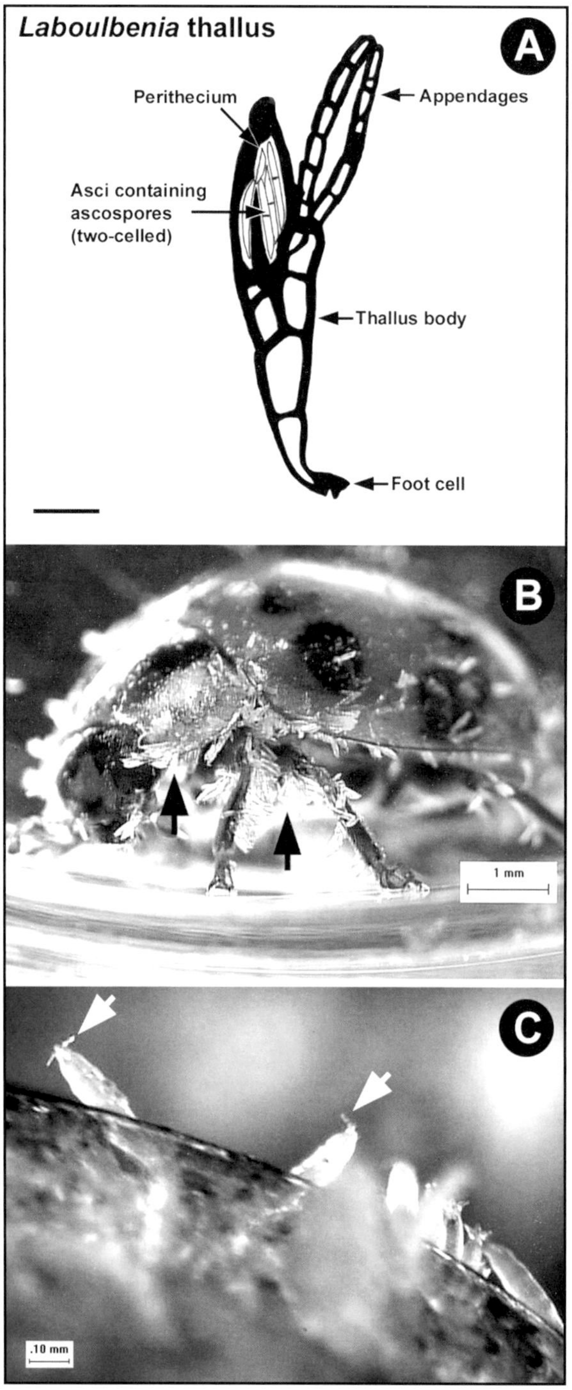

Fig. 25. Representatives of the obligately entomopathogenic fungi in the order Laboulbeniales. (A) Drawing of the *thallus* (i.e., mycelium) of *Laboulbenia* showing a perithecium producing two-celled ascospores, a conspicuous foot cell, and appendages. Bar = 50 µm. (B–C) *Hesperomyces virescens* infecting adult multicolored Asian lady beetle, *Harmonia axyridis* (Coleoptera: Coccinellidae). Note the thalli on the surface of the integument and the trigger cells (arrows) whose function is to facilitate ascospores coming into contact with an insect host. [Images B and C provided by Eric Riddick (USDA-ARS, Mississippi State, MS).]

are easily distinguished from bacteria because of their method of asexual reproduction (i.e., budding) and because they are generally larger than bacteria. Yeasts cells subjected to Gram stains appear Gram positive.

D. Protozoa and Microsporidia[2]

The majority of entomogenous protozoa- and microsporidia-insect associations produce chronic, nonlethal infections of which a common feature is a reduction in the host reproductive output (Hurd 1993). Most such infections by these organisms exhibit nonspecific signs and symptoms such as sluggishness; irregular growth; loss of appetite; malformed larvae, pupae, or adults; or adults with reduced vigor, fecundity, and longevity. Entomogenous protozoa and microsporidia are commonly transmitted on the surface of the egg or within the egg of their host. For more information on taxonomy and biology of this group of entomopathogenic organisms see the comprehensive treatment of this group by Kudo (1966), Weiser (1977), Poinar and Thomas (1984), Lee et al. (1985), Brooks (1988), Tanada and Kaya (1993), Undeen and Vávra (1997), and Boucias and Pendland (1998).

1. Apicomplexa. The phylum Apicomplexa consists of protozoa that form an *apical complex*, any of several subcellular structures (polar ring, conoid, rhoptery, microneme, and subpellicular microtubules) that are congregated at the anterior end of the cell. A subclass of Apicomplexa that includes a number of insect parasites is the Gregarinasina. The insect pathogenic gregarines are divided into two orders: the Eugregarinida and the Neogregarinida. Eugregarines have been isolated from a variety of insects, and they typically are very host specific and may be restricted to a particular tissue of a specific life stage of a single insect species (Boucias and Pendland 1998). Most eugregarines do incite overt disease, but species within the genus *Ascogregarina* can be detrimental to the host (Boucias and Pendland 1998). *Farinocystis*, *Mattesia*, and *Ophryocystis* are neogregarines that commonly produce lethal infections in insect pests in the orders Diptera, Coleoptera, and Hemiptera. Most neogregarine species infect a single or a few species of insects. However, some species such as *Farinocystis tribolli* and *Mattesia grandis* infect a large number of species (Tanada and Kaya 1993). For example, *M. grandis*, commonly infects the boll weevil, *Anthonomus gran-*

[2]Members of the phylum Microsporidia have traditionally been considered primitive protozoa. However, recent evidence indicates that they are actually highly evolved intracellular fungi. In this discussion, they are considered separate from fungi for convenience.

dis (Coleoptera: Curculionidae) (Fig. 26), but it also can infect the Heliothinae (Lepidoptera: Noctuidae) (e.g., *H. virescens* and *H. zea*) (Ignoffo and Garcia 1965). Neogregarines infect a number of organs and tissues, but the fat bodies are particularly affected in insects infected by *Farinocystis* and *Mattesia*. For example, *M. grandis* causes the progressive destruction of the fat bodies thereby severely reducing the fitness of infected individuals (McLaughlin 1965). *Farinocystis* commonly infects grasshoppers in captivity. From infected individuals, maturing cysts and liberated lemon-shaped spores can be detected (Fig. 26). *Ophryocystis* is another neogregarine that is known to afflict insects in captivity. There are over 10 species known to infect insects, and various beetle species are particularly susceptible. For example, *Ophryocystis mesnili* commonly infects the yellow mealworm, *Tenebrio molitor* (Coleoptera: Tenebrionidae) (Tanada and Kaya 1993). The Malpighian tubules of insects infected with *Ophryocystis* are particularly affected. The species *Ophryocystis elektroscirrha* is known to infect monarch butterflies, *Danaus plexippus* (Lepidoptera: Danaidae), and the Florida queen butterfly, *Danaus gilippus* (Lepidoptera: Danaidae) (McLaughlin and Myers 1970). The pathogen was reported to infect the epidermis where it remains until pupation. It completes its life cycle in tissues that become the scales of the adult butterfly and by which it is transmitted to other insects. Infection by *Ophryocystis* also has often been observed in heliothine pupae (typically insectary reared) received for diagnosis at Mississippi State University. In infected insects, large amber-colored cysts can be observed beneath the translucent pupal integument. The cysts also can be seen on the scales of the moths under a dissecting microscope, and if the moths are allowed to emerge in clean plastic cups, the cysts can be seen easily with a hand lens on the sides of the cups.

2. Cilophora, Rhizopoda, and Zoomastigina. Members of the phylum Cilophora are named for the dense hair-like cilia that cover the surface of the cells. The insect pathogenic ciliates are usually found in the larval and adult stages of dipterans, with the two most prominent genera being *Lambornella* and *Tetrahymena*. The Rhizopoda include the amoebas, and relatively few members of this phylum are parasites of invertebrates. However, *Malamoeba locustae* infects acridids (e.g., grasshoppers and locusts), and *Malpighamoeba mellificae* infects honey bees. *Malamoeba* infections are commonly encountered in acridids reared at high densities (Fig. 27). In such infections, cysts and motile vegetative states of the amoeba fill the lumen of the Malpighian tubules resulting in conspicuous swelling (Weiser 1977). Infected grasshoppers or locusts are typically lethargic. The phylum Zoomastigina is characterized by an elongate, motile trophic cell that possesses one or more *flagella* (generally whip-like structures used for locomotion, sing. *flagellum*). The majority of the Zoomastigina associated with insects are found in the order Kinetoplastida (Boucias and Pendland 1998), and most alternate between vertebrate and invertebrate hosts or between invertebrate and plant hosts. Most species do not adversely affect the health of insects. However, some are pathogens (e.g., *Blastocrithidia*), others are facultative pathogens and only invade the hemolymph under some conditions (e.g., *Leptomonas*), while others colonize the alimentary canal but only cause sublethal disease when population den-

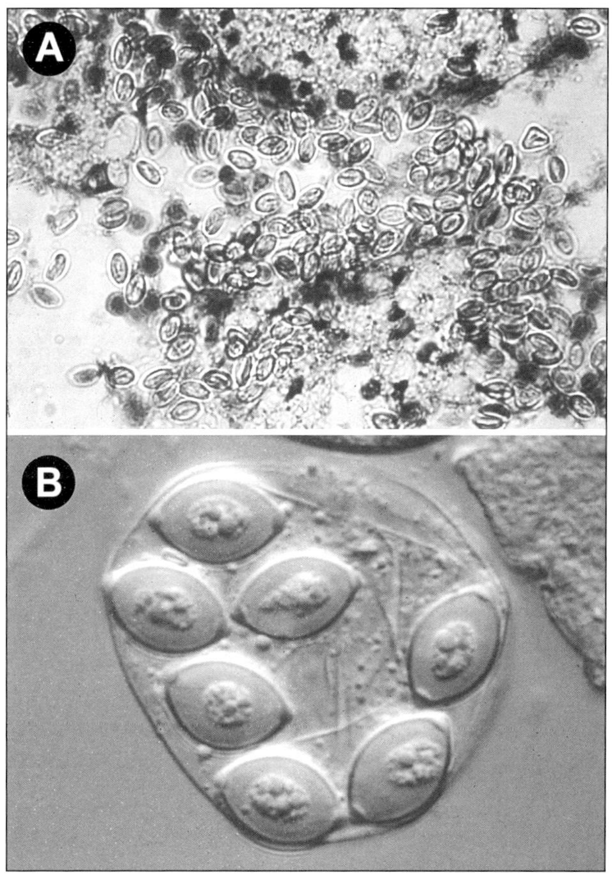

Fig. 26. Neogregarine protozoa. (A) *Mattesia* spores from an infected boll weevil, *Anthonomus grandis grandis* (Coleoptera: Curculionidae). (B) Lemon-shaped spores within a cyst of *Farinocystis* obtained from an infected grasshopper. [Image B provided by Dan L. Johnson (Agriculture and Agri-Food Canada, Lethbridge, AB, Canada).]

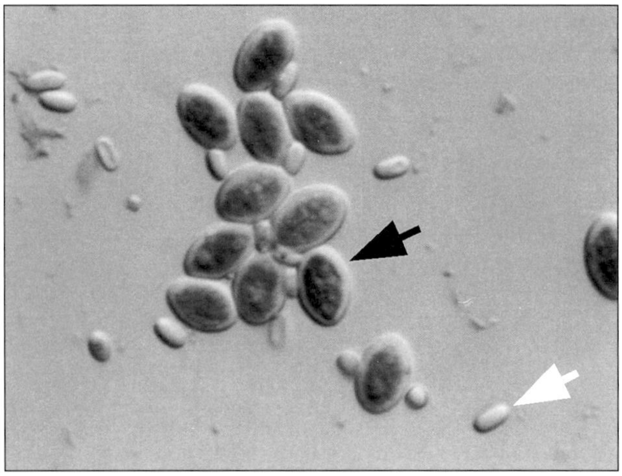

Fig. 27. A grasshopper co-infected with the rhizopodan *Malameba locustae* and the microsporidian *Paranosema locustae*. Malpighian tubules were homogenized and examined with a light microscope. Black arrow indicates spores of *M. locustae*, and white arrow indicates the smaller spores of *Par. locustae*. [Image provided by Dan L. Johnson (Agriculture and Agri-Food Canada, Lethbridge, AB, Canada).]

sities are high (e.g., *Crithidia* and *Herpetomonas*). Several species of kinetoplastids are important insect-vectored pathogens of humans (e.g., *Leishmania* and *Trypanosoma*).

3. Microsporidia. The microsporidia have traditionally been considered to be primitive protozoa. However, relatively recent evidence indicates that they are actually highly evolved intracellular fungi (Keeling and Fast 2002). Nonetheless, we discuss them here with the protozoa largely because they are traditionally handled with this phylogenetically diverse group of microorganisms. Members of microsporidia are arguably the most important entomopathogens of reared insects. Diseases caused by microsporidia are often referred to as *microsporidiosis*. Many species of microsporidia have been described (Becnel and Andreadis 1999), and approximately 600 of the approximately 1,000 named species have been reported from insects. Although microsporidia have been reported to infect all insect orders, the majority of genera occur in dipteran hosts (42 of 69 genera), and an additional 10 genera occur only in the aquatic orders Ephemeroptera (mayflies), Trichoptera (caddisflies), and Odonata (dragonflies and damselflies). Whether these genera represent a phylogenetic grouping is currently unknown. Some species of microsporidia infect a single host species, whereas other species possess a wide host range. Some insect-associated species have been shown to infect mammals. For example, *Nosema algerae* infects mosquitoes but also infects the tail and feet of immunocompromised mice, causing severe infections of muscle and connective tissue (Trammer et al. 1997).

All microsporidia are intracellular obligate parasites. They have evolved an extremely sophisticated and unique infection mechanism along with other adaptations to life inside cells (Brooks 1988, Tanada and Kaya 1993, Keeling and Fast 2002). These adaptations are primarily characterized by reductions in morphological (e.g., they are unicellular), ultrastructural (e.g., they do not produce mitochondria, Golgi apparati, centrioles or peroxisomes), biochemical, and genetic complexity (genomes are only 2.3 to 19.5 million base pairs). All microsporidia produce spores. Spores vary considerably in size and shape among the various genera, and spore characters are used in diagnosis. Microsporidian spores are among the most complex cells known to science, and they are the only easily identified stage of the parasite. The spore is an essential structure to the existence of microsporidia and functions as both a resting and infection structure. The spore is bound by two rigid extracellular walls (the exospore and endospore). Within the endospore is the *sporoplasm* (the infectious material), which includes a single nucleus or two nuclei arranged as a *diplokaryon* (two closely appressed nuclei). Also within the spore are a number of specialized structures such as the polaroplast, polar filament, and posterior vacuole (Fig. 28). The *polaroplast* is a large body of membranes occupying the anterior part of the spore. The anterior portion consists of highly organized, closely stacked membranes called the *lammellar polaroplast*, and the posterior portion is more loosely organized and is called the *vesicular polaroplast*. The *polar filament* or polar tube is derived from the Golgi apparatus and consists of a long filament 50 to 500 µm x 0.1 to 0.2 µm diam. It is attached to the apex of the spore via an umbrella-shaped structure called an *anchoring disk*. For approximately one-third to one-half the length of the spore, the polar filament is straight, and the remainder is helically coiled about the contents of the sporoplasm. The number of coils, their arrangement relative to one another, and even the angle of helical tilt are diagnostic for a particular species. To observe these structures requires the use of a TEM.

Germination of the microsporidian spore is one of the most interesting and dramatic series of subcellular events in biology (Keeling and Fast 2002). Spore germination begins with an environment trigger, and the trigger differs according to the species. Mechanical force, pH, dehydration followed by rehydration, hyperosmotic conditions, presence of cations or anions, and ultraviolet light have all been demonstrated to stimulate germination in some species. The first sign of germination is a general swelling of the spore and specific swelling of the polaroplast and posterior vacuole. This results from an increase in *osmotic pressure* (a measure of the tendency of water to move across a membrane) within the cell, but the precise mechanism is unknown. The swelling causes a rupturing of the anchoring disk and the rapid discharge of the polar filament by eversion (Fig. 29). Eversion begins at the spore apex, where the discharging polar filament breaks through the thinnest region of the two spore walls and turns itself inside out as it extends. Once the polar filament is fully everted, the sporoplasm from the spore is forced into the polar filament, which is essentially a hollow tube because its contents have been ejected. The continued pressure within the spore (resulting from the swelling of the posterior vacuole) forces the sporoplasm through the polar filament very rapidly (15–500 m/s), and the original sporoplasm membrane is left behind in the now empty spore. However, the sporoplasm obtains a brand-new membrane during transport through the polar filament from the polaroplast. The entire germination process takes < 2 sec.

Infection occurs in the midgut region of the alimentary canal following ingestion of spores; the midgut is the only portion of the insect gut that is not lined with cuticle. The polar filament is everted so rapidly that it actually pierces the peritrophic matrix barrier. In this way, the microsporidian sporoplasm can be injected into the midgut epithelial cells. Once inside a host cell, the sporoplasm is referred to as a *meront* (Fig. 30). The meront is the assimilative stage of the microsporidian, i.e., the stage of nutrient uptake. It divides mitotically; and, depending on the microsporidian and the host, infections can be localized (e.g., to a specific tissue or organ), chronic or sublethal, or systemic and lethal. How meronts spread from cell to cell is not well understood. One possibility is that the meronts infect *hemocytes* (blood cells associated with a hemocoel), and the microsporidian gains access to various regions of the hemocoel via the hemocytes. Although many tissues can be infected, fat bodies are a common site of accumulation (Fig. 28). At some point, the pathogen transforms from the meront to the sporant state, i.e. spore stage. The process is termed *sporogony*, and in most species, sporogony is also accompanied by some degree of division. Following division, the extrusion apparatus begins to develop. As the extrusion apparatus nears complete formation, the cells decrease in size, and the chitinous endospore layer develops. In some species, autoinfective spores are produced. These spores ger-

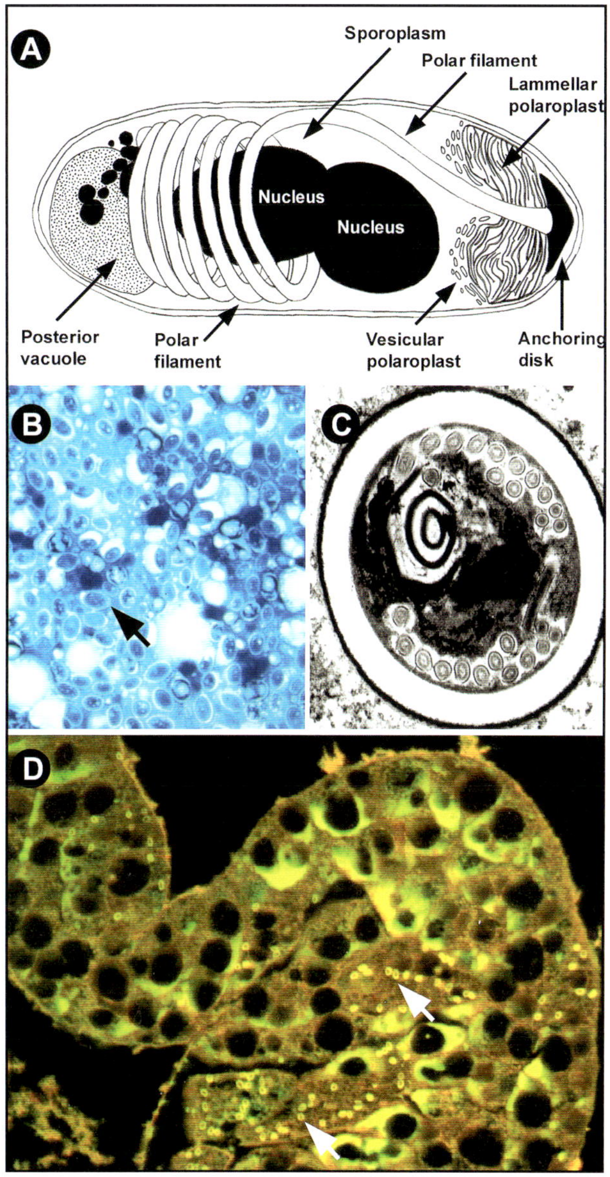

Fig. 28. An undescribed *Nosema* species infecting southwestern corn borer, *Diatraea grandiosella* (Lepidoptera: Crambidae). (A) Diagrammatic representation of a longitudinal section of a spore showing the internal structures as seen under an electron microscope. Note the presence of two nuclei and the characteristic coiled polar filament, polaroplast, and anchoring disc. (B) A light micrograph of spores (arrow) stained with buffalo black. (C) Electron micrograph of a cross-section of spore showing the coiled polar filament and lammellar polaroplast. (D) A confocal laser micrograph of a fat body showing the accumulation of spores.

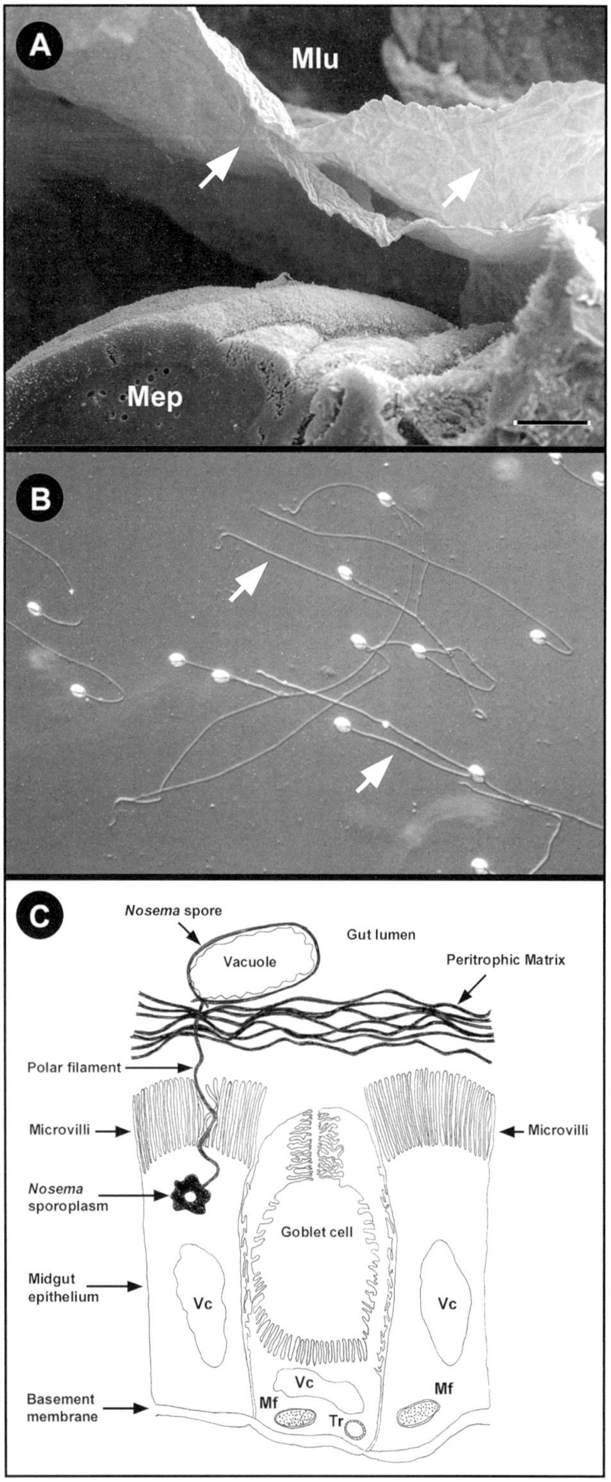

Fig. 29. The *peritrophic matrix* (PM) in relation to infection of the midgut epithelium by *Nosema* and *Paranosema*. (A) The laminar sheet-like appearance of the PM (arrows) in tobacco budworm, *Heliothis virescens* (Lepidoptera: Noctuidae). Note the spatial relationship of the PM to the midgut epithelium (Mep) and midgut lumen (Mlu). (B) Extruded polar filaments from *Paranosema locustae*. (C) Drawing illustrating the process by which the polar filament breaches the PM and injects the pathogen sporoplasm into the midgut epithelium. [Image B provided by Dan L. Johnson (Agriculture and Agri-Food Canada, Lethbridge, AB, Canada).]

minate immediately in an attempt to infect cells within the same host and cause systemic infections.

Insects infected with microsporidia can be altered in color, size, form, and activity depending on the tissues and organs infected. Malaise is a common symptom, and abnormal behavior may occur. For example, honey bees, *Apis mellifera* (Hymenoptera: Apidae), infected with *Nosema apis* are physiologically older than healthy bees; and they start guarding, foraging, dancing, and orientation flying at an earlier age than healthy bees. Furthermore, the hypopharyngeal gland typically becomes infected, and these bees are unable to produce royal jelly needed to maintain the health of the queen (Tanada and Kaya 1993). Reduced size and fecundity also are common signs of disease resulting from sublethal infection by microsporidia.

a. Nosema *and* Paranosema. These genera have more than 150 described species with hosts representing at least 12 orders of insects (Becnel and Andreadis 1999). A number of *Nosema* and *Paranosema* spp. are important pathogens of insects in rearing settings. Examples include: *Nos. algerae* (pathogen of Culicidae), *Nosema apis* (*A. mellifera*), *Nos. bombycis* (causal agent of pébrine disease in *B. mori* larvae), *Nosema fumiferana* [spruce budworm, *Choristoneura fumiferana* (Lepidoptera: Tortricidae)], *Nosema gasti* (*A. grandis*), *Nosema heliothidis* (Lepidoptera: Noctuidae: Heliothinae), and *Nosema pyrausta* (*O. nubilalis*) (Brooks 1988). *Par. locustae*, a pathogen of acridids (grasshoppers and locusts), along with *Par. grylii*, a parasite of crickets were recently transferred from *Nosema* to the new genus, *Paranosema* (synonym: *Antonospora*) (Sokolova et al. 2003). Two basic types of life cycles occur in *Nosema* and *Paranosema* species. The sporoplasm of *Nos. fumiferana*, *Nos. apis*, and *Nosema scripta* forms a binucleate meront that undergoes several cycles of division to produce a series of daughter diplokaryotic cells (Fig. 30). In other species including *Nos. algerae*, *Nos. gasti*, and *Par. locustae* the invasive sporoplasm produces an initial cycle of unikaryotic meronts (Brooks 1988). Depending on the species, meronts may be restricted to the gut or they may spread to other tissues (e.g., fat bodies, Malpighian tubules) and cause systemic infections. Infected tissues often appear abnormal (e.g., swelling). From binucleate meronts, sporonts are formed. Nuclear division within the sporont gives rise to a tetranucleate sporont, which divides to produce two *sporoblasts* (prespore stage), each of which develops into a spore.

Nosema and *Paranosema* spp. can enter the host in three ways: 1) *per os*, 2) transovarially, and 3) directly through the insect cuticle. Oral entrance includes feeding of insects on spore-contaminated food or feeding on dead or moribund insects. Transmission from adults to offspring is accomplished either by infection of hatching larvae when they eat contaminated egg chorion, or by integration of meronts, sporonts, and/or spores into the egg or embryo within the female's reproductive tract. In some instances (e.g., *H. zea*), infected females may produce 100% infected offspring (Sikorowski unpublished data). Although experimental evidence is lacking, there is anecdotal evidence that spores of *Nosema* also may be disseminated on lepidopteran scales in insectary settings.

Some species of *Nosema* and *Paranosema* can rapidly kill their host; however, the majority of species cause nonlethal, debilitating diseases. Nonlethal diseases are of particular significance to rearing quality insects (Sec. IV. B). Signs of disease are variable but may include: swelling of infected tissues, reduced insect size, delayed development, loss of appetite, inability to pupate, and various types of deformities. The pathogen can also influence the physiology of the host. For example, Thompson and Sikorowski (1979) reported that infected *H. zea* larvae accumulated fatty acids more rapidly than healthy larvae in the first 10 days, but then concentrations decreased rapidly until death. Because of the often inconspicuous evidence of infection, accurate diagnosis requires microscopic examination of susceptible tissues such as fat bodies and midgut epithelium (e.g., Fig. 3C).

b. Vairimorpha. Boucias and Pendland (1998) refer to species of *Vairimorpha* as microsporidians with a dual personality. All *Vairimorpha* spp. exhibit two sporulation sequences so that each has two morphologically distinct spore types. It is important to note that *Vairimorpha* is only one of sixteen genera of microsporidia that have polymorphic sporulation sequences (Boucias and Pendland 1998). *Vairimorpha necatrix* is the most commonly encountered species in insectaries (Fig. 31). It was first reported from armyworm, *Pseudaletia unipuncta* (Lepidoptera: Noctuidae), larvae from Hawaii (Tanada and Chang 1962); but it is now know to infect a number of lepidopterans. Although it primarily affects Lepidoptera, species from the Hymenoptera and Diptera are also susceptible (Maddox 1966, Becnel and Andreadis 1999). We have previously mentioned the polymorphic sporulation of *Vairimorpha* species. In *V. necatrix*, one spore

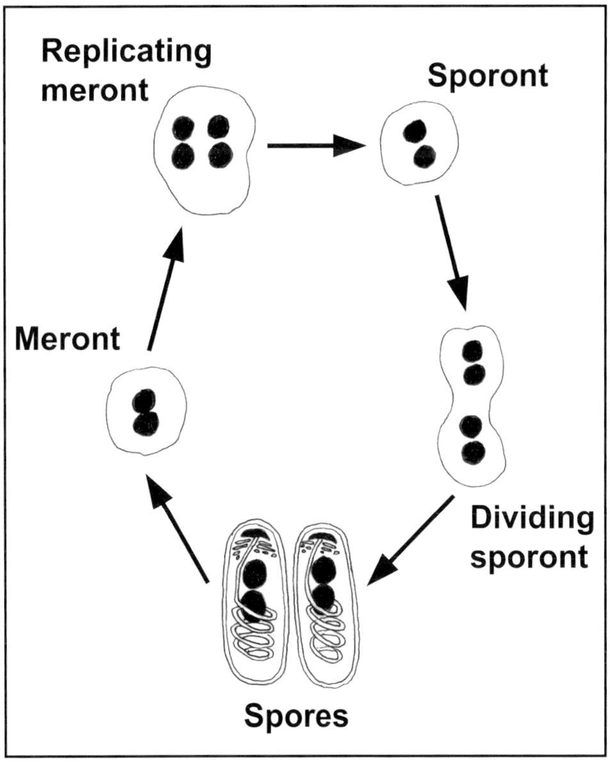

Fig. 30. Diagram of the generalized life cycle of a microsporidian. In this example, both the meronts and sporonts are dinucleate but become tetranucleate during replication. Both these stages occur intracellularly within the infected insect. In contrast, the thick-walled spores can exist outside of the host (i.e., they act as a survival structure), and they are the infectious stage of the microsporidian. Note the coiled polar filament within the spores.

form has dimensions (length, *L*; width, *W*) that are highly dependent upon temperature (at 32°C, $L = 2.4$ to 5.2 μm and $W = 2.2$ μm, but at 15°C, $L = 4.2$ to 6.9 μm and $W = 2.4$ μm) (Sikorowski and Lawrence 1997). The other spore form is smaller and less variable in size regardless of temperature ($L = 2.8$ to 4.0 μm and $W = 1.7$ to 2.4 μm).

Once ingested by the host, spores of *Var. necatrix* germinate in the midgut. A similar mechanism to that described for *Nosema* and *Paranosema* is utilized. Insects, especially young individuals exposed to large numbers of spores, often become septic (Maddox 1966). Mechanical damage to the PM and midgut epithelium due to extrusion of the polar filaments allows ingress of bacteria present in the gut lumen. Such insects typically die rapidly. In insects ingesting smaller numbers of spores, midgut tissues are initially infected, and meronts spread to hemocytes and fat bodies (Boucias and Pendland 1998; Fig. 31). These insects may not die, and they exhibit signs of chronic infection (Chu and Jaques 1979). However, the severity of infection depends on spore dosage, temperature, and larval age. In afflicted insects, feeding is usually normal at first but decreases to very little for the last few days before death. In advanced stages of infection, fat bodies are extensively affected. Afflicted fat body cells may enlarge many times, giving the fat tissue a lobed appearance. The abnormally large, white fat bodies are often conspicuous macroscopically, being evident through the cuticle and due to a dorsal swelling of the last two or three abdominal segments (Pilley 1976). Microscopic examination of the fat bodies reveals the presence of polymorphic spores. *Vairimorpha* is transmitted to the next generation both on and in the eggs of infected adults.

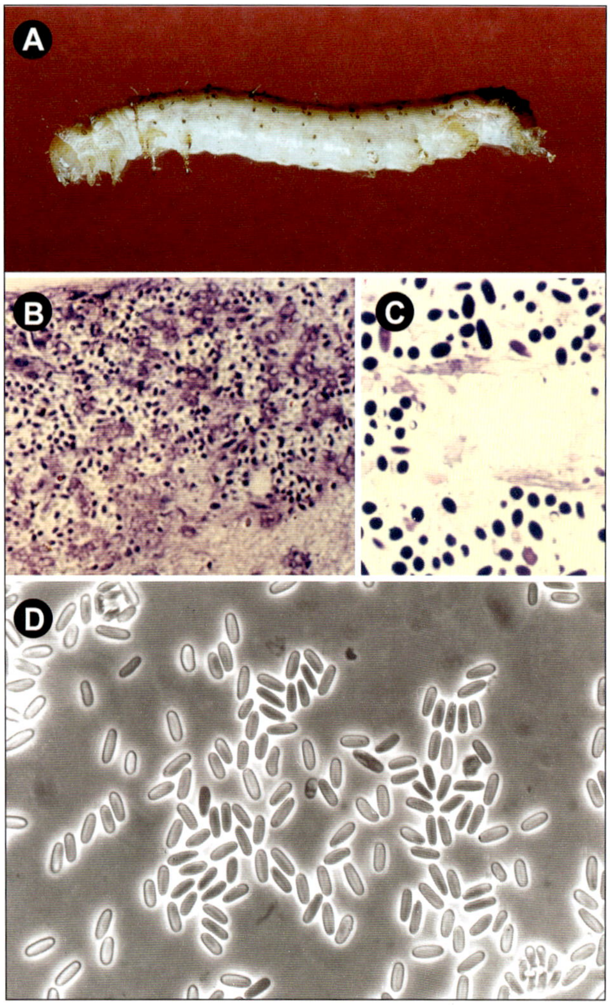

Fig. 31. The microsporidian *Vairimorpha necatrix* infecting the tobacco budworm, *Heliothis virescens* (Lepidoptera: Noctuidae). (A) Infected larva showing white hypertrophic fat bodies through the integument. (B–C) Light micrographs of sections of the fat body showing spores and total loss of cell differentiation. (D) Light micrograph of spores. [From Sikorowski and Lawrence (1997) with permission from the Mississippi Agricultural and Forestry Experiment Station, Mississippi State, MS.]

c. Other. There is a considerable diversity among microsporidian genera and species that infect insects (Undeen and Vávra 1997). Although, relatively few of these genera have been reported from insects in rearing systems, it is most likely that many of these microsporidians do infect insects in rearing settings but that they are not being detected. Factors contributing to this failure include the following: the technical nature of identifying microsporidians, that they are obligate parasites, and that infections are frequently sublethal. Characters that are used to identify microsporidia to genus include: 1) type of spore formed, 2) number of nuclei in meronts, 3) number of nuclei in sporonts, 4) presence of a sporophorous vesicle, 5) numbers of spores formed in the sporophorous vesicle, 6) spore shape, 7) spore length and width, 8) polar filament type, and 9) the tissues affected. The *sporophorous vesicle* (SPV) is a membrane-like structure that encloses spores in some species. Under light microscopy, the SPV-forming microsporidia are seen in packets containing a specific number of spores (one to many) (Undeen and Vávra 1997). However, the SPV is a delicate structure, and its presence/absence can only be definitely concluded using a TEM. Examples of microsporidia producing an SPV include *Amblyospora*, *Octospora*, and *Thelohania*. Detailed information on individual genera of microsporidia is found in Sprague et al. (1992).

E. Nematodes

Nematodes are infrequently encountered as entomopathogens of reared insects. However, insects brought from the field to establish a new colony may be infected. Currently, there are six orders and 30 families of nematodes associated with insects, but many of these are animal and plant parasites that use insects as vectors (Kaya and Stock 1997). The most common nematodes whose primary hosts are insects fall into seven families: Mermithidae, Tetradonematidae, Sphaerulariidae, Allantonematidae, Phaenopsitylenchidae, Steinernematidae, and Heterorhabditidae. Symptomatology of disease caused by nematodes includes sterility, reduced fecundity, delayed development, aberrant behavior, color changes, and host death. Some nematodes that are associated with one host are difficult to propagate on artificial medium (e.g., some terrestrial mermithids). Other less specialized forms can be mass produced on artificial media for control of agricultural pests (e.g., Steinernematidae). Here we limit our discussion to the Mermithidae, Steinernematidae, and Heterorhabditidae.

1. Mermithidae. Common genera within the Mermithidae include *Filipjevimermis*, *Gastromermis*, *Hexamermis*, *Mermis*, and *Romanomermis*. Merimithids parasitize many orders of insects as well as other classes of arthropods, including Acarina (Ixodoidea), Arachnida, and Crustacea (Kaya and Stock 1997). The vast majority of mermithids do not reach adulthood in the hemocoel of the host. They emerge from the host as post-parasitic juveniles, molt to the adult stage, mate, and produce progeny (Tanada and Kaya 1993). The parasitic juveniles are short-lived relative to the free-living stages. In most cases, the infective juveniles directly penetrate the host's exterior integument using a stylet. Once in the hemocoel, the mermithid absorbs food directly through its cuticle. Mermithids can be quite large, and can often be seen within the hemocoel of living arthropods possessing transparent cuticles (Fig. 32A). The post-parasitic juveniles of most mermithids emerge from the host to mature in the soil to free-living, nonfeeding adults. Host death occurs because of the trauma caused by the nematode exiting through the host's cuticle (Kaya and Stock 1997).

2. Steinernematidae. Steinernematids are obligate pathogens and are characterized by a mutualistic symbiosis with bacteria in the genus *Xenorhabdus* (Sec. II. B. 2. c). Steinernematids are parasites of a wide diversity of invertebrates, and a number of *Steinernema* species have been described from natural infections of insects. *Steinernema carpocapsae* has been isolated from seven different species of field-collected insects and has been experimentally transmitted to over 200 different insect species (Poinar 1979). The infective juveniles gain access to the hemocoel of susceptible insects via natural openings such as the mouth, anus, and spiracles. Infective juveniles carry bacterial cells of *Xenorhabdus* within their intestines; once the nematodes gain access to the hemocoel, the bacteria are released. Once released into the hemocoel, septicemia kills the host within 48 h. The nematodes develop rapidly to the adult stage, mate, and produce eggs. As soluble nutrients in the cadavers are depleted, the progeny of the second and third generations develop into infective juveniles, which exit from the cadaver and seek new hosts (Fig. 32C). While steinernematids can kill their host without *Xenorhabdus*, they are unable to reproduce in the absence of the bacterium (Tanada and Kaya 1993). Conversely, *Xenorhabdus* cannot invade the hemocoel of its host without the nematode.

3. Heterorhabditidae. Members of the genus *Heterorhabditis* are obligate parasites of insects and share with the steinernematids the specialized character of carrying mutualistic bacteria. Heterorhabditids are similar to steinernematids in general life cycle and gross morphology (Wouts 1984), but heterorhabditids form a symbiosis with bacteria in the genus *Photorhabdus* (Tanada and Kaya 1993). Insects killed by *Photorhabdus* typically turn dark brick red (Fig. 17). Sometimes, the cadavers are purple, orange, yellow, or green. In the dark, cadavers typically luminesce due to the presence of pigments produced by the bacterium. Another difference between steinernematids and heterorhabditids is that infective juveniles of the latter are *hermaphroditic* (i.e., having both male and

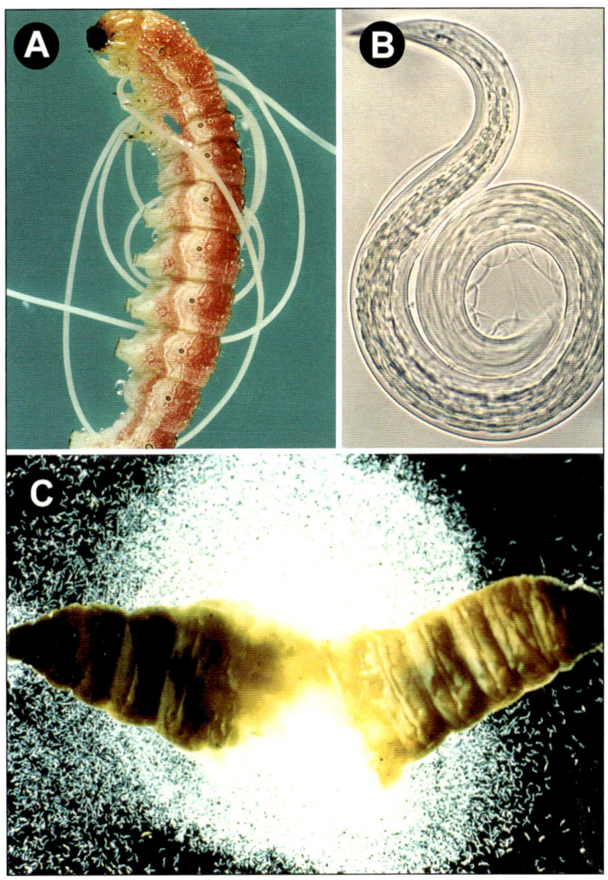

Fig. 32. Insects parasitized by entomopathogenic nematodes. (A) A pre-adult *Hexamermis* sp. emerging from a bollworm, *Helicoverpa zea* (Lepidoptera: Noctuidae), larva. (B) An infective juvenile of *Steinernema carpocapsae* from an *H. zea* larva. (C) Infective *S. carpocapsae* juveniles emerging from a greater wax moth, *Galleria mellonella* (Lepidoptera: Pyralidae), larva. [From Sikorowski and Lawrence (1997) with permission from the Mississippi Agricultural and Forestry Experiment Station, Mississippi State, MS.]

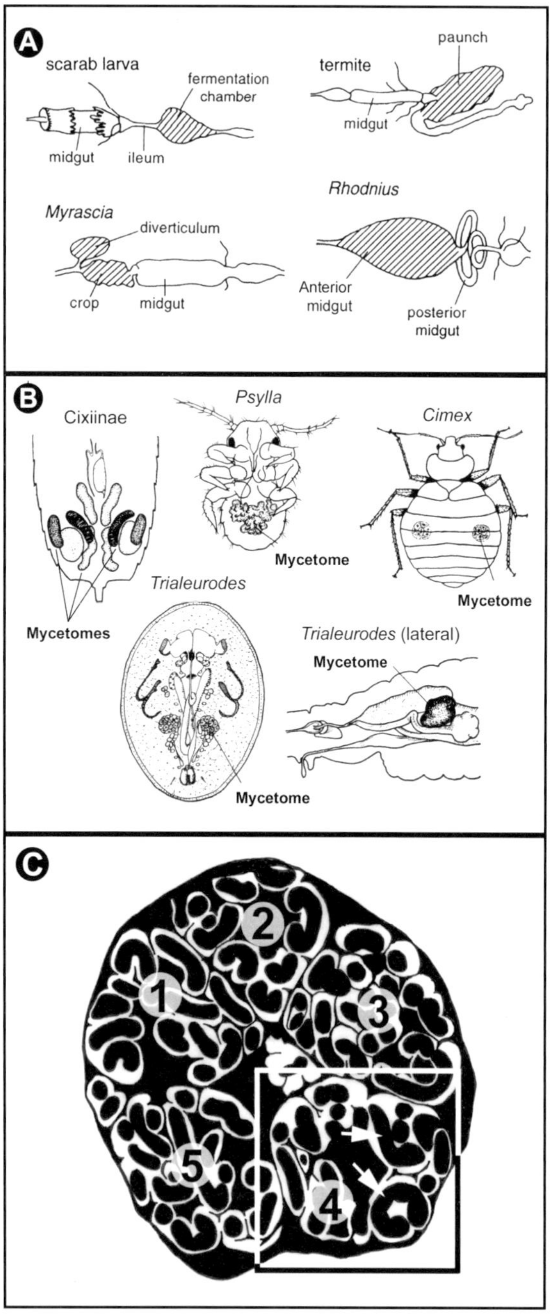

Fig. 33. Symbiotic relationships between microorganisms and insects. (A) Microorganisms within the gut lumen. This is often associated with modifications of the alimentary canal including enlarged hindgut (e.g., fermentation chambers of scarabs and paunches of termites), foregut (e.g., diverticula), and midgut. [From Chapman (1998) with permission from Cambridge University Press, Cambridge, England.] (B) Location of mycetomes in various insects. Note the four symbiote-harboring mycetomes in Cixiinae, the single mycetome in the abdomen of *Psylla buxi* nymphs, the pair of mycetomes in the bedbug, *Cimex lectularius* (Hemiptera: Cimicidae), and the pair in the greenhouse whitefly, *Trialeurodes vaporariorum* (Hemiptera: Aleyrodidae). [From Steinhaus (1949) with permission from the McGraw-Hill Book Company.] (C) A cross-section of a mycetome from the sawtoothed grain beetle, *Oryzaephilus surinamensis*) (Coleoptera: Silvanidae). Within the mycetome are five mycetocytes (1–5) containing pleomorphic bacterial cells (arrows). [Drawing based on Steinhaus 1947.]

female sexual characteristics and organs) rather than being male or female (Kaya and Stock 1997). Therefore, only one infective *Heterorhabditis* juvenile is required to infect a host. Four to five days after death of the host, the nematodes reach mature size and become giant hermaphrodite females. Eggs laid by the hermaphrodites produce juveniles that develop into males and females within the body of the host.

III. Beneficial Symbioses

Before considering the deleterious effects of disease agents on insects, the beneficial effects of associations between microorganisms and insects will be summarized briefly. A multitude of microorganisms exist in association with insects, and in many instances this relationship is essential to the normal function of the insect. When two dissimilar organisms form a relationship, regardless of the result of such an association, it is referred to as a *symbiosis*. Earlier in this chapter, *disease* was defined as an abnormal, deleterioous state resulting from a dynamic interaction between an organism and its environment. Some microbial ecologists use *parasitic symbiosis* to describe a situation in which the microorganism harms the host. This also is referred to as *pathogenism*. When the relationship is favorable to both the host and the microorganism, the relationship is referred to as a *mutualistic symbiosis*. Putative advantages to insects forming mutualistic relationships with microorganisms include: 1) the creation or maintenance of a proper environmental conditions within the alimentary canal to facilitate digestion; 2) enhanced nutrition resulting from the action of microbial enzymes on food components; 3) a source of macronutrients such as carbohydrates, protein, or lipids provided by microbial biomass; 4) a source of essential micronutrients such as vitamins or sterols; 5) production of pheromones or defensive compounds; and 6) protection from pathogens. In insects that form mutualistic relationships with microorganisms, it is often important to facilitate these relationships in reared insects. Often management strategies applied against other microorganisms can have a detrimental impact on mutualistic microorganisms. For example, the broad-spectrum antibacterial agent tetracycline included in the diet to control bacterial contaminants also may have a detrimental effect on mutualistic bacteria and thus compromise the health of the insect.

Mutualistic microorganisms can exist external to the insect (e.g., associated with food substrates) or within the alimentary canal. In insects in which microorganisms occur within the alimentary canal, the morphology of the canal often reflects the relationship (Fig. 33A). For example, in pollen- and nectar-feeding lacewing adults, the *diverticulum* (i.e., pouch associated with the foregut) is enlarged, and large numbers of yeasts are housed within the highly convoluted and folded structure (Woolfolk and Inglis 2004; Woolfolk et al. 2004; Fig. 34). Furthermore, exceptionally large tracheal trunks attach to the lateral side of the diverticulum presumably to support the high demand for gas exchange within this organ.

Microorganisms can also exist within the body cavity, inter- or intracellularly. In many instances, specialized structures termed *mycetomes* are formed to house the symbionts (Fig. 33B, C). The cells making up the mycetome and containing symbionts are known as *mycetocytes*. These terms are used by some regardless of whether they house fungi or bacteria. However, some researchers now use the term *bacteriocyte* for cells containing mutualistic bacteria. In other mutualistic relationships, microorganisms occur within cells and conspicuous structures are not formed. As mentioned earlier, the majority of bacteria cannot be cultured on microbiological media, and this holds true for many of the bacteria forming mutualistic relationships with insects. The application of molecular techniques has revolutionized the study of mutualistic symbiosis in insects. A detailed description of molecular methods that can be used is beyond the scope of this chapter, and the reader should consult any number of references for more information (e.g., Bourtzis and Miller 2003, 2006).

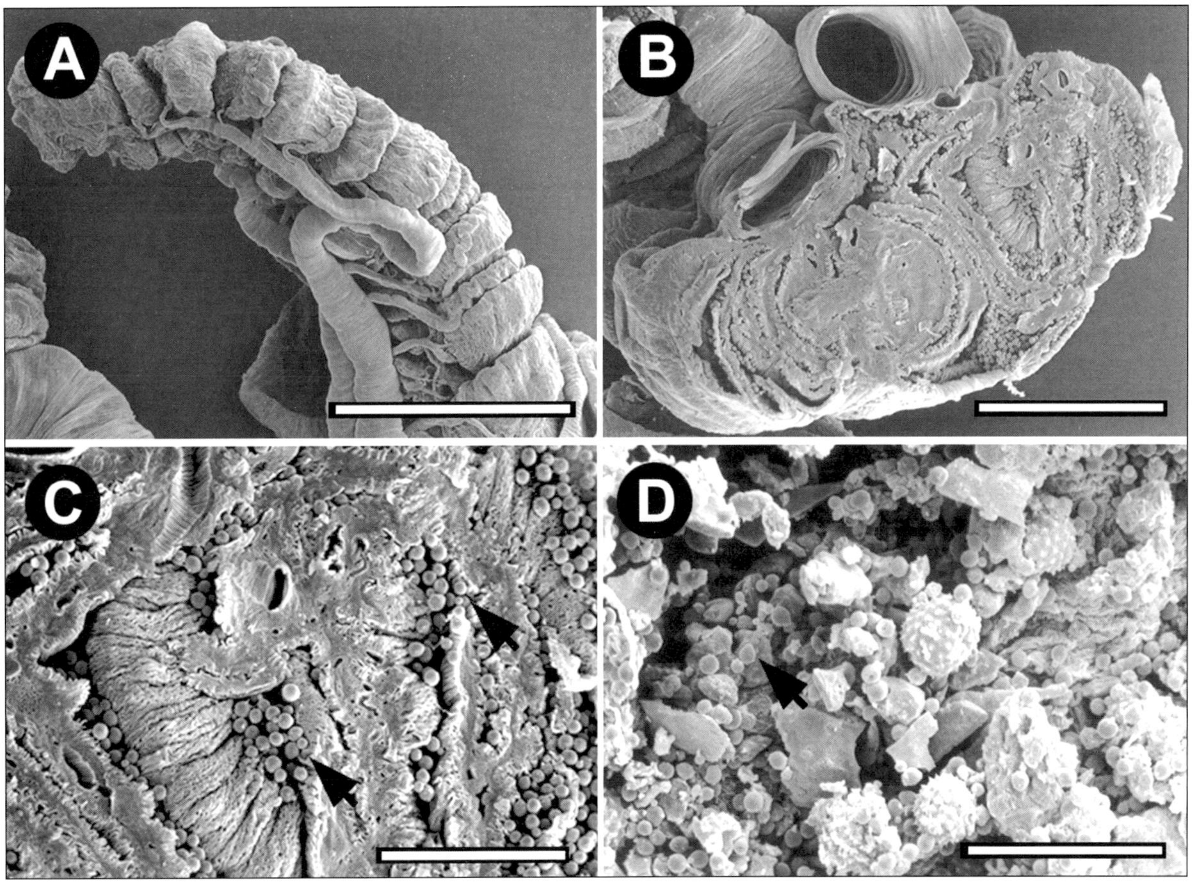

Fig. 34. Scanning electron micrographs illustrating a mutualistic symbiosis between yeasts and lacewing (*Chrysoperla rufilabris*) adults. (A) Diverticulum showing large tracheal trunk attached to the lateral surface, bar = 200 µm. (B) Cryofractured section of a diverticulum showing extensive diverticular folding and two tracheal trunks, bar = 100 µm. (C) Large numbers of yeast cells (arrows) located within the diverticulum, bar = 50 µm. (D) Pollen grains along with yeast cells in the diverticulum, bar = 50 µm. [From Woolfolk et al. (2004) with permission from Elsevier.]

IV. Disease Manifestations

A. Acute

The term *pathogenicity* refers to the ability of a pathogen to infect and cause disease and is a qualitative term. In contrast, *virulence* refers to the degree of pathogenicity and is a quantitative term. Highly virulent pathogens cause *acute disease* in which disease progression is brief but severe and frequently lethal. Many viruses (e.g., NPV), bacteria (e.g., *Bac. thuringiensis*), fungi (e.g., *Bea. bassiana* and *Met. anisopliae*), and nematodes can cause acute infections; but their virulence is often related to dose (Fig. 35). As a result of the rapidity and severity of disease, acute infections are of obvious concern in insectaries. Steinhaus (1953) suggested that if disease occurs in greater than 25% of the insects concerned, fighting the disease in the remainder of the colony is almost always useless. Under such circumstances, he suggests the complete destruction of the entire stock and beginning again with freshly collected or known disease-free insects. From a disease management perspective, the severity of acute diseases usually precludes the possibility that they escape detection by insectary personnel (Sec. V) and allows the appropriate management strategies to be employed before the incidence becomes too high.

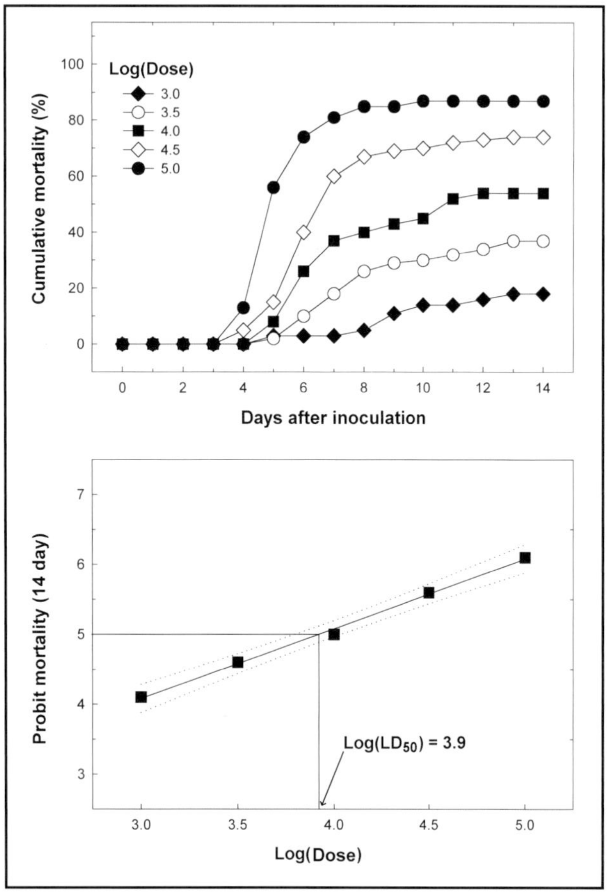

Fig. 35. A typical dose-mortality response of a virulent pathogen inciting acute disease. (A) Disease progress curves. (B) A probit mortality graph showing a log(median lethal dose) [i.e., log(LD_{50})] of 3.9 and thus a median lethal dose of 7.9 x 10^3 propagules per insect. The linear equation describing the probit-log dose regression is $y = a + bx$, where a = 1.0, b = 1.08, and r^2 = 0.99 for *y*, probit, and *x*, log(dose). Fiducial limits of the regression (95%) are presented as dotted lines.

B. Chronic

Chronic disease persists over a long time or recurs frequently and usually is nonlethal (Fig. 36). Another term that is often confused with chronic is the term *latent. Latent infection* refers to a situation where a pathogen has infected a suitable host, but the infection is dormant and has not manifested itself. While latent infections are common in mammals (e.g., HIV), they are less common in insects. Of the latent infections that do occur in insects, most are viruses and, to a lesser extent, bacteria and protozoa.

While insects do not die from chronic infections, the pathogen often reduces the fitness of the insect. For example, sublethal infections may have the following deleterious effects: 1) reduced adult size, 2) reduced food consumption and frass production, 3) delayed development, 4) reduced fecundity and egg hatch, 5) decreased anabolism (e.g., production of storage compounds such as fatty acids and glycogen), 6) increased metabolic rates (e.g., increased O_2 consump-

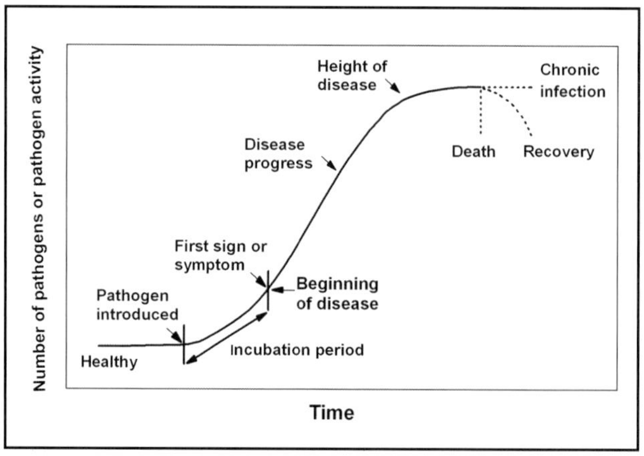

Fig. 36. Typical course of infection in an insect host. Note that there are three possible outcomes: death, recovery, and chronic infection. [From Tanada & Kaya (1993) with permission from Elsevier.]

tion), and 7) increased mortality and reduced size of offspring. Arguably, the greatest threat to insectaries does not come from highly virulent pathogens that cause conspicuous disease outbreaks but rather from less virulent pathogens that cause undetected chronic infections. Since chronically infected insects are almost always inferior to noninfected individuals, their use in field releases and research can lead to unsatisfactory or invalid conclusions, respectively. Furthermore, it should be emphasized that when chronically infected insects are exposed to a stressor (e.g., mechanical damage, environmental, or nutritional stress), outbreak of an acute disease can rapidly occur. This is a common scenario in insectaries and underlines the importance of detection and management of chronic diseases.

Examples that illustrate how a chronic infection can adversely affect the productivity of a rearing program include *H. virescens* infected with CPV (Sikorowski and Lawrence 1997) and *D. grandiosella* sublethally infected with *Nosema* (Inglis et al. 2003a). *H. virescens* larvae infected with CPV weigh only 5%

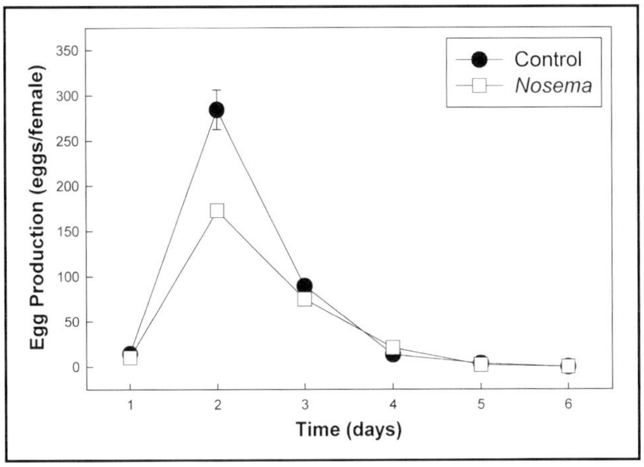

Fig. 37. Egg production (eggs/female) over time by southwestern corn borer, *Diatraea grandiosella*) (Lepidoptera: Crambidae), infected with *Nosema* and mated to healthy males. Bars represent standard errors (n = 3), and error bars not shown are within symbols. [From Inglis et al. (2003a) with permission from the Entomological Society of America.]

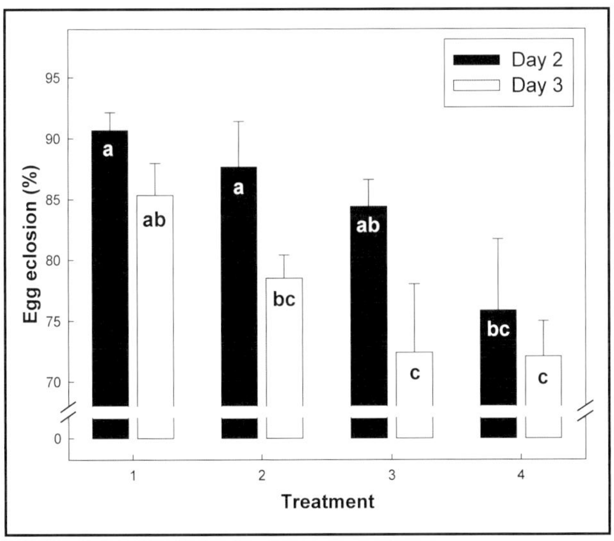

Fig. 38. Prevalence (%) hatch in southwestern corn borer, *Diatraea grandiosella*)(Lepidoptera: Crambidae) eggs produced by females assigned to the following treatments: (1) noninfected ♀ mated with noninfected ♂, (2) ♀ infected with *Nosema* mated with noninfected ♂, (3) noninfected ♀ mated with infected ♂, and (4) infected ♀ mated with infected ♂. Vertical bars represent standard errors of the mean (n = 3). Bars with different letters are significantly different (P ≤ 0.05). [From Inglis et al. (2003a) with permission from the Entomological Society of America.]

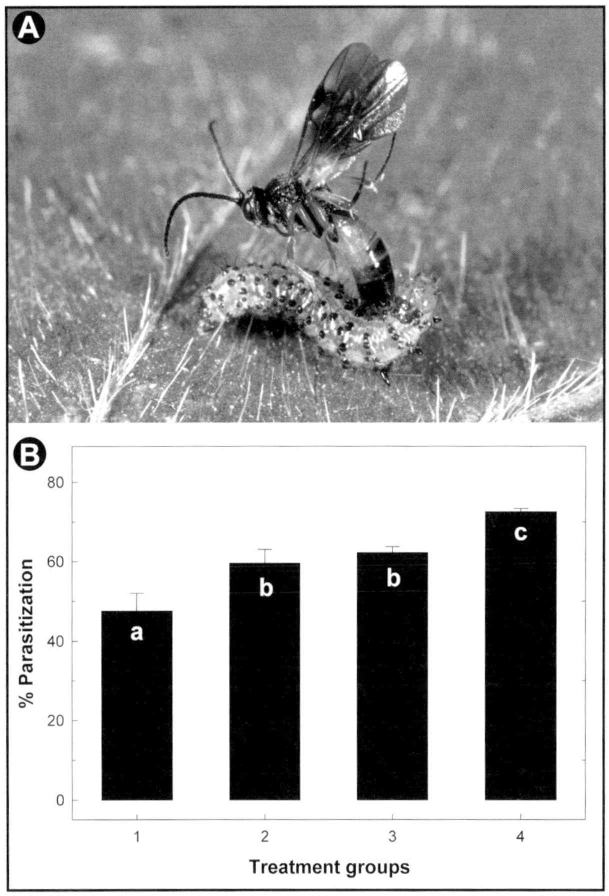

Fig. 39. Parasitism of the tobacco budworm, *Heliothis virescens* (Lepidoptera: Noctuidae), by the parasitoid *Microplitis croceipes* (Hymenoptera: Braconidae). (A) *M. croceipes* female ovipositing into a *H. virescens* larva. [Image provided by Merle Shepard, Clemson University, Clemson, SC.] (B) Effect of infection by nonoccluded baculovirus (NOBV) on parasitism of *H. virescens* larvae by *M. croceipes* females. Treatment groups represent different pairings of virus-infected and healthy parents: (1) infected ♀ and infected ♂, (2) healthy ♀ and infected ♂, (3) infected ♀ and healthy ♂, and (4) healthy ♀ and healthy ♂. Vertical bars indicate standard error of the means, and histogram bars not containing the same letter are significantly different (P ≤ 0.05). [From Sikorowski et al. (1997) with permission from Kluwer Academic Publishers.]

less than healthy larvae. However, infected larvae require 4 days longer to pupate, pupae weigh an average of 60 mg less, adult lifespan is 9 days less, and 68% fewer eggs are produced relative to noninfected individuals. In *D. grandiosella*, *Nosema* is highly infectious to larvae, but even high doses as high as 10^7 spores per larva do not result in significant mortality or reduced weight. Nevertheless, egg production of infected females is reduced by 45% (Fig. 37), and the percentage hatching is also reduced (Fig. 38).

An example of how a chronic infection can affect the quality of a mass-reared insect involves the braconid parasite *M. croceipes* infected with an NOBV (Hamm et al. 1988, 1992, Chittihunsa and Sikorowski 1995b, Sikorowski et al. 1997). *M. croceipes* is a larval endoparasitoid that attacks *H. virescens* and *H. zea*, two important agricultural pests (Fig. 39A). The Mc-NOBV (Fig. 4) is transmitted from adults to offspring on and/or in eggs and from individual to individual by ingestion of virions in frass. Furthermore, the virus can replicate in *H. virescens* larvae, and infection of *M. croceipes* can occur from infected host tissues during parasitism. Despite few indications of disease in *M. croceipes* (insects are only slightly smaller and wing deformities occur infrequently), the infection of wasps released in the field can severely decrease their effectiveness as biological control agents. Infection with Mc-NOBV has been shown to decrease the parasitism potential of adults (Fig. 39B), increase the time from parasitism to emergence of F_1 larvae; prolong larval development and pupation times for F_1 individuals; and cause decreased longevity of F_1 adult females (Fig. 40). Furthermore, the release of infected individuals can disseminate the virus to other wasps already present in the field setting. Insectaries with dense populations of wasps facilitate transmission of the virus, and a single Mc-NOBV infected wasp may contaminate the food source (frequently sugar water) with virions and thus infect the entire population in a rearing container. Hamm et al. (1992) successfully used a quarantine method to eliminate the Mc-NOBV from a laboratory colony of *M. croceipes*. See Sec. VII for more detail.

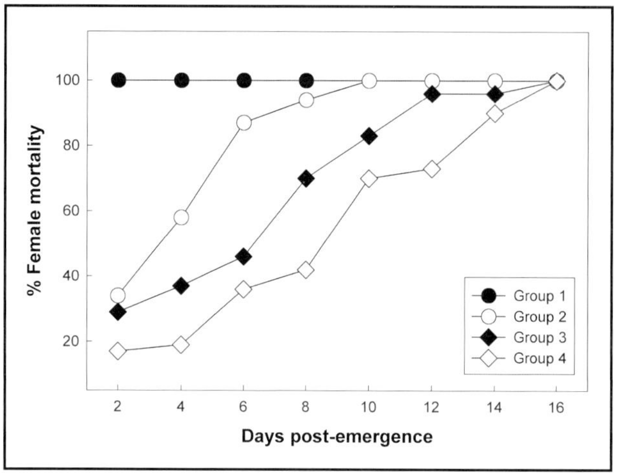

Fig. 40. Effect of nonoccluded baculovirus (NOBV) infection of the parasitoid *Microplitis croceipes* (Hymenoptera: Braconidae), on cumulative percent mortality of adult females. Treatment groups represent different pairings of virus-infected and healthy parents: (1) infected ♀ and infected ♂; (2) healthy ♀ and infected ♂; (3) infected ♀ and healthy ♂; and (4) healthy ♀ and healthy ♂. [From Sikorowski et al. (1997) with permission from Kluwer Academic Publishers.]

V. Disease Development

Disease is a dynamic process, and the discipline of *epizootiology* is concerned with the study of the occurrence of disease in time and space. Increase in disease prevalence usually takes the form of an *epizootic wave* (Fig. 41). The epizootic wave is divided into three main phases including the: 1) *pre-epizootic*; 2) *epizootic*; and 3) the *post-epizootic* (Steinhaus 1949, Tanada and Kaya 1993). The *enzootic* state is the "normal level" of disease in a population, which remains low and relatively constant for long periods of time. In contrast, epizootics are relatively sporadic, limited in time, and characterized by a sudden increase in disease prevalence. The factors that initiate and maintain an epizootic are complex and often involve a four-way interaction between the pathogen, host, environment, and time (Fig. 42). In insectaries, the insect manager attempts to disrupt one or more of the components of this interaction in an attempt to prevent disease development. Therefore, an adequate understanding of the factors controlling epizootic development is essential to effectively manage disease. Environmental factors (e.g., conditions of temperature, humidity, and light) are crucial, yet they are often overlooked. Host factors including age, density, behavior, and susceptibility (i.e., physiological status) must also be considered. As mentioned previously, exposure of insects to conditions that cause physiological stress can have devastating consequences regarding the initiation of epizootics. A stressor can be any of a number of factors, but it always acts on the host and not directly on the entomopathogen (Tanada and Kaya 1993). In contrast, a *disease incitant* is a factor that acts directly on the pathogen. It has long been recognized that stressed insects are more susceptible to entomopathogens and less likely to survive infection. For example, workers associated nutritional stress with outbreaks of diseases in honey bees and silkworms in the early 1800s (Tanada and Kaya 1993). In modern times, Steinhaus (1958, 1959) experimentally observed that a number of factors (e.g., crowding, abrasives, and chemicals) enhanced the efficacy of the entomopathogenic bacterium *S. marcescens*. It is often not possible to have a pathogen-free colony, and therefore it is imperative that good husbandry methods are used to minimize physiological stress and prevent disease outbreaks.

Regarding the pathogen component of the interaction, salient factors influencing disease development include transmission, virulence, ability to persist, capacity to disperse, and dose. There are two basic types of transmission: 1) vertical and 2) horizontal. *Vertical transmission* is transmission from parent to offspring. Most commonly, transmission occurs from the female parent to her offspring. Transmission from the male parent does occur but it is relatively rare. Vertical transmission is common among the viruses, bacteria (e.g., mutualistic symbionts), and microsporidia but can occur with other entomopathogens as well. There are two primary types of vertical transmission of significance to insect rearing: 1) *transovum* refers to the transmission of a microorganism on the egg; and 2) *transovarial* refers to the transmission of a microorganism within the egg (i.e., the microorganism enters the egg while it is still in the ovary). In *D. grandiosella* infected with *Nosema*, spores are commonly detected on the surface of the chorion, within the ovarian tissues, and within the eggs themselves (Fig. 43). Transovarial transmission represents the most difficult type of vertical transmission to manage in insectaries (Sec. VII).

For *horizontal transmission*, the route of infection is an important consideration. Viruses, bacteria, protozoa, most nematodes, and some fungi infect *per os*; and infective propagules must be ingested. In contrast, most fungi infect through the integument and most nematodes via natural openings including spiracles and the anus. In insectaries, infectious propagules may become airborne and contaminate diets, cages, and insects themselves. In communal rearing systems, a single contaminated and/or infected insect can provide enough inoculum of *Ser. marcescens* to infect all the individuals within the same enclosure (e.g., Sikorowski and Lawrence 1998). Consumption of infested frass and food and cannibalism can all contribute to the horizontal transmission of entomopathogens. Insectary design and rearing systems have a major role in preventing horizontal transmission (Chap. 2 *Insectary Design and Construction* and Chap. 11 *Insect Rearing Production Systems, A Case Study: The Southwestern Corn Borer*.

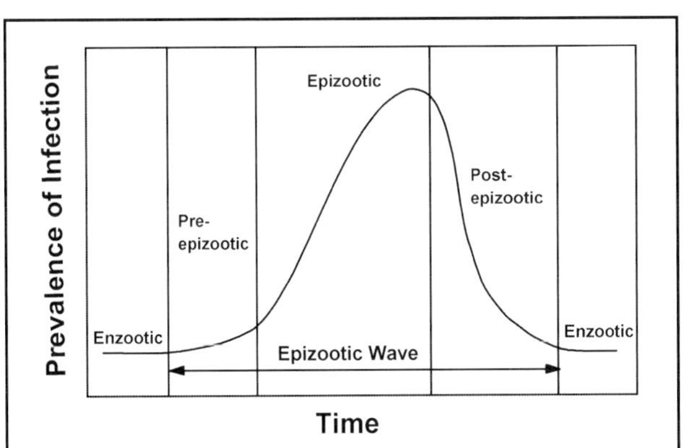

Fig. 41. Typical temporal pattern of change in infection during an epizootic. The curve is divided into the *enzootic* (low prevalence of disease), *pre-epizootic* (a slow increase in the prevalence of disease), *epizootic* (an exponential increase in disease), and *post-epizootic* phases (prevalence of disease decreases back to enzootic levels). [From Tanada and Kaya (1993) with permission from Elsevier.]

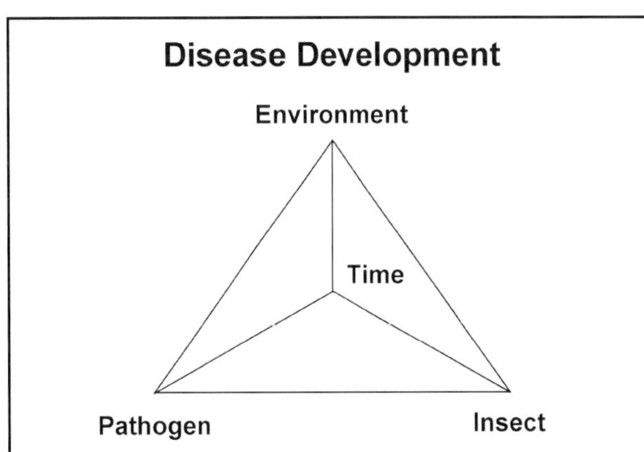

Fig. 42. The disease tetrad or "theory of predisposition model." Disease can only occur in the presence of a susceptible host, a pathogenic pathogen, and a suitable environment. Since all diseases develop through time, a temporal component is added.

To illustrate factors that contribute to disease outbreaks in reared insects, we use the example of the bacterium *Ser. marcescens,* which in some circumstances can be devastating. The bacterium is ubiquitous, present in air, water, and diet ingredients (McLaughlin and Keller 1964, Sikorowski and Lawrence 1994), in the alimentary canal, on and within eggs, and in the hemolymph of insects (Bucher 1963, Bell 1969, Sri-Arunotai et al. 1975, Grimont and Grimont 1978). *Ser. marcescens* can be transmitted vertically, albeit at low frequencies (Sikorowski and Lawrence 1998). Furthermore, it is readily transmitted horizontally. Introducing a single pair of *H. virescens* moths inoculated with *Ser. marcescens* into a cage can result in a very high prevalence of septicemia in moths that were previously free of *Ser. marcescens* (Sikorowski et al. 2001; Fig. 44). In addition to its ubiquitous occurrence, a factor that greatly contributes to the transmission of *Ser. marcescens* in insectary settings is its ability to grow saprotrophically. For example, the bacterium is able to proliferate in honey and sugar solutions commonly used to feed adult Lepidoptera (Sikorowski et al. 2001). It is important to note that *Ser. marcescens* is able to grow in/on artificial diets containing many commonly used antibacterial agents (Sikorowski and Lawrence 1998). For example, *Ser. marcescens* cells can be excreted in the frass of grasshoppers treated with sulfanilamide in sufficient numbers to give the frass a red coloration (Fig. 45), and bacterial cells in frass that are ingested by other grasshoppers can initiate infections.

A number of environmental factors have been shown to influence both the persistence and infection of arthropods by *Ser. marcescens*. Conditions of high humidity are thought to enhance the persistence of the bacterium, and temperature conditions have been shown to predispose the host. For example, susceptibility of the predatory mite *Metaseiulus occidentalis* (Acarina: Phytoseiidae) to *Ser. marcescens* increased significantly if they were exposed to a preinoculation high-temperature pulse (Lighthart et al. 1988). High RH also was found to interact with temperature to increase the sus-

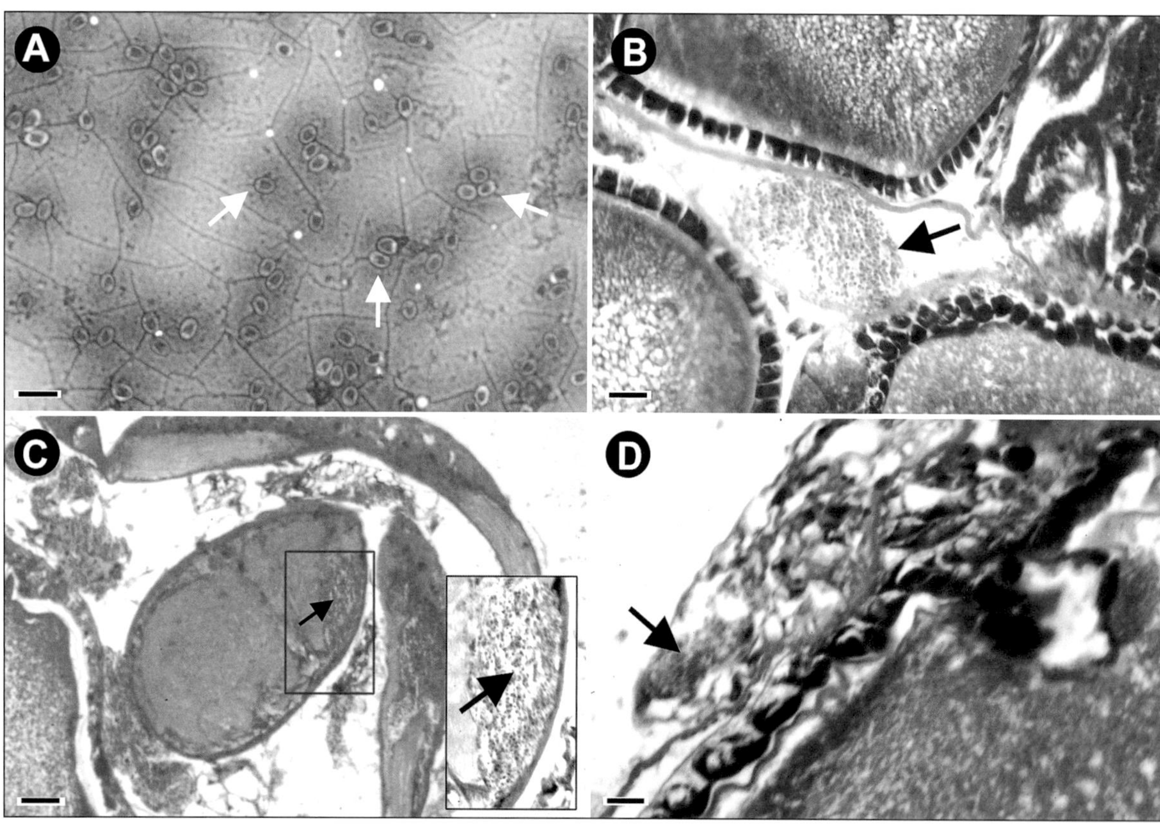

Fig. 43. ***Nosema*** **spores (indicated by arrows) associated with various products, tissues, and structures of the southwestern corn borer,** ***Diatraea grandiosella*** **(Lepidoptera: Crambidae). (A) Homogenate of meconium obtained from an infected individual upon pupal eclosion. Bar = 10 µm. (B) Within ovarian tissues. Bar = 50 µm. (C) Within an egg. Bar = 100 µm (bar = 50 µm within box). (D) On the chorion of an egg. Bar = 25 µm. [From Inglis et al. (2003a) with permission from the Entomological Society of America.]**

ceptibility of the mites to the bacterium. A variety of host factors such as housing density and age can influence disease caused by *Ser. marcescens* (Lighthart et al. 1988, Sikorowski and Lawrence 1998). Typically, high host densities increase the severity of epizootics (e.g., viral infections), but Lighthart et al. (1988) observed that increasing the area of containment (i.e., decreasing the density of mites) increased the impact of *Ser. marcescens*, possibly because the mites had to spend more time searching for food.

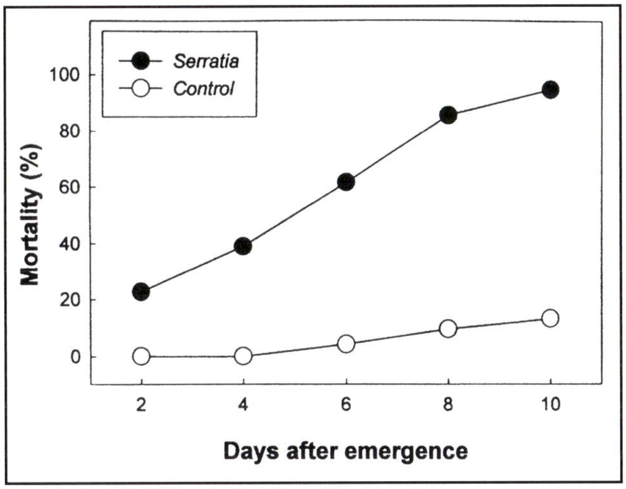

Fig. 44. Prevalence of mortality in tobacco budworm, *Heliothis virescens* (Lepidoptera: Noctuidae), adults (four pairs) exposed to a pair (♀ and ♂) of moths that eclosed from pupae that were submerged in a suspension containing 4 x 10^8 cells/ml of *Serratia marcescens* or to adults from pupae that were not exposed to the bacterium (control). Each value represents the mean of 10 replicates (a total of 50 pairs of moths per treatment). Comparisons at each observation time between the *Serratia* and control treatments were conducted using a Student's *t*-test, and all comparisons were significantly different ($P < 0.01$). [From Sikorowski and Lawrence (1998) with permission from Entomological Society of America.]

Fig. 45. Frass from grasshoppers infested with *Serratia marcescens* (arrows). Heavily colonized frass is red in color. [Frass provided by Dan L. Johnson and image provided by Grant Duke, Agriculture and Agri-Food Canada, Lethbridge, AB, Canada.]

VI. Disease Diagnosis

Rapid and accurate diagnosis is complex but an essential component in the management of disease in insectaries. Steinhaus (1963) emphasized that an accurate diagnosis is required before a disease can be properly studied, controlled, or suppressed. As mentioned previously, the detection of chronic infections is essential in insectaries. There are four basic types of diagnostics (Tanada and Kaya 1993): 1) *differential diagnosis*, which compares and contrasts the signs and symptoms and postmortem changes of different diseases in a systematic manner to distinguish one disease from another; 2) *preliminary diagnosis*, which is the first cursory examination of a diseased insect; 3) *tentative diagnosis*, made after general macroscopic and microscopic examinations have been conducted and usually involves routine laboratory tests; and 4) *definitive diagnosis* in which a final conclusion is made based on all pertinent data.

The data upon which a diagnosis is based can be categorized into three main areas (*sensu* Steinhaus 1963): 1) history of disease; 2) physical examination; and 3) ancillary examination. A diagnosis may be done in-house; however, due to the complexity of some of the observations and analyses and the need for specialized equipment, many may decide to submit specimens for diagnosis by an outside expert. These options will be considered in turn.

A. History

Taking a history to diagnose insect disease is comparable to taking a medical history to diagnose human disease. The diagnostician must prepare the history of disease based on their observations within the insectary. Examples of data collected include the past occurrence of disease in the insectary, environmental conditions including occurrences of abnormal events (e.g., temperature increases/decreases), observations of symptoms and/or signs of disease, host density and housing conditions, nutritional information, and exposure to potentially harmful substances (e.g., insecticides). An example of a form used for recording pertinent history information is provided in Appendix A.

B. Physical Examination

The first step of the physical examination is to inspect the surface of the insect's body. This is followed by a macroscopic examination of the internal organs and tissues with the naked eye, magnifying glass, or with the aid of a stereomicroscope. This can provide valuable information on the identity of the pathogen but is usually insufficient alone to make a conclusive identification. Therefore, more detailed observations involving microscopic examination of suspect tissues are often required. Finally, molecular-level examination or analysis and other observations may be necessary.

1. Macroscopic. An example of a form used for recording pertinent macroscopic information is provided in Appendix B. When performing an external examination, record any observed stunting, discoloration, and textural, or morphological abnormalities (e.g., wing deformities). This is followed up by an internal gross dissection under a stereomicroscope or with the aid of a magnifying device. To facilitate observations of the internal anatomy, the insect is usually cut open with a midventral incision and laid out flat with pins (this is termed an open dissection). In most instances it is necessary to keep the tissues moist to prevent damage or disruption to the tissues. Typically, a saline solution (e.g., 0.85% NaCl) is employed, but more sophisticated solutions may be required for specialized tissues (Becnel 1997). It is often advantageous to dissect a healthy insect of approximately the same age to serve as a comparison. Of particular significance are abnormalities in organ or tissue size (e.g., *hypertrophy*, which is an increase in size due to cell enlargement, in contrast to *hyperplasia*, which is an increase in size due to an increase in cell numbers), colors, or texture (e.g., liquefaction) (Fig. 3). Where possible, conspicuous odors should also be recorded. Many bacteria and fungi are saprotrophs, so many nonentomopathogens may be present in the cadaver, especially if the insect has been dead for a period of time before examination. While many entomopathogens are too small to observe at low magnification, in some instances it will be possible to detect them when they occur *en masse*. For example, nematode adults as well as hyphae of fungi that erupt through the intersegmental folds of insect cadavers under conditions of high humidity can be conspicuous.

Some keys have been developed based exclusively on the macroscopic signs of disease. In all instances, these keys are accurate only under certain conditions and for certain insects. Goodwin (1984) presented a key to the major pathogen groups of reared insects based solely on symptoms and signs of diseases as observed with the naked eye or dissecting microscope. He specified that the key was designed for use with living insects in an insectary or under conditions where continuous observation of living specimens is possible. Starvation, asphyxiation, overheating, drowning, wounding (including gut or integument compromise), and poisoning or other chemical assaults must be ruled out before using the key. With experience, sign- and symptom-based keys can be developed for diseases commonly encountered in a particular insectary, and the development of such keys will greatly facilitate diagnosis and subsequent management of diseases.

2. Microscopic. An example of a form used for recording pertinent microscopic information is provided in Appendix C. Following the macroscopic examination, microscopic observations are made using light and/or electron microscopes. The easiest procedure is to make a wet-mount of the entire insect (including the gut and internal organs) or of specific tissues (e.g., alimentary canal, fat bodies, or Malpighian tubules). This is most commonly accomplished by gently crushing the tissue in water on a glass microscope slide by exerting pressure on a glass cover slip. However, for intact insects and tougher tissues, it may be necessary to homogenize the tissue prior to placement on the slide. This can be accomplished using mechanical homogenizers (e.g., tissue tearers) or pestles. An alternate homogenization method involves the use of polypropylene micropestles in microcentrifuge tubes (Fig. 46). The goal of tissue homogenization is to disrupt tissues sufficiently to facilitate observations of microorganisms in and/or on the tissues.

Many entomopathogens can be observed easily using bright field, phase-contrast, differential interference contrast (DIC or Nomarski), or epifluorescence light microscopy. To visualize entomopathogens *in situ* using light microscopy (e.g., in fat bodies), tissues are usually embedded in paraffin and then sectioned (4- to

8-μm-thick sections) with a microtome. In some instances, it may be possible to freeze specimens and section them with a cryostatic microtome (i.e., a microtome that operates at subfreezing temperatures), which can save considerable time because the fixation and dehydration steps are not required. For specimens for which a greater level of resolution is required, plastic embedding is commonly used to obtain sections < 4 μm. Histological sections are usually required to observe some bacteria (e.g., rickettsias), occluded and nonoccluded viruses, and some protozoa. The major drawback of light microscopy is inadequate resolution to detect nonoccluded viruses and small viral occlusion bodies (e.g., granulosis virus) with any certainty.

For small entomopathogens, it is often necessary to use TEM. The most rapid preparatory strategy for electron microscopy is to use a negative staining method, e.g., the method published by Sikorowski and Lawrence (1997) for detection of Mc-NOBV in *M. croceipes* and its hosts. In some instances, it may be necessary to fix, embed, and section the specimen for observation with TEM. For more information on techniques used for preparation and observation of histological sections at the light and electron levels, the reader is referred to Becnel (1997), Evans and Shapiro (1997), and references therein.

While fungi, bacteria, and protozoa can be observed at the light level without the aid of stains using phase-contrast or DIC, the visualization of these and other entomopathogens is often enhanced by the use of differential stains and bright field or epifluorescence microscopy. There is a vast diversity of simple to complex staining procedures, but the most commonly used stains include: 1) haematoxylin, 2) Giemsa stain, 3) buffalo black, 4) acid fuchsin, and 5) lactophenol cotton blue. Haematoxylin is one of the most common and useful histological stains, and a number of different formulas have been used depending on the specific objectives of the study (Sheehan and Hrapchak 1987). Giemsa is a polychromatic stain that is often used to visualize some bacteria, fungal propagules, microsporidia, and viral occlusion bodies in wet smears (Becnel 1997, Evans and Shapiro 1997). For the majority of entomopathogens, Giemsa stain is a negative stain (i.e., the pathogen tissues remain unstained while the host tissues stain in blues and reds); however, entomopoxvirus occlusion bodies appear blue (Evans and Shapiro 1997). Buffalo black is another stain used for tissue preparations, and this stain is commonly used in diagnostic facilities for routine examination of specimens for bacteria, microsporidia, fungi, and viral occlusion bodies. Buffalo black is a positive stain in that the entomopathogens are stained varying shades of blue-black while the insect host tissue remains light blue in color. Both Giemsa and buffalo black are relatively simple to use because they do not require a complex fixation step, and both are virtually permanent, enabling specimens to be stored in a light-tight container for many years without significant loss of detail. An advantage of buffalo black over Giemsa stain is the dissolution of uric acid crystals by the latter, which can obscure visualization of entomopathogens. Evans and Shapiro (1997) describe a staining method to distinguish between occlusion bodies of NPV, CPV, and EPV within the same tissue: using various prefixatives, it is possible to change the characteristics of CPV to take up stain (they remain colorless under normal Giemsa staining) while leaving NPV occlusion bodies colorless and EPV occlusion bodies stained blue. Fungal structures (e.g., hyphae, hyphal bodies, and spores) are often relatively conspicuous, but a number of nonspecific stains can be used to aid differentiation from host tissues, including acid fuchsin and cotton blue.

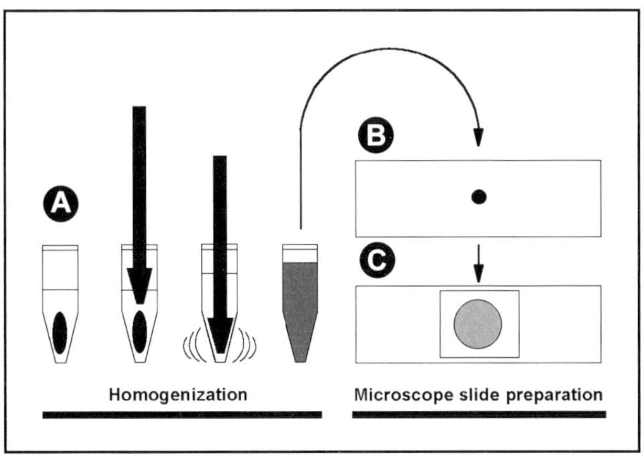

Fig. 46. Preparation of an insect for examination with a light microscope. The diseased insect is placed at the bottom of a 1.5-ml microcentrifuge tube containing 500 μl of sterile water or buffer and homogenized using a sterile plastic pestle. A drop of the homogenate is then placed on a glass microscope slide and covered with a glass cover slip. The specimen is then examined using phase-contrast microscopy. If bright field microscopy is used, the homogenate is mixed with an appropriate stain prior to being covered with a cover slip.

A number of keys are available for the microscopic identification of major entomopathogen groups (e.g., Weiser and Briggs 1971, Poinar and Thomas 1984, Lacey and Brooks 1997). We present the key published by Lacey and Brooks (1997) (Table 1). Microscopic identification of entomopathogens at lower taxonomic levels requires specialized knowledge and techniques (e.g., isolation) beyond the intent of this chapter. For introductions to these more detailed procedures, one should consult any number of appropriate references. Some examples include the following: viruses (Poinar and Thomas 1978, Padhi 1985, Volkman 1985, Maramorosch 1987, Tompkins 1991, Adams and Bonami 1991, Ausubel et al. 1991); bacteria (Poinar and Thomas1978, Thiery and Frachon 1997); fungi (Harris 1986, Samson et al. 1988, Goettel and Inglis 1997, Papierok and Hajek 1997, Humber 1997); protozoa (Brooks 1988, Undeen and Vávra 1997, Becnel and Andreadis 1999); and nematodes (Kaya and Stock 1997). In instances where it is not possible to make a

Table 1. A key to the major groups of entomopathogens based primarily on microscopic signs (Lacey & Brooks 1997)

1.	Distinct external growth on insect	2
1.	No external growth on insect	3
2.	Mass of wormlike (nonsegmented) organisms over the surface of the insect which may be red or cream to light brown in color, many with two, distinct body coverings	*Nematodes*
2.	Growth or organisms on the surface of the insect, often floccose (i.e., cottony-like), powdery (white, green, or red), and sometimes limited to intersegmental areas including clublike growths emanating from these areas	*Fungi*
3.	Insect usually normal in appearance but may be stunted or slightly malformed; upon dissection the body cavity may contain one or more multicellular organisms with many segments and a distinct to indistinct head region with mandibles, possessing a tracheal system marked by various types and arrangements of spiracles; in some cases the body of the organism may protrude through the host's integument or even may be feeding externally on the host tissues	*Parasitoids*
3.	Insect may be normal or abnormal in appearance but upon dissection may contain wormlike, nonsegmented organisms or small, unicellular particles represented by motile or nonmotile rod-shaped to spherical microorganisms or various life-cycle states of other microorganisms including spores, cysts, occlusion bodies or hyphal-like structures	4
4.	Aquatic or other insect with essentially transparent cuticle (integument)	5
4.	Terrestrial insect or those with basically nontransparent cuticle	12
5.	Insect iridescent or specific tissues (especially fat bodies) iridescent	6
5.	Insect not iridescent	7
6.	Insect (e.g., scarabaeid grubs) white-bluish to blue-grayish in coloration, containing crystals and minute bacterial-like forms, often *pleomorphic* (multiple forms) in shape and that are just visible by light microscopy	*Rickettsias*
6.	Insects orange to green to blue in coloration, infectious agent not visible with the light microscope	*Viruses*
7.	Hemolymph (blood) as seen through the cuticle milky in coloration, rod-shaped, motile cells often with refringent spores giving footprint appearance under phase-contrast microscopy	*Bacteria*
7.	Hemolymph clear, essentially normal in appearance	8
8.	Wormlike or rapidly motile, ciliated organisms visible through the cuticle at low microscopic magnification	9
8.	Nematodes or ciliated organisms absent	10
9.	Organisms wormlike and elongated, nonsegmented	*Nematodes*
9.	Organisms pear-shaped and motile by cilia	*Protozoa*
10.	Digestive tract abnormally opaque white, particles polyhedral in shape, visible with phase microscopy in cytoplasm of gut cells	*Viruses*
10.	Digestive tract essentially normal in appearance	11
11.	Abnormal whitish masses in hemocoel associated with various tissues (e.g., fat bodies) or in hemolymph itself, masses composed of ovoid to pyriform spores refringent under phase microscopy	*Protozoa* or *Microsporidia*
11.	Hemocoel (especially of mosquito larvae) filled with hyaline (i.e., clear) hyphal-like bodies or rust-colored oval spores with sculptured walls	*Fungi*
12.	Insect body usually hardened, mummified, and cheesy in consistency, filled with hyaline hyphae, hyphal-like bodies, or spherical resting spores	*Fungi*
12.	Insect not hardened or cheesy, may be stunted or with malformed body parts	13
13.	Insect flaccid and usually discolored, integument may be fragile	15
13.	Insect may be stunted or malformed, integument usually normal in appearance	14
14.	Various body tissues and cells containing refringent, nonmotile spores or cysts (oval, pyriform, navicular, or spherical in shape) best visualized by phase-contrast microscopy; in some cases the digestive tract contains flagellated, large, slow moving, septate organisms in pairs or chains	*Protozoa* or *Microsporidia*
14.	Various body tissues and cells containing minute, nonmotile, bacterial- to pleomorphic-like cells that are just visible by light microscope, exhibit Brownian motion, are highly refringent, and occur in pairs or chainlike structures; crystals may or may not be present	*Rickettsias*
15.	Wormlike, nonsegmented organisms in body tissues which may be liquefied and creamy, grayish to reddish in coloration	*Nematodes*
15.	Nematodes not present in body tissues	16
16.	Insect often with putrid odor; usually brown, black, or reddish in coloration; and may be liquefied with rod-shaped, motile organisms that may contain refringent spores evident under phase microscopy	*Bacteria*
16.	Insect flaccid and discolored, integument may be very fragile, liquefied body tissues filled with refringent spherical to polyhedral-shaped occlusions which may also occur in the cytoplasm or nuclei of intact cells, occlusions usually dissolved by a weak solution of NaOH, caterpillar cadavers may be attached to the host plant in an inverted v-shaped manner hanging by abdominal prolegs	*Viruses*

diagnosis, specimens should be submitted to experts for diagnostic testing (Sec. VI. A).

3. Molecular. Molecular methods have been used extensively to diagnose diseases and/or detect pathogens in plants and vertebrate animals and in many instances are the methods of choice. At present, there are two primary strategies employed, including immunological (polyclonal or monoclonal antibodies) and nucleic acid (RNA or DNA) based methodologies. Diagnosis of infection in mammals often relies on the detection of *antibodies* (immunoglobulin proteins produced by the host in response to the pathogen) each of which has an affinity to a particular *antigen* (a protein on the surface of the pathogen). This strategy does not work for insects because they do not produce antibodies. Therefore, immunological detection of entomopathogens relies on the use of polyclonal or monoclonal antibodies produced in mammals (e.g., rabbits) or birds to antigens of the entomopathogen itself. Given the small size of virions and our inability to detect them by light microscopy, antibodies are frequently used to detect entomopathogenic viruses. Methods such as enzyme-linked immunosorbent assay (ELISA) are commonly applied. In this method, the breakdown of a substrate by an enzyme bound to an antibody causes a color change. Antibodies are typically conjugated with enzymes such as horseradish peroxidase, glucose oxidase, or alkaline phosphatase.

Direct ELISA is primarily used to detect viral or bacterial entomopathogens (Fig. 47). For this method, antibodies specific to the entomopathogen of interest are produced in a vertebrate and bound to the wall of a microtiterplate. The plate is then washed to remove any excess antibody, and a solution containing the potential antigen (e.g., an insect homogenate) is added to the plate. If the antigen matches the antibody, it binds. A second antibody with an attached enzyme ligand is added, and the plate is washed to remove any unbound antibody. Finally, the substrate for the enzyme is added, and any color reaction is detected visually or quantitatively using a spectrophotometer.

Indirect ELISA is typically used in human and veterinary medicine to detect the production of antibodies by an animal to a specific antigen. As indicated above, insects do not produce immunoglobulins, and we mention indirect ELISA only for comparative purposes (Fig. 47).

One of the major problems with immunological methods is poor sensitivity due to low amounts of antigen or cross-reactivity with nontarget antigens. To increase specificity and enhance the sensitivity of detection, *monoclonal antibodies* (an antibody that is mass produced in the laboratory from a single clone and that recognizes only one antigen) can be used. For example, monoclonal antibodies have been used successfully to detect pathogens in a number of insects including *Par. locustae* in grasshoppers (Knoblett and Yousseff 1996). The disadvantage of monoclonal antibodies relates to the difficulty in producing them and that they can be so specific that intraspecific variants of an entomopathogen may not be detected.

The application of nucleic acid–based methodologies that utilize polymerase chain reaction (PCR) has revolutionized the detection of pathogens. These methods can be much more sensitive than immunological methods, and they are relatively easy to conduct and robust. Initially, hybridization methods including North-

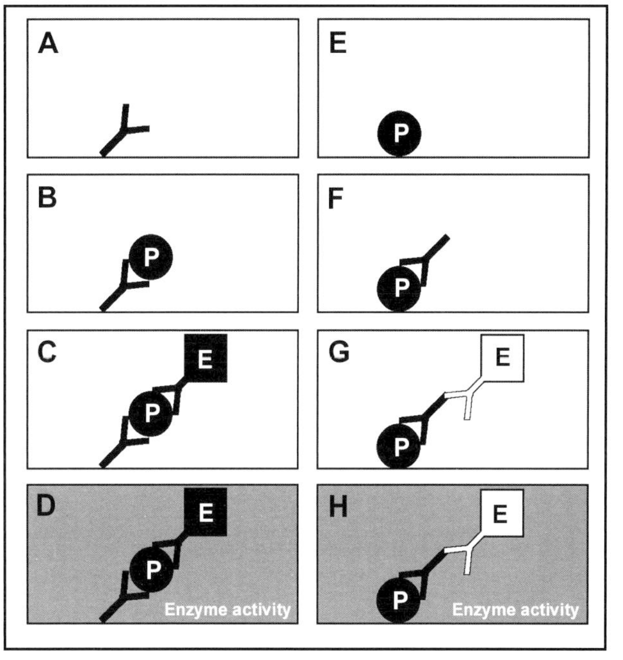

Fig. 47. Enzyme-linked immunosorbent assay (ELISA) used to detect entomopathogenic microorganisms. (A–D) Direct ELISA. An antibody (Y) to a particular entomopathogen is bound to the wall of a microtiter plate. A homogenate of the insect to be tested is applied to the plate; if the homogenate contains the antigen of interest (P), some will bind to the antibody. A second antibody specific to the antigen and conjugated to an enzyme that catalyzes a colored product in the presence of a specific substrate is added, and the excess rinsed away. Finally, the conjugated enzyme's substrate is added, and any appearance of the colored product is noted. (E–H) Indirect ELISA. The antigen (P) is bound to the wall of a microtiter plate, an antibody specific to the antigen, an antibody specific to the first antibody and conjugated to an enzyme, and a color-producing substrate of the enzyme are added in order.

ern (i.e., RNA) and Southern (i.e., DNA) blots were used to qualitatively detect viruses, bacteria, fungi, and various protozoa in insects (St. Leger and Joshi 1997 and references therein). Another method is fluorescence *in situ* hybridization (FISH), which is used not only to detect microorganisms but also provides a measure of tissue specificity and spatial distribution of microorganisms within tissues (Heddi et al. 1999). However, the greatest advance in pathogen detection using molecular methods involves the use of the polymerase chain reaction (PCR), and this technology is being extensively used to detect and characterize infectious agents in insects. DNA coding for a protein specific to a particular entomopathogen occurs in very small quantities. PCR is a technique to create exact copies of the DNA in sufficiently large quantities that it can be detected (Fig. 48). The enormous body of literature dealing with PCR-based detection of pathogens is reviewed by Innis et al. (1990), Persing (1996), Caetano-Anollés and Gresshoff (1997), and St. Leger and Joshi (1997).

4. Nondestructive Sampling and Specific Pathogen-free Insects. A major limitation of all techniques used to detect entomopathogens in insectary settings is the lack of nondestructive sampling strategies. While there are currently few viable alternatives to destructive sampling, two possibilities are the nonlethal removal of quantities of hemolymph and quantification of entomopathogens excreted in substances such as frass or meconia. In some insects, it is possible to remove relatively large quantities of hemolymph without causing mortality. For example, half of the hemolymph can be removed from *P. japonica* larvae without killing them (Beard 1949). Even if relatively large volumes of hemolymph can be removed, this strategy assumes that the entomopathogens will be present in the blood in detectable quantities. This is often not the case.

Many *holometabolous* insects (i.e., insects that interpose a pupal stage between the final larval stage and the adult) produce meconium. Very little is known

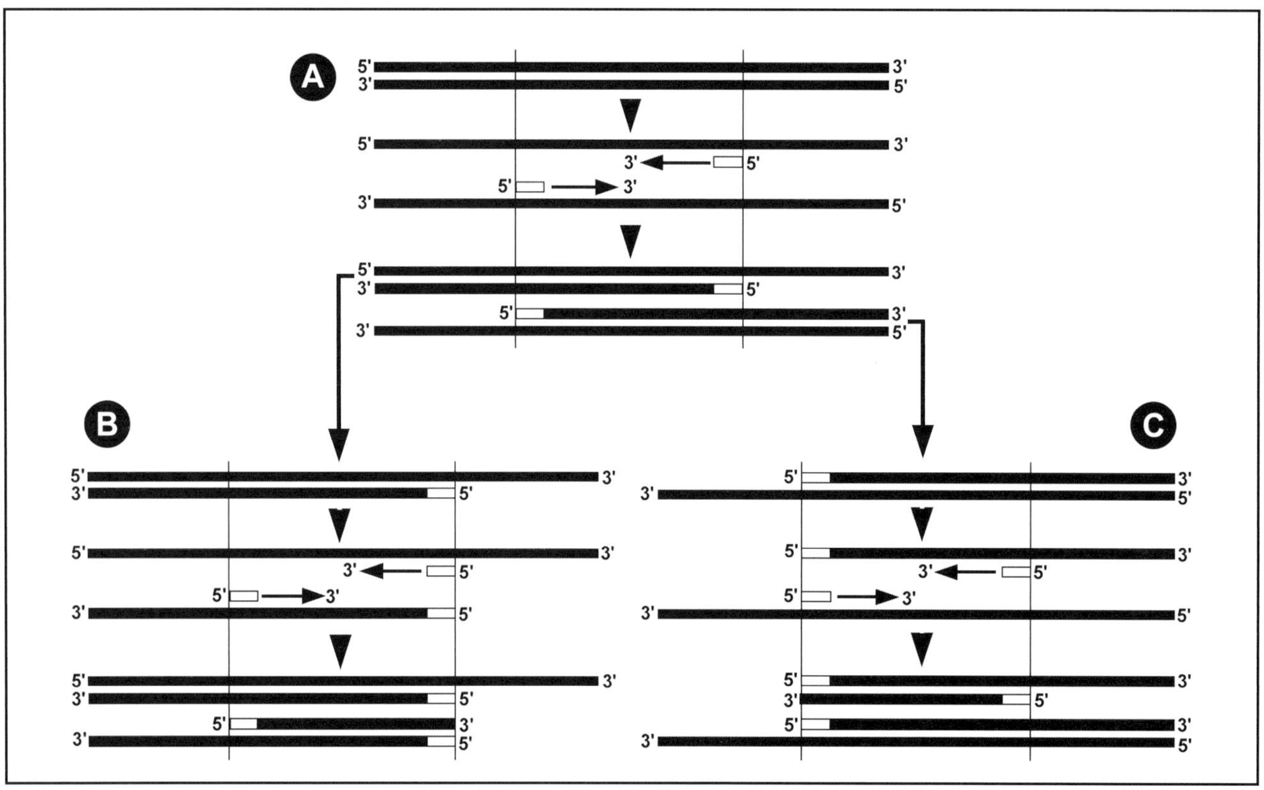

Fig. 48. Polymerase chain reaction (PCR) to identify entomopathogenic microorganisms. (A) The first of a series of 20–30 similar cycles: denaturation at high temperature (94°C) of the double-stranded DNA coding for a protein specific to the entomopathogen, annealing at a lower temperature (ca. 55°C) of a primer (open rectangle) to a specific location on each of the template DNA strands, and extension of the primers by the action of *Taq* polymerase, an enzyme that polymerizes DNA at 72°C. (B–C) In the second cycle, 4 copies (the 2 original and the 2 new copies) serve as templates, and the steps of denaturation, annealing, and extension are repeated. At the end of C cycles, the number of copies present is given by $N = N_0 \cdot 2^C$, for N_0, the initial number of copies present. For example, if one starts with two copies and applies 20 PCR cycles, then $N = 2 \cdot 2^{20} = 2.1 \times 10^6$ copies.

about the ontogeny of meconium, but it is believed to contain the waste products of pupal metabolism such as uric acid (Chapman 1998). A number of entomopathogens are discharged in the meconium of Lepidoptera and other orders of insects. A method has been developed to detect viral OBs, microsporidian spores, and *Ser. marcescens* cells in meconia of lepidopteran insects (Sikorowski et al. 1971, Inglis et al. 2003b). To stain viral occlusion bodies and microsporidian spores in meconium, dissolve buffalo black (0.1 g) in a solution of 100% methyl alcohol (50 ml), glacial acetic acid (30 ml), and distilled water (20 ml), and filter the stain to remove particulate matter (the stain should be used within 1 month of preparation). Aseptically homogenize the meconia in a small quantity of sterile distilled water (approximately 50 μl), smear the homogenate onto a glass slide, and allow it to air dry. Then heat the slide to 40°C, and cover the dried homogenate with a solution of buffalo black for 5 min with care taken not to allow the staining solution to dry on the slide during this period. Remove the slide from the slide warmer and dip the slide in tap water for 5 sec, allow the specimen to dry, and examine without a cover slip at appropriate magnification. Meconium containing *Nos. heliothidis* stained using this method is shown in Fig. 49.

We have found that the method is logistically feasible. For example, two experienced technicians, one preparing meconium smears and the other examining them with a compound microscope, can process approximately 50 samples per h (Sikorowski and Lawrence 1997). Regarding accuracy, using a heliothine model, Inglis et al. (2003b) found that the meconium method was very accurate in detecting infections by CPV and somewhat less accurate for *Ser. marcescens* and *Nos. heliothidis* (Fig. 50). False negatives (i.e., falsely concluding that an insect is free of an entomopathogen) are of paramount concern to insect rearers. Based on the prevalence of false negatives, the meconium method was found to be > 99% effective in detecting infection by CPV, 95% effective for *Ser. marcescens*, and 77% effective for *Nos. heliothidis* (Inglis et al. 2003b). These levels of predictability will greatly facilitate the selection of pathogen-free individuals in rearing programs for these species.

Whether insect hemolymph or meconia will prove useful in the detection of infections will no doubt be dependent on the pathogen and host. Therefore, experimentation on an insect-specific basis is necessary before these strategies may be successfully implemented.

Mammals (e.g., rabbits, mice, and rats) purchased for scientific reasons are certified specific pathogen free (SPF). SPF mammals are maintained under quarantine conditions, and sentinel animals are used to determine if they are free of the specified pathogens (e.g., various types of viruses, bacteria, and parasites including protozoa). To accomplish this, tissue, feces, and blood samples are obtained on a defined schedule; and the presence of antibodies in the blood is tested extensively. If the sentinel animals are deemed to be free of the specified pathogens, the colony is designated as SPF.

Currently there are no sources of certified SPF insects; however, the cultured shrimp industry provides a model for the development of certified SPF arthropods. Disease, primarily incited by a number of viruses, threatened the cultured white and tiger shrimp industry in the 1980s and 1990s. Outbreaks of disease forced the destruction of entire populations. Like viral diseases of insects, there are no medications efficacious against these pathogens. In response, a program to provide SPF brood cultures of shrimp was initiated by public research groups in the USA. This program has largely now been transferred to private industry. The same model could be utilized for economically important insects. Quarantine facilities utilized in combination with "selection methods" using sentinel insects and sound diagnostic principles, including the

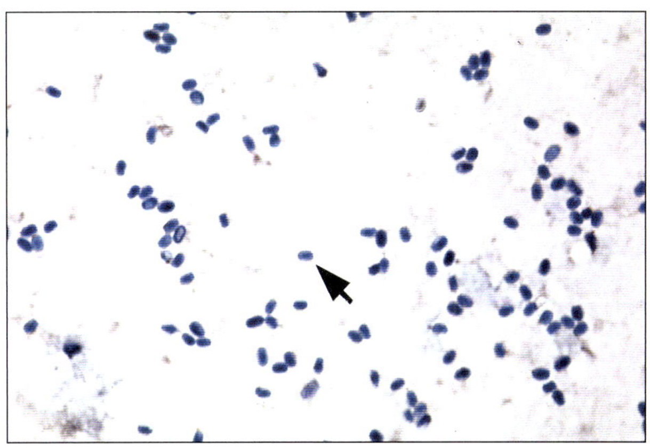

Fig. 49. Spores of *Nosema* in the meconium from a southwestern corn borer, *Diatraea grandiosella* (Lepidoptera: Crambidae). The spores were stained with buffalo black (using the meconium method) and examined with a light microscope under bright field illumination. The arrow points to a *Nosema* spore (stained blue).

application of molecular and statistical methods, could be used to designate insects as "SPF." Such insects could then be used to initiate colonies.

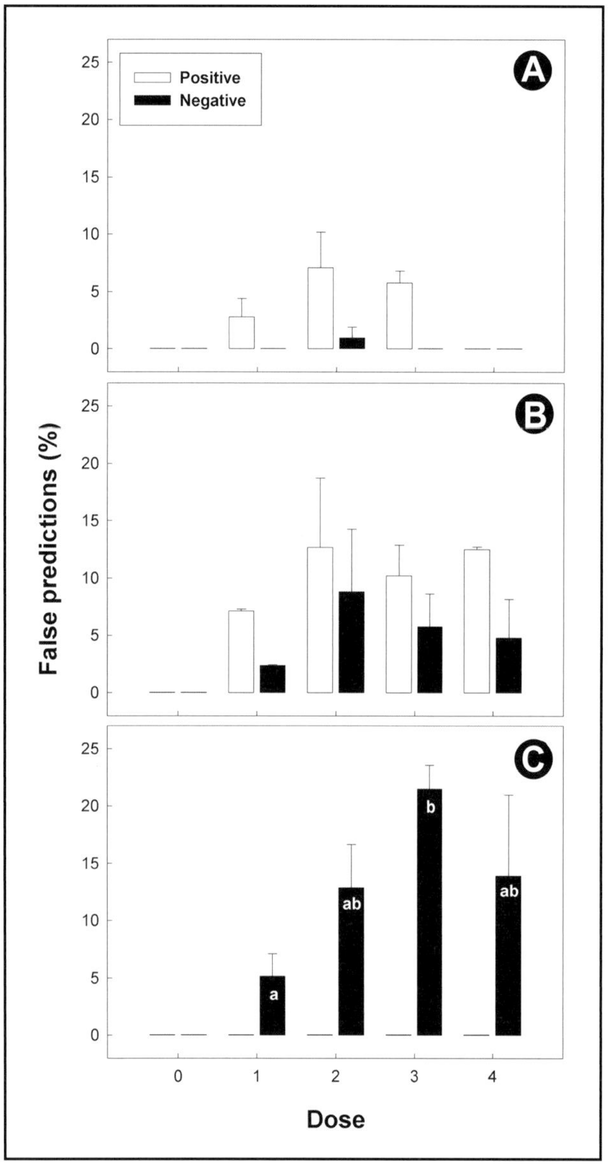

Fig. 50. The accuracy of using meconium to nondestructively detect entomopathogens of some moths of the Heliothidinae. False positives occur when the meconium is positive for the pathogen but the adult is negative, and false negatives are the reverse. (A) Tobacco budworm, *Heliothis virescens* (Lepidoptera: Noctuidae), larvae ingested 0 (control), 10^1 (dose 1), 10^2 (dose 2), 10^3 (dose 3), and 10^4 (dose 4) cytoplasmic polyhedrosis virus (CPV) occlusion bodies. (B) *H. virescens* larvae ingested 0 (control), 10^1 (dose 1), 10^2 (dose 2), 10^4 (dose 3), and 10^6 (dose 4) *Serratia marcescens* cells. (C) Bollworm, *Helicoverpa zea* (Lepidoptera: Noctuidae), larvae ingested 0 (control), 10^3 (dose 1), 10^4 (dose 2), 10^5 (dose 3), and 10^6 (dose 4) *Nosema heliothidis* spores. Error bars represent standard errors (n = 3), and histogram bars not labeled with the same letter or unlabeled are significantly different ($P \leq 0.05$). [From Inglis et al. (2003b) with permission from the Entomological Society of America.]

C. Ancillary Examination

Frequently, additional, more detailed examinations are required to confirm the diagnosis made based on the physical examination. An example of a form used for recording pertinent information on the final diagnosis is provided in Appendix D. This type of examination is sometimes referred to as *ancillary examination*. This may involve isolation, culture, and further characterization using microscopic, physiological, and/or molecular methods. It may also be necessary to study various aspects of a pathogen's infectivity (e.g., using bioassay) and to ensure that the isolated pathogen is the actual causal agent of disease. Just because an entomopathogen is isolated from a diseased individual does not prove that it is the cause of disease. In cases where the pathogen has a history of inciting disease in insects maintained in a particular insectary, then its isolation alone can be used to make a final diagnosis. However, when a pathogen is isolated from a diseased insect but the diagnosis is uncertain, the following steps, known as *Koch's postulates*, are taken to confirm that the isolated microorganism is the causal agent of the disease (modified from Agrios 1978): 1) the pathogen must be isolated from the every diseased insect examined, and the signs and/or symptoms of the disease recorded; 2) the pathogen must be grown in axenic culture on a nutrient medium (nonobligate pathogens) or in a susceptible insect (obligate pathogens), and it must be identified and/or characterized; 3) the pathogen must be inoculated on or in healthy insects of the same or a similar species to the original, and signs and symptoms of disease must be the same; 4) the pathogen must be isolated in axenic culture again from the inoculated insect, and its characteristics must be exactly like those observed in the second step above; and 5) given a situation where entomopathogens cannot be recovered in artificial culture, a revision of Koch's postulates based on recovery of nucleic acids (e.g., PCR amplification products) has been proposed by Relman and Persing (1996).

It must be stressed that many diseases are not caused by entomopathogenic microorganisms, and assays for abiotic agents such as poisons, toxins, and metabolic diseases cannot be overlooked. Such assays often require detailed biochemical or molecular analysis of tissues. In such a situation, it is often necessary to consult a skilled diagnostician or submit a specimen for expert diagnosis.

D. Submission for Diagnosis

An important consideration in obtaining an accurate diagnosis is the proper submission of specimens. A number of laboratories at universities and other institutions at the state, federal, and private levels may conduct such diagnoses. The following information is provided by Tanada and Kaya (1993).

It is preferable to submit fresh specimens for diagnosis. In any event, they should not be fixed in preservatives, such as formaldehyde or ethanol, unless they are to be sectioned for histological examination. It is usually not possible to make an accurate diagnosis from dead, dried specimens. If available, uninfected individuals, individuals at an early stage of infection, at a late stage of infection, and moribund specimens should be submitted. The specimens should be placed individually in clean (preferably sterile) containers, and if alive, the containers should be stoppered with cotton or non-airtight covers. In most instances, living specimens should be placed individually in containers with only small portions of food in an attempt to minimize mechanical damage. The specimens should be shipped by the fastest available means—usually a courier service. There may be quarantine regulations in place in some instances (e.g., between countries), and it will be necessary to obtain the necessary import/export permits before shipping. The insectary manager should check with the diagnostician regarding such regulations.

The specimens should be shipped with pertinent data to facilitate the diagnosis including: 1) common and scientific names of the insect; 2) name of the collector and address including a contact number or e-mail address; 3) name and address of the submitter if different from that of the collector; 4) date and location of collection; 5) nature of the environment in which the diseased insect was maintained, including conditions of temperature, relative humidity, light, insect density, and nutrition; 6) extent of the disease outbreak or epizootic; 7) conditions under which the disease outbreak occurred (e.g., an extreme temperature fluctuation); 8) abnormal behavior or appearance of the insect, if any; 9) whether or not the insect was exposed to potentially harmful nonbiotic agents (e.g., chemicals); and 10) past history of similar outbreaks or previous observations and experiences.

VII. Disease Management

The primary goal of disease management is prevention. However, it is often an unrealistic goal to completely eliminate entomopathogens from a colony, and therefore the objective should be to maintain a "disease-free" colony. To obtain and maintain a disease-free colony, a number of methods are employed, but none is more important than early detection. Once an outbreak occurs in a colony, therapeutic approaches may be implemented with some probability of success depending on the type of entomopathogen causing the disease, but these should always be considered as a secondary strategy. The appropriate strategy to be used can be simple to complex and will vary according to the situation. For example, if the insect colony is used for the production of entompathogens (e.g., viruses or protozoa), parasites, or heterologous proteins, it is often sufficient and, in some cases desirable, to maintain a colony with limited genetic diversity. However, if a virulent pathogen is introduced into a colony possessing limited genetic diversity, devastating epizootics can result. Another example, also involving genetic diversity, is the following: if the goal of the rearing program is to produce insects for field release, it is desirable that insects represent field populations as closely as possible; so a number of researchers infuse genes into the population in an attempt to maintain genetic diversity. However, this practice results in a much greater risk of introducing new and potentially virulent pathogens into the colony.

A. Prevention

1. Pathogen-Free Individuals. In establishing a colony, starter stock should be free of disease and, if possible, entomopathogens as well. This often involves obtaining stock cultures from an established colony that has a history of being disease free. Since no standards exist for SPF insects, it is often necessary to start a new colony from field populations, which has obvious consequences for the potential introduction of entomopathogens in the rearing system. When using field-collected insects, it is useful to possess knowledge of the diseases commonly associated with the species. For example, *M. croceipes* adults that live for 7 days or longer should be selected for incorporation into or establishment of a colony because Mc-NOBV–infected wasps usually live less than seven days (Hamm et al. 1988).

A second example of a strategy in which knowledge of the host-pathogen relationship is instrumental in selecting pathogen-free individuals is provided by Hamm et al. (1996). They found that reductions in GSV infection of *H. zea* colonies could be accomplished using eggs only from the first or second night of oviposition. The prevalence of progeny infected with GSV increased with the day of oviposition from 5% for day 1, to 17% for day 2, to 63% for day 3. Thus the presence of GSV in the *H. zea* colony is substantially reduced by using eggs from only the first or second night of oviposition. While the routine implementation of this method did not rid the colony of the pathogen, it effectively reduced the prevalence and impact of the GSV. A third example is provided by Boohene et al. (2003). They found that development of pupae of the parasitoid *Muscidifurax raptor* (Hymenoptera: Pteromalidae) infected with *Nosema muscidifuracis* was significantly longer than that of healthy individuals. The difference in development time provided a narrow window for obtaining pathogen-free adults. After 30 days at 20°C, 60% of uninfected females had emerged compared to only 10% of infected adults.

The quarantine or Pasteur method is one of the oldest and best methods for preventing the introduction of pathogens into a colony. This involves quarantine of field-collected insects followed by selection of healthy individuals. A generalized method that can be applied to decrease the likelihood of introducing pathogens into the colony is provided in Table 2. However, the procedure will have to be modified based on the particular characteristics of the insects and pathogens involved (e.g., Fig. 51).

2. Sanitation. Implementation of the appropriate sanitation and decontamination measures is a prerequisite for managing disease in laboratory colonies of insects. Although a number of entomopathogens are transmitted transovarially and cause chronic infections, the vast majority of vertically transmitted entomopathogens are transmitted transovum. For information on the sterilization and pasteurization of diets, see Chap. 8 *Microbial Contamination and Insect Rearing*.

a. Surface Sanitation of Insects. A key method for eliminating or reducing entomopathogens on the surface of eggs involves the use of surface sanitizers. For example, eggs from *Ser. marcescens*–infected parents may be externally contaminated, and surface sanitation of the eggs is an excellent management strategy for this entomopathogen. Formaldehyde and sodium hypochlorite are two commonly used sanitizers, but a

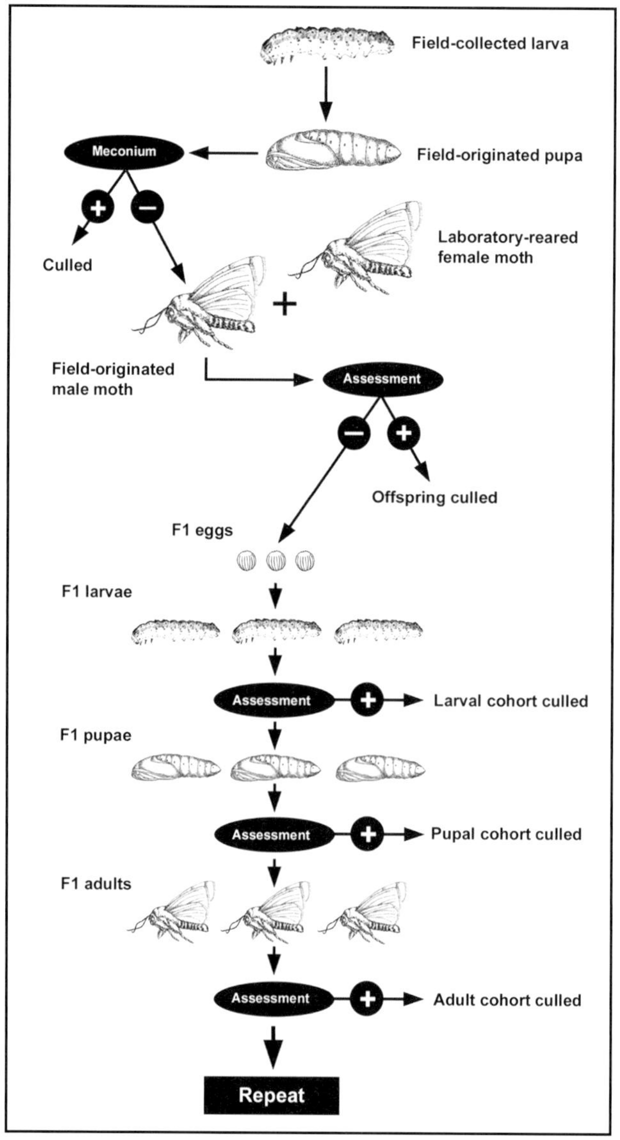

Fig. 51. A flowchart describing the basic quarantine method for a lepidopteran insect. In this scenario, an immature male insect is collected from an apparently healthy field population. The larva is isolated and its health is monitored visually throughout the development period. Once it pupates, the meconium is checked for evidence of microsporidian spores, viral occlusion bodies, or bacterial cells. If negative, it is mated with a laboratory-reared female. After mating, the moth is destructively examined for signs of infection and tested for the presence of entomopathogens. Standard fitness quality control standards are applied. If the moth is positive for infection, all of its offspring are destroyed. If it does not exhibit signs or symptoms of infection, periodic checks for infection in its offspring are conducted. The process is then repeated.

number of other chemicals (e.g., quaternary ammonium compounds) have demonstrated their utility with various pathogens and insects. Such surface sanitation procedures are not benign and can cause considerable damage to eggs (e.g., destruction of the egg chorion; Fig. 52) and loss of viability (e.g., Leppla et al. 1974). It is imperative that the efficacy of the chemical treatment as a function of time and concentration be determined for each insect species. To increase the efficacy of sanitation, agitation is usually employed to increase the degree of contact between the sanitizer and the egg surface. Following the sanitation step, the sanitizing chemical must be removed, and this is usually accomplished by rinses in sterile water, taking care to ensure that the washing and drying environment are free of entomopathogens. The deactivating chemical *antichlor* (a solution of sodium or ammonium thiosulfate) is sometimes used to deactivate residual sodium hypochlorite. After submersion in sodium hypochlorite, the insect is rinsed in an aqueous solution of antichlor (generally 2 to 5% sterile sodium thiosulfate) followed by a final rinse in water. When working with eggs, special care must be taken to prevent excessive desiccation. Varying degrees of mechanization have been developed to expedite the process and reduce labor inputs (Chap. 11 *Insect Rearing Production Systems, A Case Study: The Southwestern Corn Borer*). For more details on the surface sanitiza-

Table 2. A generalized quarantine strategy for field-collected insects or for eliminating a pathogen from a diseased colony

1. Collect insects from an apparently healthy field population or collect apparently healthy individuals from a diseased colony (in most instances, immature stages are preferred).
2. Isolate insects if possible (e.g., in individual cells), provide a suitable diet and rearing conditions, and allow the insects to complete development. Remove those insects that die. Check for the presence of entomopathogens in cadavers.
3. Check the meconium of eclosed adults for evidence of microsporidian spores, occlusion bodies, or bacteria (see description in text).
4. Allow insects to mate in as controlled a situation as possible. Where possible, individual pairings of apparently healthy ♂♂ and ♀♀ is desirable. A preferable strategy is to mate wild ♂♂ with laboratory-reared ♀♀, since vertical transmission from adult males to offspring is a relatively rare occurrence.
5. After oviposition is completed, examine the female for signs of infection, and test for the presence of entomopathogens. In some instances, it is beneficial to apply other criteria such as initiation of oviposition, fecundity, pupal weights, and pupation period (i.e., to facilitate assessments of infection).
6. If the female is positive for infection or if she and/or her offspring exhibit abnormal traits, destroy all her offspring.
7. Keep offspring originating from each female separate.
8. Rear the offspring of uninfected females only, and conduct periodic checks for infection in cohorts. After checking cohorts for infection by a nonoccluded baculovirus, Hamm et al. (1992) noted that the colony appeared to be free of the baculovirus. However, they continued to monitor the population and found viruses again in the next generation. After repeating the selection process, disposing of the original virus-infected colony, and cleaning the rearing facility extensively, the colony was deemed free of the baculovirus. This example stresses the importance of vigilance in monitoring colonies for entomopathogens and disease.
9. An adequate number of insects must be used in the selection process to ensure the genetic vigor of the colony. For example, Hamm et al. (1971) successfully eliminated *Nosema* from a colony of the bollworm, *Helicoverpa zea* (Lepidoptera: Noctuidae); but this did not improve the vigor and productivity of the colony. They attributed this to the small number of insects used in the selection process. They subsequently repeated the process with a larger number of insects and successfully obtained a "standardized colony."
10. The selection of uninfected individuals is made easier when the entomopathogen is not transmitted transovarially. In such cases, the application of the appropriate sanitation methods (e.g., removal of dead adults and treatment of the egg surface to remove entomopathogens) is usually sufficient. Using these methods, Briese and Milner (1986) were able to clear a colony of the eyed ladybird, *Anatis efformata* (Lepidoptera: Geometridae), of the protozoan *Pleistophora schubergi* in only one generation. For more information on quarantine, see the previous section on selection of pathogen-free individuals.
11. If possible, individuals from the field should be maintained in a separate room or facility from established laboratory colonies, and appropriate sanitation practices should be exercised. A method that has not been extensively used to select for pathogen-free insects is the deliberate exposure of insects to conditions that will induce physiological stress. Since physiological stress often predisposes insects to disease (Vago 1963), it may be possible to utilize prescribed stressors to induce disease in insects with latent or chronic infections.
13. The appropriate quality control parameters (e.g., time of oviposition, fecundity, weight, pupation period) should be applied at all assessment steps indicated in Fig. 51 to facilitate assessment of infection.
14. If the new colony started in this way is uninfected, destroy the contaminated colony, and sterilize everything to be used in the new colony.
15. It is essential that appropriate sanitation measures be used in conjunction with quarantine methods. In addition to standard sanitation, special attention should be paid to the sterilization of any object that is in proximity to the insects, including all rearing equipment, counter tops, walls, ceiling, glassware, dissection instruments, and even pencils and pens. In addition, dead individuals should be removed frequently, especially before surface sanitation of eggs.

tion of insects, see Chap. 8 *Microbial Contamination and Insect Rearing*.

The surface sanitation of pupae prior to eclosion can prevent the transmission of entomopathogens (e.g., viruses and bacteria) from the surface of the pupa to the adult. However, surface sanitation of pupae is often not warranted if eggs are surface-sanitized and appropriate measures have been taken to prevent contamination of larvae. Sodium hypochlorite is the most commonly used surface sanitizer for pupae.

b. General Sanitation. The implementation of appropriate sanitation procedures in the daily activities of an insectary are of critical importance in minimizing the occurrence of disease. Thorough and regular sanitization of the rearing facility is a prerequisite. Along with appropriate sanitation methods, compartmentalization of insect individuals is favored over communal rearing methods. By separating insects, the chance of horizontal transmission (e.g., ingestion of infected cadavers) is reduced substantially, and dead or moribund insects can be readily removed from the colony. Moth scales represent an allergenic concern for insectary personnel (Chap. 4 *Health and Safety Issues in the Rearing of Arthropods*), and it is possible that they can serve as an airborne carrier for microsporidian spores (e.g., *Nosema*) and microbial contaminants. The most desirable situation would be to locate the moth processing facility in a separate building from the main insectary and to utilize a proper scale handling system (Chap. 2 *Insectary Design and Construction* and Chap. 11 *Insect Rearing Production Systems, A Case Study: The Southwestern Corn Borer*). A number of authors have detailed appropriate sanitation programs including the use of air filtration systems (Steinhaus 1953, Davis 1976, Martignoni and Iwai 1977, Bell et al. 1981, Davis and Jenkins 1995). More detail on sanitation strategies is provided in Chap. 8 *Microbial Contamination and Insect Rearing*.

B. Therapeutic Measures

The term *therapeutic* is defined as the treating or curing of disease. If an entomopathogen-caused disease is detected in the colony, it may be possible, depending on the entomopathogen involved, to employ therapeutic treatments to eliminate or manage the disease. However, availability of the various therapeutic treatments does not preclude the use of disease prevention strategies (e.g., sanitation programs, adequate insectary design, and rearing strategies).

1. Quarantine. We have previously provided a generalized scheme for quarantine of field-collected individuals to reduce the chance of introducing pathogens into a colony (Table 2). The same procedures can be applied to select healthy individuals from a colony in the event that a pathogen inciting chronic disease is detected. In his pioneering work in the 1880s, Louis Pasteur utilized selection methods to rid a silkworm colony of the microsporidium *Nos. bombycis*, the causal agent of pébrine disease. He determined that the organism was transmitted from mother to progeny though eggs. Using this knowledge in combination with quarantine and selection of healthy progeny, he was able to initiate new and uninfected colonies and save the sericulture industry in Europe.

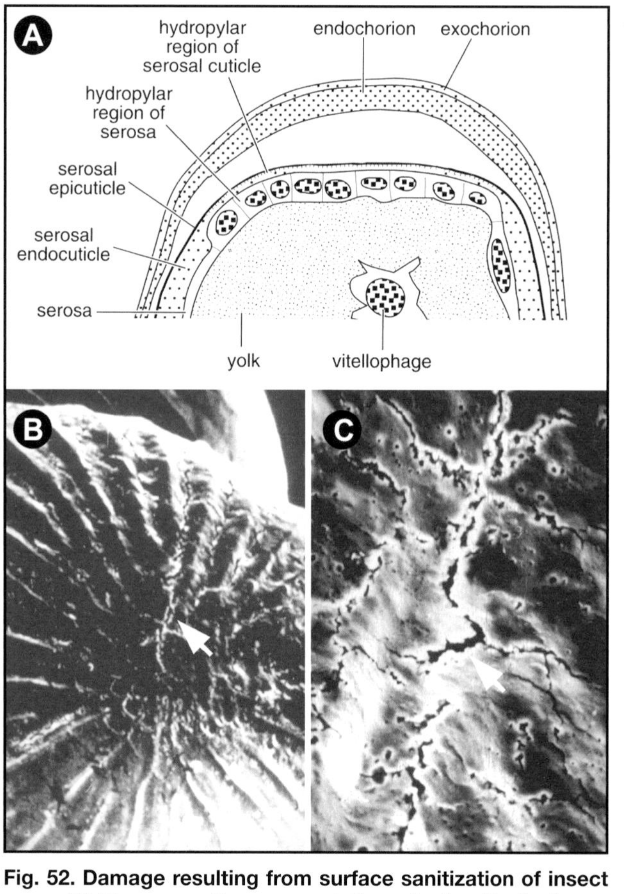

Fig. 52. Damage resulting from surface sanitization of insect eggs. (A) Diagrammatic representation of a longitudinal section through the posterior pole of a grasshopper egg. [Image A from Chapman (1998) with permission from Cambridge University Press.] (B) Micropylar erosion (arrow) following treatment of cabbage looper, *Trichoplusia ni* (Lepidoptera: Noctuidae), eggs with sodium hypochlorite in a mechanical washing machine. (C) Enlargement of a chorionic fissure (arrow). [Images B and C from Norman C. Leppla (USDA-APHIS, Gainesville, FL).]

2. Disinfection. *a. Chemotherapy.* A number of chemicals have been tested to control various entomopathogens of reared insects, with varying levels of success. Chemicals can have adverse effects on insects, usually as a function of dose; and for this reason, the application of most chemotherapies has typically taken a therapeutic rather than a prophylactic approach. Before implementing treatment with a chemotherapeutic agent, it is critical that comprehensive experimentation be conducted to determine the effects of the agent on both pathogen and host. As a result of the detrimental effects of many antimicrobial drugs on insects, the insect manager must weigh the detrimental effects on the host against the therapeutic effects.

Numerous antibacterial agents have been tested for effectiveness in protecting insects from bacterial pathogens. Once an insect ingests an antibacterial agent, a certain portion of the agent may end up in the hemolymph and impart activity against the pathogen. Sikorowski and Kent (1988) found that although tetracycline diffused readily between hemolymph and tissues of *H. zea* larvae apportionment of tetracycline among pupal tissues was uneven, and the agent reduced biosynthesis of high molecular weight proteins in the hemolymph. In another study, Sikorowski and Thompson (1985) observed that ingested streptomycin sulfate was present in the tissues and hemolymph of *H. virescens* larvae, but as the insects aged, concentrations of the antibacterial agent decreased.

Perhaps the best commercial example of using antibiotics to control an entomopathogen is the use of various antibacterial agents (e.g., oxytetracycline, tylosin, erythromycin, and sodium sulfathiazole) against *Pae. larvae*, the causal agent of American foulbrood disease of honey bees. While antibacterial agents can provide good control of American foulbrood, their extended use has led to the development of resistance in *Pae. larvae* populations (e.g., Alippi 1997). Another example where antibacterial agents have been used to successfully control bacterial infections in reared insects is the use of tetracycline against rickettsial infections. Both *Rickettsia* and *Wolbachia* can be successfully treated using tetracycline-class antibacterial agents. In contrast, *Rickettsiella popilliae*, the causal agent of blue disease in larvae of the Japanese beetle, can be cured with streptomycin but not with tetracycline treatments (Boucias and Pendland 1998). The nontarget effects of antibacterial agents on mutualistic bacteria should be considered. For more information on antibacterial agents, see Chap. 8 *Microbial Contamination and Insect Rearing*.

Fungicides have been shown to adversely influence the efficacy of fungal entomopathogens against insect pests in field environments, but relatively limited work has focused on the use of fungicides to treat fungal disease outbreaks in reared insects. An outbreak of *Conidiobolus thromboides* in cultures of the Russian wheat aphid, *Diuraphis noxia* (Homoptera: Aphididae), in a greenhouse environment was suppressed by applying the fungicide Mancozeb (1.6%) to the aphids and soil, in combination with other therapeutic and preventive measures (Hatting et al. 1999).

Most research on chemotherapeutic agents against microsporidia has focused on laboratory colonies of insects, but they have been used as a preventative in a few instances (e.g., honey bees). The drug, fumagillin (Fumidil B®) has been used extensively to manage infections by *Nos. apis* in honey bees (see reviews by Goetze and Zeutzschel 1959, Gochnauer et al. 1979, Furgala and Mussen 1978). Fumidil B inhibits meronts in the gut tissues, but it is not effective against spores or resting stages (Bailey 1953). Fumidil B has also been used against microsporidia of laboratory-reared insects including various species of mosquitoes; *O. nubialis*; *A. grandis*; *C. fumiferana*; the red flour beetle, *Tribolium castaneum* (Coleoptera: Tenebrionidae); the black blow fly, *Phormia regina* (Diptera: Calliphoridae); the forest tent caterpillar, *Malacosoma disstria* (Lepidoptera: Lasiocampidae); and *Drosophila* spp. (Brooks 1988). In most instances, the drug does not eliminate the entomopathogen from the colony, but it can provide significant suppression of infection. One of the major limitations of using Fumidil B at therapeutic doses is that adverse side effects such as increased larval period and mortality and reduced pupal weights, fecundity, and egg hatch are common (Brooks 1988). Fumidil B has also been used to treat cell lines to eliminate *Nosema* infections, providing excellent control in some cases. Although Fumidil B was originally shown to have amoebicidal activity, little effort has been made to evaluate its activity against entomopathogenic protozoa.

Another compound that has demonstrated therapeutic efficacy against microsporidia is the fungicide benomyl (Benlate®). For example, Hsiao and Hsiao

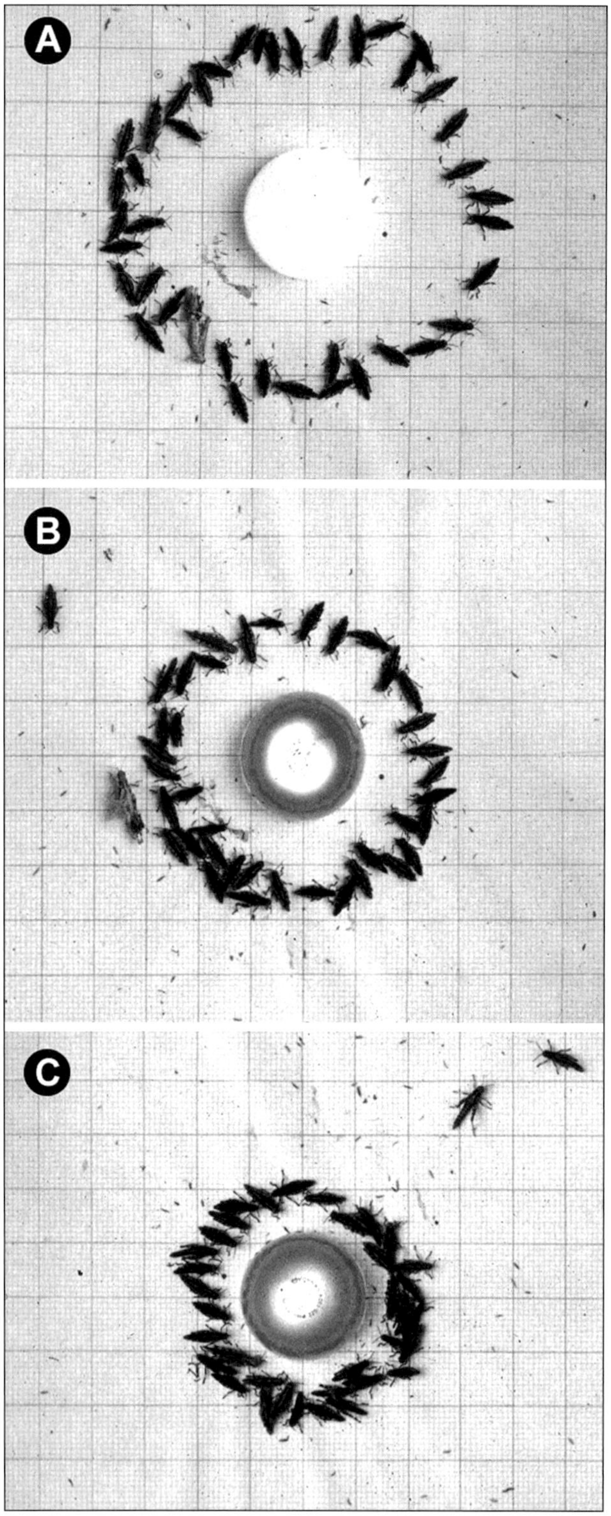

Fig. 53. Basking thermoregulatory behavior of the migratory grasshopper, *Melanoplus sanguinipes* (Orthoptera: Acrididae). Grasshoppers aggregated around incandescent light bulbs of various wattages. (A) 100 W. (B) 40 W. (C) 25 W. In all instances, the internal body temperature of the grasshoppers was determined to be 40°C, the optimal body temperature for many grasshopper species. [Images provided by Derek Lactin, Agriculture and Agri-Food Canada, Lethbridge, AB, Canada.]

(1973) found that alfalfa weevil, *Hypera postica* (Coleoptera: Curculionidae), larvae from adults infected with *Nosema* were completely free of the microsporidian 4 and 6 days after hatching when they ingested diet containing benomyl (250 ppm), compared to ≥ 90% infection prevalence in control individuals. This prompted Brooks et al. (1978) to test the effects of benomyl (500 and 1,000 ppm) on *Nos. heliothidis* infecting *H. zea*. They found that consumption of benomyl caused aberrant morphology in both meronts and spores of *Nos. heliothidis* in larvae. However, the microsporidian survived in various tissues within the larvae and proliferated rapidly in pupae and adults. Others have also determined that, like Fumidil B, the suppressive action of benomyl may be only temporary. For example, Harvey and Gaudet (1977) reported strong resurgence of microsporidiosis during the pupal and adult stages of insects reared as larvae on benomyl-treated diet. In addition, benomyl at high doses may be toxic to the host insect and is ineffective against some microsporidioses (e.g., Mussen and Furgala 1975). A number of other anti-microsporidial compounds have been tested. However, of these only the compound buquinolate (ethyl 4-hydroxy-6, 7-diisobutoxy-3-quinoline-carboxylate) has shown any promise as a chemotherapeutic agent of microsporidia in invertebrates (Brooks 1988). Most of the antibiotics used to control bacterial and fungal contaminants in artificial diets have not demonstrated any activity against either spores or vegetative cells of microsporidia.

In addition to microsporidioses, several antibiotics have shown promise for therapeutically managing amoebic diseases of reared grasshoppers. For example, thiapyrimeth (a triple sulfa drug containing 3% sulfamethazine sodium, 6% sulfathiozole sodium, and 4% sulfapyridine sodium) has been recommended as the most effective means for control of *Malameba locustae* in laboratory colonies of grasshoppers (Henry 1968, Henry and Oma 1975). This drug has been applied both therapeutically and propylactically.

b. Thermotherapy. The application of thermotherapy or thermal shock treatment has provided varying levels of success in curing insects of infection by fungi, bacteria, microsporidia, and viruses. For example, Koch (cited in Steinhaus 1947) showed that maintaining sawtoothed grain beetles, *Oryzaephilus surinamensis* (Coleoptera: Silvanidae), at

36°C eliminated symbionts from the beetles. Heat-treating potato tuberworm, *Phthorimaea operculella* (Lepidoptera: Gelichiidae), eggs in a hot water bath at 47°C for 20 min reduced the incidence of *Nosema* infection (Allen and Brunson 1947). Exposure of *H. zea* larvae to 35°C for ≥ 4 days appeared to eliminate *Nos. heliothidis* from the larvae (Hamm et al. 1971). However, heat treatment of larvae did not prevent vertical transmission of *Nosema*. Since the results were quite variable and inconsistent, the authors concluded that elevated temperatures alone could not be relied upon to eliminate these pathogens from a colony (Hamm et al. 1971). Heat shock of eggs infected with *Nosema muscidifuracis* reduced infection by 35–93%, but the temperature required to achieve this effect (47°C for 60 min) resulted in high host mortality (Geden et al. 1995). In the 1960s, Raun (1961) reported the complete elimination of *Nos. pyrausta* in *O. nubialis* by treating eggs at 43°C for 30 min. However, Lewis and Lynch (1970) observed that after several years of use, Raun's heat treatment had gradually lost its effectiveness, presumably due to the selection of heat-resistant *Nos. pyrausta*. Against viral infections, thermal therapy has demonstrated considerable promise in some instances. In other instances, disease has actually been exacerbated by exposure to high temperatures, presumably as a result of causing physiological stress in the insects. Exposure to temperatures ≥ 35°C has been shown to inhibit replication of CPV (Tanada and Chang 1968, Kobayashi and Kawase 1980) in cell culture and within the insect hosts, and Tanada (1967) observed that silkworm larvae reared at 37°C were less susceptible to both NPV and CPV. In a recent study, Kyei-Poku et al. (2003) demonstrated that continuous rearing of the parasitic wasp *Urolepis rufipes* (Hymenoptera: Pteromalidae) at 34°C eliminated *Wolbachia* from experimental populations after six generations. While thermal therapy shows considerable promise, the mechanisms by which it works are not well understood; and while it can be a valuable management tool, it generally cannot be used exclusive of other strategies.

In most thermotherapy strategies, insects are forced to maintain their body temperatures at a specific temperature. A different thermal therapy approach utilizes a phenomenon called *behavioral fever*. It has been shown that some insects infected with a pathogen will increase their body temperatures higher than noninfected insects through basking. Such basking has been documented in the migratory grasshopper, *Melanoplus sanguinipes* (Orthoptera: Acrididae); the Madagascar hissing cockroach, *Gromphadorhina portentosa* (Blattodea: Blaberidae); the cricket, *Gryllus bimaculatus* (Orthoptera: Gryllidae); and the house fly, *Musca domestica* (Diptera: Muscidae) (Boorstein and Ewald 1987, Bronstein and Conner 1984, Inglis et al. 1996, Louis et al. 1986, Watson et al. 1993). Inglis et al. (1996) demonstrated behavioral fever in *M. sanguinipes*. These grasshoppers possess an optimal body temperature of 38 to 40°C and, like other insects, elevate their body temperatures by basking (Fig. 53). In a communal rearing environment, *M. sanguinipes* infected with the entomopathogenic fungus *Bea. bassiana* were allowed to elevate their body temperature by basking *ad lib.* adjacent to an incandescent light bulb for various periods of time per day. The impact of this treatment on disease development was dramatic. In excess of 95% of the individuals that were not allowed to bask died within 10 days, while the prevalence of mortality was 59, 13, 2, and 0% in those allowed to bask for 1, 2, 4 and ≥ 6 h/d, respectively (Fig. 54).

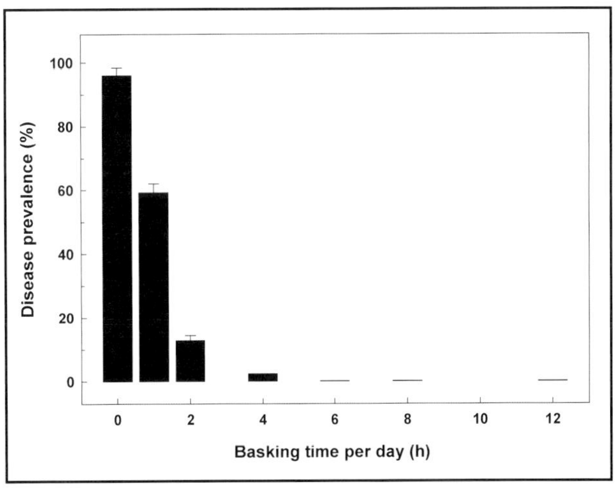

Fig. 54. Mortality of the migratory grasshopper, *Melanoplus sanguinipes* (Orthoptera: Acrididae), caused by *Beauveria bassiana* as a function of basking time by the grasshoppers. In grasshoppers allowed to bask for 1 h and 2 h per day, the prevalence of disease was reduced by 40% and 90%, respectively. No disease was detected in grasshoppers permitted to bask for 6 or more h per day. [From Inglis et al. 1996 with permission from Elsevier.]

VIII. Conclusions

The efficacious management of entomopathogens in insectaries is a prerequisite to successful insect rearing. Disease involves a complex interaction among the pathogen, the host, the environment, and time. It can manifest itself in mortality or debilitation of insects. To efficaciously manage disease, it is essential that a multilevel infrastructure be implemented. Most importantly, it is imperative that insectary management and workers are cognizant of the potential seriousness of entomopathogens in insectaries. Prevention is the primary objective. This involves aspects of insectary design, implementation of appropriate sanitation programs, and vigilance in quality control, diagnostics, and detection. Despite the best preventive measures, it is inevitable that diseases will eventually be encountered, and when they are, measures must be taken to deal with them. The consequences of ignoring the problem, misdiagnosing the problem, or implementing inappropriate management strategies are dire and could lead to the failure of the rearing program. Insectary personnel must not only possess the appropriate equipment, but they must also possess the technical ability and vigilance to rapidly diagnosis disease. In many instances, the appropriate strategy will be dependent on the insect reared and insectary environment and may require in-house experimentation to determine the best therapeutic method(s). Once a diagnosis has been made, the appropriate management strategies must be rapidly implemented and the efficacy of the strategies employed monitored.

Acknowledgments

We would like to acknowledge Amanda Lawrence for her extensive contribution to the Insect Pathology program at Mississippi State University. Amanda is responsible for many of the micrographs presented in this chapter. We also would like to thank the following persons who supplied images or specimens for this chapter: James Becnel (USDA-ARS, Gainesville, FL), Richard Brown (Department of Entomology & Plant Pathology, Mississippi State University, Mississippi State, MS), John Cunningham (Forestry Canada, Sault St. Marie, ON, Canada), Grant Duke (Agriculture & Agri-Food Canada, Lethbridge, AB, Canada), Brian Federici (University of California, Riverside, CA), John Hamm (USDA-ARS, Tifton, GA), Dan Johnson (University of Lethbridge, Lethbridge, AB, Canada), Norm Leppla (USDA-APHIS, University of Florida, Gainesville, FL), Joe MacGown (Department of Entomology & Plant Pathology, Mississippi State University, Mississippi State, MS), Eric Riddick (Biological Control and Mass Rearing Research Unit, USDA-ARS, Mississippi State, MS), Merle Shepard (Clemson University, Clemson, SC), Eloise Styer (Veterinary Diagnostic Laboratory, University of Georgia, Athens, GA), Sandra Woolfolk (Department of Entomology & Plant Pathology, Mississippi State University, Mississippi State, MS), and Jay Yanke (Agriculture & Agri-Food Canada, Lethbridge, AB, Canada). We are grateful to Mark Goettel, George Kyei-Poku, Jay Yanke, and Amanda Lawrence for providing critical reviews of the chapter. This is Lethbridge Research Centre contribution 03104.

References[3]

Adams, J. R., and J. R. Bonami. 1991. Preparation of invertebrate viruses and tissue for examination, pp. 9–30. *In* J. R. Adams and J. R. Bonami (eds.), Atlas of invertebrate viruses. CRC Press, Boca Raton, FL.

Agrios, G. N. 1978. Plant pathology, 2nd ed. Academic Press, New York, NY.

Alexopoulos, C.J, C.W. Mims, and M. Blackwell 1996. Introductory mycology. 4th ed. John Wiley & Sons, Inc., New York, NY.

Alippi, A. 1997. Background on American foulbrood in Argentina. Bee World 78: 92–95.

Allen, H. W., and M. H. Brunson. 1947. Control of *Nosema* disease of potato tuber worm, a host used in the mass production of *Macrocentrus ancylivorus*. Science 105: 394–395.

Atlas, R. M. 1993. Handbook of microbiological media, L. C. Parks (ed.). CRC Press, Boca Raton, FL.

Ausubel, F. M., R. Brent, R. E. Kingston, D. D. Moore, J. G. Seidman, J. A. Smith, and K. Struhl. 1991. Current protocols in molecular biology. Wiley Interscience, New York, NY.

[3]**Key references are indicated by "*".**

Bailey, L. 1953. Effect of fumagillin upon *Nosema apis* (Zander). Nature 171: 212.

Bando, H., T. Hayakawa, S. Asano, K. Sahara, M. Nakagaki, and T. Iizuka. 1995. Analysis of the genetic information of a DNA segment of a new virus from silkworm. Arc. Virol. 140: 1147–1155.

Beard, R. L. 1949. Physiological effects of induced hemorrhage in Japanese beetle larvae. J. N. Y. Entomol. Soc. 57: 79–91.

Becnel, J. J. 1997. Complementary techniques: Preparations of entomopathogens and diseased specimens for more detailed study using microscopy, pp. 338–353. *In* L. A. Lacey (ed.), Manual of techniques in insect pathology. Academic Press, San Diego, CA.

Becnel, J. J., and T. G. Andreadis. 1999. Microsporidia in insects, pp. 447–501. *In* M. Wittner (ed.), The Microsporidia and microsporidiosis. American Society for Microbiology, Washington, D.C.

Bell, J. V. 1969. *Serratia marcescens* found in eggs of *Heliothis zea*: tests against *Trichoplusia ni*. J. Invertebr. Pathol. 13: 151–152.

Bell, J. V., E. G. King, and J. Hammalle. 1981. Some microbial contaminants and control agents in a diet and larvae of *Heliothis* spp. J. Invertebr. Pathol. 37: 243–248.

Berns, K. 1990. Cytoplasmic polyhedrosis viruses—Reoviridae. Adv. Virus Res. 37: 173–206.

Bird, F. T., C. J. Sanders, and J. M. Burke. 1971. A newly discovered virus disease of the spruce budworm, *Choristoneura biennis* (Lepidoptera: Tortricidae). J. Invertebr. Pathol. 18: 159–161.

Björnson, S., and C. Schütte. 2003. Pathogens of mass-produced natural enemies and pollinators, pp. 133–165. *In* J. C. van Lenteren (ed.), Quality control and production of biological control agents: theory and testing procedures. CAB International, Wallingford, UK.

Bong, C. F. J., and P. P. Sikorowski. 1991. Histopathology of cytoplasmic polyhedrosis virus (Reoviridae) infection in corn earworm *Helicoverpa zea* (Boddie) larvae (Insecta: Lepidoptera: Noctuidae). Can. J. Zool. 69: 2121–2127.

Boohene, C. K., C. J. Geden, and J. J. Becnel. 2003. Development of microsporidia-infected *Muscidifurax raptor* (Hymenoptera: Pteromalidae) at different temperatures. Biol. Contr. 26: 1–7.

Boorstein, S. M., and P. W. Ewald. 1987. Costs and benefits of behavioral fever in *Melanoplus sanguinipes* infected by *Nosema acridophagus*. Physiol. Zoo. 60: 586–595.

*__Boucias, D. G., and J. C. Pendland. 1998.__ Principles of insect pathology. Kluwer Academic Press, Boston, MA.

Boucias, D. G., J. E. Maruniak, and J. C. Pendland. 1987. Characterization of an iridovirus isolated from the southern mole cricket, *Scapteriscus vicinus*. J. Invertebr. Pathol. 50: 238–245.

Boucias, D. G., J. C. Pendland, and J. P. Latge. 1991. Attachment of mycopathogens to cuticle: the initial event of mycoses in arthropod hosts, pp. 101–128. *In* G. T. Cole and H. C. Hoch (eds.), The fungal spore and disease initiation in plants and animals. Plenum Press, New York, NY.

Bourtzis, K, and T. A. Miller (eds.). 2003. Insect symbiosis, vol. 1. CRC Press, Boca Raton, FL.

Bourtzis, K, and T. A. Miller (eds.). 2006. Insect symbiosis, vol. 2. Francis and Taylor, Boca Raton, FL.

Briese, D. T., and R. J. Milner. 1986. Effect of the microsporidian *Pleistophora schubergi* on *Anaitis efformata* (Lepidoptera: Geometridae) and its elimination from a laboratory colony. J. Invertebr. Pathol. 48: 107–116.

Bronstein, S. M., and W. E. Conner. 1984. Endotoxin-induced behavioral fever in the Madagascar cockroach, *Gromphadorhina portentosa*. J. Insect Physiol. 30: 327–330.

Brooks, W. M. 1988. Entomogenous protozoa, pp. 1–149. *In* C. M. Ignoffo and N. B. Mandava (eds.), Handbook of natural pesticides, vol. V, microbial pesticides, part A. CRC Press, Boca Raton, FL.

Brooks, W. M., J. D. Cranford, and L. W. Pearce. 1978. Benomyl: effectiveness against the microsporidian *Nosema heliothidis* in the corn earworm, *Heliothis zea*. J. Invertebr. Pathol. 31: 239–245.

Bucher, G. E. 1957. Disease of the larvae of tent caterpillars caused by a sporeforming bacterium. Can. J. Microbiol. 3: 695–709.

Bucher, G. E. 1961. Artificial culture of *Clostridium brevifaciens* n. sp. and *C. malacosomae* n. sp., the caused of brachytosis of tent caterpillars. Can. J. Microbiol. 7: 641–655.

Bucher, G. E. 1963. Nonsporulating bacterial pathogens, pp. 117–146. *In* E. A. Steinhaus (ed.), Insect pathology, vol. 2. Academic Press, New York, NY.

Caetano-Anollés, G., and P. M. Gresshoff (eds.). 1997. DNA markers: protocols, applications, and overviews. Wiley-VCH, New York, NY.

Carey, G. P., T. Lescott, J. S. Robertson, L. K. Spencer, and D. C. Kelly. 1978. Three African isolates of small iridescent viruses: Type 21 from *Heliothis armigera* (Lepidoptera: Noctuidae), Type 23 from *Heteronychus arator* (Coleoptera: Scarabaeidae), Type 28 from *Lethocerus columbiae* (Hemiptera: Belostomatidae). Virology 85: 307–309.

Carter, J. B. 1973. The mode of transmission of *Tipula* iridescent virus II: route of infection. J. Invertebr. Pathol. 21: 136–143.

Chapman, R. R. 1998. The insects: structure and function. Cambridge University Press, Cambridge, UK.

Chittihunsa, T., and P. P. Sikorowski. 1995a. Occurrence of nonoccluded baculovirus (Baculoviridae) in *Microplitis croceipes* (Hymenoptera: Braconidae). Biol. Contr. 5: 622–628.

Chittihunsa, T., and P. P. Sikorowski. 1995b. Effects on nonoccluded baculovirus (Baculoviridae) infection on *Microplitis croceipes* (Hymenoptera: Braconidae). Environ. Entomol. 24: 1708–1712.

Chu, W. H., and R. P. Jaques. 1979. Pathologie d'une microsporidiose de l'arpenteuse du chou, *Trichoplusia ni* [Lep.: Noctuidae], par *Vairimorpha necatrix*. Entomophaga 24: 229–235.

Chukhri, M. G. 1987. An atlas of the ultrastructure of viruses of lepidopteran pests of plants. Amerind Publishing Co., New Delhi, India.

Cooksey, K. E. 1971. The protein crystal toxin of *Bacillus thuringiensis*: Biochemistry and mode of action, pp. 247–274. *In* H. D. Burges and N. W. Hussey (eds.), Microbial control of insects and mites. Academic Press, London, UK.

Davis, F. M. 1976. Production and handling of eggs of southwestern corn borer for host plant resistance studies. Tech. Bull. 74. Mississippi Agric. For. Exp. Stn., Mississippi State, MS.

Davis, F. M., and J. N. Jenkins. 1995. Management of scales and other insect debris: occupational health hazard in a lepidopterous rearing facility. J. Econ. Entomol. 88: 185–191.

Day, M. F. 1965. *Sericesthis* iridescent virus infection of the adult *Galleria*. J. Invertebr. Pathol. 7: 102–105.

Derkson, A. C. G., and R. R. Granados. 1988. Alteration of a lepidopteran peritrophic membrane by baculoviruses and enhancement of viral infectivity. Virology 167: 242–250.

Eilenberg, J., J. Bresciani, and J. P. Latge. 1986. Ultrastructural studies of primary spore formation and discharge in the genus *Entomophthora*. J. Invertebr. Pathol. 48: 318–324.

Evans, H., and M. Shapiro. 1997. Viruses, pp. 18–53. *In* L. A. Lacey (ed.), Manual of techniques in insect pathology. Academic Press, San Diego, CA.

Federici, B. A. 1993. Viral pathobiology in relation to insect control, pp. 81–99. *In* N. E. Beckage, S. N. Thompson, and B. A. Federici (eds.), Parasites and pathogens of insects, vol. 2: pathogens. Academic Press, New York, NY.

Federici, B. A., J. J. Hamm, and E. L. Styer. 1991. Ascoviridae, pp. 339–349. *In* J. R. Adams and J. R. Bonami (eds.), Atlas of invertebrate viruses. CRC Press, Boca Raton, FL.

Frutos, R., B. A. Federici, B. Revet, and M. Bergoin. 1994. Taxonomic studies of *Rickettsiella*, *Rickettsia*, and *Chlamydia* using genomic DNA. J. Invertebr. Pathol. 63: 294–300.

Furgala, B., and E. C. Mussen. 1978. Protozoa, pp. 1–62. *In* R. A. Morse (ed.), Honey bee pests, predators, and diseases. Comstock, Ithaca, NY.

Garrity, G. M. (ed.). 2005. The Proteobacteria, vol. 2: Bergey's manual of systematic bacteriology, 2nd ed.. Springer Verlag, New York, NY.

Geden, C. J., S. J. Long, D. A. Rutz, and J. J. Becnel. 1995. *Nosema* disease of the parasitoid *Muscidifurax raptor* (Hymenoptera: Pteromalidae): prevalence, patterns of transmission, management, and impact. Biol. Contr. 5: 607–614.

Gochnauer, T. A., B. Furgala, and H. Shimanuki. 1979. Diseases and enemies of the honey bee, pp. 617–619. *In* J. M. Graham (ed.), The hive and the honey bee. Dadant and Sons, Hamilton, IL.

Goettel, M. S., and G. D. Inglis. 1997. Fungi: Hyphomycetes, pp. 213–249. *In* L. A. Lacey (ed.), Manual of techniques in insect pathology. Academic Press, San Diego, CA.

Goetze, G., and B. Zeutzschel. 1959. *Nosema* disease of honeybees, and its control with drugs: review of research work since 1954. Bee World 40:217.

*__Goodwin, R. H. 1984.__ Recognition and diagnosis of diseases in insectaries and the effects of disease agents on insect biology, pp. 96–129. *In* E. G. King and N. C. Leppla (eds.), Advances and challenges in insect rearing. USDA, New Orleans, LA.

Gothama, A. A. A., P. P. Sikorowski, and M. R. McLaughlin. 1998. Replication of nonoccluded baculovirus associated with the parasitoid *Microplitis croceipes* (Hymenoptera: Braconidae) in *Heliothis virescens* (Lepidoptera: Noctuidae). Biol. Contr. 12: 103–110.

Granados, R. R. 1973. Entry of an insect poxvirus by fusion of the virus envelope with the host cell membrane. Virology 52: 305–309.

Greany, P. D., G. E. Allen, J. C. Webb, J. L. Sharp, and D. L. Chambers. 1977. Stress-induced septicemia as an impediment to laboratory rearing of the fruit fly parasitoid *Biosteres* (*Opius*) *longicaudatus* (Hymenoptera: Braconidae) and the Caribbean fruit fly *Anastrepha suspensa* (Diptera: Tephritidae). J. Invertebr. Pathol. 29: 153–161.

Grimont, P. A. D., and F. Grimont. 1978. The genus *Serratia*. Annu. Rev. Microbiol. 32: 221–248.

Hall, D. W. 1985. Pathobiology of invertebrate icosahedral cytoplasmic deoxyriboviruses (Iridoviridae), pp. 163–196. *In* K. Maramorosch and K. E. Sherman (eds.), Viral insecticides for biological control. Academic Press, New York, NY.

Hamm, J. J., R. L. Burton, J. R. Young, and R. T. Daniel. 1971. Elimination of *Nosema heliothidis* from a laboratory colony of the corn earworm. Ann. Entomol. Soc. Am. 64: 624–627.

Hamm, J. J., D. A. Nordlund, and O. G. Marti. 1985. Effects of a nonoccluded virus of *Spodoptera frugiperda* (Lepidoptera: Noctuidae) on the development of a parasitoid, *Cotesia marginiventris* (Hymenoptera: Braconidae). Environ. Entomol. 14: 258–261.

Hamm, J. J., E. L. Styer, and W. J. Lewis. 1988. A baculovirus pathogenic to the parasitoid *Microplitis croceipes* (Hymenoptera: Braconidae). J. Invertebr. Pathol. 52: 189–191.

Hamm, J. J., E. L. Styer, and W. J. Lewis. 1992. Three viruses found in the braconid parasitoid *Microplitis croceipes* and their implications in biological control programs. Biol. Contr. 2: 329–336.

Hamm, J. J., J. E. Carpenter, and E. L. Styer. 1996. Oviposition day effect on incidence of agonadal progeny of *Helicoverpa zea* (Lepidoptera: Noctuidae) infected with a virus. Ann. Entomol. Soc. Am. 89: 266–275.

Harris, J. L. 1986. Modified method for fungal slide culture. J. Clinc. Microbiol. 24: 460–461.

Harvey, G. T., and P. M. Gaudet. 1977. The effects of benomyl on the incidence of microsporidia and the development performance of eastern spruce budworm (Lepidoptera: Tortricidae). Can. Entomol. 109: 987–993.

Hatting, J. L., T. J. Poprawski, and R. M. Miller. 1999. Managing the entomopathogenic fungus *Conidiobolus thromboides* in Russian wheat aphid cultures. Southwest. Entomol. 24: 99–106.

Hawksworth, D. L., P. M. Kirk, B. C. Sutton, and D. N. Pegler. 1995. Ainsworth and Bisby's dictionary of the Fungi, 8th ed. CAB International, Wallingford, UK.

Heddi, A., A. M. Grenier, C. Khatchadourian, H. Charles, and P. Nardon. 1999. Four intracellular genomes direct weevil biology: nuclear, mitochondrial, principal endosymbiont, and *Wolbachia*. Proc. Natl. Acad. Sci. 12: 6814–6819.

Henry, J. E. 1968. *Malameba locustae* and its antibiotic control in grasshopper cultures. J. Invertebr. Pathol. 11: 224.

Henry, J. E., and E. A. Oma. 1975. Sulphonamide antibiotic control of *Malameba locustae* (King and Taylor) and its effect on grasshoppers. Acrida 4: 217.

Herzog, G. A., and J. R. Phillips. 1982. Manifestation of an abnormal reproductive system in a laboratory strain of the bollworm *Heliothis zea*. J. Georgia Entomol. Soc. 17: 506–513.

Hofte, H., and H. R. Whitely. 1989. Insecticidal crystal proteins of *Bacillus thuringiensis*. Microbiol. Rev. 53: 242–255.

Houston, A. S. 1991. Detection of symptomatic and asymptomatic iridescent virus infection in *Helicoverpa zea*. M.S thesis. Mississippi State University, Mississippi State.

Hsiao, T. H., and C. Hsiao. 1973. Benomyl: a novel drug for controlling a microsporidian disease of the alfalfa weevil. J. Invertr. Pathol. 22: 303–304.

Humber, R. A. 1997. Fungi: identification, pp. 154–185. *In* L. A. Lacey (ed.), Manual of techniques in insect pathology. Academic Press, San Diego, CA.

Hurd, H. 1993. Reproductive disturbances induced by parasites and pathogens of insects, pp. 87–93. *In* N. E. Beckage, S. N. Thompson, and B. A. Federici (eds.), Parasites and pathogens of insects, vol. 1 parasites. Academic Press, Inc., San Diego, CA.

Ignoffo, C. M. 1966. Susceptibility of the first-instar of the bollworm, *Heliothis zea*, and the tobacco budworm, *Heliothis virescens*, to *Heliothis* nuclear-polyhedrosis virus. J. Invertebr. Pathol. 8: 531–536.

Ignoffo, C. M., and C. Garcia. 1965. Infection of the cabbage looper, bollworm, tobacco budworm, and pink bollworm with spores of *Mattesia grandis* McLaughlin collected from boll weevils. J. Invertebr. Pathol. 7: 260–262.

Ignoffo, C. M., N. L. Marston, D. L. Hostetter, B. Puttler, and J. V. Bell. 1976. Natural and induced epizootics of *Nomuraea rileyi* in soybean caterpillars. J. Invertebr. Pathol. 27: 191–198.

Inglis, G. D., and A. M. Lawrence. 2001. Effects of *Serratia marcescens* on the F_1 generation of laboratory-reared *Heliothis virescens*. J. Econ. Entomol. 94: 362–366.

Inglis, G. D., D. L. Johnson, and M. S. Goettel. 1996. Effects of temperature and thermoregulation on mycosis by *Beauveria bassiana* in grasshoppers. Biol. Contr. 7: 131–139.

Inglis G. D., A. M. Lawrence, and F. M. Davis. 2003a. Impact of a novel species of *Nosema* on southwestern corn borers (*Diatraea grandiosella*). J. Econ. Entomol. 96: 12–20.

Inglis G. D., A. M. Lawrence, and P. P. Sikorowski. 2003b. The use of meconia to non-destructively detect sublethal infections in heliothines (Lepidoptera: Noctuidae). J. Econ. Entomol. 96: 272–279.

Innis, M. A., D. H. Gelfand, J. J. Sninski, and T. J. White. 1990. PCR protocols: a guide to methods and applications. Academic Press, San Diego, CA.

Kawase, S. 1985. Pathology associated with denosviruses, pp. 197–231. *In* K. Maramorosch, and K. E. Sherman (eds.), Viral insecticides for biological control. Academic Press, New York, NY.

Kaya, H. K., and S. P. Stock. 1997. Techniques in insect nematology, pp. 281–324. *In* L. A. Lacey (ed.), Manual of techniques in insect pathology. Academic Press, San Diego, CA.

Keeling, P.J., and N. M. Fast. 2002. Microsporidia: biology and evolution of highly reduced intracellular parasites. Annu. Rev. Microbiol. 56: 93–116.

Knoblett, J. N., and N. N. Yousseff. 1996. Detection of *Nosema locustae* (Microsporidia: Nosematidae) in frozen grasshoppers (Orthoptera: Acrididae) by using monoclonal antibodies. J. Econ. Entomol. 89: 841–847.

Kobayashi, M., and S. Kawase. 1980. Absence of detectable accumulation of cytoplasmic polyhedrosis viral RNA in the silkworm *Bombyx mori*, reared at a supraoptimal temperature. J. Invertebr. Pathol. 35: 96–98.

Krieg, A. 1987. Diseases caused by bacteria and other prokaryotes, pp. 323–354. *In* J. R. Fuxa and Y. Tanada (eds.), Epizootiology of insect diseases. Wiley, New York, NY.

Kurstak, E., and S. Garzon. 1977. Entomopoxviruses (poxviruses of invertebrates), pp. 29–66. *In* (K. Maramorosch, ed.), The atlas of insect and plant viruses. Academic Press, New York, NY.

Kudo, R. 1966. Protozoology. C. C. Thomas, Springfield, IL.

Kyei-Poku, G. A., K. D. Floate, B. Benkel, and M. S. Goettel. 2003. Elimination of *Wolbachia* from *Urolepis rufipes* (Hymenoptera: Pteromalidae) with heat and antibiotic treatments: implications for host reproduction. Biocontr. Sci. Technol. 13: 341–354.

*__Lacey, L. A. (ed.). 1998.__ Manual of techniques in insect pathology. Academic Press, San Diego, CA.

Lacey, L. A., and W. M. Brooks. 1997. Initial handling and diagnosis of diseased insects, pp. 1–15. *In* L. A. Lacey (ed.), Manual of techniques in insect pathology. Academic Press, San Diego, CA.

Lee, J. L., S. H. Hunter, and E. C. Bovee. 1985. An illustrated guide to the Protozoa. Allen Press, Lawrence, KS.

Leppla, N. C., S. L. Carlyle, and T. C. Carlysle. 1974. Effects of surface sterilization and automatic collection on cabbage looper eggs. J. Econ. Entomol. 67: 33–36.

Lewis, L. C., and R. E. Lynch. 1970. Treatment of *Ostrinia nubilalis* larvae with Fumidil B to control infections caused by *Perezia pyraustae*. J. Invertebr. Pathol. 15: 43–48.

Lighthart, B., D. Sewall, and D. R. Thomas. 1988. Effect of several stress factors on the susceptibility of the predatory mite, *Metaseiulus occidentalis* (Acari: Phytoseiidae), to the weak bacterial pathogen *Serratia marcescens*. J. Invertebr. Pathol. 52: 33–42.

Lindberg, R. B. 1974. Culture and identification of commonly encountered Gram negative bacilli: *Pseudomonas, Klebsiella-Enterobacter, Serratia, Proteus, and Providencia*, pp. 19–35. *In* J. E. Prier and H. Friedman (eds.), Opportunistic pathogens. University Park Press, Baltimore, MD.

Linley, J. R., and H. T. Nielson. 1968. Transmission of a mosquito iridescent virus in *Aedes taeniorhynchus*. I. Laboratory experiments. J. Invertebr. Pathol. 12: 7–16.

Louis, C., M. Jourdan, and M. Cabanac. 1986. Behavioral fever and therapy in a rickettsia-infected Orthoptera. Am. Physiol. 250: R991–R995.

Lynch, R. E., L. C. Lewis, and T. A. Brindley. 1976. Bacteria associated with eggs and first instar larvae of the European corn borer. Identification and frequency of occurrence. J. Invertebr. Pathol. 27: 229–237.

MacLeod, D. M. 1963. Entomophthorales infections, pp. 189–231. *In* E. A. Steinhaus (ed.), Insect pathology: an advanced treatise, vol. 2. Academic Press, New York, NY.

MacLeod, D. M., and E. Muller-Kogler. 1973. Entomogenous fungi: *Entomophthora* species with pear-shaped to almost spherical conidia (Entomophthorales: Entomophthoraceae). Mycologica 65: 823–893.

Maddox, J. V. 1966. Studies on a microsporidiosis of the armyworm, *Pseudaletia unipuncta* (Haworth). Ph.D. dissertation, University of Illinois, Urbana.

Maramorosch, K. E. 1987. Biotechnology in invertebrate pathology and cell culture. Academic Press, New York, N.Y.

Martignoni, M. E., and P. J. Iwai. 1977. Sanitation program for diet preparation. Standard operating procedures, RWU-2203, Forestry Sciences Laboratory. U.S. Forest Service, Corvallis, OR.

Martignoni, M. E., and P. J. Iwai. 1986. A catalogue of viral disease of insects, mites, and ticks. Gen. Tech. Rep. PNW-195. U.S. Dept. Agric., For. Serv., Pac. Northwest Res. Stn., Corvallis, OR.

McCoy, C. W., R. A. Samson, and D. G. Boucias. 1988. Entomogenous fungi, pp. 151–236. *In* C. M. Ignoffo and N. B. Mandava (eds.), Handbook of natural pesticides, vol. V, microbial insecticides, part A. CRC Press, Boca Raton, FL.

McLaughlin, R. E. 1965. Some relationships between the boll weevil, *Anthonomus grandis* Boheman, and *Mattesia grandis* McLaughlin (Protozoa: Neogregarinida). J. Invertebr. Pathol. 7: 464–473.

McLaughlin, R. E., and J. C. Keller. 1964. Antibiotic control of an epizootic caused by *Serratia marcescens* Bizio in the boll weevil, *Anthonomus grandis* Boheman. J. Invertebr. Pathol. 6: 481–485.

McLaughlin, R. E., and J. Myers. 1970. *Ophryocystis elektroscirrha* sp. n., a neogregarine pathogen of the monarch butterfly *Danaus plexippus* (L.) and the Florida queen butterfly *D. gilippus berenice* Cramer. J. Protozool. 17:300–305.

Merrill, W. 1980. Theory and concepts of plant pathology. Department of Plant Pathology, The Pennsylvania State University, University Park, PA.

Mohamed, A. K. A., P. P. Sikorowski, and J. V. Bell. 1978. Histopathology of *Nomuraea rileyi* in larvae of *Heliothis zea* and *in vitro* enzymatic activity. J. Invertebr. Pathol. 31: 345–352.

Murphy, F. A., C. M. Fauquet, D. H. L. Bishop, S. A. Ghabrial, A. W. Jarvis, G. P. Martelli, M. A. Mayo, and M. D. Summers (eds.). 1995. Virus taxonomy; classification and nomenclature of viruses. Sixth Report of the International Committee on Taxonomy of Viruses. Springer-Verlag, New York, NY.

Murray, P. R., E. J. Baron, J. H. Jorgensen, M. A. Pfaller, and R. H. Yolken. 2003. Manual of clinical microbiology, vol. 1. ASM Press, Washington, DC.

Mussen, E. C., and B. Furgala. 1975. Benomyl ineffective against *Nosema apis*. Am. Bee J. 115: 478.

Neter, E. 1974. Opportunistic pathogens: immunological aspects, pp. 37–48. *In* J. E. Prier and H. Friedman (eds.), Opportunistic pathogens. University Park Press, Baltimore, MD.

Ohba, M. 1996. *Bacillus thuringiensis* populations naturally occurring on mulberry leaves: a possible source of the populations associated with silkworm-rearing insectaries. J. Appl. Bacteriol. 80: 56–64.

Padhi, S. B. 1985. Viral proteins for the identification of insect viruses, pp. 55–78. *In* K. Maramorosch and K. E. Sherman (eds.), Viral insecticides for biological control. Academic Press, New York, NY.

Papierok, B., and A. E. Hajek. 1997. Fungi: Entomophthorales, pp. 187–212. *In* L. A. Lacey (ed.), Manual of techniques in insect pathology. Academic Press, San Diego, CA.

Persing, D. H. (ed.). 1996. PCR protocols for emerging infectious diseases. ASM Press, Washington, DC.

Pilley, B. M. 1976. A new genus, *Vairimorpha* (Protozoa: Microsporidia), for *Nosema necatrix* Kramer 1965: pathogenicity and life cycle in *Spodoptera exempta* (Lepidoptera: Noctuidae). J. Invertebr. Pathol. 28: 177–183.

Pitt, J. I. 2000. A laboratory guide to common *Penicillium* species, 3rd ed. Food Science Australia, North Ryde, NSW, Australia.

Poinar, G. O. 1979. Nematodes for biological control of insects. CRC Press, Boca Raton, FL.

***Poinar, G. O., and G. M. Thomas. 1978.** Diagnostic manual for the identification of insect pathogens. Plenum Press, New York, NY.

***Poinar, G. O., and G. M. Thomas. 1984.** Laboratory guide to insect pathogens and parasites. Plenum Press, New York, NY.

Raina, A. K., and J. R. Adams. 1995. Gonad-specific virus of corn earworm. Nature 374: 770.

Raun, E. S. 1961. Elimination of microsporidiosis in laboratory-reared European corn borers by the use of heat. J. Insect Pathol. 3: 446–468.

Relman, D. A., and D. H. Persing. 1996. Genotypic methods for microbial identification, pp. 3–32. *In* D. H. Persing (ed.), PCR protocols for emerging infectious diseases. ASM Press, Washington, DC.

Rothman, L. D., and J. H. Myers. 1996. Debilitating effects of viral diseases on host Lepidoptera. J. Invertebr. Pathol. 67: 1–10.

St. Leger, R. J., and L. Joshi. J. 1997. The application of molecular techniques to insect pathology with emphasis on entomopathogenic fungi, pp. 367–394. *In* L. A. Lacey (ed.), Manual of techniques in insect pathology. Academic Press, San Diego, CA.

***Samson, R. A., H. C. Evans, and J. -P. Latgé. 1988.** Atlas of entomopathogenic Fungi. Springer-Verlag, Berlin.

Sheehan, D. C., and B. B. Hrapchak. 1987. Theory and practice of histotechnology. Battelle Press, Columbus, OH.

Sikorowski, P. P. 1985. Pecan weevil pathology, pp. 87–101. *In* W. W. Neel (ed.), Pecan weevil: research perspective. Quail Ridge Press, Jackson, MS.

Sikorowski, P. P. and A. D. Kent. 1988. Distribution of tetracycline in haemolymph and tissue of all stages of corn earworm (*Heliothis zea* Boddie). Comp. Biochem. Physiol. 91: 437–441.

Sikorowski, P. P., and A. M. Lawrence. 1994. Microbial contamination and insect rearing. Am. Entomol. 40: 240–253.

Sikorowski, P. P., and A. M. Lawrence. 1997. Major diseases of *Heliothis virescens* and *Helicoverpa zea* in Mississippi field and insectaries. Tech. Bull. 218. Miss. Agric. and For. Exp. Stn., Mississippi State, MS.

Sikorowski, P. P., and A. M. Lawrence. 1998. Transmission of *Serratia marcescens* (Enterobacteriaceae) in *Heliothis virescens* colonies (Lepidoptera: Noctuidae). J. Invertebr. Pathol. 12: 50–55.

Sikorowski, P. P., and A. C. Thompson. 1985. Residual quantities of streptomycin sulfate in haemolymph and tissue of *Heliothis virescens* reared on treated diet. Pharmacol. Biochem. Behav. 23: 525–528.

Sikorowski, P.P, and G. E. Tyson. 1984. Per os transmission of iridescent virus of *Heliothis zea* (Lepidoptera: Noctuidae). J. Invertr. Pathol. 44: 97-102.

Sikorowski, P. P., J. R. Broome, and G. L. Andrews. 1971. Simple methods for detection of cytoplasmic polyhedrosis virus in *Heliothis virescens*. J. Invertebr. Pathol. 17: 451–452.

Sikorowski, P. P., G. L. Andrews, and J. R. Broome. 1973. Transovum transmission of a cytoplasmic polyhedrosis virus of *Heliothis virescens* (Lepidoptera: Noctuidae). J. Invertebr. Pathol. 21: 41–45.

Sikorowski, P. P., J. E. Powell, and A. M. Lawrence. 1992. Effects of bacterial contamination on development of *Microplitis croceipes* [Hym: Braconidae]. Entomophaga 37: 475–481.

Sikorowski, P. P., A. A. A. Gothama, and A. M. Lawrence. 1997. Potential consequences of field releasing virus infected endoparasitoids. Int. Pest Mngmt. Rev. 2: 113–124.

Sikorowski, P. P., A. M. Lawrence, and G.D. Inglis. 2001. Effects of *Serratia marcescens* on rearing of tobacco budworm, *Heliothis virescens* (Lepidoptera: Noctuidae). Am. Entomol. 47: 51–61.

Sokolova, Y. Y., V. V. Dolgikh, E. V. Morzhina, E. S. Nassonova, I. V. Issi, R. S. Terry, J. E. Ironside, J. E. Smith, and C. R. Vossbrinck. 2003. Establishment of the new genus *Paranosema* based on the ultrastructure and molecular phylogeny of the type species *Paranosema grylli* Gen. Nov., Comb. Nov. (Sokolova, Selezniov, Dolgikh, Issi 1994), from the cricket *Gryllus bimaculatus* Deg. J. Invertebr. Pathol. 84: 159–172.

Sprague, V, J. J. Becnel, and E. I. Hazard. 1992. Taxonomy of the phylum Microspora. Crit. Rev. Microbiol. 18: 285–295.

Sri-Arunotai, S., P. P. Sikorowski, and W. W. Neel. 1975. Study of pathogens of the pecan weevil larvae. Environ. Entomol. 4: 790–792.

***Steinhaus, E. A. 1947.** Insect microbiology. Comstock Publishing Company, Ithaca, NY.

***Steinhaus, E. A. 1949.** Principles of insect pathology. McGraw-Hill, New York, NY.

Steinhaus, E. A. 1953. Diseases of insects reared in the laboratory or insectary. Calif. Agric. Exp. Stn. Ext. Serv. Leafl. 9.

Steinhaus, E. A. 1958. Crowding as a possible stress factor in insect disease. Ecology 39: 503–514.

Steinhaus, E. A. 1959. *Serratia marcescens* Bizio as an insect pathogen. Hilgardia 28: 351–380.

***Steinhaus, E. A. 1963.** Background for the diagnosis of insect diseases, pp. 549-589. *In* E. A. Steinhaus, (ed.), Insect pathology: an advanced treatise. vol. 2. Academic Press, New York, NY.

Steinhaus, E. A., and M. E. Martignoni. 1970. An abridged glossary of terms used in invertebrate pathology. Pacific Northwest Forest and Range Experiment Station, U.S. Forest Service, Corvallis, OR.

Styer, E. L., and J. J. Hamm. 1991. Report of a densovirus in a commercial cricket operation in the southeastern United States. J. Invertebr. Pathol. 58: 283–285.

Suzuki, A., S. Kuyama, Y. Kodair, and S. Tamura. 1966. Structural elucidation of destruxin A. Agric. Biol. Chem. (Tokyo) 30: 517–518.

Suzuki, A., H. Taguchi, and S. Tamura. 1970. Isolation and structure elucidation of three new insecticidal cyclodepsipeptides, destruxins C and D and desmethyldestruxin B, produced by *Metarrhizium anisopliae*. Agric. Biol. Chem. (Tokyo) 34: 813–816.

Suzuki, A., K. Kawakami, and S. Tamura. 1971. Detection of destruxins in silkworm larvae infected with *Metarrhizium anisopliae*. Agric. Biol. Chem. (Tokyo) 35: 1641–1643.

Tanada, Y. 1967. Effects of high temperatures on the resistance of insects to infectious diseases. J. Seric. Sci. Japan 36: 333–339.

Tanada, Y., and G. Y. Chang. 1962. An epizootic resulting from a microsporidian and two virus infections in the armyworm, *Pseudaletia unipuncta* (Haworth). J. Invertebr. Pathol. 4: 128–131.

Tanada, Y., and G. Y. Chang. 1968. Resistance of the alfalfa caterpillar, *Colias eurytheme*, at high temperatures to a cytoplasmic-polyhedrosis virus and thermal inactivation point of the virus. J. Invertebr. Pathol. 10: 79–83.

*__Tanada, Y. and H. K. Kaya. 1993.__ Insect pathology. Academic Press, New York, NY.

Tanaka, S. 1971. Cross transmission of cytoplasmic polyhedrosis viruses, pp. 201–207. In H. Aruga and Y. Tanada (eds.), The cytoplasmic-polyhedrosis virus of silkworm. University of Tokyo Press, Tokyo, Japan.

Thiery, I., and E. Frachon. 1997. Identification, isolation, culture and preservation of entomopathogenic bacteria, pp. 55–77. *In* L. A. Lacey (ed.), Manual of techniques in insect pathology. Academic Press, San Diego, CA.

Thomas, G. M., and G. O. Poinar, Jr. 1973. Report of diagnoses of diseased insects 1962–1972, University of California Division of Agricultural Sciences. Hilgardia 42: 320–322.

Thompson, A. C., and P. P. Sikorowski. 1979. Effects of *Nosema heliothidis* on fatty and amino acids in larvae and pupae of the bollworm *Heliothis zea*. Comp. Biochem. Physiol. 63A: 325–328.

Tompkins, G. J. 1991. Purification of invertebrate viruses, pp. 31–40. *In* J. R. Adams and J. R. Bonami (eds.), Atlas of invertebrate viruses. CRC Press, Boca Raton, FL.

Trammer, T., F. Dombrowski, M. Doehring, W. A. Maier, and H. M. Seitz. 1997. Opportunistic properties of *Nosema algerae* (Microspora), a mosquito parasite, in immunocompromised mice. J. Euk. Microbiol. 44: 258–262.

*__Tzean, S. S., L. S. Hsieh, and W. J. Wu. 1997.__ Atlas of entomopathogenic fungi from Taiwan. Council of Agriculture, Executive Yuan, Taiwan.

Undeen, A. H., and J. Vávra. 1997. Research methods for entomopathogenic protozoa, pp. 117–151. *In* L. A. Lacey (ed.), Manual of techniques in insect pathology. Academic Press, San Diego, CA.

Vago, C. 1963. Predispositions and interrelations in insect diseases, pp. 339–370. *In* E. A. Steinhaus (ed.), Insect pathology, an advanced treatise. Academic, New York, NY.

Volkman, L. E. 1985. Classification, identification, and detection of insect viruses by serologic techniques, pp. 27–53. *In* K. Maramorosch and K. E. Sherman (eds.), Viral Insecticides for biological control. Academic Press, New York, NY.

Watson, D. W., B. A. Mullens, and J. J. Petersen. 1993. Behavioral fever response of *Musca domestica* (Diptera: Muscidae) to infection by *Entomophthora muscae* (Zygomycetes: Entomophthorales). J. Invertebr. Pathol. 61: 10–16.

Webster, J. 1980. Introduction to fungi, 2nd ed. Cambridge University Press, Cambridge, UK.

*__Weiser, J. 1977.__ An atlas of insect diseases. Academia, Prague, Czechoslakia.

Weiser, J., and J. D. Briggs. 1971. Identification of pathogens, pp. 13–66. *In* H. D. Burges (ed.), Microbial control of insects and mites, 1971. Academic Press, New York, NY.

Williams, T. 1996. The iridoviruses. Adv. Virus Res. 46: 345–412.

Wood, H. A., and R. R. Granados. 1991. Genetically engineered baculoviruses as agents for pest control. Annu. Rev. Microbiol. 45: 69–87.

Woolfolk, S. W., and G. D. Inglis. 2004. Microorganisms associated with field-collected *Chrysoperla rufilabris* (Neuroptera: Chrysopidae) adults with emphasis on yeast symbionts. Biol. Contr. 29: 155–168.

Woolfolk, S. W., A. C. Cohen, and G. D. Inglis. 2004. Morphology of the alimentary canal of *Chrysoperla rufilabris* (Neuroptera: Chrysopidae) adults in relation to microbial symbionts. Ann. Entomol. Soc. Am. 97:796–808.

Wouts, W. M. 1984. Nematode parasites of lepidopterans, pp. 655–696. *In* W. R. Nickle (ed.), Plant and insect nematodes. Marcel Dekker, New York, NY.

Appendix A

History of disease data sheet. An example of an accession form used to record the preliminary history of diseased insects received for diagnosis. Information pertaining to the principal manifestation, the past history, companion history, and history of the present condition may be recorded in the spaces toward the bottom. [From Steinhaus (1963) with permission from Elsevier.]

ACCESSION AND HISTORY CARD

Date received.............................. Collector's Number.............................. Accession No..............................

Scientific name of insect..............................

Common name of insect..............................

Condition of specimens..............................

Method of submittal or shipment..............................

No. of specimens: Dead.............. Ailing.............. Healthy.............. Stage.............. Instar..............

Males.............. Females.............. Time held prior to submittal.............. Temp..............

Abundance or prevalence of insect..............................

Host plant or animal..............................

Extent of disease outbreak..............................

....

Environmental conditions (including temperature and humidity)..............................

....

....

Possibility of insecticide treatment..............................

Name of collector.............................. Address of collector..............................

....

Name of submitter.............................. Address of submitter..............................

....

Cover letter dated.............................. Date collected.............................. Locality..............................

....

Additional description or history provided by collector or submitter:

Additional history provided by diagnostician:

Recorder..............................

UC DIP Form 1A

Appendix B

Physical examination data sheet. An example of a form on which may be recorded observations made during the course of the physical examination of diseased insects. [From Steinhaus (1963) with permission from Elsevier.]

PHYSICAL EXAMINATION

(Observations made macroscopically or with dissecting microscope)

Date.. Insect.. Accession No.............................

A. Abnormal movements and behavior:

B. General external appearance; general condition and odor (relate to healthy specimen)

Pre-mortem: Post-mortem:

C. Signs and symptoms: (Record according to anatomical system or tissue, correlating description with appropriate number.)

1. Integument
2. Circulatory system
3. Alimentary system
4. Excretory system
5. Glandular system
6. Muscular system
7. Respiratory system
8. Nervous system and sense organs
9. Reproductive system
10. Fat body (adipose tissue)

Photographs..

Examiner...

UC DIP Form 2A

Appendix C

Laboratory examination data sheet. An example of a form on which may be recorded preliminary microscopic findings and which may be used to indicate the type of additional examinations that might be required. [From Steinhaus (1963) with permission from Elsevier.]

LABORATORY EXAMINATION SCAN SHEET

Date........................ Insect........................ Accession No........................

Microscopic findings: (Record according to anatomical system or tissue, correlating description with appropriate number.)

1. Integument
2. Circulatory system
3. Alimentary system
4. Excretory system
5. Glandular system
6. Muscular system
7. Respiratory system
8. Nervous system and sense organs
9. Reproductive system
10. Fat body (adipose tissue)

Sketch:

Photographs........................

Examined further for: (Check square ☒ if Special Sheet is used.)

☐ Viruses	☐ Protozoa	☐ Nutritional disturbances	☐ Chemical injuries
☐ Rickettsiae	☐ Nematodes	☐ Genetic anomalies	☐ Hematology
☐ Bacteria	☐ Insect parasites	☐ Nongenetic teratologies	☐ Serology
☐ Yeasts	☐ Neoplasms	☐ Mechanical injuries	☐ Histopathology
☐ Fungi	☐ Deranged metabolism	☐ Physical injuries	☐ Biochemistry

☐ Other anomalies, disturbances, or pathological conditions

General microbiology and miscellaneous:

Method of examination: (Method of external sterilization, types of preparations, stains used, etc.)

Examiner........................

UC DIP Form 2B-1m-5,'61(B9695s)9915

Appendix D

Final diagnosis data sheet. An example of a form depicting the manner in which the diagnosis may be recorded. A brief statement as to the reasons for the particular conclusions reached should be included in the entry under *Final Diagnosis*. [From Steinhaus (1963) with permission from Elsevier.]

DIAGNOSIS

Date diagnostic examination begun.. Date completed..

Date receipt card sent.................................. Signed by.. How sent..................................

Preliminary diagnosis:

By:..

Tentative diagnosis:

By:..

FINAL DIAGNOSIS

By:.. Confirmed by:..

Material referred to.. Date..

Date report sent to collector or submitter.. How?.................................. By..................................

Additional specimens needed?.............................. Kind needed?..

Disposition of material..

Microscope slides or photographs..

New record?.............................. New pathogen?.............................. New host?.............................. New area?..............................

Previous record or reference (if not common)..

Published?.............................. Reference..

Department of Insect Pathology
University of California, Berkeley
UC DIP Form 1B-5c-4,'61(B9315s)9138

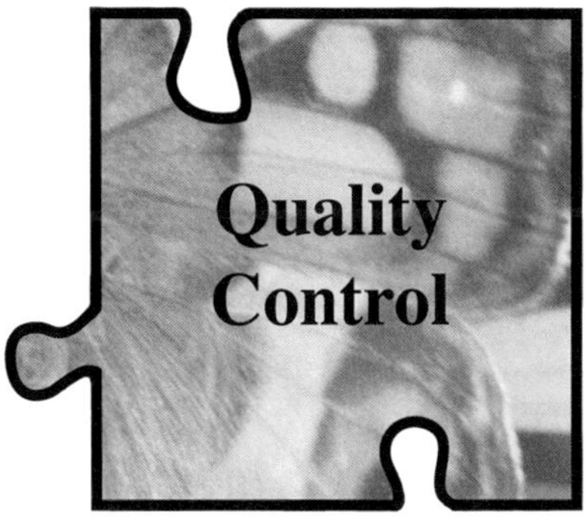

Chapter 10

The Basics of Quality Control for Insect Rearing

NORMAN C. LEPPLA
Professor and IPM Coordinator
Department of Entomology & Nematology
University of Florida, Institute of Food and Agricultural Sciences
Gainesville, FL

I. Introduction

Quality control has become an established component of insect rearing systems, regardless of their purposes or complexities. This is because it helps to assure the reliable and consistent production of insects that satisfy required specifications and standards. As mass rearing capabilities were developed to support pest management programs in the second half of the twentieth century (Cohen 2004), production failures became extremely costly and difficult to overcome. A means was needed to organize and monitor all of the components essential for rearing the insects and troubleshooting the systems to identify and correct problems. Unfortunately, existing industrial quality control procedures developed for nonliving products did not fit the production of living organisms. This is because, unlike metal and plastic parts, insects immediately adjust to both intentional and unintentional changes in the environment. Due to differential survival and reproduction, genetic adaptation occurs over generations accompanied by exponential growth of the survivors. A dynamic relationship thus exists between quality control monitoring and the underlying insect rearing procedures. Quality control for insect rearing systems was consequently designed to incorporate variables in both the environment and biological products.

Basic principles of production and quality control emerged as insect rearing became a well-organized and professional activity. Skill in rearing based on training and experience remained indispensable, the so-called "brown thumb" analogous with the gardener's green thumb. Insect quality is still monitored subjectively by the direct observation and scrutiny of experts. There is economy of scale to an optimum threshold size, and then insect rearing facilities should be modular and redundant so that environmental quality is easier to maintain and backup is provided in the event of catastrophic production failures. Even in relatively small colonies, division of processes and associated labor increases precision and enables work to be performed in isolated physical locations. This isolation can be especially useful for preventing microbial contamination of clean areas or for separating closely related species. Traffic patterns that cross from dirty to clean areas eventually result in the spread of diseases that reduce the quality of the insects. Thus, mass rearing systems are not merely the sum of their individual parts, rather each step along the production chain influences production efficiency and insect quality.

Quality control in insect rearing depends on effectively establishing and maintaining colonies that meet expectations relative to their intended purposes (Fig. 1). There are many important uses for insects, including pest management, aesthetics, and biological products. Once established, considerable expertise is required to maintain a rearing system and monitor colonies for consistent quality. Quality control in insect rearing can be divided into production, process, and product control (Leppla and Ashley 1989). In production control, monitoring assures the performance of standard operating procedures (SOPs) (Chap. 3 *The Insectary Manager*). Process and product control are based on standardized tests developed to monitor insect characteristics predictive or indicative of quality. Insect stages are sampled, various characteristics are measured, and the resulting data are analyzed and plotted on process control charts. Information from these three types of monitoring is used to evaluate colonies and decide when changes are warranted.

It is advisable to effectively review rearing programs periodically and make necessary changes rather than when there is a production crisis. To support these reviews, there are quality control manuals for individual production facilities and access to insect rearing expertise. There also are two primary, international quality control systems: the International Organization for Standardization (IOS) that emphasizes SOPs and participation by both workers and management, and the American Society for Testing and Materials (ASTM) that focuses on product performance. Additionally, the International Organization for Biological Control (IOBC) has developed specific guidelines to assure the quality of commercially produced natural enemies (van Lenteren 2003). Total quality control (TQC) for insect rearing has been adapted from an industrial model that defines the generic elements and provides a structure for organizing production systems (Leppla and Fisher 1989).

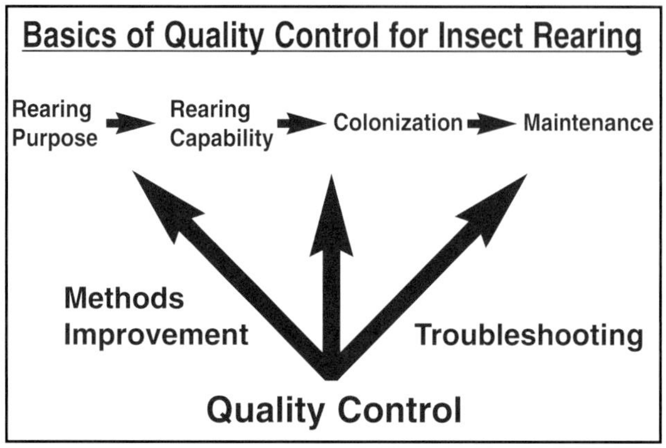

Fig. 1. The basics of quality control for insect rearing include a purpose for producing the insects, a rearing capability, a suitable founder population to establish the colony, and maintenance of the colony to assure stable survival and reproduction. Quality control tests are used to improve methods and troubleshoot and correct problems.

II. Colony Establishment and Maintenance

Quality control principles are applicable for any size insect rearing program. For small colonies, quality control is intuitive and subjective because a single person performs virtually all of the rearing processes. On a large scale, however, rearing involves the separation of tasks but with interdependence and shared responsibility for meeting a specified standard of insect quality. These tasks can be accomplished in different places and times for a single colony or involve geographically distinct colonies from which insects are pooled. For example, the screwworm, *Cochliomyia hominivorax* (Diptera: Calliphoridae), has been produced in Mexico and used in Africa; and Mediterranean fruit flies, *Ceratitis capitata* (Diptera: Tephritidae), from Guatemala, Mexico, and Hawaii have been released in California. These insect mass rearing and associated area-wide pest management programs have become very expensive: for the screwworm, about $40 million per year at its peak (Leppla 1989). By 2004, Mediterranean fruit fly production exceeded two billion insects per week at Metapa, Mexico. Regardless of the size of an insect rearing effort, application of quality control principles and practices more than pay for themselves by increasing production efficiency and reliability and insect quality.

A. Purposes for Rearing Insects

Insects are reared for many purposes, including education, personal collections, stationary exhibits, insect zoos, butterfly houses, household pets, human and animal food, agricultural and medical research, pest management, and pollination (Leppla 2002). They are used in the form of biochemical products, body parts, individuals, and small to large populations. Therefore, requirements for quality vary widely—from large size to complex behaviors performed in the field. Parasites and predators released for biological control of agricultural pests probably have the most complex requirements. For example, *Encarsia formosa* (Hymenoptera: Aphelinidae) is widely sold to control greenhouse whitefly, *Trialeurodes vaporariorum* (Homoptera: Aleyrodidae), and must survive the prevailing environmental conditions, fly and orient to crop plants, locate its host, and reproduce (van Lenteren and Woets 1988; Bigler 1989, 1991). Consequently, rearing procedures and quality control standards must be matched to the purpose for which the insects are reared (Chambers 1975, Boller and Chambers 1977a).

B. Types of Insect Rearing Systems

There are three basic types of insect rearing: small scale, single species; medium scale, multiple species; and mass rearing, single species (Leppla 2004). In small-scale, single-species rearing, usually both immature and adult stages are fed on host plants or animals, although artificial diets are sometimes substituted to improve production efficiency and consistency. Natural environments and oviposition substrates are duplicated or slightly modified from nature, and a single person can perform all of the rearing processes (Fig. 2).

Medium-scale, multiple-species insect rearing programs are usually maintained in centralized facilities by a small staff (Leppla et al. 1978). There is economy of scale in these programs because diet preparation, egg treatment, larval rearing, harvesting of pupae, and adult colony maintenance can be combined for similar species (Fig. 3). However, isolation must be ensured when closely related species or multiple strains are reared in the same facility. This type of rearing is common in research laboratories, such as those used for insecticide or transgenic crop development and in commercial natural enemy production facilities (Penn et al. 1998).

Single-species mass rearing primarily supports biologically-based pest management and is accomplished in factory-like facilities with controlled environments, artificial diets and oviposition substrates, mechanized equipment, and different processes performed by separate work units. The largest facilities, those used to rear the screwworm and the Mediterranean fruit fly, employ several hundred workers during three shifts per 24-hour day, seven days per week throughout the year (Fig. 4). Insect mass rearing is not merely a multiplication of small-scale rearing (Peterson 1964); rather it is uniquely designed to fit factory-like production (Leppla and Ashley 1989).

The easiest insects to rear are relatively small, plant-feeding terrestrial species that have continuous, short generations and no unusual environmental requirements. Examples include crickets, mealworms, certain flies, and a variety of stored product pests. Because of their relatively uncomplicated life histories, insects that have been colonized and reared for multiple generations include 300+ species of moths and butterflies (Lepidoptera) (Leppla et al. 1982), 200+ species of beetles (Coleoptera), nearly 200 species of flies (Diptera), 100+ species of bugs (Hemiptera), and 100+ species of bees and wasps (Hymenoptera) (Edwards et al. 1987). Grasshoppers and katydids (Orthoptera), lacewings (Neuroptera), cockroaches (Dictyoptera), termites (Isoptera), and fleas (Siphonaptera) are also reared routinely but only about 10–20 species. Undoubtedly, many more species could be reared for multiple generations using techniques developed for close relatives, and single-stage or single-generation rearing of some species could be expanded to support continuous production.

Fig. 2. Materials for rearing larvae of a blister beetle, *Nemógnatha* spp. (Coleoptera: Melòidae). Although this is small-scale, single-species rearing, the quality control concepts are the same as for medium-scale and mass rearing.

C. Options for Colonizing Insects

The genetic variability of a founding population determines its potential for adapting to a rearing environment and producing subsequent generations of high quality insects (Hoy 1979; Bartlett 1984, 1985). Genetic bottlenecks are minimized when insect survival and reproduction are maximized (Mackauer 1972, 1976). Most of the individuals should develop at a normal rate, survive to the adult stage, and produce viable offspring. A general rule-of-thumb for achieving an acceptable level of genetic variability is to obtain a total of at least 200 individuals from five to ten locations in the field, hold them individually to eliminate diseases and natural enemies, isolate pairs of adults to verify that they reproduce, and use surface-sterilized eggs to infest larval rearing containers. Considerable experience may be required to select and handle founders and adjust the rearing system so that it promotes their survival (Fig. 5; Leppla 1994). Regardless, within a few generations, laboratory-reared insects typically become selected for rapid development and maturation, minimal locomotion, and reduced behavioral discrimination (Boller 1972). If bottlenecks cause excessive mortality, it may be necessary to improve the rearing methods and establish a new colony. Also, because of low yields during colonization of insects from the field, colonies often are initiated by using eggs from established insectaries. Where to obtain a new insect strain, and when and how it should eventually be replaced, has been deliberated most intensively for *C. hominivo-*

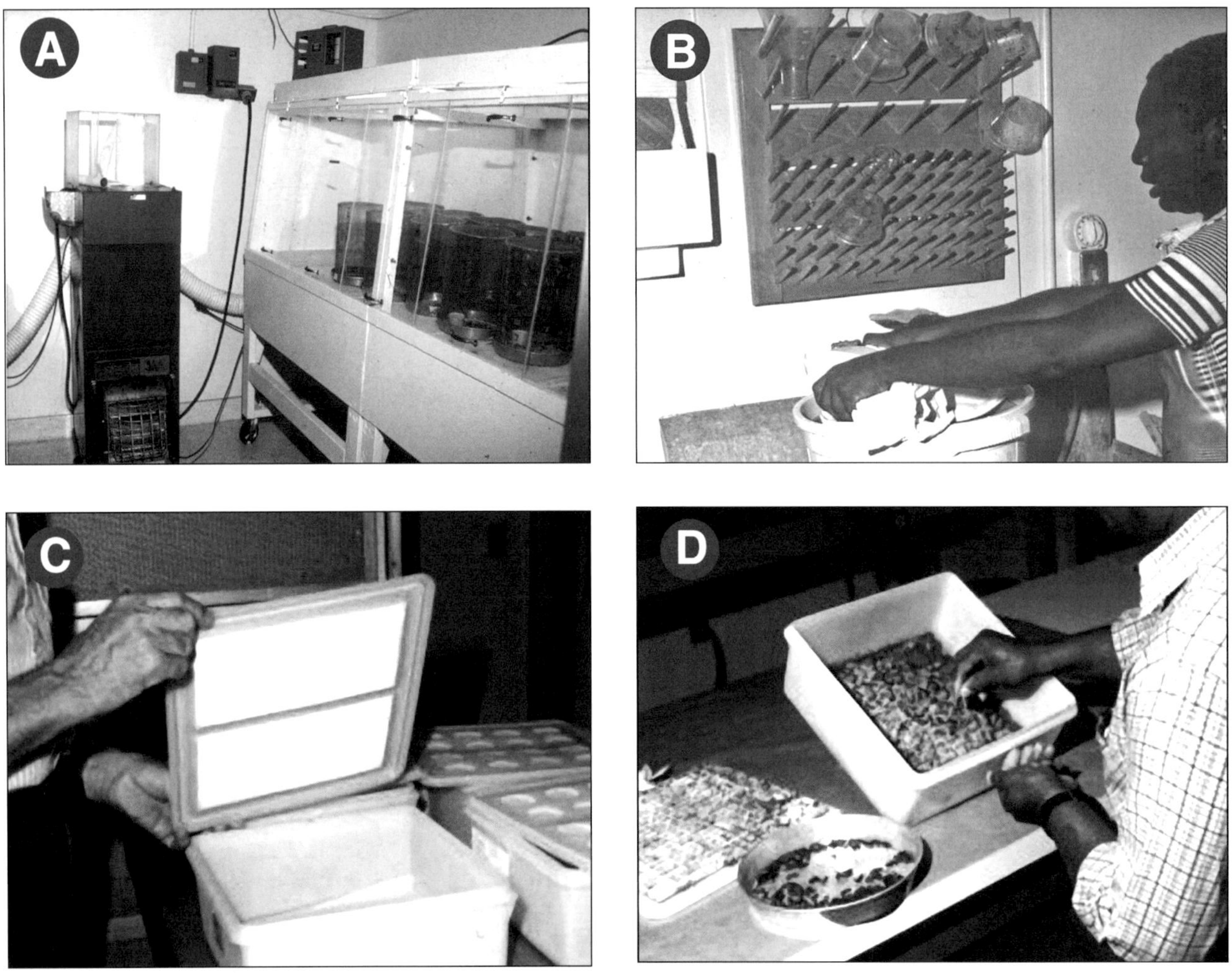

Fig. 3. Rearing the fall armyworm, *Spodoptera frugiperda* (Lepidoptera: Noctuidae), in a medium-scale, multiple-species rearing system. (A) Maintenance of adult cages. (B) Surface sterilization of eggs on an oviposition substrate. (C) Diet infestation with eggs on container lids. (D) Harvesting pupae.

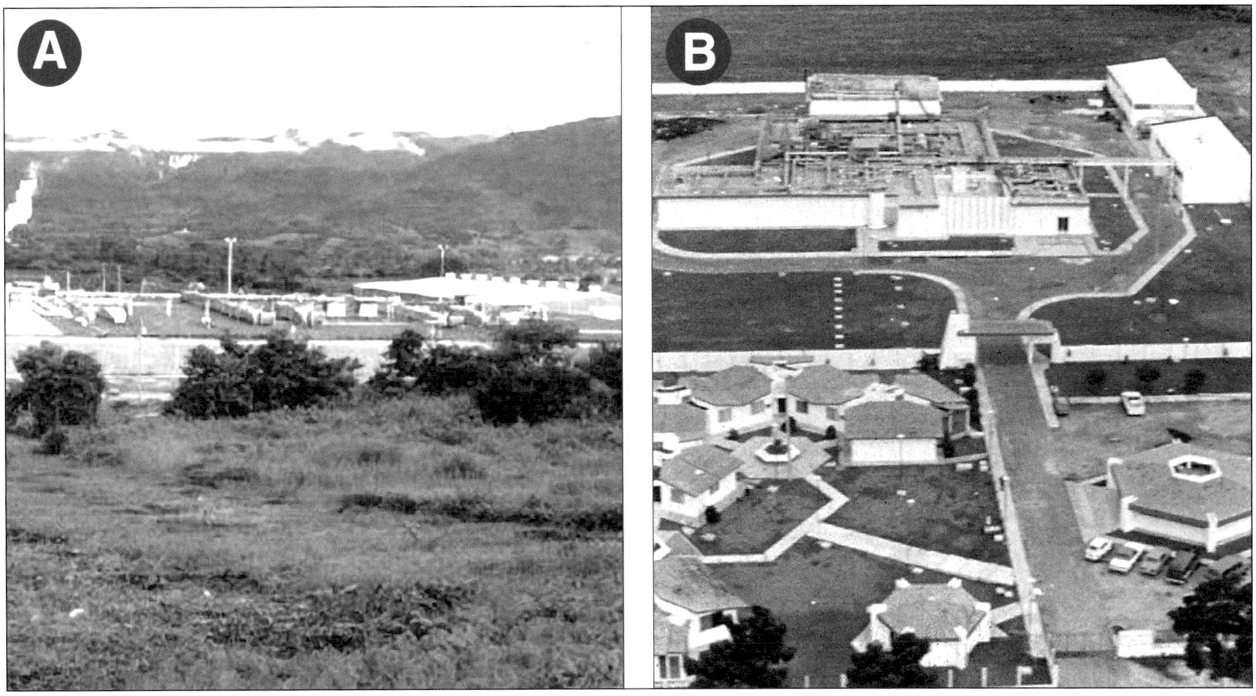

Fig. 4. Two of the largest, well-established insect production facilities in the world. (A) Screwworm, *Cochliomyia hominivorax* (Diptera: Calliphoridae), mass rearing facility at Tuxtla Gutierrez, Mexico. (B) Mediterranean fruit fly, *Ceratitis capitata* (Diptera: Tephritidae), mass rearing facility at Metapa, Mexico.

rax (Marroquin 1985, Meyer 1987, Mangan 1992). See Chap. 5 *Genetic Considerations and Strategies for Rearing High Quality Insects* for additional information.

D. Maintaining the Quality of Established Colonies

The quality of reared insects depends on determining the conditions and procedures that optimize insect performance and then on maintaining the rearing system as designed (Smith 1966; Singh and Moore 1985; Minno and Holler 1987). In practice, declining insect production and quality almost always result from defective environmental controls or failure to perform SOPs. Performance of each procedure is therefore monitored for conformance to prescribed SOPs, e.g., the number of viable eggs used to establish each rearing container, because insect quality is directly affected (Fig. 6). This helps to assure that desirable criteria such as large size and uniformity are maintained. Similarly, environmental conditions and materials are monitored to keep them within established tolerances. Monitoring is based on criteria that can be used to evaluate the colony and make corrective changes, thus stabilizing

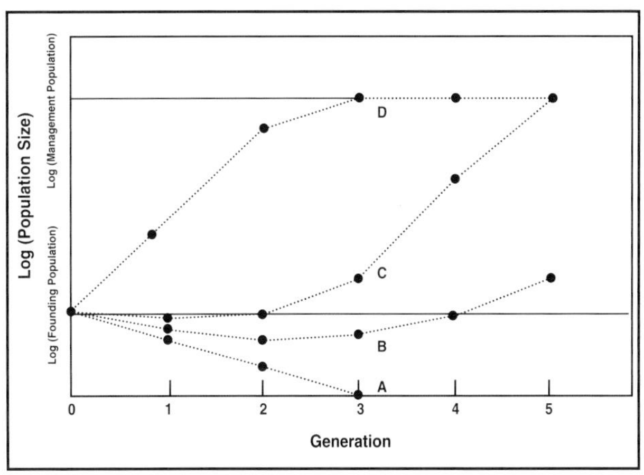

Fig. 5. Typical patterns of change observed in population size over generations following the founding of phytophagous insect colonies from the field. Colonies *A-D* illustrate increasing levels of preadaptation to laboratory conditions and decreasing genetic bottlenecking during establishment. The degree of preadaptation is reflected in the initial population growth rate (difference in log[population size] between generations 0 and 1). The degree of genetic bottlenecking is reflected in the minimum population size experienced. Colony *A* was lost in three generations due to deficient rearing capability and absence of laboratory adaptation. Colony *B* established after significant laboratory adaptation accompanied by a potentially severe genetic bottleneck. Colony *C* established after limited laboratory adaptation and a minor genetic bottleneck. Colony *D* established with no apparent laboratory adaptation and no genetic bottleneck beyond that imposed by the size of the founding population.

product quality at acceptable levels (Boller and Chambers 1977b, Fisher 1983, Chap. 5 *Genetic Considerations and Strategies for Rearing High Quality Insects*, Chap. 11 *Insect Rearing Production Systems, A Case Study: The Southwestern Corn Borer*).

Insects adapt to very subtle changes in their environments, including temperature and humidity fluctuations, altered dietary ingredients (Odell 1992) and oviposition substrates, and shifts in their densities. Therefore, critical aspects of insect development and behavior must be monitored to assure that colonies do not deviate from expectations over time. In insects such as the Mediterranean fruit fly, significant laboratory adaptation occurs as the colony diverges from the source population, regardless of the rearing regimen (Leppla et al. 1983). A colony derived from mass production adapted at about the same rate as one established from the field when both were subjected to mass rearing materials and procedures in a new environment (Fig. 7). However, another colony established with field-collected flies but given guava branches with leaves for courtship sites and papaya fruit for oviposition reproduced at a greater rate until the 6th generation when they too were subjected to the mass rearing regimen. The change in regimen caused an immediate decline in yields of pupae followed by adaptation at about the same rate as the other two experimental colonies. Fortunately, the field performance of colonized

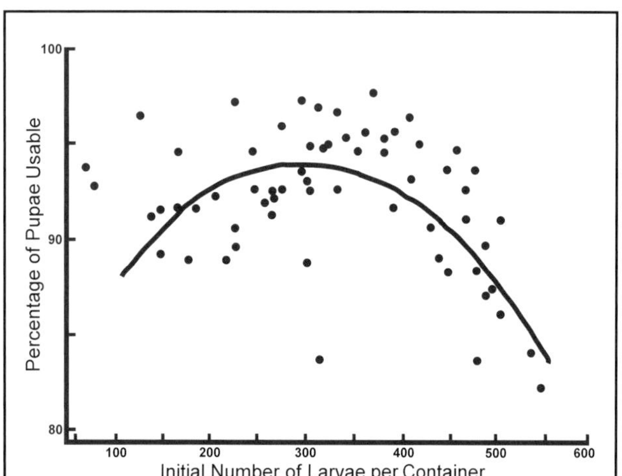

Fig. 6. Regression of usable fall armyworm, *Spodoptera frugiperda* (Lepidoptera: Noctuidae), pupae produced per rearing container from an initial number of developing larvae (Fisher 1983).

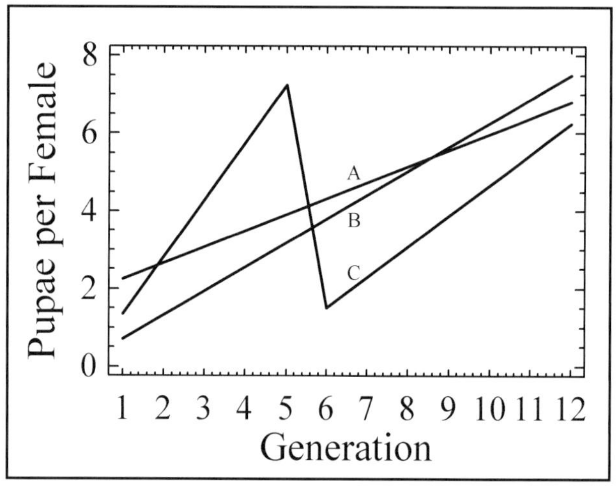

Fig. 7. Regressions of numbers of Mediterranean fruit fly, *Ceratitis capitata* (Diptera: Tephritidae), pupae produced by three colonies, all founded with flies collected from peaches in the field but subsequently reared for 12 generations as follows: Colony *A* was continuously reared under nearly natural conditions, including guava branches for mating substrate and papaya fruit for oviposition substrate; Colony *B* was continuously reared under the mass production regime, including artificial mating and oviposition substrates; Colony *C* was reared for five generations under somewhat natural conditions, including papaya fruit for oviposition substrate, and then switched to the mass rearing production regime (Leppla et al. 1983).

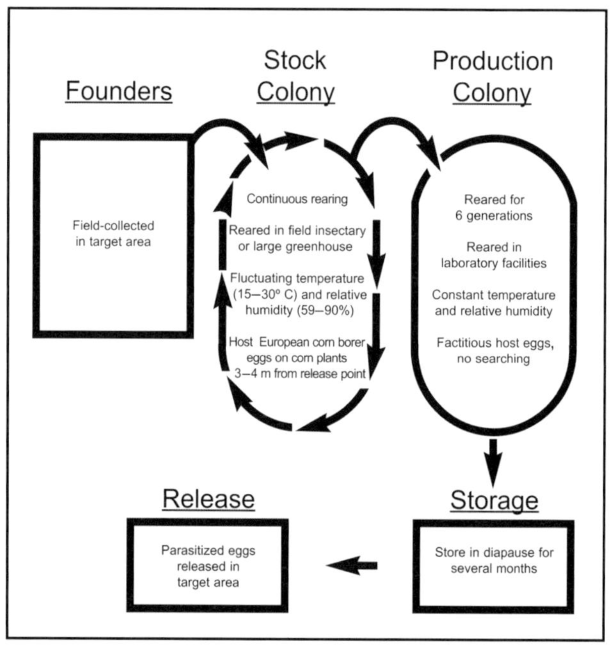

Fig. 8. Mass rearing system for *Trichogramma maidis* (Hymenoptera: Trichogrammatidae) released to control the European corn borer, *Ostrinia nubialis* (Lepidoptera: Crambidae), in Switzerland and other European countries. Founders from the target area were used to establish a carefully reared stock colony that supported a separate mass production colony to provide insects for storage and subsequent release.

Mediterranean fruit flies apparently is not compromised by this laboratory adaptation (Cayol 2000, Cayol et al. 2002).

To minimize divergence of a laboratory colony of the parasitoid *Trichogramma evanescens* (formerly *T. maidis*) (Hymenoptera: Trichogrammatidae) from field populations, a stock colony was maintained under seminatural conditions (Fig. 8). This wasp is an egg parasitoid used to control the European corn borer, *Ostrinia nubilalis* (Lepidoptera: Crambidae), in France, Switzerland, and Germany. The release colony is derived from the stock colony and maintained relatively inexpensively under artificial conditions on factitious host eggs (Bigler 1989).

III. Monitoring Quality

Insect colonies are monitored for life history and behavioral parameters with the minimum intensity and frequency required to make decisions about their condition (Chambers 1977). Monitoring is expensive, so it is imperative to separate "need-to-know" from "nice-to-know" when designing a sampling procedure. A critical parameter that is correlated with and predictive of product quality for a colonized species, such as variability in the rate of larval development, must be monitored intensively. It is necessary to sample larvae from several rearing containers arranged horizontally and vertically on shelves in holding rooms. The rate of larval development can be analyzed by position on a shelf and shelf location or for the entire room, depending on what information is needed. Typically, sampling is more intense while a new rearing system is being established or when problems occur.

The amount of sampling is based on the specifications and standards established for the insect products (Boller and Chambers 1977a). A more exacting standard requires more frequent and extensive sampling and results in more below-standard batches of insects being detected and eliminated. Due to a low standard for size and related behavioral performance, very limited sampling is warranted to reliably produce insects for animal food or fish bait. Conversely, natural enemies and sterile insects must be evaluated carefully to assure that they perform as specified in the field. A *specification* is defined as "the document that prescribes the requirements with which the product or service has to conform" (ANSI/ASQC 1987). A specification is written at the level of quality needed, i.e., a *product grade* or *product standard*. A typical example would be pupae supplied with a specified minimum average weight and variability and at least 90% emergence of adults (standard). If field performance is critical, e.g., for a predator, the product standards could include minimum levels of dispersal and maximum impact on the target pest.

A. Quality Control Criteria and Standardized Tests

Quality control tests are selected or developed on the basis of how well they measure or are correlated with the specified insect characteristics and the amount of effort that is required to conduct them routinely (Fisher 1983). There are standardized tests for the screwworm because these insects are mass produced in several locations and transported great distances for release (IAEA 2003). These tests include the size and weight of pupae; timing and percentage of emerging adults; and adult sex ratio, flight ability, and longevity under stress. Some of these tests are also conducted after irradiation at the origin and again at the destination. Also, due to the importance of assuring that sterile males mate with wild females, periodic tests are conducted on mating compatibility, dispersal, and survival in the field. Mating tests in Hawaii indicated that mass reared sterile males were about 30% as effective as wild males in mating with wild females (McInnis et al. 1996).

The quality control characteristics generally tested for mass reared natural enemies are rate of adult emergence, sex ratio, fecundity, longevity, flight ability, and parasitism/predation (van Lenteren et al. 1993, van Lenteren 2003). Laboratory tests have been developed for at least 26 species of natural enemies although they are often not conducted because feedback on field performance has indicated that quality is adequate. Quality criteria for the lacewing predator, *Chrysoperla carnea* (Neuroptera: Chrysopidae), are indicative of the potential complexity of these kinds of tests. When lacewings are shipped as eggs, the product standards tested include the following: hatchability (> 65% within 5 days, weekly test) and rate of development (> 65% develop to second instar larvae within 4 days, yearly test or when the rearing system is changed). Lacewings shipped as second instar larvae must meet the following product standards: rate of development (> 65% develop to third instar larvae within 5 days, yearly test or when the rearing system is changed) and rate of prey consumption [> 25 potato aphids, *Macrosiphum euphorbiae* (Homoptera: Aphididae), or > 50 green peach aphids, *Myzus persicae* (Homoptera: Aphididae), consumed within 4 days, annual test]. Performance of parasitoids may also be monitored in the field. For example, *Trichogramma brassicae* (= *T. maidis*) (Hymenoptera: Trichogrammatidae) has consistently achieved about 75% parasitism of European corn borer larvae as a result of quality control monitoring and associated adjustments to the rearing system (Bigler 1994).

B. Sampling for Consistent Quality

Although a colony must be sampled for a set of characteristics that directly measure or are predictive of product quality, data acquisition is minimized because it is expensive (Wajnberg 2003). There is no universal sampling method (Southwood and Henderson 2000), but the required number, size, and frequency of samples must be determined. Each unique sampling plan is designed to answer specific questions normally asked about the condition of the colony. Is the yield of pupae from larval rearing containers optimal, i.e., within the range below or above which production becomes unacceptable? Is development synchronized, i.e., do most of the insects reach the stage of development to be supplied in an acceptable time frame? Does the average weight of pupae meet the established product specification, and does the colony yield the normal number of adults? Most importantly, perhaps, do adults engage in normal courtship, mate, and oviposit viable eggs? The characteristics to be sampled and the amount of sampling required to answer these kinds of questions can be determined by tracking means and their variability and correlating characteristics indicative of quality. A relatively high level of variability over time in a stable colony indicates that a characteristic is sensitive to the environment and, therefore, useful for rapidly detecting changes.

Life history, morphological, and behavioral characteristics (variables) used to monitor insect quality usually have a normal or "bell-shaped" distribution. Estimates of the mean (a measure of central tendency) and standard deviation (a measure of spread) of the observed distribution of a quality control variable (X) can be made by calculating AVE[X] = SUM[X_i]/N (for X_i, the ith of N observed values of the quality control variable) and SD[X] = SQRT[(SUM [(X_i - AVE[X])^2]/(N -1)]), respectively. Although these formulas appear complicated, means and standard deviations are easily calculated using spreadsheets, such as Microsoft Excel. In a fictitious example, the numbers of cabbage looper, *Trichoplusia ni* (Lepidoptera: Noctuidae), pupae produced per larval rearing container were counted from five containers (N = 5) removed at the same time from each of six vertically distributed shelves in a larval holding room. The average and standard deviation of yields were calculated for the containers on each shelf (Table 1).

From the shelf averages, it is apparent that containers on the lower and upper shelves produced fewer pupae than did those nearer the middle. The room environment might have had a gradient of temperature and relative humidity that supported suboptimal larval development on the lower and upper shelves. Improving air circulation within the room could increase production uniformity among the shelves and insect quality. The variability among containers within shelves also was higher for the less productive upper and lower shelves. Larvae probably developed over a wider time period, and more of the pupae typically would have been smaller than the specified size.

A decision must be made about the frequency and size of samples needed to estimate quality control vari-

Table 1. Fictitious data of the number of cabbage looper, *Trichoplusia ni* (Lepidoptera: Noctuidae), pupae produced per larval rearing container on each of six shelves in a holding room.

	Shelf Number[1]					
Container No.	1	2	3	4	5	6
1	175	190	190	191	180	142
2	167	191	191	190	180	176
3	144	191	190	192	189	156
4	166	191	192	190	176	138
5	136	190	191	190	174	181
Average	157.6	190.6	190.8	190.6	179.8	158.6
Standard Deviation	16.7	0.5	0.8	0.9	5.8	19.4

[1]Shelf 1 lowest.

ables. The precision of these estimates is limited by cost; however, uncertainty in the estimates must be within acceptable limits (Southwood and Henderson 2000). In most large-scale insect rearing facilities, much more quality control data are generated than are needed to monitor the colonies with acceptable precision. An adequate sample size, *n*, can be estimated by using one of the following formulae involving the permissible error in AVE[*X*] expressed as *E*, a fraction of AVE[*X*]; α, the probability of exceeding *E*; and $\nu = N - 1$, degrees of freedom of the *t*-distribution (Walpole and Myers 1972):

$$n = \left[\frac{t(\alpha/2,\nu)\cdot \mathrm{SD}[X]}{E\cdot \mathrm{AVE}[X]}\right]^2, N < 30 \qquad \text{Eq. 1}$$

$$n = \left[\frac{Z(\alpha/2)\cdot \mathrm{SD}[X]}{E\cdot \mathrm{AVE}[X]}\right]^2, N \geq 30 \qquad \text{Eq. 2}$$

Values for $t(\alpha/2,\nu)$ and $Z(\alpha/2)$ can be obtained from standard statistical tables, e.g., Ott and Longnecker (2001). For $N < 30$ (e.g., N = 5) and $\alpha = 0.1$, 0.05, and 0.01, one has: $t(0.05, 4) = 2.13$, $t(0.025, 4) = 2.78$, and $t(0.005, 4) = 4.60$, respectively. For $N \geq 30$ and $\alpha = 0.1$, 0.05, and 0.01, one has: $Z(0.05) = 1.64$, $Z(0.025) = 1.96$, and $Z(0.005) = 2.57$, respectively.

For example, the number of containers that must be sampled to estimate the average number of cabbage looper pupae produced per container on shelf 1 (Table 1, $N = 5$, AVE[*X*] = 157.6, SD[*X*] = 16.7) with an error of 5% of the mean ($E = 0.05$) all but 5% of the time ($\alpha = 0.05$) is estimated to be 35: $n = [(2.78 \times 16.7)/(0.05 \times 157.6)]^2 = 34.7$. If an estimate subject to an error of only 1% of the mean ($E = 0.01$) with 99% confidence ($\alpha = 0.01$) is needed, then 432 containers would have to be sampled. Fortunately, this level of precision and degree of confidence is almost never warranted. When more than 30 containers are sampled, SD[*X*] can be calculated and the sample size estimated by using the formula for $N \geq 30$. However, the number of samples must not be too great due to cost and the possibility of disrupting the entire rearing operation.

The sampling units for small colonies are the individual rearing containers or host plants, and pupae or adults are usually pooled to establish a new generation. In order to detect sources of variability and identify specific causes of rearing problems, these sampling or rearing units should be evaluated individually. In mass rearing facilities, however, yields from immature and adult rearing units are typically pooled for an entire workday shift, and this batch becomes the sampling unit. Such pooling obscures sources of variation so that, for example, reduced yields can result either from relatively small reductions in most of the rearing units or large declines in only a few. Considering the example in Table 1, yields of pupae were highly variable and on average lower from the containers on the upper and lower shelves. Perhaps the room could be modified to retain high yields in mid-level containers and improve the others. Efforts to increase yields would be approached differently if containers with both high and low yields were found on all the shelves.

C. Production, Process, and Product Control

Monitoring the SOPs of production, characteristics of the developing insects, and final products can be divided into the three respective categories: production, process, and product control (Leppla and Fisher 1989; Fig. 9). These categories are separated and combined in different ways depending on the structure of the insect production system, which work unit conducts the monitoring, and how the data are managed. For example, production and process control are often combined because they monitor closely related variables, such as temperature and rate of larval development, that determine or are correlated with the quality of insects. Rearing processes are monitored by examining the insects at critical points in their production and handling, such as egg treatment, container infestation, larval holding, harvesting and sexing pupae, specialized treatments, and packaging for shipment. In mass rearing facilities, the production staff normally monitors materials, personnel, and SOPs by using check sheets and keeping accurate records. They may also be responsible for process control, but a separate quality control staff usually handles product control in terms of emergence, behavior, and survival. Key life history, morphological, and behavioral characteristics are defined by the species being reared. Regardless of the organizational structure for producing the insects, the same kinds of information are always required. The insectary manager often performs all of the quality control functions in small, single-species and medium-size, multiple-species rearing programs.

1. Production Control. Proficiency and consistency in performing established insect rearing procedures are essential to the production of high quality insects. SOPs, such as those developed for the Caribbean fruit fly, *Anastrepha suspensa* (Diptera: Tephritidae), must be performed as specified (Minno and Holler 1987). These

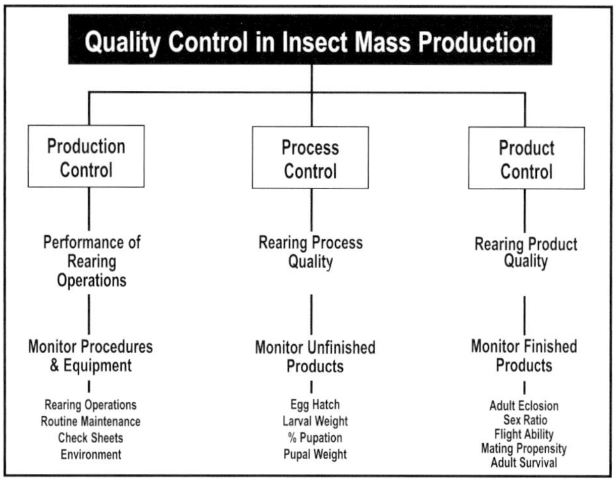

Fig. 9. The three types of quality control used for monitoring insect production and assuring that the insects meet established standards: production control, process control, and product control. These three types can be organized into different combinations depending on the size and structure of the insect rearing program (Leppla and Fisher 1989).

"recipes" include employee selection and training, facility and equipment maintenance, sanitation and monitoring for microbial contamination, rearing procedures, and quality control tests. Sources of commonly used materials and equipment, checklists, and data sheets are also included. Details of rearing processes, such as preparation of diets for adults and immatures and holding the insects in appropriate environments, are prescribed (Singh and Moore 1985). It is also important that rearing personnel have an adequate understanding of how the biology of the insects and associated pests and pathogens affect product quality. Once established, rearing processes become routine and individual procedures are easily forgotten. A diet ingredient may be omitted or degraded by overheating, eggs may be accidentally desiccated following surface sterilization, closely related species may be mixed unintentionally, or larval densities may be allowed to deviate from the specifications. Temporary loss of temperature control can destroy all the insects in a rearing room, pathogens can be devastating, and parasitoids and predators must be detected and eliminated. In practice, declining insect production and quality almost always results from defective environmental controls or failure to perform SOPs (Chap. 11 *Insect Rearing Production Systems, A Case Study: The Southwestern Corn Borer*).

2. Process Control. Quality control data can be presented as simple tables, histograms, pie charts, or graphs; however, one of the most useful methods is to use *process control charts* (Chambers and Ashley 1984). For a *means chart*, AVE[X] is plotted on the vertical axis and time or sample number on the horizontal axis. The expected mean or performance standard, PS, is established by experience to meet the required rearing goals, and control limits are set no wider than $PS \pm 3$ standard deviations of the mean (SEM[X]). In an example using the weight of cabbage looper pupae (Chambers and Ashley 1984), PS was set to be the grand mean (AVE[AVE[X]] = 253.5 mg) of 10 five-pupa samples removed randomly from all of the pupae reared on a particular day, and the upper, UCL, and lower, LCL, control limits were calculated using the standard deviation of the mean (SEM[X] = 20.3 mg): $UCL = PS + 3$ SEM[X] = 314.4 mg and $LCL = PS - 3$ SEM[X] = 192.6 mg, respectively. Thereafter, for each sampling day, the observed mean weight for a five-pupa sample was plotted and compared visually with the control limits for the performance standard (Fig. 10). Means usually increase with improvements in rearing proficiency until they stabilize, so subsequent major directional changes signal serious rearing problems. Similarly, an increasing standard deviation indicates that container yields are becoming less consistent and that the cause should be identified and corrected. A variable is usually considered under control if the sample mean varies randomly and is less than two or three standard deviations from the performance standard. The range of a variable can also be plotted for samples on a *range chart*, but this has not gained acceptance for insect rearing.

3. Product Control. Quality control tests are conducted to assure that insect products meet acceptable quality specifications and standards when they leave the production facility. However, products typically are pack-

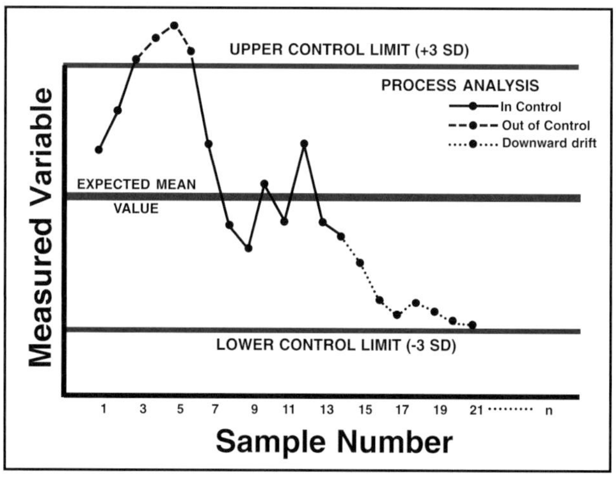

Fig. 10. A hypothetical process control chart based on an expected mean of a measured parameter, illustrating acceptable and unacceptable trends. Process control charts are also used (SD = SEM[X]) for product quality control (Chambers and Ashley 1984).

aged and shipped long distances before being used, and tests repeated at the delivery site usually indicate that insect performance has diminished. Product control must therefore extend into utilization control by conducting the same tests at critical points between production and use of the product. Results of these tests identify points at which performance declines and often enable significant improvements to be made in handling procedures. For example, screwworm pupae are transported in bulk by refrigerated trucks from the main production facility to remote distribution centers, repackaged into small boxes containing adult food, and loaded onto airplanes for release. The company Gardens Alive purchases and resells the convergent lady beetle, *Hippodamia convergens* (Coleoptera: Coccinellidae), after feeding them a "special conditioning diet" for two weeks (http://www.gardensalive.com/). Additionally, this company recommends that water be provided when the beetles are received. Survival, predation, and egg production are thus improved. Feeding on diets containing sugar plus protein is essential for optimum survival of the Caribbean fruit fly (Teal et al. 2004). This assures the effectiveness of the sterile males in the field. Process control charts can be used to plot and assess product control characteristics.

IV. Colony Evaluation and Management

Insect colonies are evaluated constantly, both by means of quality control procedures and informally through experience. Process and product control variables indicate if a colony is producing insects that meet the established specifications and standards (Leppla 2003). Additionally, experienced rearing personnel gain a sense of the condition of the production system and can determine the health of a colony by inspection. Substandard materials and microbial contamination are common in poorly managed systems. These are easy to identify and correct in small, single-species rearing programs; but, in more complex situations, structured diagnostic procedures may be necessary. When a quality control variable for an insect product is not under control, e.g., size falls below the lower control limit, operations are checked for compliance with all SOPs, including environmental conditions and materials. Rearing problems originating from these sources, including microbial contamination, can be corrected by improving production control. However, sources of physiological and behavioral problems, such as delayed and highly variable egg hatch or larval development, incomplete molting, poor survival, and reduced incidence of mating and associated fertility, although considerably less common, are much more difficult to identify.

A. Structured Diagnostic Procedures

A structured diagnostic procedure is usually required to determine the specific causes of insect rearing problems (Leppla and Fisher 1989; Fig. 11). This is illustrated by the sudden decline in a colony of the spruce budworm, *Choristoneura fumiferana* (Lepidoptera: Tortricidae), that had been in large-scale production since the mid-1970s (van Frankenhuyzen 2004). Production declined from ca 15,000 larvae per 300 mated pairs to a few hundred, and the colony was on the verge of collapse. After considerable scrutiny of the rearing processes and resulting insect stages, the cause was determined to be a microsporidian of unknown source. The infection could be eliminated by using high concentrations of therapeutic drugs, but the insects suffered unacceptable side effects. Consequently, a microsporidian-free spruce budworm colony was established by screening females from individual mating pairs and using clean eggs. Several rearing protocols were changed to prevent reinfection: 1) spatially separating the colony from other insect species, 2) substituting wax paper for field-collected foliage as an oviposition substrate, 3) using disposable plastic bags rather than wire mesh mating cages, 4) splitting the colony into isolated subcolonies or cohorts, and 5) instituting rigorous quality control, including microscopic and molecular examination of adults from each cohort. Thus, production of the spruce budworm has been reestablished and maintained at 50,000–100,000 uninfected larvae per week.

B. Quality Control versus Methods Improvement

Making corrections to reestablish production control is not to be confused with continuous methods improvement and optimization. Existing quality control data and rapid bioassays are used to determine the causes of unanticipated rearing problems. For instance, a dietary ingredient or egg handling procedure can cease to be effective because a supplier modifies a previously acceptable product. New materials or methods must be substituted quickly to avoid major bottlenecks and ultimately the loss of the colony. If necessary, bioassays are conducted on a variety of alternatives to assure that the substituted materials and methods are effective and cause minimal disruption of the established production system. This emergency response is approached very differently than the continuous process of intentionally deciding to make changes in a fully functioning system to improve production efficiency and decrease costs. In this case, careful testing is required for each material or method, one at a time and outside of the production system, before modifications are implemented. A stable rearing system is particularly important when the production strain is replaced, replenished, or restored. It must be demonstrated that a new strain will perform as well in production and be as effective as the old one and that there is some advantage to the replacement (Chap. 5 *Genetic Considerations and Strategies for Rearing High Quality Insects*). Colonies of insects are maintained according to SOPs unless a component fails; then quality control assessments identify the cause, and stability is restored with minimal change. In contrast, optimization by means of research and methods development is a separate experimental process that carefully alters the established production system.

C. Periodic Review versus Crisis Review

Periodic reviews of an insect rearing system are inexpensive relative to a failure and usually more than pay for themselves by involving experienced consultants who can recommend significant improvements. For large production programs, structured reviews should be performed at least annually as a routine activity and not just when there is a crisis. When there is a production failure, per-

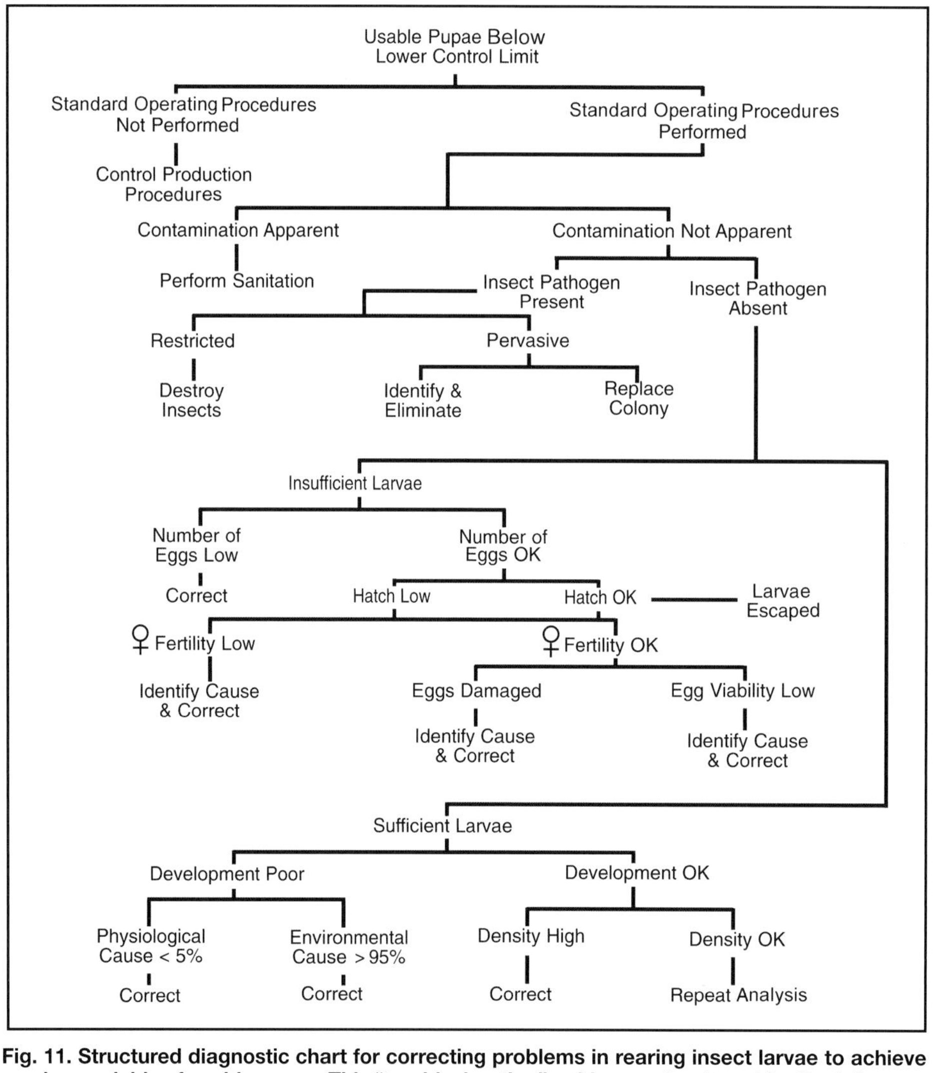

Fig. 11. Structured diagnostic chart for correcting problems in rearing insect larvae to achieve maximum yields of usable pupae. This "troubleshooting" guide was developed for the fall armyworm, *Spodoptera frugiperda* (Lepidoptera: Noctuidae), and incorporates both production and process control (Leppla and Fisher 1989).

sonnel are stressed and expecting a review to find the cause and assign blame. Indeed, most failures are caused by a breakdown in SOPs that can be traced to specific personnel. Examples from experience include the improper handling of eggs and use of outdated and deficient dietary ingredients, e.g., diet-induced straggling in gypsy moth, *Lymantria dispar* (Lepidoptera: Lymantriidae), larvae (Odell et al. 1997). Both of these production control failures caused a lack of developmental synchrony, reduced survival of larvae, and a high proportion of unacceptably small adults. Conversely, noncrisis reviews provide opportunities to brainstorm potential improvements in rearing facilities, production equipment and procedures, colonies, and employee training. Valuable reviews have been conducted and associated unpublished reports produced for rearing the following (among others): screwworm; pink bollworm, *Pectinophora gossypiella* (Lepidotera: Gelechiidae); boll weevil, *Anthonomus grandis* (Coleoptera: Curculionidae); gypsy moth; spruce budworm; codling moth, *Cydia pomonella* (Lepidoptera: Tortricidae); tobacco budworm, *Heliothis virescens* (Lepidoptera: Noctuidae); tsetse fly, *Glossina* spp. (Diptera: Glossinidae); and several species of fruit flies (Diptera: Tephritidae), butterflies, and natural enemies.

V. Guidelines for Quality Systems

Quality control guidelines and programs have been developed for mass rearing single or closely related species of insects for sterile insect technology and for rearing a wide range of natural enemies used in commercial biological control (Moore et al. 1985). Manuals and other publications that describe life history and behavioral characteristics have been published for tropical fruit flies in general (IAEA 2003); the Mediterranean fruit fly (Brazzel et al. 1986); Caribbean fruit fly (Minno and Holler 1987); and melon fly, *Bactrocera cucurbitae* (Diptera: Tephritidae) (Nakamori and Soemori 1981, Nakamori et al. 1992); and several others. Most of these publications are not widely distributed, often being intended only for use within a single rearing facility. The Association of Natural Bio-control Producers (ANBP, http://www.anbp.org/) included quality control as a central part of its mission when it was founded in 1990. The European counterpart to ANBP, the International Biocontrol Manufacturer's Association (IBMA, http://www.ibma.ch/), has a similar interest in assuring the quality of commercial natural enemies. Individual companies in both organizations have used generic guidelines for establishing quality standards (van Lenteren and Tommasini 1999), including IOS, ASTM, and IOBC. These quality control programs overlap in species and tests but have different approaches. IOS attempts to engage rearing personnel and management in designing and controlling production so that the product is "automatically" of high quality. ASTM is a post-production product quality control program that establishes and publishes standards available to customers. IOBC emphasizes product control standards and tests to assure that products are effective (Boller 2002, Boller and Leppla 2006, van Lenteren et al. 2003). Quality standards are ultimately determined by the users or customers who accept or reject the product. Additionally, regulatory authorities may require accurate identification and purity of natural enemies, and assurance that the products are effective in controlling specific pests. The IOBC Working Group on Arthropod Mass Rearing and Quality Control (AMRQC, http://www.amrqc. org/) in concert with the ANBP and IBMA have attempted to harmonize the requirements of the marketplace and international plant protection organizations (Leppla et al. 2002, Grenier et al. 2003).

VI. Total Quality Management

Total quality control (TQC) is defined as "an effective system for integrating the quality-development, quality-maintenance, and quality-improvement efforts of the various groups in an organization so as to enable marketing, engineering, production, and service at the most economical levels which allow for full customer satisfaction" (Feigenbaum 1983). As applied to insect rearing, TQC provides a means of organizing and managing not only mass rearing programs but also, by combining the divisions, rearing programs on any scale (Leppla 1994). The TQC structure incorporates all of the interacting variables that must be considered in the persistent search for ways to improve all of the science, technology, and management issues associated with insect rearing. The divisions of TQC as applied to insect rearing are *Management, Methods Development, Material, Production, Research,*

Utilization, Personnel, and *Quality Control* (Leppla and Fisher 1989; Fig. 12).

Management is responsible for establishing and administering the overall rearing program. SOPs are developed in close collaboration with employees in each of the other divisions. These procedures maximize the efficiency of insect production and result in product quality that meets prescribed minimum performance standards. Diagnostic tests and feedback loops are designed to monitor the system and, if necessary, trigger efforts to correct deficiencies.

Methods Development continuously strives to improve the rearing program's design by testing new technologies, whether invented or transferred. It bridges the gap between research and implementation. Strain development, one of its primary areas of responsibility, depends on the specific purpose, required specifications, and rearing capabilities for the colonized insects. Process capability studies, evaluations of alternative system designs, and pilot tests are conducted by methods development.

Material, as the name implies, emphasizes the reception and storage of materials that conform to specified quality and economy. This can involve both incoming materials and those produced in one area of an organization and transferred to another internal location. Required specifications and standards for materials are documented as part of the purchasing process, usually by administrative personnel. Although vendors are responsible for maintaining the quality of purchased materials, periodic verification is necessary.

Production assures the specified consistency, timing, and numbers of insects. The production of high quality insects depends on the appropriate use of suitable facilities and equipment and on the accurate performance of rearing processes, i.e. SOPs. These rearing processes include the following: maintenance of adult colonies, collection and treatment of eggs, preparation of diets for immature and adult insects, rearing immature stages, harvesting the stages to be supplied, and associated quality control.

Research is the source of new technologies for obtaining, rearing, and monitoring insects within a produc-

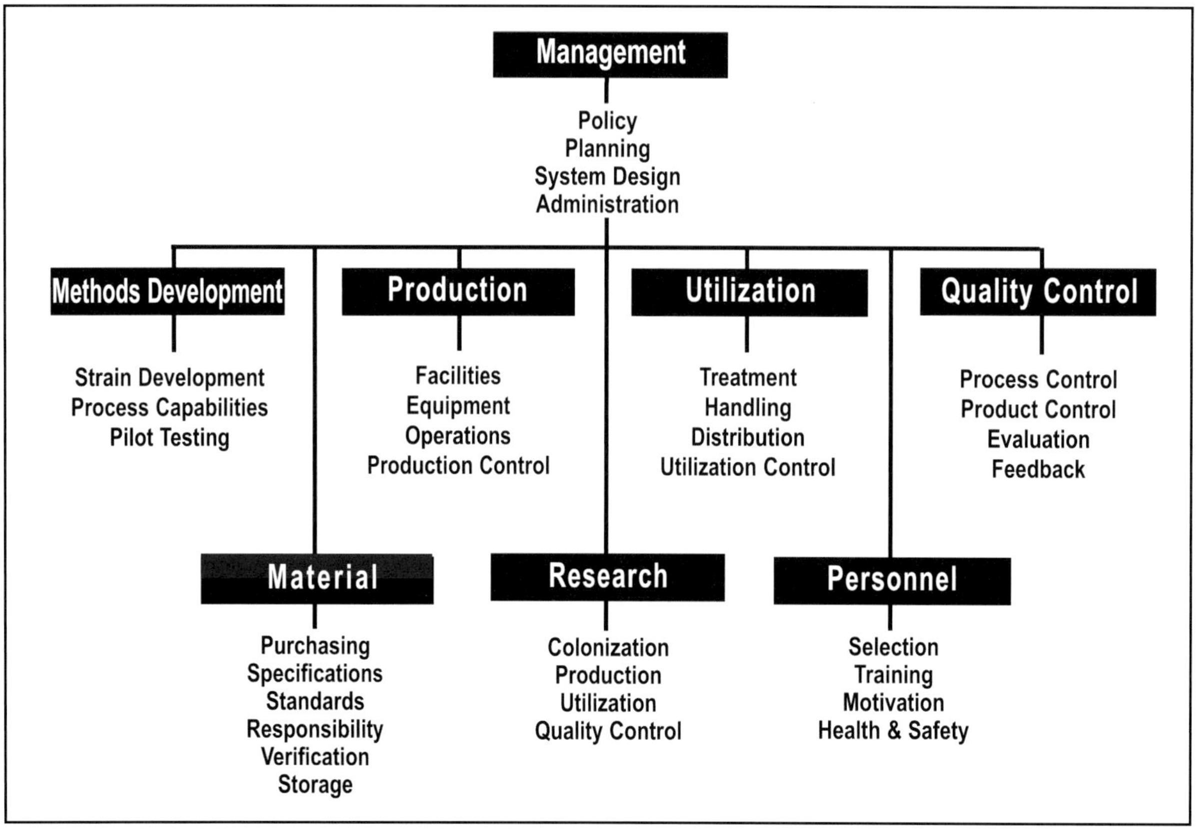

Fig. 12. Generic divisions of total quality control for insect production. Regardless of size or complexity, all insect rearing systems incorporate some form of *Management, Methods Development, Material, Production, Research, Utilization, Personnel,* and *Quality Control* (Leppla and Fisher 1989).

tion system and for product evaluation (Dame 1989). In practice, *Research* and *Methods Development* often involve the same work although Research should contribute fundamental knowledge whereas methods development focuses on technology improvement. A separate, on-site research component is essential for advancing mass rearing programs and helping to correct problems. It also supports field operations and overall decision making for the associated pest management effort.

TQC often includes a *Utilization* division, the post-production chain of delivery, use, and evaluation, particularly if there is customer involvement and feedback. Users monitor the performance of the insects relative to their intended purpose and report success or failure. Often the quality control criteria and tests at delivery are the same as those used for product control at the end of production.

The selection, training, supervision, and motivation of personnel within the *Personnel* division have only recently been recognized formally as being critical to success in insect rearing (Cohen 2001). Professionals must have accurate position descriptions, clear lines of authority, a sense of the importance of their work, individual and group goals, incentives, recognition, safe environments, and an understanding of the organization's mission. Highly skilled and diligent personnel can often achieve success in an otherwise marginal insect rearing program.

The *Quality Control* division is responsible for monitoring both unfinished and finished insect products to assure that they conform to required specifications and standards. This monitoring often extends to testing along the supply chain to assure that products do not deteriorate unacceptably before being used. Information is obtained, evaluated, and reported to other divisions that are affected and to *Management*. Although independent, *Quality Control* shares responsibility with the other divisions for preventing or correcting problems throughout an insect rearing system that could compromise production of high quality products.

A TQC system can be designed for managing the quality of all inputs for insect production, the quality of the insect products, and the effectiveness of the insects relative to their intended purpose (Fig. 13). The system organizes the divisions and their respective functions in a sequence ending with use of the insects. Thus, the TQC system begins with *Management* orchestrating development of an insect rearing program that can reliably produce high quality insects. The design phase involves all of the divisions of the system, usually outside consultants,

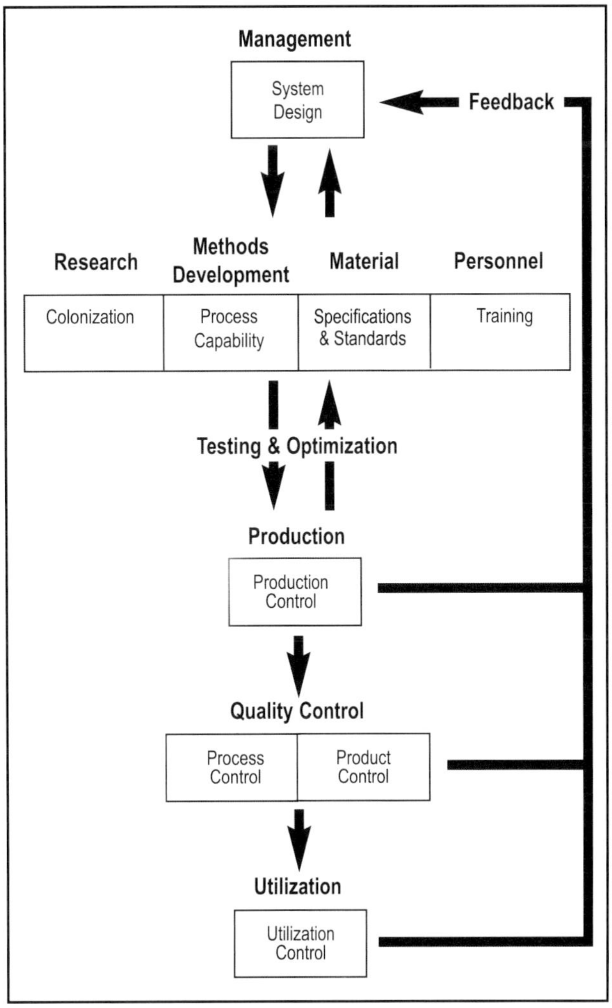

Fig. 13. A total quality control system for insect production. Management establishes the system based on available research, methods development, materials, and personnel. Insects are produced, monitored for quality, used, and evaluated relative to their purpose. Management receives feedback on product quality from *Production, Quality Control,* and *Utilization* enabling decisions to be made about adjusting the system.

and virtually every aspect of quality control. The initial design is tested and optimized by *Research, Methods Development, Material,* and *Personnel* in close cooperation with *Production*. This pilot testing phase usually requires considerable reevaluation of optional facilities, equipment, materials, and procedures. Once insect production is established, it is maintained by *Production* and the quality of both developing and mature insect products is monitored by *Quality Control*. The same product control tests are conducted again at critical points in the post-production handling and use of the insects. This additional testing is particularly important if the insects are altered and transported in ways that could diminish their quality. Typical examples of the latter are Mediterranean fruit flies produced in Guatemala, irradiated, and shipped to California for release and commercial natural enemies from California used in Florida. Feedback on the effectiveness of the insects is provided to management in a form that can be evaluated easily and disseminated to appropriate divisions to assure that the TQC system is successful (Leppla 1994).

VII. Conclusions

The basics of quality control for insect rearing include methods for establishing and maintaining appropriate colonies, monitoring them for consistent quality, correcting deficiencies if necessary, and making planned improvements. Insect rearing systems are designed to produce insects for specific purposes according to prescribed specifications and standards. Requirements for the insects range from minimal concern about performance to very exacting demands for essential behavior. Also, the specified number of insects must be delivered consistently according to a defined schedule. Colonies are established and monitored to assure that they satisfy these goals. This monitoring incorporates SOPs and tests that extend from production through evaluation of the product and its use. Structured diagnostic procedures based on SOPs are used to identify and eliminate causes for declining production. Optimization of the rearing system is accomplished by allied research and methods development, and through periodic review. Guidelines have been adapted from industrial processes and organized into TQC for insect rearing systems.

Acknowledgments

This paper would not have been written without the collegial encouragement of Frank Davis and the students of the insect rearing workshop he so ably developed and conducted. David Dame and Robert McSorley provided exceptionally helpful reviews of the manuscript, and Mike Sanford and Denise Thomas prepared the figures. This research was supported by the Florida Agricultural Experiment Station and approved for publication as Journal Series No. N-0274.

References

(ANSI/ASQC) American National Standards Institute/American Society for Quality Control. 1987. Quality systems terminology. Publication A3-1987. ANSI/ASQC, Milwaukee, WI.

Bartlett, A. C. 1984. Genetic changes during insect domestication, pp. 2–8. *In* E. G. King and N. C. Leppla (eds.), Advances and challenges in insect rearing. Agric. Res. Serv., USDA, U.S. Government Printing Office, Washington, DC.

Bartlett, A. C. 1985. Guidelines for genetic diversity in laboratory colony establishment and maintenance, pp. 7–17. *In* P. Singh and R. F. Moore (eds.), Insect rearing handbook, vol. 1. Elsevier, Amsterdam, The Netherlands.

Bigler, F. 1989. Quality assessment and control in entomophagous insects used for biological control. J. Appl. Entomol. 108: 390–400.

Bigler, F. (ed.) 1991. Fifth Workshop of the International Organization of Biological Control Working Group on Quality Control of Mass-Reared Arthropods/West Palaearctic Regional Section. Wageningen, The Netherlands. (http://users.ugent.be/~padclerc/AMRQC/ proceedings.html).

Bigler, F. 1994. Quality control in *Trichogramma* production, pp. 93–114. *In* E. Wajnberg and S. A. Hassan (eds.), Biological control with egg parasitoids. CABI Publishing, Wallingford, United Kingdom.

Boller, E. F. 1972. Behavioral aspects of mass-rearing of insects. Entomophaga 17: 9–25.

Boller, E. 2002. History of quality control in mass-reared insects, pp. 1–5. *In* N. C. Leppla, K. A. Bloem, and R. F. Luck (eds.), Proceedings, 8th and 9th Workshops of the International Organization of Biological Control Working Group on Quality Control of Mass-Reared Arthropods. Santa Barbara, CA and Cali, Colombia, respectively. (http://users.ugent.be/~padclerc/AMRQC/ proceedings.html).

Boller, E. F., and D. L. Chambers. 1977a. Concepts and approaches, pp. 4–13. *In* E. F. Boller and D. L. Chambers (eds.), Quality control, an idea book for fruit fly workers. International Organization of Biological Control/West Palaearctic Regional Section Publication Series No. 1977/5.

Boller, E. F., and D. L. Chambers. 1977b. Quality aspects of mass-reared insects, pp. 219–235. *In* R. L. Ridgway and S. B. Vinson (eds.), Biological control by augmentation of natural enemies, insect and mite control with parasites and predators. Plenum, New York, NY.

Boller, E. F. and N. C. Leppla. 2006. Global Working Group on Arthropod Mass Rearing and Quality Control (1968-present), pp 129-134. In E. F. Boller, J. C. van Lenteren and V. Delucchi (eds.) International Organization for Biological Control of Noxious Animals and Plants, history of the first 50 Years (1956-2006). Ponsen & Looijen, Wageningen, The Netherlands.

Brazzel, J. R., C. O. Calkins, D. L. Chambers, and D. B. Gates. 1986. Required quality control tests, specifications, and shipping procedures for laboratory produced Mediterranean fruit flies for sterile insect control programs. Publication 81-51, U.S. Dept. Agric., Animal and Plant Health Inspection Service, Beltsville, MD.

Cayol, J. P. 2000. Changes in sexual behavior and life history traits of tephritid species caused by mass-rearing processes, pp. 843–860. *In* M. Aluja and A. L. Norrbom (eds.), Fruit flies (Tephritidae): phylogeny and evolution of behavior. CRC Press, Boca Raton, FL.

Cayol, J. P., P. Coronado, and M. Taher. 2002. Sexual compatibility in medfly (Diptera: Tephritidae) from different origins. Florida Entomol. 85: 51–57.

Chambers, D. L. 1975. Quality in mass-produced insects, definition and evaluation, pp.19–32. *In* Controlling fruit flies by the sterile-insect technique. Publication PL-582/3. International Atomic Energy Association, Vienna, Austria.

Chambers, D. L. 1977. Quality control in mass rearing. Annu. Rev. Entomol. 22: 289–308.

Chambers, D. L., and T. R. Ashley. 1984. Putting the control in quality control in insect rearing, pp. 256–260. *In* E. G. King and N. C. Leppla (eds.), Advances and challenges in insect rearing. USDA, Agric. Res. Serv., U.S. Government Printing Office, Washington, DC.

Cohen, A. C. 2001. Formalizing insect rearing and artificial diet technology. Am. Entomol. 47: 198–206.

Cohen, A. C. 2004. Insect diets: science and technology. CRC Press, Boca Raton, FL.

Dame, D. A. 1989. The relationship of research to total quality control with special reference to sterile insect technique. J. Appl. Entomol. 108: 476–482.

Edwards, D. R., N. C. Leppla, and W. A. Dickerson. 1987. Arthropod species in culture. Entomological Society of America, College Park, MD.

Feigenbaum, A. V. 1983. Total quality control, 3rd ed. McGraw-Hill, New York, NY.

Fisher, W. R. 1983. Quality control of mass-reared Lepidoptera using the fall armyworm, *Spodoptera frugiperda* (J. E. Smith), as a model. Ph.D. dissertation, University of Florida, Gainesville, FL.

Grenier, S., P. DeClercq, and N. C. Leppla (eds.). 2003. Proceedings, 10th Workshop of the International Organization of Biological Control Global Working Group on Arthropod Mass Rearing and Quality Control. Global IOBC Bulletin No. 2.

Hoy, M. A 1979. The potential for genetic improvement of predators for use in pest management programs, pp. 106–115. *In* M. A. Hoy and J. J. McKelvey, Jr. (eds.), The use of genetics in insect management. Rockefeller Foundation, New York, NY.

(IAEA) International Atomic Energy Agency. 2003. Product quality control, irradiation and shipping procedures for mass-reared tephritid fruit flies for sterile insect release programs. (http://www-naweb.iaea.org/nafa/ipc/ public/d4_pbl_5.html).

Leppla, N. C. 1989. Laboratory colonization of fruit flies, pp. 91–103. *In* A. S. Robinson and G. Hooper (eds.), World crop pests, vol. 3B, Fruit flies, their biology, natural enemies and control. Elsevier, Amsterdam, The Netherlands.

Leppla, N. C. 1994. Principles of quality control in mass-reared arthropods, pp. 1–11. *In* G. Nicoli, M. Benuzzi, and N. C. Leppla (eds.), Proceedings, 7th Workshop of the International Organization of Biological Control Global Working Group on Quality Control of Mass-Reared Arthropods. Rimini, Italy. (http://users.ugent.be/~padclerc/AMRQC/ proceedings.html).

Leppla, N. C. 2002. Rearing of insects, pp. 975–979. *In* V. H. Resh and R. Carde (eds.), Encyclopedia of insects. Academic Press, San Diego, CA.

Leppla, N. C. 2003. Guidelines for quality control of commercially produced natural enemies, pp. 19–24. *In* J. C. van Lenteren (ed.), Quality control and production of biological control agents, theory and testing procedures. CABI Publishing, Cambridge, MA.

Leppla, N. C. 2004. Rearing of insects, pp. 2–8. *In* J. L. Capinera (ed.), Encyclopedia of entomology. Kluwer, Dordrecht, The Netherlands.

Leppla, N. C., and T. R. Ashley. 1989. Quality control in insect mass production: a review and model. Bull. Entomol. Soc. Amer. 35(4): 33–44.

Leppla, N. C. and W. R. Fisher. 1989. Total quality control in insect mass production for insect pest management. J. Appl. Entomol. 108: 452–461.

Leppla, N. C., S. L. Carlyle, C. W. Green, and W. J. Pons. 1978. Custom insect rearing facility, pp. 66–70. *In* N. C. Leppla and T. R. Ashley (eds.), Facilities for insect research and production. USDA Tech. Bull. 1576. US Government Printing Office, Washington, DC.

Leppla, N. C., W. R. Fisher, J. R. Rye, and C. W. Green. 1982. Lepidopteran mass rearing: an inside view, pp. l23–133. *In* Proceedings, Symposium on sterile insect technique and radiation in insect control. Publication SM-255/14. International Atomic Energy Agency, Vienna, Austria.

Leppla, N. C., M. D. Huettel, D. L. Chambers, T. R. Ashley, D. H. Miyashita, T. T. Y. Wong, and E. J. Harris. 1983. Strategies for colonization and maintenance of the Mediterranean fruit fly. Entomol. Exp. Appl. 33:89–96.

Leppla, N. C., K. A. Bloem, and R. F. Luck (eds.). 2002. Proceedings, 8th and 9th Workshops of the International Organization of Biological Control Working Group on Quality Control of Mass-Reared Arthropods. Santa Barbara, CA and Cali, Colombia, respectively. (http://users.ugent.be/ ~padclerc/AMRQC/ proceedings.html).

Mackauer, M. 1972. Genetic aspects of insect production. Entomophaga 17: 27–48.

Mackauer, M. 1976. Genetic problems in the production of biological control agents. Annu. Rev. Entomol. 21: 369–385.

Mangan, R. L. 1992. Evaluating the role of genetic change in insect colonies maintained for pest management, pp. 269–288. *In* T. E. Anderson and N. C. Leppla (eds.), Advances in insect rearing for research and pest management. Westview Press, Boulder, CO.

Marroquin, R. 1985. Mass production of screwworm in Mexico, pp. 31–36. *In* O. H. Graham (ed.), Proceedings, Symposium on eradication of the screwworm from the United States and Mexico. Misc. Publ. Entomol. Soc. Amer., MPEAAL 62: 1-68. Entomological Society of America, College Park, MD.

Meyer, N. L. 1987. History of the Mexico-United States screwworm eradication program. Vantage Press, New York, NY.

McInnis, D. O., D. R. Lance, and C. G. Jackson. 1996. Behavioral resistance to the sterile insect technique by Mediterranean fruit fly (Diptera: Tephritidae) in Hawaii. Ann. Entomol. Soc. Amer. 89: 739–744.

Minno, M. C., and T. C. Holler. 1987. Procedures manual for mass-rearing the Caribbean fruit fly, *Anastrepha suspensa* (Loew) (Diptera: Tephritidae). Internal Publication. Florida Department of Agriculture and Consumer Services, Div. of Plant Industry, Gainesville, FL.

Moore, R. F., T. M. Odell, and C. O. Calkins. 1985. Quality assessment in laboratory-reared insects, pp. 107–135. *In* P. Singh and R. F. Moore (eds.), Insect rearing handbook, vol. 1. Elsevier, Amsterdam, The Netherlands.

Nakamori, H., and H. Soemori. 1981. Comparison of dispersal ability and longevity for wild and mass-reared melon flies, *Dacus cucurbitae* Coquillett (Diptera: Tephritidae), under field conditions. Appl. Entomol. Zool. 16: 321–327.

Nakamori, H., H. Kakinohana, and M. Yamagishi. 1992. Automated mass production system for fruit flies based on the melon fly, *Dacus cucurbitae* Coquillett (Diptera: Tephritidae), pp. 441–454. *In* T. E. Anderson and N. C. Leppla (eds.), Advances in insect rearing for research and pest management. Westview Press, Boulder, CO.

Odell, T. M. 1992. Straggling in gypsy moth production strains: a problem analysis for developing research priorities, pp. 325–350. *In* T. E. Anderson and N. C. Leppla (eds.), Advances in insect rearing for research and pest management. Westview Press, Boulder, CO.

Odell, T. M., M. A. Keena, and R. B. Willis. 1997. Dietary influence of iron formulation on the development of gypsy moth (Lepidoptera: Lymantriidae) in laboratory colonies. Ann. Entomol. Soc. Am. 90: 149–154.

Penn, S. L., R. L. Ridgway, G. T. Scriven, and M. N. Inscoe. 1998. Quality assurance by the commercial producer of arthropod natural enemies, pp. 202–230. *In* R. L. Ridgway, M. P. Hoffmann, M. N. Inscoe, and C. S. Glenister (eds.), Mass-Reared natural enemies: application, regulation and needs. Thomas Say Publications in Entomology. Entomological Society of America, Lanham, MD.

Peterson, A. 1964. Entomological techniques, how to work with insects. Entomological Reprint Specialists, Los Angeles, CA.

Singh, P., and R. R. Moore (eds.). 1985. Handbook of insect rearing, vols. 1 & 2. Elsevier, Amsterdam, The Netherlands.

Smith, C. N. 1966. Insect colonization and mass production. Academic Press, New York, NY.

Southwood, T. R. E., and P. A. Henderson. 2000. Ecological methods, 3rd ed. Blackwell Science, Oxford, United Kingdom.

Ott, R. L., and M. Longnecker. 2001. An introduction to statistical methods and data analysis, 5th ed.. Thompson Learning, Duxbury, Pacific Grove, CA.

Teal, P. E. A., J. M. Gavilanez-Slone, and B. D. Dueben. 2004. Effects of sucrose in adult diet on mortality of males of *Anastrepha suspensa* (Diptera: Tephritidae). Florida Entomol. 87: 487–491.

van Frankenhuyzen, K., P. Ebling, B. McCron, T. Ladd, D. Gauthier, and C. Vossbrinck. 2004. Occurrence of *Cystosporogenes* sp. (Protozoa, Microsporidia) in a multispecies insect production facility and its elimination from a colony of the eastern spruce budworm, *Choristoneura fumiferana* (Clem.) (Lepidoptera: Tortricidae). J. Invert. Pathol. 87: 16–28.

van Lenteren, J. C. 2003. Quality control and production of biological control agents, theory and testing procedures. CABI Publications, Cambridge, MA.

van Lenteren, J. C., and M. G. Tommasini. 1999. Mass production, storage, shipment and quality control of natural enemies, pp. 276–294. *In* R. Albajes, M. L. Gullino, J. C. van Lenteren, and Y. Elad (eds.), Integrated pest and disease management in greenhouse crops. Kluwer Academic Publishing, Dordrecht, The Netherlands.

van Lenteren, J. C., and J. Woets. 1988. Biological and integrated control in greenhouses. Annu. Rev. Entomol. 33: 239–269.

van Lenteren, J. C., F. Bigler, and C. Waddington. 1993. Quality control guidelines for natural enemies, pp. 222–230. *In* G. Nicoli, M. Benuzzi and N. C. Leppla (eds.), Proceedings, 7th Workshop of the International Organization of Biological Control Global Working Group on Quality Control of Mass-Reared Arthropods. Rimini, Italy. (http://users.ugent.be/~padclerc/AMRQC/proceedings.html).

van Lenteren, J. C., A. Hale, J. N. Klapwijk, J. van Schelt, and S. Steinberg. 2003. Guidelines for quality control of commercially produced natural enemies, pp. 265–303. *In* J. C. van Lenteren (ed.), Quality control and production of biological control agents, theory and testing procedures. CABI Publishing, Cambridge, MA.

Wajnberg, E. 2003. Basic statistical methods for quality control workers, pp. 305–314. *In* J. C. van Lenteren (ed.), Quality control and production of biological control agents, theory and testing procedures. CABI Publishing, Cambridge, MA.

Walpole, R. E., and R. H. Myers. 1972. Probability and statistics for engineers and scientists. Macmillan, New York, NY.

Chapter 11

Insect Rearing Production Systems, A Case Study: The Southwestern Corn Borer

FRANK M. DAVIS
Emeritus Adjunct Professor of Entomology
Department of Entomology and Plant Pathology
Mississippi State University
Mississippi State, MS

I. Introduction

The production system of an insect rearing program, which I equate to "rearing system," includes the housing, provisioning, and handling of the insects in their various growth stages. This involves development and utilization of standard operation procedures (SOPs), supplies, and equipment for operations such as preparing diet and presenting it in a manner that is acceptable, containing the insects, and harvesting the various growth stages.

All nine insect rearing program components, illustrated as parts of the puzzle introduced in the preface, are essential to the development of a successful rearing program and are very much interrelated. The success of the production system to rear high quality insects is strongly dependent upon how well the other components function. Continuous production of high quality insects requires that all components function properly and efficiently.

Insect rearing programs can be characterized by the size of their production potential: small (hundreds of adults produced per week), midsized (thousands of adults produced per week), or mass (millions of adults produced per week). Many of the basic principles and procedures used in these different size programs are the same or very similar. However, small-scale operations involve the production of one or a few species for use in

Fig. 1. A form-fill-seal machine for mass rearing lepidopterous insects at the USDA Gast Insect Rearing Laboratory located at Mississippi State University, Mississippi State, MS.

research or education at a relatively high cost per insect, midscale operations involve the production of a few to several species for public and private rearing ventures at a lower cost per insect, and mass rearing operations usually involve the production of a single species for use in pest management at very low cost per insect. Another major difference between large-scale rearing and small-to midscale rearing is the issue of disposal of refuse (e.g., fecal material, leftover diet and disposable rearing containers) created during the rearing process. Small and midsized rearing programs do generate refuse that must be removed; however, the amount is normally small enough that it can be handled using plastic garbage bags that can be taken to a landfill. Large programs produce huge amounts of refuse that must be dealt with according to environmental regulations. The disposal of huge amounts of refuse normally requires the use of lagoons or special types of landfills on the facility property. This is a significant additional rearing cost.

Regardless of size, insect rearing programs and their components tend to evolve over time. Changes occur as programs face new challenges such as significantly increased demand, the need to reduce production costs, the occurrence of problems (e.g., diet contamination by microbes), the need to improve the quality of the insects reared, and the addition of new species to be reared.

Many insect rearing production systems initially use unsophisticated equipment and materials for preparing, dispensing, and infesting the diet with insects: hot plates, metal cooking pots, and quart- or gallon-sized blenders for preparing diets; a plastic beaker or squeeze bottle for dispensing the diet into rearing containers; and an artist's brush or a salt shaker to infest the diet with neonates (recently hatched, first instar larvae) or eggs (Davis 1976). Over time, many small-sized programs evolve into midsized ones; and their production system utilizes equipment designed specifically for insect rearing, e.g., a steam-jacketed kettle with an attached variable speed mixer for diet preparation, specially designed and fabricated equipment for dispensing the diet semiautomatically into rearing containers, and a hand-held device for rapidly infesting each container or rearing cell with eggs or neonates (Davis 1989). Some large programs have progressed to the sophistication level of having a single piece of equipment that can form the insects' rearing containers, dispense the diet (prepared and flash sterilized by adjacent equipment) into the containers, quickly cool the diet using a chilling chamber, infest the rearing cells with the desired insect stage, place the cover on the rearing container, and separate the containers into groups of the desired size in a continuous process (e.g., Sparks and Harrell 1976; Tillman et al. 1997). Such equipment is referred to

Fig. 2. Twin screw extruder equipment for preparing and dispensing diet for mass rearing the pink bollworm at the USDA/APHIS Pink Bollworm Rearing Facility in Phoenix, AZ.

as an "in-line, form-fill-seal machine" (Fig. 1). The latest piece of equipment used in some large-scale rearing programs [e.g., the United States Department of Agriculture/Animal and Plant Health Inspection Service (USDA/APHIS) pink bollworm Sterile Insect Technique (SIT) rearing program in Arizona] is referred to as an "extruder" (Edwards et al. 1996). This equipment (Fig. 2) can be used to cook, form, mix, texturize, shape, and dispense food products under conditions that favor quality retention, high productivity, and low cost (Riaz et al. 1996).

Much of the progress in sophistication of equipment for producing insects has followed the progress made by human food technologists. Therefore, consulting with food technologists for information on their latest technological advances is important to those trying to increase automation potential for their insect rearing production system.

One insect production system is presented here in detail as a case study to illustrate the evolution and development of a midsized program for rearing four pest species of Lepidoptera. It is a rearing program I initiated in the late 1960s. The case study approach provides me the opportunity to share with the insect rearing community my experience of the continuous building of a successful rearing program for providing insects for the Mississippi State USDA/Agricultural Research Service (ARS) corn and cotton host plant resistance programs in which I worked as a team member as an entomologist for 35+ years. The principles, procedures, and equipment used in this program can and have been used for developing and improving rearing programs for other insects.

The rearing program described here began in the late 1960s with the rearing of one species: the southwestern corn borer, *Diatraea grandiosella* (Lepidoptera: Crambidae) (SWCB). In the mid 1970s, the fall armyworm, *Spodoptera frugiperda* (Lepidoptera: Noctuidae) (FAW), was added; and in the 1980s, two species were added: the cotton bollworm, *Helicoverpa zea* (Lepidoptera: Noctuidae) (CBW), and the tobacco budworm, *Heliothis virescens* (Lepidoptera: Noctuidae) (TBW).

In the beginning, it was a small-sized program with the production system based on simple, hand labor procedures. As the demand for significantly more SWCB for research grew and with the successive addition of three new insect species to be reared, the program evolved into a multispecies, midsized program with the production system component becoming more and more mechanized. Additionally, other components of the rearing program evolved in an effort to improve support and protection of the production system component and to improve the safety and health issues for the rearing staff. During the years from 1965 to 1999, the opportunity to design and construct or renovate facilities to produce insects occurred three times.

Fig. 3. The "Little Black Table," located in the break room of the rearing facility, around which the team approach for solving problems was practiced.

The goals for the USDA/ARS program at Mississippi State (see Acknowledgements for a partial list of team members) have been the following: 1) produce high quality insects capable of performing their intended purpose for resident host plant resistance programs; 2) place a minimum of stress on the laboratory-reared insects in order to maintain healthy, robust colonies; 3) produce the number of insects required by the research program in a reliable and timely manner at the lowest possible cost; and 4) provide rearing personnel with a safe and healthy environment in which to work. The accomplishment of these goals allows the research units served to receive maximum benefit from the insects reared at a reasonable cost.

A team approach was utilized to the fullest to accomplish the above goals. The team members varied according to the problem to be solved. On occasion, local maintenance personnel plus others from the public and private sectors were brought in to work with me and my staff. A space for team problem solving was designed in the break room, which featured a table referred to as the "Little Black Table" (Fig. 3). Also, a fully equipped shop was provided to design and fabricate new rearing equipment (e.g., adult cages).

At the end of each peak production period, staff involved in the day-to-day rearing were asked what problems needed to be solved and what new equipment was needed to increase the efficiency of the system. Problem solving and development of new equipment, rearing containers, and adult cages occurred mainly during the non-field season when rearing production was in a low or maintenance mode.

The SWCB is a major pest of corn/maize (*Zea mays*) in the southern part of the United States. In the mid 1960s, a team of USDA scientists located at Mississippi State, MS, was charged with creating a program for developing corn germplasm with resistance to this pest. I and my agronomist team member recognized early in the program that a rearing program for this insect would be absolutely necessary for successful identification, development, and release of resistant corn germplasm (Davis and Guthrie 1992). The laboratory-reared insects' primary role is to damage susceptible plants grown under field conditions in the same manner and to the same degree as their feral counterparts. The keys to successful screening are that each plant within an experiment should have an equal opportunity to be damaged and that the plants in the susceptible control plots should all be heavily damaged. The technique developed for artificially infesting corn plants with SWCB (and FAW) neonates is described by Davis et al. (1989). The success of the USDA host plant resistance program for identifying resistant corn germplasm and the development of these selections into resistant populations and inbred lines for release to the public have been documented by Williams and Davis (1997).

The technologies associated with this production system have been transferred to the rearing community using a variety of ways including publications, oral presentations, visits to research sites around the world, visits to the rearing facility by interested parties, and making equipment and supplies available to others in the insect rearing community by developing special relationships with vendors. The following publications demonstrate the evolution of the production system discussed here: Davis (1976, 1980, 1982a, 1982b, 1983, 1989, 1997), Davis et al. (1978, 1985, 1990), Davis and Oswalt (1979), Ng et al. (1987), Davis and Guthrie (1992), Davis and Jenkins (1995), and Jenkins et al. (1995).

In the following account, the principles, procedures, equipment, and supplies used to rear all growth stages of the SWCB are described. The production system today is relatively simple and typical of many multispecies, mid-sized rearing programs. It relies on a well-defined set of SOPs for handling and rearing each insect growth stage, excellent larval diets, a plastic multicellular larval rearing container, a few pieces of specialized equipment to increase the efficiency of the operation, and large adult cages. Where appropriate, comments are made concerning the three other species when they differ from the SWCB, e.g. in SOPs, equipment, supplies, or cages. Examples of real insect rearing problems and their solutions will be presented. This chapter overlaps broadly with others in this book; however, in order to fully describe the SWCB production system, the context of the other components of the rearing program must be provided.

II. Developmental Stages

A. Egg

SWCB eggs are small and ellipsoidal with a flat bottom. Eggs are glued to the oviposition substrate by an adhesive produced by the female. In the wild, females normally oviposit 1–5 eggs per mass in a fish scale pattern. Under laboratory conditions, the female oviposits in variable-sized masses of up to 50 eggs when oviposition on a wax paper substrate is unrestricted within a large screened cage.

SWCB eggs go through three color phases before eclosion of the neonate. The first is the "yellow-green phase." After about 36 h, the color changes to white with 3–4 red horizontal stripes. At this point, the egg is commonly referred to as being in the "candy stripe phase." A day before eclosion, the dark head capsule of the developing larva can easily be seen through the chorion, so the egg is referred to as being in the "blackhead phase."

1. Environment. The following environmental regime was determined to be optimal for the development of eggs: 27.6°C, 50–60% relative humidity (% RH), and a 16:8 L:D [(photophase duration in hours):(scotophase duration in hours)] photoperiod cycle. Under these conditions, egg hatch occurs in 5 d, which is the same as that for eggs of the wild population developing under field conditions during the summer. Details concerning the rooms for holding the egg and larval stages during their development and the equipment for providing the temperature/humidity/photoperiod regimes are described in Sec. II. B. 1.

To delay egg hatch or to coordinate hatch of eggs oviposited on different nights, eggs can be held for varying time periods in a selected suboptimal environment of 21.1 ± 0.6°C, 55 ± 5% RH, and continuous light (Ng et al. 1987). Eggs held exclusively under the suboptimal environment require 9 d to hatch. This coordination procedure has proven to be very useful when scheduling availability of unusually large numbers of neonates or for preventing hatch on the weekend. SWCB eggs oviposited on as many as five consecutive nights can be made to hatch on the same day by coordinating their development using various combinations of periods of time at these two temperatures. This technique has also been used effectively for coordinating egg hatch of the FAW, CBW, and TBW.

In the early 1990s, the photoperiod under which the SWCB eggs and larvae were reared was changed from continuous light to 16:8 (L:D) to provide the insects with a more normal light/dark cycle. Earlier, continuous light was used because some SWCB larvae were observed to enter the diapause stage at a 14:10 (L:D) regime. See Chap. 6 *Environmental Biology of Insect Rearing* for a discussion of diapause. The change in photoperiod from continuous light to 16:8 (L:D) did not elicit larval diapause or have any noticeable effects on egg hatch synchronization. However, quality of the lighting apparently does have some effect on eggs. Under natural conditions, wild SWCB eggs in the candy stripe phase are strikingly beautiful with very distinct red stripes against a pure white background. In contrast, eggs produced under laboratory conditions have indistinct red stripes, and the background color is abnormal. Interestingly, when eggs produced by females reared under laboratory conditions are exposed to field environmental conditions, they develop the coloration seen in wild eggs. The above is an example of where there was concern that something was wrong with the laboratory-reared SWCB.

Another false alarm due to a similar phenomenon was experienced with the FAW. Wild females ovipositing under field conditions produce gray-brown eggs in masses, which they cover with a layer of scales from their bodies. Artificially-reared females ovipositing under laboratory conditions produce eggs that vary in color from gray-brown to green to red and which are not covered with a layer of scales. The fact that these differences are environmental in origin was discovered when laboratory-reared FAW moths were released into large cages containing corn plants in the field to screen for adult nonpreference resistance. Surprisingly, the laboratory-reared females oviposited normal colored eggs in masses covered with scales.

2. Harvest and Surface Sterilization. Each morning during the 7-day week, wax paper sheets with SWCB eggs attached are harvested by pulling the sheets vertically through slits in the top of the oviposition cage (Sec. II. D). Eggs used for colony maintenance and for some laboratory experiments are surface-sterilized to kill potential disease and diet contaminating microorganisms (Davis 1989). Eggs used for infesting plants in the field or greenhouse do not require egg surface sterilization.

The equipment and materials necessary for surface sterilization of the eggs include the following: a timer, a gallon (3.8-liter) plastic or metal beaker, a rocker (Hoefer Scientific Instruments, San Francisco, CA), a laminar flow hood, a simple rack for suspending wax paper strips, bleach (Clorox®: 5.25% sodium hypochlorite), tap water, two pieces of 3/8-in (1-cm) thick honeycomb panel with 3/8-in (1-cm) openings (Hexcel Corp., Arlington, TX) cut to fit inside the beaker, sanitized pint (0.47-liter) or quart (0.95-liter) glass canning jars, and autoclaved paper (Kim-Wipes®). Eggs are normally surface-sterilized when they reach the blackhead stage. Ten to twelve wax paper strips (cut from the 15.24 cm- or 30.48 cm-wide by 60.96 cm-long wax paper) with attached eggs are placed in the gallon (3.8–liter) beaker containing an aqueous solution of 0.2% sodium hypochlorite (2,020 ml of tap water + 80 ml of bleach). Two circular pieces of honeycomb panel are placed on top of the wax paper strips to keep them submerged. The contents of the beaker are agitated on the mechanical rocker for 5 min, the sodium hypochlorite solution is discarded, and the beaker is flushed with cold water directly from the tap for 15 min. The wax paper strips are removed from the beaker and sniffed to insure that the odor of sodium hypochlorite has been removed completely. If not, the strips are returned to the beaker and washed with tap water for another 5 min. After no odor of sodium hypochlorite is detectable, the strips are hung vertically on a drying rack for 20–30 min in a laminar flow hood. Approximately 7 strips of wax paper with eggs are placed in each sanitized 1-quart (0.95-liter), wide-mouth glass jar. Fewer egg strips per jar may be necessary because high egg densities result in difficulty handling the neonates during infestation due to tangling of the neonates in silk webbing. The mouth of the jar is covered with an autoclaved Kim-Wipe® secured by the screw-on metal ring.

There are several critical considerations when surface-sterilizing eggs. Overexposure to sodium hypochlorite can result in dechorionation of the eggs and mortality due to dehydration. Consequently, the 5-min period of exposure and the 0.2% concentration should not be exceeded. The concentration of sodium hypochlorite in generic brands of bleach can vary from that labeled (P. Sikorowski, unpub.) and should be avoided. Even in name brands of bleach, different formulations can be encountered, e.g., Ultra Clorox (6% sodium hypochlorite). Underexposure to sodium hypochlorite can result in transmission of diseases due to inadequate surface sterilization. Because sodium hypochlorite is chemically unstable, purchase quart rather than gallon size containers, keep the caps on the bottles tightly secured, and use only freshly prepared solutions.

B. Larva

The SWCB has two larval forms: nondiapausing ("summer form") and diapausing ("winter form"). Summer form larvae are characterized by having large black, dorsal spots on a white background, whereas winter form larvae lack black spots (referred to as "immaculate") and are light yellow in color.

At an age of about 7 days, female larvae begin to consume more diet than male larvae, and their efficiency of conversion of food to body mass is significantly higher than that for males (Ng et al. 1993). There are 5–6 instars in both sexes of both larval forms. Female pupae weigh about 1/3 more than male pupae.

The larvae of both forms are highly cannibalistic in the late instars, so only one pupa can be obtained per rearing cup or cell. The CBW is also highly cannibalistic in the larval stage. The FAW and TBW are less cannibalistic, and two or more pupae may be produced from a single rearing cup or cell.

1. Environment. SWCB larvae are reared using four different environments, depending upon production goals. For normal uninterrupted development and for diapause termination, an environment of 26.7 ± 0.6°C, 55 ± 5% RH, and 16:8 (L:D) is used. To slow normal development, an environment of 21.1 ± 0.6°C, 55 ± 5% RH, and 16:8 (L:D) is used. A combination of periods in these two temperature regimes is used to coordinate larval growth rates—similar to the technique described for egg development (Ng et al. 1987). To induce diapause, an environment with both lower than optimum temperature and short photophase duration is used: 21.1 ± 0.6°C, 55 ± 5% RH, and 12:12 (L:D). The environment for storing or keeping the larvae in diapause is 10.0 ± 0.6°C, 55 ± 5% RH, and 12:12 (L:D) photoperiod. To terminate diapause, the larvae are returned to the environment for normal development.

To allow the rearing staff winter holiday leave while still satisfying research needs, a schedule was developed for placing SWCB and the other Lepidoptera into the 26.7 and 21.1°C, 55 ± 5% RH, and 16:8 (L:D) non–diapause inducing environments during November and the first half of December so that during the last two weeks of December no eggs or larvae need to be handled. Only a few hours of work are necessary to harvest pupae during this period so that eggs can be available the first week of January. Careful planning of development within the two temperature regimes allows synchronized eclosion of a sufficient number of adults weekly in January to restore the maintenance colony age distribution. This system has been used successfully for the last 17 years without losing any Lepidoptera colonies or creating any problems for researchers. Prior to implementation of this system, the staff worked 365–366 days a year.

The developmental environments are provided by two rooms, each of which has its own environmental control system. The major advantage of this arrangement is that these critical environmental regimes can be maintained within close tolerances without interference from building environmental policies, equipment, and repairs. Also, by having two large rooms, insects can be moved from one to the other when repairs are necessary.

Variation in temperature should be held to a range of ± <1°C. Large fluctuations in temperature can be particularly deleterious because they cause moisture droplets to form inside closed rearing containers. Free water can favor growth of microbial contamination, and a procedure has never been identified whereby this moisture problem could be corrected once it occurs.

Lighting within each room is provided by four banks of fluorescent bulbs (two standard cool white, 1.2 m (4 ft), T-12, 30-watt bulbs per bank) mounted in the ceiling. Photoperiod is controlled using an electromechanical timer (Dayton Model 2E021, Dayton Electric Mfg. Co., Chicago, IL).

Humidification is an important environmental factor and should be held year round at 50–60% RH. This requires humidification equipment that has the capacity to maintain favorable humidity levels even during the winter months when these humidity levels are difficult to obtain. Currently, the facility uses evaporative humidifiers (Aprilaire®, Research Products Corp., Madison, WI; Fig. 4), which operate by blowing air through a mat through which warm water percolates. Humidifiers that operate by atomization or by steam generation require potentially burdensome reservoir maintenance unless the water supply is substantially free of dissolved minerals or unless the expense of softening or deionizing the water supply is incurred. Humidifiers that operate by steam generation also introduce a significant source of heat into the rearing room that may interfere with temperature control. Dehumidification can be a problem in the summer months—depending on the climate. Consequently, an upright dehumidifier (Whirlpool Inc., Benton Harbor, MI) is operated in the 27°C rearing room.

Air circulation and sanitation is provided within these developmental rooms by air filtration equipment similar to that used to control scales from the adults in the "moth house" (Sec. II. D. 3). This circulation appears to be sufficient to eliminate any significant, vertical stratification of temperature. Stratification of temperatures can result in variation in developmental rates of the insects (Chap. 6 *Environmental Biology of Insect Rearing*).

Because environments are so important to the success of rearing, they are monitored at least daily and even more frequently during peak production periods. Dial thermometers and hygrometers are mounted at sites within each room and environmental chamber where staff and workers can easily see if any changes from normal are occurring. Additionally, hygrothermographs are operated in each developmental room to record variation in temperature and humidity. These records indicate when problems occur and to what extent.

2. Rearing Container. From the late 1960s to the late 1980s, the rearing container used by the USDA/ARS rearing program at Mississippi State was a 1-ounce (29.6-ml), clear plastic cup with a paperboard insert cap, coated on one side with a polysaran compound to reduce cap porosity. The reasons for choosing this container were the following: 1) suitability for rearing the highly cannibalistic SWCB larvae individually, 2) minimization of spread of diseases and diet-contaminating microbes, 3) ready availability, and 4) low cost.

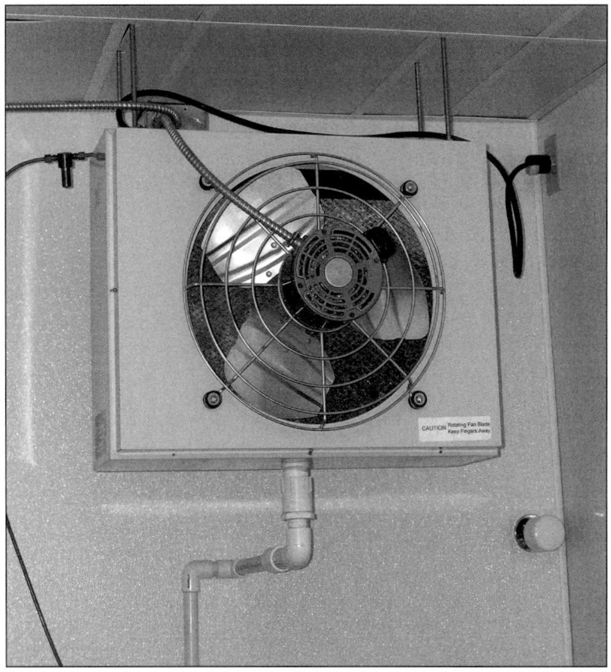

Fig. 4. An Aprilaire® humidifier used to provide preferred humidity in the rearing facility.

The following is an example of how even minute procedural details can have a very significant effect on success of a rearing effort. Late instar FAW larvae were observed to be dying in the plastic cups at an unacceptably high rate. The first hypothesis was that this mortality was due to a disease organism; however, only saprophytic bacteria could be detected in the cadavers. Observation that the diet in the cups with dead larvae was abnormally wet and that wet diet was associated with orientation of the polysaran-coated side of the caps to face the diet led to an experiment demonstrating that orientation of the coated side of the caps to face away from the diet allowed the diet to dry faster and increased larval survival. This result prompted a change in the SOP for capping the cups for FAW.

As demand for SWCB and other species of Lepidoptera grew, the cup and cap system became too inefficient, both in terms of labor and space. In addition, the direct cost of cups and caps had significantly increased. Similar to the approach taken by Mihm (1983), large plastic, circular and rectangular containers with inserted plastic grids to separate the larvae were tried as potentially more efficient rearing containers for the SWCB and FAW. However, reliable production using these containers was never obtained because of microbial contamination and larval cannibalism.

In 1985, technical staff from the USDA/ARS Gast Insect Rearing Laboratory (Mississippi State, MS) expressed an interest in rearing Lepidoptera in a 32-cell, plastic tray fabricated on-site using an in-line, form-fill-seal machine (Fig. 1; Sparks and Harrell 1976) obtained from a USDA/ARS facility located at Tifton, GA. Personnel from the Gast facility came to the author's research unit to learn about rearing Lepidoptera because their experience had centered on rearing the boll weevil, *Anthonomus grandis grandis* (Coleoptera: Curculionidae). The opportunity to share knowledge and technology for rearing of Lepidoptera and to work in developing a new technology was welcomed.

A major problem with the Sparks and Harrell 32-cell rearing tray was that the porosity of the lid material was too high, resulting in excessive drying of the diet before the larvae could complete their development when reared under preferred environmental conditions. When the diet became too dry, many of the larvae escaped by chewing a hole through the lid material or through the plastic tray. In addition, the gray color of the plastic used to fabricate the tray made it difficult to monitor the survival and growth of the larvae and the incidences of disease and microbial-contaminated diet.

After some searching, a clear, 15-mil polyvinyl chloride (PVC) plastic sheet was located that could be used to thermo-form the tray. The overall dimensions of the tray and the configuration and size of individual rearing cells were modeled after the 32-cell tray created by Sparks and Harrell (1976) because the depth and volume of their rearing cell was adequate not only to hold sufficient food for normal growth and development of the four species of Lepidoptera of interest, but also provided sufficient space between the diet surface and the lid material to allow for adequate exchanges of gases.

The problem of finding a satisfactory lid material was much more challenging. In addition to the effects of the lid material on larvae discussed above, it had to be sealable to the tray in a manner that was durable for the several-week period of larval development but easily removed when desired. To satisfy these criteria, personnel from Oliver Products Co. (Grand Rapids, MI) were invited to assist in the research and development effort. Lid materials of various types were tested, including a series of polyester materials of varying thicknesses. The polyester film Mylar® showed the greatest potential. Mylar® of 1-, 2-, 3-, and 5-mil thicknesses were tested. The larvae of all four species could easily penetrate 1-mil-thick Mylar® but not the 3- or 5-mil. SWCB were the only larvae observed to cut through the 2-mil Mylar® but only at the end of their development; and, even then, a high percentage of these larvae did not leave the cell or, if they left, re-entered it to pupate. When testing lid material of various thicknesses for larval penetration, the lid material was simply glued onto the tray bottoms using common glue, and a few pin holes were punched through the lid covering each rearing cell to allow gas exchange.

The next step was to have the staff at Oliver Products place a heat-sensitive adhesive on one side of the Mylar® that would adhere to the plastic tray for at least 6 to 8 weeks, would allow removal of lid from tray with only minimum effort, and would not adversely affect the insects if eaten. They responded quickly with an adhesive that met the above criteria.

Since Mylar® is impermeable to gases, the next step was to test various densities of perforations to achieve appropriate porosity for normal development of the larvae to the pupal stage. A pinhole every 5 mm (referred to by Oliver Products as "single punched") created a microenvironment resulting in normal larval development and pupation in our midsized production system. This porosity allowed the diet to gradually dry down during the 20-day development period for FAW, CBW, and TBW and the 26-day period for SWCB, resulting in high pupal productivity and ease of harvest. The Mylar® lid material for the trays formed and processed by the in-line, form-fill-seal-machine required holes punched every 2.5 mm (referred to by Oliver Products as "double punched") to produce similar production results. A possible reason for the differences in pinhole spacing may be as follows. In the midsized rearing system, after the diet is dispensed into the trays, the trays are placed under a laminar flow hood to allow the diet to cool and dry for a preselected time period prior to infestation (normally 1 to 2 h). In the case of the in-line, form-fill-seal system, the diet is cooled as the trays pass quickly through a chilling chamber, so there is no extended period of cooling and drying (Tillman et al. 1997). This diet may begin with much higher moisture content and require a lid material with greater porosity to produce the same suitable microenvironment.

An experiment comparing the four species reared in the newly created 32-cell rearing containers (Fig.

5) with those reared in the cup and cap system using the midsized rearing system's equipment indicated that the new rearing container was a suitable replacement for the old (Davis et al. 1990).

Oliver Products decided to market the Mylar® lid material (2 and 3 mil) and to design, fabricate, and market equipment for semiautomatically sealing their lid material (single or double punched and in rolls or individually cut to fit the 32-cell tray). Their first model of the sealing equipment was found to have design problems, and I and my staff assisted Oliver Products personnel in designing an improved model that is the one presently being marketed.

I negotiated with a private company (formerly Dixon Paper in Lubbock, TX; now owned and operated by Mayfield Paper, San Angelo, TX) to produce and market the thermo-formed tray. Today, the tray, lid material, and sealing equipment can be purchased from Oliver Products. Bio-Serv, Inc. (Frenchtown, NJ) markets the Oliver Products lid material plus a 32-cell PVC plastic tray with rearing cells that are slightly deeper than the ones sold by Oliver Products or Mayfield Paper.

The above is an example of the development of a mutually beneficial relationship between insectary personnel and vendors that has resulted in the availability of a unique rearing container with associated equipment to the insect rearing community. I believe strongly in developing these types of relationships.

The Corn Research Unit has used the 32-cell rearing tray for over 15 years to rear SWCB, FAW, CBW, and TBW. The savings of using this rearing container rather than the individual cup with cap have been substantial: 1) significant reduction in container cost; 2) reduced time and number of personnel required to dispense diet into cells, infest rearing cells with neonates, and harvest pupae from cells; and 3) reduced space required for housing the insects (Davis 1997). The Gast Laboratory has also used the multicellular tray produced in their own in-line, form-fill-seal machine to successfully rear several noctuid species (Tillman et al. 1997).

a. Diet Ingredients. The diet used to rear SWCB larvae to pupation is a modified version of the commercially available wheat germ-casein-based Vanderzant-Adkisson diet for Lepidoptera (Davis 1989). Major modifications included the following: substitution of sterilized, 60-mesh corn cob grits for cellulose fiber; substitution of stabilized wheat germ for raw wheat germ, and increase in the amount of ascorbic acid in the vitamin mix to compensate for possible losses during diet cooking. This diet has proven to be highly suitable for rearing both the SWCB and FAW and serves as a cornerstone for the production system. The individuals produced are comparable in size, morphology, and fecundity to feral individuals obtained from late-instar larvae collected in corn fields. The diet for CBW and TBW is similar to the SWCB/FAW diet except Nutrisoy, a soy protein, is substituted for casein, and the antibiotic aureomycin is substituted for neomycin sulfate (Jenkins et al. 1995). SWCB reared on diet containing aureomycin are diminutive compared to those reared on corn, their preferred host, or on diet containing neomycin sulfate. The SOPs for preparing the two diets are similar, with the same equipment used for preparation and dispensing.

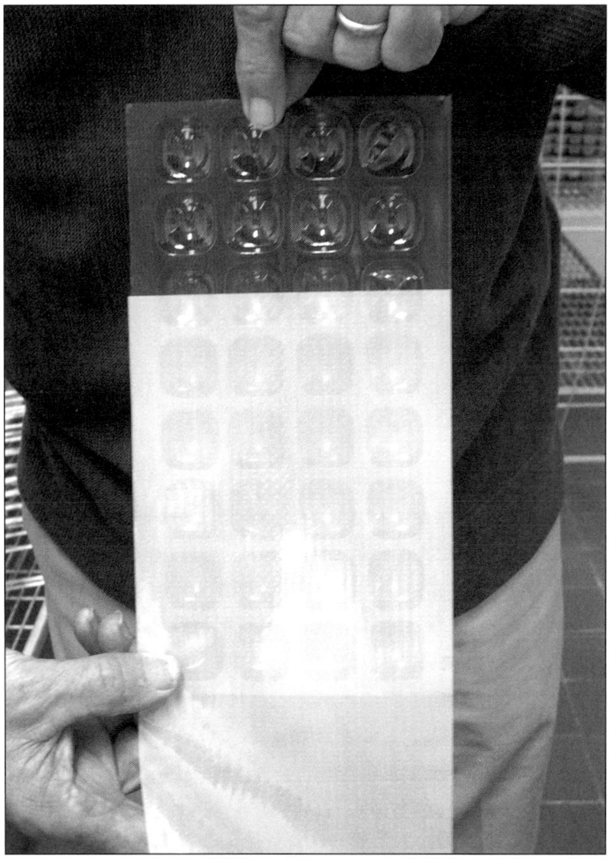

Fig. 5. A 32-cell rearing container consisting of plastic tray and Mylar® lid.

To avoid having to purchase and stock the numerous individual diet ingredients, reduce the time required to weigh the ingredients separately, and reduce the possibility of preparation errors, Bio-Serv, Inc. (Frenchtown, NJ) agreed to market the SWCB/FAW and

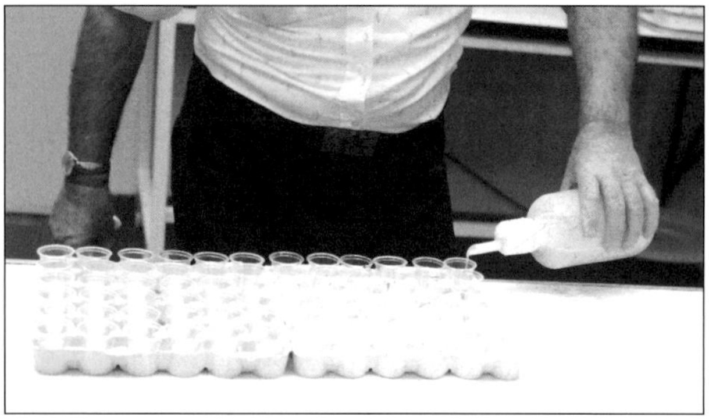

Fig. 6. Dispensing larval diet into 29.6-ml (1-oz), plastic cups using a hand-held, plastic squeeze bottle.

TBW/CBW diets as three packs, each containing one or more of the 23 ingredients. Pack A (Bio-Serv #7060) contains agar, a gelling agent. Ultimately, I decided to purchase agar from a variety of vendors depending on unit cost. However, before buying large quantities of agar or any other gelling agent, one should request a sample to be tested for suitability. Pack B (Bio-Serv #F0635; also referred to as the "dry mix") contains nine, non–heat labile diet ingredients and is ground to 60 mesh to be compatible with the dispensing equipment described below. Pack C (Bio-Serv #F0635; also referred to as the "vitamin mix") contains 13 heat-labile ingredients. Additional ingredients added separately include the antibiotic neomycin sulfate [available from Sigma (St. Louis, MO) and Bio-Serv] and water. Packs B and C and neomycin sulfate are stored at about 40°F (4.4°C) to increase shelf life.

Factors that should be considered when selecting vendors to supply diet ingredients include: 1) quality, consistency, and availability of diet ingredients; 2) price; and 3) other vendor services such as ability to sterilize selected diet ingredients upon request and to provide the results of assays for pathogens in diet ingredients.

b. Preparation and Dispensing. Great advances have been made in the procedures and equipment for preparing and dispensing diet. Initially, the equipment used included a hot plate; a 1-quart (0.95-liter)-capacity, household blender; and a 1,000-ml squeeze bottle from which the diet was dispensed individually by hand into plastic cups (Fig. 6). Today, the diet is prepared in 20-liter steam-jacketed kettles (Groen, Elk Grove Village, IL) with a variable-speed mixer (Model PM6015, Eastern Mixers, Clinton, CT) attached to the side (Fig. 7). The diet is moved by a pump (Model 2971-1 FSV-260, National Instrument Co., Baltimore, MD; described by Tillman et al. 1997) directly from the kettle into a custom-designed and fabricated, electro-pneumatic dispenser that delivers an adjustable quantity of diet simultaneously into all 32 cells of the multicell tray (Fig. 8). This equipment can dispense four gallons of diet in <10 min.

The kettles and diet dispenser are on carts. Standard laboratory carts are used for the kettles, allowing them to be moved easily in front of a large stainless steel sink for cleaning. The diet dispenser and associated pump are mounted on a custom-designed cart allowing staff to move the dispenser from storage to a position adjacent to the kettles and then back when the task is completed.

The SOP for preparing approximately one gallon (3.8 liter) of diet is as follows:

- Weigh the dry ingredients: Pack A (84.3 g), Pack B (472.0 g), Pack C (29.6 g), and neomycin sulfate (2.0 g).
- Dissolve Pack A in 700 ml of cool tap water. (Note: If agar is placed directly into boiling water, it adheres to the surface of the mixing container.)
- Bring 480 ml of tap water to a boil in the kettle.
- Add Pack B and the dissolved agar to the kettle with the mixer on, and allow to "cook" for 3 min.
- Turn off the heating element in the kettle and add 1,464 ml of 10°C distilled water. (Note: This will immediately bring the temperature down to about 80°C, a temperature low enough to prevent harm to the heat-labile ingredients to be added next.)
- Add Pack C and the neomycin sulfate, and mix the diet for 2–3 min before dispensing.

Fig. 7. Cooking and mixing larval diet in a steam-jacketed kettle with variable-speed mixer attached.

Fig. 8. A custom-designed and fabricated piece of equipment to dispense diet simultaneously into all cells of 32-cell rearing trays.

Before dispensing the diet, hot water plus an antimicrobial solution of ammonium chloride is run through the dispenser to heat the metering head and to resanitize the system prior to placing the supply hose into the kettle. As trays are filled with diet, they are placed in custom-designed holding racks (Davis et al. 1990) under a laminar flow hood for 1 to 2 h to cool and dry before infestation. When the trays of diet are to be infested with FAW, they are allowed to dry about 30 min longer than for infestation with the other species. As discussed above (Sec. II. B. 2), FAW larvae require a drier diet than other species. After dispensing the diet, the equipment is sanitized again using hot water plus ammonium chloride and the pump and metering head are disassembled, cleaned, sanitized, serviced, and reassembled.

4. Infestation. For many years, neonates were transferred one at a time onto diet using a moistened artist's brush. In the fall of 1978, Dr. John A. Mihm, International Center for Wheat and Maize Breeding (CIMMYT) in Mexico, visited the USDA/ARS laboratory at Mississippi State and demonstrated a hand-held, plastic device referred to as the "Bazooka" that he and his colleagues had designed and fabricated for infesting corn plants with neonate larvae mixed in corn cob grits (Mihm et al. 1978). It was obvious from his demonstration that the Bazooka was a simple and highly useful piece of equipment that could be used to quickly infest plants or rearing containers with neonate larvae.

Working from the Bazooka's operational principle, a version was developed that was referred to as the "Inoculator" (Davis and Oswalt 1979). Others also developed versions of the Bazooka around this time, as reported by Wiseman et al. (1980). Today, one can purchase a "Davis Inoculator" from Bio-Serv, Inc. or a "Mihm Bazooka" from Custom Bio-Products (Maxwell, IA).

Before adopting the Inoculator, experiments were conducted to compare SWCB larval survival and development using the new infestation technique with the manual transfer technique. The results indicated that there were no significant differences between the two techniques. Also, it was noted that the need to mechanically score the surface of the diet to provide a thigmotactic stimulus facilitating initiation of feeding was eliminated by the presence of corn cob grits.

The best sized corn cob grits for use in this device is # 2040 (0.4-0.8 mm) (Wiseman et al. 1980). This particular grit can be obtained from Anderson Cob Grit Co. (Maumee, OH) or Bio-Serv, Inc. Before the grits can be used for rearing purposes, they must be sterilized by autoclaving or by exposing the grits to gamma radiation, a service provided by Bio-Serv, Inc. To prevent subsequent contamination, the fungicide Captan (0.11% w/w) and the antibiotic neomycin sulfate (0.03% w/w) are added to the grits and mixed in a jar roller mill (U.S. Stoneware, Akron, OH) for 2 to 3 hours. In order to minimize microbial contamination, the treated grits are divided into small containers

Fig. 9. Infestation of 32-cell rearing trays with neonate Lepidoptera mixed in sterile corn cob grits using an Inoculator. In this case, the bottom of the plastic bottle has been removed to facilitate recharging from a larger container in which the grits and larvae are frequently mixed.

Fig. 10. Sealing of infested 32-cell rearing trays with Mylar® lid material using the piece of custom equipment produced by Oliver Products, Inc. The unit uses heat (107°C) and pressure [1.3 x 10^{-2} Pa (90 psi)] and requires electricity (110 v) and a source of compressed air.

(e.g., plastic, self-seal bags) and only sanitized spoons or scoops are used to dispense them.

Typically the Inoculator is used by screwing a 1,000-ml plastic bottle containing the grit/neonate mixture into the top of the device, tumbling the contents of the bottle gently to uniformly mix the neonates and grits, placing the exit tube of the Inoculator over the target, and moving its handle from one side to the other to deliver a precise quantity of grits and an approximate number of neonates to the target (Fig. 9). The goal is to infest the rearing cells with an average of 2–3 neonates per Inoculator delivery. Fewer neonates than this and a significant number of cells receive 0 neonates; more than this and competition and excessive moisture build-up become problematic for the only moderately cannibalistic TBW and FAW. To achieve 2–3 neonates per Inoculator delivery, one goes through the following, simple calibration procedure:

1. Into the glass jar in which the neonates have hatched, pour a quantity of corn cob grits that is expected—through experience—to result in >3 neonates per Inoculator delivery.
2. Remove the wax paper oviposition strips after gently tapping them to remove neonates from their surface, and gently swirl the jar to mix the neonates and corn cob grits.
3. Pour the neonate-grit mixture into a sanitized 1,000-ml plastic bottle.
4. Attach the bottle to an Inoculator, and gently tumble the neonate-grit mixture again.
5. Make five Inoculator deliveries, and determine the average number of neonates per delivery.
6. Continue to add small amounts of grits to the source bottle, tumbling the mixture after each addition of grits, and recalculate the average number of neonates per five deliveries until the desired number of neonates is obtained.

Once calibration of the neonate-grit mixture has been achieved, infestation of rearing cells should begin immediately. The infestation process is conducted under a laminar flow hood to protect the diet from contamination. A skilled technician can infest a 32-cell tray in approximately 10 sec, which is >20 times faster than manual transfer using an artist's brush.

After infestation, the trays are immediately covered with lid material using the Oliver Products' semi-automatic sealer (Fig. 10). The steps for sealing the lid material are as follows:

1. Pull the sealer's tray out and place two infested trays in the sealer's rearing cell webs.
2. Wipe the top of each tray with a sanitized sponge moistened with tap water to reduce electrostatic attraction of the corn cob grits to the lid material.
3. Place precut pieces of lid material on top of the trays with the adhesive side down.
4. Push trays forward under the sealing mechanism.
5. Activate the sealer by simultaneously pressing buttons on both sides of the unit (an operational safety feature).
6. Remove the sealed trays and repeat.

Following sealing, the trays are labeled (species, date of infestation, and projected date to harvest pupae), placed on racks, and the racks transferred to species-specific shelves within the larval development room. Finally, all utensils and equipment are sanitized for subsequent use.

C. Pupa

Female SWCB pupae are about 1/3 larger than male pupae because of the greater feeding rate and higher growth efficiency of female larvae (Sec. II. B). Sexing of pupae is readily achieved by observing morphological differences on the ventral side of the last few abdominal segments. Similar morphological structures are used to sex FAW, CBW, and TBW pupae.

1. Environment. Prior to harvest, SWCB pupae are held under the same environmental conditions as described for larvae (Sec. II. B. 1). Following harvest, pupae are taken to a separate building that also houses the adults and is consequently referred to as the "moth house." Temperature and humidity conditions in the moth house are the same as those in the holding rooms for larvae (26.7 ± 0.6°C, 55 ± 5% RH), but the photoperiod is 14:10 (L:D) instead of 16:8 (L:D). This photoperiod was chosen because it is close to the average photoperiod for the four-month period centered on the summer solstice at 33°N latitude.

The number of days from pupation to adult eclosion at 26.7 ± 0.6°C is about 10 d. A small percentage of male pupae invariably produce adults the day before the first females eclose. Peak adult eclosion of both sexes occurs 30–31 days after infestation of diet with neonates (Ng et al. 1993).

Just prior to pupation, SWCB larvae prepare a cell within the diet. After completion of this cell, the larvae enter the prepupal stage characterized by inactivity and physical shortening followed by pupation within a few hours. Until a pupa's integument hardens (within 48 h), it is highly vulnerable to injury and should not be disturbed.

2. Harvest. SWCB pupae are harvested 26 d after introduction of neonates to diet. Harvesting at this time allows all individuals that are undergoing normal development to pupate, reduces pupal injury to a negligible level, and avoids adult eclosion in larval rearing cells.

Once, without consulting me, the rearing technicians made a decision to harvest SWCB pupae early because of an upcoming holiday. The pupae ranged between about 1 and 4 days old. Their action resulted in a high percentage of pupal mortality and a subsequent disruption of egg production during a critical research period. Personnel who do not have the knowledge to predict the consequences of not following an SOP must never do so without approval by their supervisor. In this case, a lesson was learned and a repeat never occurred.

Very little silk is produced by an SWCB larva when it is preparing its pupation cell, and none is attached to the pupa itself. Such pupae are referred to as "naked" and are easy to harvest, especially when the diet is slightly moist but not wet. The non-naked pupae of species such as the European corn borer are more difficult to harvest (Guthrie et al. 1965). Fortunately, FAW, CBW, and TBW pupae are also devoid of silk and can be easily harvested 20 d after infestation when reared in the optimum developmental environment.

When 1-ounce (29.6-ml) cups were used for rearing the SWCB, equipment was designed and fabricated that split the cups in half, dumped the contents onto a conveyer belt, and moved the contents onto a mechanical shaker, where pupae were separated from the cup parts and large pieces of diet (Davis 1982b). When the 32-cell rearing container was created, the shaker portion of this equipment was adapted for harvesting the pupae. Eventually, a cleaner and simpler method was adopted. The new method consists of culling cells that show signs of microbial contamination prior to harvest (Sec. IV), removing the lid material by hand from the tray (Fig. 11), dislodging the contents into a plastic tub by gentle tapping, and simply picking the pupae out of the debris by hand. With practice, workers can quickly remove the pupae without injuring them.

After harvest, the pupae are washed gently with tap water for several seconds to remove traces of diet and fecal material that may be a source of microbial contamination. Pupal harvest is a dirty task during which pathogens and diet-contaminating microorganisms can contaminate the rearing facility. Therefore, this task is conducted in a room separate from the clean rooms used to prepare the larval diet, place neonates on diet, and rear the larvae. Sanitation procedures for clean and dirty rooms are discussed in Sec. IV.

Fig. 11. Removal of the Mylar® lid from a 32-cell rearing tray to harvest southwestern corn borer, *Diatraea grandiosella* (Lepidoptera: Crambidae), pupae.

D. Adult

Eclosion of SWCB adults occurs primarily during late photophase to early scotophase. It requires about 30 min for these newly emerged adults to expand their wings and void their gut of *meconium*, a waste material that accumulates during the pupal stage. Shortly after wing expansion, a high percentage of the females begin releasing pheromone to attract males. Mated females begin ovipositing the second night after emergence, with most egg production occurring the following two nights. Based on spermatophore counts, females typically mate two to three times (Davis 1968) and oviposit 300–400 eggs during their life span of five to six days.

Because SWCB adults do not feed, the diet ingested during the larval stage must be sufficient to provide for all metabolic needs of the pupal and adult stages. The other Lepidoptera require an adult diet. FAW adults are fed beer ("regular", not "light"), and TBW and CBW adults are fed a 10% (w:v) sucrose solution (Jenkins et al. 1995).

Fig. 12. A wire mesh, mating/oviposition cage with a capacity of about 200 southwestern corn borer, *Diatraea grandiosella* (Lepidoptera: Crambidae), adults.

1. Environment. The environment for maintaining adults is based on a typical, early summer day: 27.6°C, 50–60% RH, and 14:10 L:D photoperiod. These conditions are maintained in a separate, well-insulated building referred to as the "moth house". The moth house is composed of two large rooms (6.1 m x 7.3 m x 2.4 m). One room is used for SWCB and FAW and the other for CBW and TBW. The rooms are separated by a wall and a door. The reasons for the moth house being separate from the main facility are to minimize exposure of personnel to scales, to eliminate contamination of larvae by microbial organisms associated with adults, and to have sufficient space for the large mating/oviposition cages. Heating and cooling in the moth house is provided by a 3.6-kW (4-ton) heat pump utilizing a thermostat that automatically switches between heating and cooling modes. Humidification and lighting are provided by the same equipment as used for the larval development rooms (Sec. II. B. 1). The reasons for using basic, readily available environmental control equipment for the moth house are that a stable environment is maintained at a very reasonable cost and that facilities maintenance staff or private service personnel can fix or replace equipment quickly.

Hygrothermometers are placed in prominent positions within each room of the moth house so that workers can detect even small deviations from normal. During peak production periods, all rooms are checked twice daily for any changes in temperature and relative humidity.

The standard for relative humidity is to keep it above 50%. During the winter months when outside air with low relative humidity is being heated, the demand for humidification is at its maximum. When purchasing a humidifier, a critical consideration is that its capacity must be sufficient to meet this winter peak demand.

To stress the importance of monitoring environments regularly, two examples are presented. In the first example, while repairs were being made in the moth house on a Friday afternoon, the service personnel found the room's temperature uncomfortable, so they reduced the temperature significantly. The thermostats were not readjusted when the job was completed. In the following days, a decline in the numbers of adults eclosing from pupae in the oviposition cages resulted in a significant reduction in egg production before the lower temperature was detected. This

Fig. 13. A wire mesh, mating/oviposition cage with the capacity to house up to 1,500 southwestern corn borer, *Diatraea grandiosella* (Lepidoptera: Crambidae), adults.

episode resulted in the posting of signs above the thermostats warning against adjustment by unauthorized personnel and the placement of hygrothermometers within each room for regular monitoring of temperature and relative humidity.

In the second example, a former graduate student reported that egg production had suddenly dropped to near zero in his rearing facility. I suggested that he first check the adult holding environment to ensure that all factors were being maintained as desired. If this did not solve the problem, I asked him to provide some of his female adults to determine if a disease microbe could be detected. He discovered that the "lights off" tripper on the electromechanical timer for the lights had come loose so that the moths were being held in continuous light. Moths are nocturnal and only oviposit in the dark phase of the light/dark cycle. With the tripper reinstalled properly, the adults resumed normal egg production. This illustrates an important lesson: check your equipment periodically to insure proper operation.

2. Cage. Initially, a simple, cylindrical cage made of 1/8-in (0.32-cm) hardware cloth was used to contain adults (Fig. 12; Davis 1976). This cage could accommodate ca. 150 adults. Pupae were placed in a walk-in screen cage for adult eclosion, and adults were collected using a modified hand-held vacuum for transfer to the cylinder cages.

A sheet of wax paper was secured around the outside of the cage to provide an oviposition substrate. By forcing females to oviposit through the wire mesh, the eggs occurred in discrete masses on one side only of the oviposition substrate. This allowed small wax paper discs to be punched out, each containing a group of eggs of approximately constant number. These discs were pinned to plants in the field to screen genotypes of corn for resistance to SWCB (Davis 1976).

The Davis Inoculator revolutionized the technique used for infesting plants in the field. Instead of pinning groups of eggs to plants, neonates mixed in corn cob grits could be delivered into the whorl or leaf axils of plants (Davis and Williams 1980). The flexibility of this new infestation technique permitted consideration of cage designs that allowed more efficient collection and handling of eggs. At the same time, several new functional goals were considered for the adult cages: (1) *in situ* eclosion of adults from pupae, (2) accommodation of much larger numbers of adults per cage, (3) prevention of the build up of shed scales, and (4) facilitation of the removal of moribund and dead moths. A number of prototypes were made before one (Fig. 13) satisfactorily met all of the functional goals (Davis 1982a, 1989). Design features of the SWCB cage in use today include: 1) 0.32-cm mesh hardware cloth walls and inserts covering the cage frame to restrict oviposition to the wax paper substrate provided and to allow shed scales to exit the cage; 2) slits in the cage top for hanging wax paper sheets, the slits being wide enough to allow safe retrieval of the sheets with eggs but not wide enough to allow moth escape (Fig. 14); 3) a door to permit introduction of pupae into the cage—feasible because of docile behavior of SWCB adults during photophase (Fig. 15);

Fig. 14. View of the top of mating/oviposition cage showing slits through which wax paper oviposition sheets are introduced and later retrieved with southwestern corn borer, *Diatraea grandiosella* (Lepidoptera: Crambidae), eggs.

Fig. 15. Front door used to place southwestern corn borer, *Diatraea grandiosella* (Lepidoptera: Crambidae), pupae into cage. The docile nature of the moths during the light phase allows this procedure with a minimum of adult escape.

4) a horizontal 5.08 x 10.16-cm mesh wire floor (for holding wire trays containing pupae) above a downward-slanting subfloor with a trap door at the bottom of the cage through which moribund and dead moths can be removed; and 5) casters to facilitate moving the cage. Presently, a cage containing up to 1,500 SWCB moths can be serviced in a few minutes. Each weekday during the peak production period, 300–350 pupae are placed into each cage.

Presently, up to 10 wax paper sheets (30.5 cm wide x 61 cm long) are inserted into each cage as oviposition substrate (earlier, wax paper sheets 15.24 cm wide x 61 cm long were used). The wax paper used is stiffer than household grade and is waxed on both sides (Papertec, Inc., Elizabeth, NJ; product number CC-40/55w2sbl) This special wax paper has been purchased through the years from several sources as vendors have gone out of business. Each time this occurs, samples of new vendors' paper are requested to test for female acceptance. The wisdom of the general practice of testing samples of a new product before purchasing large quantities was illustrated in an incident involving wax paper. A researcher who was having a problem with FAW egg production compared oviposition on the type of wax paper sheets in use at Mississippi State with those he was using and found that his paper was significantly less preferred.

The resulting SWCB cage was such a success that one was designed and fabricated for FAW (Davis et al. 1985). In addition, researchers in the USDA/ARS laboratory at Mississippi State were assisted in designing and fabricating a similar cage for CBW and TBW (Parrott et al.1986, Jenkins et al.1995). Although, these cages were patterned after the SWCB cage, some changes in design were needed to accommodate differences in adult behaviors. For example, the docile behavior of the SWCB allowed a front door for resupplying the cage with pupae, but the "excitable" nature of the FAW, TBW, and CBW moths required a cage design and SOPs that minimize adult escape while servicing them.

The SOPs used for managing FAW cages are similar to those for SWCB cages. Pupae are added to the cage periodically throughout the week, and each cage remains in service for 2–3 months before sanitizing and reusing it. The main difference between the two species is that FAW adults require food to maximize adult longevity and reproduction. They are fed beer in circular, plastic dishes (15.2 cm diameter and 1.3 cm deep), which contain a disc of foam, air conditioner filter to provide a surface on which the moths can rest while feeding. During peak production periods, up to four plastic dishes are placed on the bottom screen of each cage by slightly raising a side panel to allow introduction and removal of dishes (Fig. 16; Davis et al. 1985). This minimizes adult escape as FAW is easily disturbed and will exit the cage. Because adults feed only during the scotophase and because the container with beer is a source of microbial contamination, beer is provided from 5:00 p.m. to 8:00 a.m. only. The following day the plastic containers and the foam discs are removed from the cage, sanitized using an ammonium chloride compound, and reused.

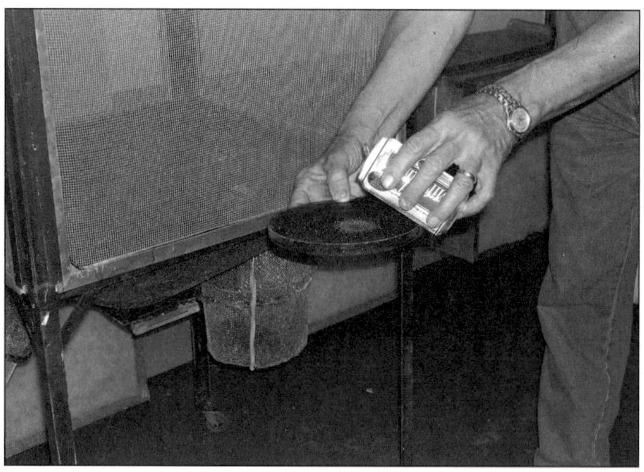

Fig. 16. Beer being poured into a feeding dish before placement into a fall armyworm, *Spodoptera frugiperda* (Lepidoptera: Noctuidae), mating/oviposition cage.

The cage design and SOPs for the CBW/TBW cage differ somewhat from those for the SWCB and FAW. A group of pupae (up to 2,000) is placed into each cage. Eggs are collected from this cohort of adults for a period of seven days. Each day, the cages are serviced by removing the top and four side panels on which the eggs have been oviposited, while simultaneously replacing these panels with fresh ones (Fig. 17). This process minimizes adult escape. After seven days, the cage is sanitized using an ammonium chloride agent and reused. Each cage has 15-watt lights mounted on two sides and the top. These lights simulate moonlight during the scotophase to promote mating and uniform oviposition around the cage.

To feed adult TBW and CBW, four or five household synthetic sponges (7.6 cm x 12.7 cm x 1.9 cm) are placed on the top panel of the cage, and a 10% sucrose solution is poured over each sponge to saturation. To reduce the evaporation rate, a square plastic dish is placed over the sponge. The moths are fed at about 9:00 a.m. each day and, during peak production of eggs, again at about 5:00 p.m. Both species feed primarily during the scotophase but do feed during the photophase—especially the CBW. To avoid significant microbial contamination of the sponges, they are washed in hot water each day and autoclaved before subsequent use.

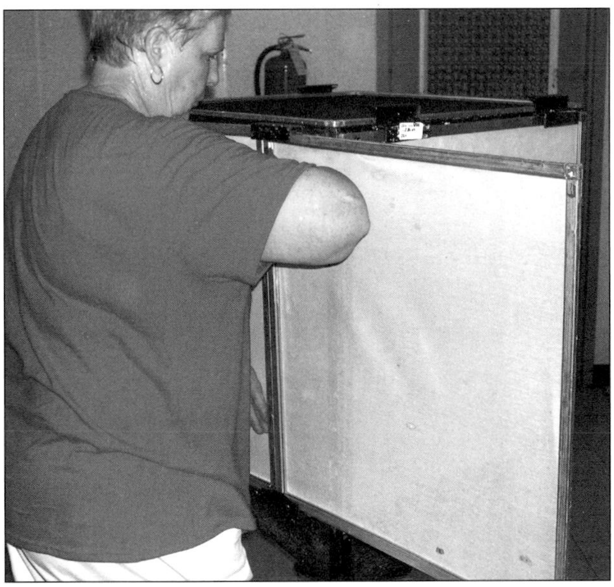

Fig. 17. Replacing a cage panel on which bollworm, *Helicoverpa zea* (Lepidoptera: Noctuidae), or tobacco budworm, *Heliothis virescens* (Lepidoptera: Noctuidae), eggs have been oviposited with a fresh panel without allowing adults to escape. The new panel is slid into mounting tracks as the old panel is slid out.

To remove the TBW and CBW eggs from the panels, a custom-designed machine referred to as the "egg harvester" was fabricated (Fig. 18; Parrott and Jenkins 1992, Jenkins et al. 1995). The eggs are separated from the screen panels by washing them with a 0.18% sodium hypochlorite solution. The eggs sink to the bottom of the wash tank, are caught in a 60-mesh sieve, and washed with tap water to remove any traces of the sodium hypochlorite. This procedure also surface-sterilizes the eggs.

3. Scale Collection. Moth scales are a serious human allergenic hazard (Chap. 4 *Health and Safety Issues in the Rearing of Arthropods*) and have been a major concern for years. Early on, air filters purchased from local hardware stores were used to collect scales from air in the space housing cages for adults (Davis 1976, 1989). In the early 1990s, with the assistance of representatives from Air Filter Sales and Service, Inc. in Jackson, MS, filter specialists from Freudenberg Nonwovens L.P. (Chelmsford, MA) recommended using a combination of two different prefilters and one primary filter for scale removal. The outer prefilter is a 50.8 x 50.6 x 2.5-cm P5 polyester pad (Moldan Air Filters, Pineville, NC). The inner prefilter is a 50.8 x 50.8 x

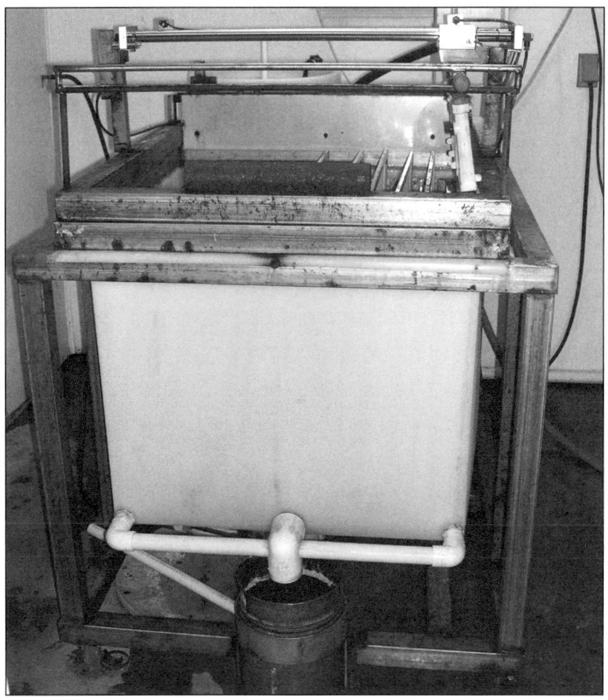

Fig. 18. The custom-designed and fabricated piece of equipment used to harvest cotton bollworm, *Helicoverpa zea* (Lepidoptera: Noctuidae), and tobacco budworm, *Heliothis virescens* (Lepidoptera: Noctuidae), eggs from mating/oviposition cage panels.

10.2-cm DQP pleated filter (Quality Filters, Robertsdale, AL). The purpose of these first two filters is to protect the high efficiency third and primary filter by removing the larger particles from the air flow. The primary filter is a Viledon MF-90 pocket filter (Freudenberg Nonwovens L.P.). A team was assembled consisting of specialists from the private sector and personnel from the Mississippi State USDA/ARS research unit and facility maintenance staff to design and fabricate a sheet metal housing for a direct drive blower and the filters. Four complete housing systems were constructed, one for each of the Lepidoptera species; and each system has six intake stations—one for each of six large cages (Fig. 19). The cages are positioned in close proximity to their respective intake sites to allow scales to be drawn directly into the filtering system without settling to the floor, a major key to the success of this system. This system removes 95–100% of the particles ranging from 0.5 to 5.0 μm (Davis and Jenkins 1995). Persons who could not enter the moth house because of allergies associated with scales before this system was installed can now enter the building and work.

The moth scale collection system is reasonable in cost and does not require a burdensome level of maintenance. The primary filter can be replaced at greater than yearly intervals if the polyester prefilters are cleaned daily during the peak production period and 2–3 times per week during the low maintenance production period using a modified vacuum cleaner system and changed regularly (every 3 to 4 months). The unit costs of the prefilters as of January 2008 are $0.98 for the polyester pad and $7.98 for the DOP pleated filter. The cost of the primary filter MF-90 is $215.25.

Fig. 19. The custom-designed and fabricated scale removal system for the moth house.

III. Colony Genetics

At the end of each year, individuals from wild SWCB populations are crossed with individuals from the current laboratory colony to generate the colony reared during the next production season. Guthrie et al. (1974) found that continuous rearing of the European corn borer, *Ostrinia nubialis* (Lepidoptera: Pyralidae), on an artificial diet in the laboratory yielded an insect that could no longer establish as readily on or inflict as much damage to corn plants as could the wild type. Given that the primary purpose for rearing the SWCB is to screen corn for resistance to attack by SWCB larvae, such laboratory adaptation had to be avoided. See Chap. 5 *Genetic Considerations and Strategies for Rearing High Quality Insects* for a general discussion of a laboratory adaptation.

A two-pronged approach is used to maintain the genetic diversity and wild-type genetic makeup of the SWCB colony. To maintain genetic diversity during the production season, the breeding stock is prevented from falling below 1,000 moths even during low production periods. This policy is followed for FAW, CBW, and TBW as well. To maintain the genetic fidelity of the cultures relative to wild populations, cultures are outcrossed to wild populations annually. In the case of SWCB, small fields of corn are planted late in the season at three locations in Mississippi. From each location, 250 to 500 SWCB overwintering larvae are collected during October and early November when they have finished feeding and are in the immaculate (spotless) stage. The larvae are stored individually in empty, 29.6-ml (1-ounce) plastic cups for about 2 weeks at 21.6°C, 12:12 (L:D), and 50-60% RH in order to cull weak and injured larvae. The remaining, healthy-appearing larvae are placed individually in 29.6 ml (1-ounce) plastic cups containing 8–10 ml of 2% agar plus the mold inhibitors sorbic acid and methyl paraben (Davis 1983). The top of each cup is closed with a polysaran-coated paperboard insert cap and placed in a cold storage environment (Sec. II.B.1).

In mid-August, approximately 1,500 colony larvae are reared under diapause-inducing conditions (Sec. II.B.1) in

29.6-ml (1-ounce) cups containing a quantity of diet sufficient to allow the larvae to develop to the immaculate stage, at which time they are placed in the cold storage environment. Transfer of the larvae to cups containing the 2% agar plus mold inhibitors had no effect on survival. Given the 50- to 55-day period required for the colony larvae to reach diapause, both the colony and wild larvae are ready for cold storage at about the same time. Normally, diapausing laboratory and wild larvae are placed in a diapause termination environment (Sec. II.B.1) in three equal-size groups about a week apart in February. This allows for availability of neonates from the feral by laboratory adult mating over a period of about 4 weeks. Median time to pupation under these conditions is ca. three weeks. As the pupae reach harvestable age, they are culled for morphological defect, sexed, and separated by sex into two adult cages in the moth house. In one cage, colony males are crossed with wild females; and in the other, the reverse cross is arranged. Eggs from both cages are produced over a period of ca. four weeks and combined to begin the colony used for the new production season.

The new SWCB colony is observed for normality (survival, development times, egg production, and egg hatch) for at least two generations before the old colony is discarded. This practice is very important because one can not assume that the new colony will always be as fit as the old one that it is replacing. In the case of SWCB, no diseases have been introduced into the colony over the three decades of using the above procedure.

For FAW, CBW, and TBW, only wild males are used for crossing to the colony because females of these species are known to transmit viruses and protozoa to the next generation through their eggs. As a consequence, the only major microbial-caused disease problem with any of the insect colonies has been a gonad-specific virus in the CBW (Raina and Adams 1995), which can be transmitted by males as well as females (Sec. IV. F).

IV. Microbial Management

For many years, diet contamination by bacteria and fungi (especially *Aspergillus niger*) posed a constant threat to all four species produced in the USDA/ARS rearing program at Mississippi State. As described in Sec. III, only one significant disease has been encountered and that was in the CBW colony. A multitactic approach to prevent and minimize disease due to entomopathogens and microbial contamination of the insects' diet has evolved over the years and has been greatly influenced by the assistance of Dr. Peter P. Sikorowski and Dr. G. Douglas Inglis, insect pathologists; Amanda Lawrence, research associate, in the Department of Entomology and Plant Pathology at Mississippi State University; and by Tom Riddell, President, Air Filter Sales and Services, Jackson, MS, who has been instrumental in adding air filtration equipment to each of the insect rearing rooms. Today, this rearing program experiences a very low incidence of diet contamination by microbes and has no known problems with diseases (Chap. 8 *Microbial Contamination and Insect Rearing* and Chap. 9 *Entomopathogens and Insect Rearing*).

A. Standard Operating Procedures

An important and basic tactic for combating disease and diet contamination is development of a strict and clear set of SOPs, including: 1) personnel hygiene—a major source of microbes; 2) sanitation of floors, walls, and work surfaces; 3) sanitation of equipment and rearing supplies; 4) sterilization of the surface of eggs; 5) addition of the antimicrobial agents sorbic acid and methyl paraben and the antibiotic neomycin sulfate to the diet; 6) cooking the diet in a steam-jacketed kettle (Sec. II. B. 3.); 7) control of visits by nonrearing personnel to contamination–sensitive areas; 8) routing of rearing personnel traffic from "clean" to "dirty" areas rather than the reverse; 9) removal of microbial-contaminated rearing cells prior to harvest of pupae to prevent fungal spores from entering the air; 10) proper disposal of rearing waste materials; and 11) proper use and maintenance of laminar flow hoods and air filtration equipment. An up-to-date manual of SOPs is kept for use by the rearing supervisor and staff, and any change in an SOP is authenticated with a signature of authorization and the date. Adherence to SOPs is monitored frequently to ensure compliance by personnel as there may be a tendency by some to "short cut" instructions.

B. Insectary Design

In 1976, the opportunity to renovate the insectary facilities housed in the USDA/ARS, Boll Weevil Research Laboratory arose. Rooms were assigned functions so that the diet preparation room, the one requiring the cleanest environment, was located as far as possible from the pupal harvest room, the dirtiest room. Doors isolate each room in the facility; and recently, devices have been installed on the doors so they close automatically. This feature has improved air purification by compartmentalizing the areas to be cleaned and minimizing air infiltration from outside sources. Also, the adults are maintained in a separate building to avoid microbial problems associated with moth cadavers and insect scales. See Chap. 2 *Insectary Design and Construction* for detailed information on this subject.

C. Air Filtration

After the success of the insect scale collection system described above, a decision was made to invest in a similar system for each of the rearing rooms in the main building to remove dust and other particulates including microbes (especially *Aspergillus* spp. spores) from the air. The air filtration system in each of the rooms is based on the use of prefilters and a primary filter. The housings of the air filtration systems in the diet preparation room and the two developmental rooms are extensions of the heating and cooling units. Those in the egg sterilization room and the pupal harvest room are independent units placed over work areas, such as the work bench where pupal harvest occurs (Fig. 20). Each system continuously draws air from the room through the prefilters and pushes it through a primary filter back into the room. Addition of air filtration units within each room has played a key role in minimizing microbial contamination of the insects' diets, especially *A. niger* contamination.

D. Hygiene and Sanitation

Supplies for personnel hygiene and for sanitizing facilities and equipment are usually purchased from janitorial supply businesses that service hospitals. The surgical/germicidal soap Stoko Clear Antimicrobial Hand Cleaner (Stockhausen, Inc., Greensboro, NC) has been used for sanitizing hands and is placed in dispensers above each sink. Also, over-the-counter, antibacterial hand soaps have been used. Ammonium chloride antimicrobial compounds such as Virex 128 (S. C. Johnson Commercial Markets, Inc., Sturtevant, WI) have been and continue to be used to sanitize floors, walls, work benches, equipment, and adult holding cages. This disinfectant has activity against a wide range of microbes, including *Aspergillus* spp., a major fungal contaminant (Chap. 8 *Microbial Contamination and Insect Rearing*). Virex 128 is used as directed at a 1:128 dilution [29.6 ml (1 oz.) per 3.8 liter (1 gallon) of water]. Occasionally, a freshly prepared solution of 10% Clorox is used to sterilize equipment, utensils, and work areas. An autoclave is available in the rearing laboratory to sterilize corn cob grits, paper toweling, glass jars, and similar materials; and an oven is used to keep glass containers sanitized until used.

E. Stress Reduction

As a complement to the multitactic approach to microbial problems, efforts have been made to create rearing conditions that minimize stress on the reared insects. Stressed insects are more vulnerable to diseases than are nonstressed insects (Soares 1992, Sikorowski and Lawrence 1994; Chap. 8 *Microbial Contamination and Insect Rearing* and Chap. 9 *Entomopathogens and Insect Rearing*). A common stress is significant variation in temperature caused by heating and cooling equipment failures. Overheating not only stresses the insects but creates an environment in rearing containers that is favorable for microbial growth. Other factors that can cause stress are overcrowding of insects in a rearing container and not providing sufficient food for completion of larval development.

Fig. 20. An air filtration system above a work bench where pupae are harvested. The system cleans the air of microbes that may be released into the air during harvest.

F. Control of the CBW Gonad-Specific Virus

Symptoms of the gonad-specific virus in CBW include protrusion of a white, waxy plug from the female's vulva, a significant drop in egg production, and abnormal adult behavior, including reduced mating, feeding activity, and mobility. The existence of the CBW gonad-specific virus (Herzog and Phillips 1982; Raina and Adams 1995; Hamm et al. 1996; Sikorowski and Lawrence 1997) was known as was the fact that wild males infected with the virus could transmit it to laboratory females when they were crossed each fall. Sikorowski and Lawrence (1997) and Hamm et al. (1996) provided useful information on ways to control and eliminate the virus from laboratory colonies. In particular, Hamm et al. (1996) made the recommendation that eggs from only the first two nights of oviposition be used to obtain neonates for colony maintenance. An attempt was made to follow this recommendation; however, eggs from the third night of oviposition had to be used to obtain sufficient numbers of larvae.

In 1998, incidence of the gonad-specific virus began to increase in the CBW colony. Because the TBW, SWCB, and FAW colonies are housed with the CBW colony, various stages of these three species were sent to Dr. A. K. Raina (USDA-ARS, Beltsville, MD) to determine if they were also susceptible to this disease. Fortunately, his results were negative. A decision was made to eliminate the CBW colony and start over with pupae from a colony maintained by Monsanto Co. (Union City, TN) that was verified by dissection of adults to be free of the virus.

Before initiating the new CBW colony, the moth house and the large cages used for the CBW were thoroughly sanitized to eliminate the virus. All surfaces of the furniture, equipment, floor, ceiling, and walls were washed with either 10% Clorox® solution or Virex 128. The cages were sanitized using the following procedure: the cage frames were washed with a disinfectant several times, spray painted, and placed outside in direct sunlight before reusing. All cage panels were soaked in 10% Clorox® solution before reusing.

Until the outbreak of gonad-specific virus, the plywood walls of the moth house were repainted annually during the December holiday break in an attempt to reduce microbial contamination. Following the outbreak, it was decided to cover the walls of the moth house with fiberglass reinforced polymer (FRP) panels. This nonporous, plastic-like surface is similar to ceramic tile in that it can be cleaned with a damp cloth, hosed down, or even steam cleaned but costs a small fraction of ceramic tile. In 2000, once the above measures had been taken and the necessity of strictly following all SOPs related to sanitation reinforced with personnel, the new colony of CBW was initiated. In 2001, as a further measure to control the gonad-specific virus, the annual, end-of-season mating of laboratory females with wild males was discontinued. The CBW colony has been monitored about twice a month for the gonad-specific virus by inspecting 20 females per sample for vulvular plugs and dissecting them to check for abnormal ovaries. No infected CBW have been observed since the new colony was established.

Annual mating of female TBW from the laboratory colony with wild males was also discontinued in 2000. In this case, extremely low population densities of wild TBW in northeast Mississippi made it difficult to collect a sufficient number of wild males to perform the crosses. Apparently, the high percentage of cotton acreage planted to Bt-transgenic varieties of cotton, which are extremely effective in controlling the TBW, has greatly suppressed the density of wild TBW populations.

V. Quality Control

Quality control is an essential component of any size rearing program. Large-scale rearing programs have staff designated to monitor quality of the insects as they are being reared and to evaluate their performance, such as successfully competing with wild males for mating with wild females. Midsize and small rearing programs must rely on staff members with multiple responsibilities to monitor on a regular basis for abnormalities in the insects being produced. See Chap. 10 *The Basics of Quality Control for Insect Rearing* for more detailed information.

The primary goal of the USDA/ARS Lepidoptera rearing program at Mississippi State has been and continues to be production of high quality insects suitable for plant resistance research and to produce them in sufficient numbers at the times required. Rates of success of achieving each developmental stage and percentage normality of the resulting individuals (assumed to correlate with high quality) are monitored on a daily basis during both maintenance and peak production periods. These evaluations are done primarily visually by rearing staff that have developed, through training and experience, the skill to recognize significant declines in success rates and increases in frequencies of abnormal individuals. When a decline in production rate is detected, efforts are made to correct the problem as quickly as possible. For SWCB, remedial efforts are triggered when egg hatch, pupation rate, or adult eclosion falls to < 80%.

When an increase in the frequency of abnormal individuals indicates a decline in quality, troubleshooting is initiated to determine the cause. Some examples of normal versus abnormal SWCB are described. Normal SWCB neonates are grayish in color and very active, in contrast to abnormal neonates, which are noticeably shorter, reddish, and sluggish. Low egg hatch is generally associated with occurrence of these abnormal neonates. This abnormality is thought to be caused by microbial contamination of the parents' diet. A common abnormality for pupae is a whitish band between the thorax and abdomen. This abnormality has been observed to result in poor adult emergence and is thought to also be caused by larval development on microbial-contaminated diet. Adults are observed for twisted wings, nonseparation of the pupal exuvium from the adult, and inability of pairs to separate following pairing. These adult abnormalities become a concern when they reach 20%.

Observations on production rates and quality are routinely made for FAW, CBW, and TBW when the pupae are harvested. Also, visual estimation of the number of empty pupal cases versus dead pupae within the adult cages provides useful information on quality of the insects being produced. Monitoring for adult activity of CBW is very important because adults suffering from the gonad-specific virus are noticeably sluggish.

Each year, small field trials are conducted to showcase the inbred lines of corn with leaf feeding resistance to SWCB and FAW larvae developed by USDA/ARS at Mississippi State. These trials also provide an opportunity to confirm that the insects reared to infest the plots are performing well in the field on the highly susceptible inbred lines included in the trials. If the susceptible controls sustain high levels of damage, it is concluded that the reared insects are performing as desired (Fig. 21A), however, if low levels of damage are observed, then—along with many other potential factors—low quality of the insects used for infestation is considered as an explanation. Fig. 21A when contrasted to Fig. 21B illustrates the ease of distinguishing resistant from susceptible genotypes when the plants are infested with reared SWCB neonates that are essentially identical to wild neonates.

Fig. 21. Damage to corn plants when infested with 30 southwestern corn borer, *Diatraea grandiosella* (Lepidoptera: Crambidae), neonates per plant. (A) Susceptible control corn plants. (B) Resistant test plants.

VI. Insectary Pests

Through the years, a variety of pests have been encountered in the USDA/ARS rearing facilities at Mississippi State, including rodents, ants, spiders, egg parasitoids, and escaped moths. These animals can damage supplies, prey upon or parasitize reared insects, and cause contamination of eggs harvested from one colony with eggs from another.

Rodents, particularly mice, have occasionally invaded the facilities, especially during the fall and winter months. They have been observed to chew through the rearing trays and consume larvae and pupae. Mice have also been observed to disturb moths in their cages resulting in poor oviposition. The best control tactic for mice has proven to be spring traps baited with peanut butter.

Control of roaches in the main rearing facility, where use of chemical insecticides are prohibited, is achieved using "Roach Prufe" (98% boric acid) (Cooper Brite, Inc. Santa Barbara, CA). The mode of action for this compound is not well understood, but death results from dehydration. The white powder is sprinkled periodically on the floor near the walls where roaches have been observed. Roaches are considered to contribute to filth and most likely are carriers of problem microbes.

Ants, especially the red imported fire ant, *Solenopsis invicta* (Hymenoptera: Formicidae), and the black imported fire ant, *Solenopsis richteri* (Hymenoptera: Formicidae), enter the moth house during the spring and summer months and attack both pupae and adults. Insecticides containing pyrethrins (e.g., the ant and roach killer Enforcer®; Acuity Brands, Cartersville, GA) when sprayed on surfaces around facility doors kill ants during their ingress and egress. The insecticide bait Amdro® (AMBRANDS, Atlanta, GA) is used to treat fire ant mounds near the building.

Managing a trichogrammatid egg parasitoid, *Trichogramma* sp. (Hymenoptera: Trichogrammatidae), attacking SWCB in the moth house has been particularly challenging. Fortunately, this species of *Trichogramma* parasitizes only SWCB eggs. Fumigation of the moth house during the December holiday period when no adult Lepidoptera are present and efforts to minimize escape of SWCB moths from their cages to reduce the supply of hosts for the parasitoids have been the main tactics for management of this pest. When egg masses are oviposited on the walls by escaped moths, they are manually destroyed. Frame surfaces exposed inside the cages are covered with wire mesh to prevent SWCB oviposition there, thus removing another supply of hosts. Annual fumigation using an indoor fogger [e.g., Four Hour Fogger® (active ingredient Tetramethrin (0.20%) + Sumithrin (0.40%)); Enforcer Products, Carterville, GA] for the egg parasitoid also controls roaches and spiders.

Escaped CBW and TBW moths can oviposit on surfaces of cages housing the other species. As a result, eggs of both species are harvested together, causing a species separation problem. Each day, escaped adults are killed using a fly swatter; and during periods of high production, sticky fly paper is suspended from the ceiling to help capture them. An anecdote related to the use of fly paper illustrates the need for supervisors to communicate properly with assistants and the need for all concerned to carefully read labels on products before using them. A graduate student, who was responsible for the adult TBW and CBW colonies, sent some of his temporary helpers to purchase fly paper. Instead of fly paper, they purchased strips of Vapona, a fumigant insecticide and hung them up around the CBW/TBW room. The next day most of the moths in both holding rooms were dead. Fortunately, the pupae were not harmed, and field infestations for host plant resistance studies had just been completed. If this incident had occurred earlier in the summer, a major interruption in the supply of insects would have occurred.

Another arthropod pest that can invade a rearing facility and cause substantial problems bears mention. In 1980 I was temporarily assigned by USDA/ARS to the International Rice Research Institute in the Philippines to set up a rearing program for Lepidoptera. After rearing of several species of Lepidoptera on artificial diet was begun, mite infestations were observed within the larval rearing containers. It was noted that large mite populations could build up on the artificial diet in only a few days. Four procedures were implemented to prevent the mite problem. All rearing containers with mite infestations were discarded without opening the containers. All work bench surfaces were wiped daily with alcohol to kill the various stages of the mite. Since mites have no wings, the legs of all furniture (especially rearing container racks) in the rearing facility were coated with the adhesive Stick-em® (J. T. Eaton & Co., Inc., Twinsburg, OH) commonly used in sticky traps for insects and rodents. The final procedure involved determining how the mites were entering the facility and developing an SOP to prevent mite entry. The mites were observed to be originating in rice straw that had accumulated in close proximity to the entrance door of the rearing facility and to be simply crawling under the door into the facility. The problem was solved by removing the straw and informing maintenance personnel that the area around the rearing facility should be kept free of straw.

VII. Planning and Record Keeping

Proper planning and recordkeeping have been essential to management of the USDA/ARS insect rearing program at Mississippi State. The success of the program has depended upon the ability of the rearing manager and staff to plan for variation in production demands in a timely manner and to provide excess production (up to 20%) to cope with unexplained variation in production output. It has also depended upon the maintenance of information on SOPs, equipment (including maintenance supplies and schedules), supplies, vendors, insect production, insect growth by instars within different environmental regimes, incidents of diet contamination and diseases, insect abnormalities, environmental failures, diet ingredient stocks, and dates of delivery of diet components and other rearing materials.

A list of alternate vendors for critical supplies and equipment is maintained, including current contact information. Over the years, vendors have gone out of business, and replacements have had to be found on short notice. Also, vendors have been asked to make sure that they send notification when modification of their product is anticipated so that compatibility of the modified product with the production system can be tested and an alternate vendor found if necessary.

Over time, improvements have been made in planning, recording essential rearing information, and running the rearing program in a more businesslike manner. Computers have helped enormously with these tasks. Also, communications have improved between rearing staff and users of the insects with respect to procedures for requesting insects and the best procedures for handling orders so that good research results can be achieved. See Chap. 3 *The Insectary Manager* for detailed information.

VIII. Equipment

Equipment of various types is essential to the various components of any rearing program but especially the production system. Criteria used by the USDA/ARS rearing group at Mississippi State for selection of equipment purchased or fabricated have been: 1) task appropriateness (e.g., if equipment involves handling the insect, it should cause minimal stress or mechanical injury to the insects); 2) ease and efficiency of operation; 3) good durability; 4) ease of servicing, cleaning, and sanitizing; 5) safety features; and 6) reasonable cost.

To keep equipment running properly, staff and facility maintenance personnel have fully understood the critical importance of this equipment to the well being of the overall rearing program. For example, the equipment used to provide the insects with suitable environments is absolutely critical to the production system's success; and the consequences of its failure can cause severe disruption of the production of quality insects. Such equipment requires regular maintenance check ups and repairs. If environmental control equipment fails, the response—to repair or to replace—needs to be immediate. Also, when equipment is replaced, consideration should be given to retaining some equipment if operative to serve as a back up (e.g. a manual balance for weighing out diet ingredients).

For equipment such as automatic balances and laminar flow hoods, professional maintenance service contracts have been utilized to ensure their continued proper operation and to extend the life of the equipment.

The air filtration system in each room is serviced routinely by facility maintenance personnel replacing filters as recommended by the vendor. See Chapter 3 *The Insectary Manager* for general information regarding equipment choices and care.

IX. Conclusions

The insect rearing production system described here for SWCB, FAW, CBW, and TBW has evolved from a small-scale system based primarily on hand labor to a midsized system utilizing semiautomatic equipment for preparing and dispensing artificial diet and infesting rearing cells with neonates mixed in corn cob grits, a 32-cell rearing container for rearing neonate larvae to the pupal stage, and large cages for holding up to 2,000 adults for oviposition. Today, only two, full-time USDA technicians and three, part-time workers are required to produce up to 39,000 high quality pupae per week. The success of the system is based on three pillars: 1) commercially available or locally designed and fabricated equipment is used extensively to substantially reduce the amount of hand labor; 2) as the system developed, the rearing staff utilized a team approach to solve problems and to make system improvements; and 3) experts in both the public and private sectors have been invited to contribute to the system and subsequently have been treated as team members.

Additional important elements for developing a successful rearing program include the following:

1. Know the insect reared thoroughly, including its biology, behavior, physiology, food, and environmental needs and preferences.
2. Develop or acquire a highly suitable diet to serve as the cornerstone of the production system.
3. Develop precise goals for improving a rearing component prior to its modification.
4. Conduct thorough investigations to ensure that proposed changes accomplish desired goals without harmful side effects.
5. Create, implement, and enforce sound and clear SOPs for handling each growth stage; preparing, dispensing, and infesting diet; harvesting insects; using and maintaining equipment; and minimizing microbial problems.
6. Form a working relationship with an insect pathologist to develop a multitactic approach for preventing microbe-caused disease and diet contamination.
7. Implement quality control monitoring to ensure production of high quality insects and to be able to respond quickly to solve rearing problems.
8. Create mutually beneficial relationships with vendors of diets, supplies, and equipment.
9. Develop a conscientious staff well trained in all of the essential areas of rearing and who take pride in producing high quality insects.
10. Review all of the pieces of the insect rearing puzzle (Frontispiece) at least annually, make modifications where necessary, and recombine the pieces into a well-integrated whole.

Acknowledgments

The insect rearing production system presented here as a case study is truly the product of contributions made by a host of special people. They include Thomas Oswalt, Maggie Collier, Vivian Crowley, Susan Wolf, and Eloise Sloan (USDA/ARS technicians); Stan Malone and his staff (USDA facility maintenance personnel); Peter Sikorowski and Doug Inglis [Mississippi State University (MSU) faculty]; Amanda Lawrence (MSU staff); and Sen Seong Ng (MSU graduate student). Vendors of rearing equipment and supplies, visiting scientists, and others who have helped throughout the years are too numerous to name. Their hard work, creativity, and willingness to contribute to a team effort is recognized and greatly appreciated. Johnnie Jenkins and Paul Williams of the Crop Science Research Laboratory (USDA/ARS) are acknowledged for their permission to use archival photos of the production system during its development starting in 1965. Susan Wolf and Amanda Lawrence kindly contributed photographs of the system as it currently exists. Amanda Lawrence and Sen Seong Ng provided valuable reviews of the manuscript.

References

Davis, F.M. 1968. Morphology of the reproductive systems of the southwestern corn borer, *Diatraea grandiosella*. Ann. Entomol. Soc. Amer. 61: 1143–1147.

Davis, F.M. 1976. Production and handling of eggs of southwestern corn borer for host plant resistance studies. Mississippi Agric & Forestry. Exp. Stn. Tech. Bull. 74. Mississippi State University, Mississippi State, MS.

Davis, F M. 1980. A larval dispenser-capper machine for mass rearing the southwestern corn borer. J. Econ. Entomol. 73: 692–693.

Davis F. M. 1982a. Southwestern corn borer: Oviposition cage for mass production. J. Econ. Entomol. 75: 61–63.

Davis. F. M. 1982b. Mechanically removing southwestern corn borer pupae from plastic rearing cups. J. Econ. Entomol. 75: 393–395.

Davis F. M. 1983. Simple technique for storing diapausing southwestern corn borers (Lepidoptera: Pyralidae). J. Econ. Entomol. 76: 1191–1192.

Davis F. M. 1989. Rearing the southwestern corn borer and fall armyworms at Mississippi State, pp. 27–36. *In* J. Keyser and N. Russell (eds.), Toward insect resistant maize for the third world. Proceedings of the International Symposium on Methodologies for Developing Host Plant Resistance to Maize Insects, 9–14 March 1987, El Baton, Mexico. International Maize and Wheat Improvement Center (CIMMYT), Mexico City, Mexico.

Davis F. M. 1997. Improved technologies for rearing lepidopterous pests for plant resistance research, pp. 184–188. *In* J. A. Mihm (ed.), Proceedings of the International Symposium on Insect Resistant Maize: Recent Advances and Utilization, 27 November–3 December 1994, El Baton, Mexico. International Maize and Wheat Improvement Center (CIMMYT), Mexico City, Mexico.

Davis, F. M., and W. D. Guthrie. 1992. Rearing Lepidoptera for plant resistance research, pp. 211–227. *In* T. E. Anderson and N. C. Leppla (eds.), Advances in insect rearing for research and pest management. Westview Press, Boulder, CO.

Davis, F. M., and J. N. Jenkins. 1995. Management of scales and other insect debris: occupational health hazard in a lepidopterous rearing facility. J. Econ. Entomol. 88: 185–191.

Davis, F. M., and T. G. Oswalt. 1979. Hand inoculator for dispensing lepidopterous larvae. USDA-SEA-AAT-S-9. US Government Printing Office, Washington, DC.

Davis, F. M., and W. P. Williams. 1980. Southwestern corn borer: comparison of techniques for infesting corn for plant resistance studies. J. Econ. Entomol. 73: 704–706.

Davis, F. M., T. G. Oswalt, and J. C. Boykin. 1978. Insect diet dispenser for medium-size rearing programs. USDA-ARS-S-182. US Government Printing Office, Washington, DC.

Davis, F. M., T. G. Oswalt, and Sen-Seong Ng. 1985. Improved oviposition and egg collection system for the fall armyworm (Lepidoptera: Noctuidae). J. Econ. Entomol. 78: 725–729.

Davis, F. M., W. P. Williams, and B. R. Wiseman. 1989. Methods used to screen for and to determine mechanisms of resistance to southwestern corn borer and fall armyworm, pp. 101–108. *In* J. Keyser and N. Russell (eds.), Toward insect resistant maize for the third world. Proceedings of the International Symposium on Methodologies for Developing Host Plant Resistance to Maize Insects, 9–14 March 1987, El Baton, Mexico. International Maize and Wheat Improvement Center (CIMMYT), Mexico City, Mexico.

Davis, F. M., S. Malone, T. G. Oswalt, and W. C. Jordan. 1990. Medium-sized lepidopterous rearing system using a multicellular rearing tray. J. Econ. Entomol. 83: 1535–1540.

Edwards, R. H., E. Miller, R. Becker, A. P. Mossman, and D. W. Irving. 1996. Twin screw extrusion processing of diet for mass rearing the pink bollworm. Trans. Amer. Soc. Agric. Engineers 39: 1789–1797.

Guthrie, W. D., E. S. Raun, F. F. Dicke, G. R. Pesho, and S. W. Carter. 1965. Laboratory production of European corn borer egg masses. Iowa State J. Sci. 40: 65–83.

Guthrie, W. D., Y. S. Rathore, D. F. Cox, and G. L. Reed. 1974. European corn borer: virulence on corn plants of larvae reared for different generations on a meridic diet. J. Econ. Entomol. 67: 605–606.

Hamm, J. J., J. E. Carpenter, and E. L. Styer. 1996. Oviposition day effect on incidence of agonadal progeny of *Helicoverpa zea* (Lepidoptera: Noctuidae) infected with a virus. Ann. Entomol. Soc. Amer. 89: 266–275.

Herzog, G. A., and J. R. Phillips. 1982. Manifestation of an abnormal reproductive system in a laboratory strain of the bollworm *Heliothis zea*. J. Georgia Entomol. Soc. 17: 506–513.

Jenkins, J. N., J. C. McCarty, and M. S. Moghal. 1995. Rearing tobacco budworm and bollworm for host plant resistance research. Mississippi Agric. & Forestry Exp. Stn. Tech. Bull. 208. Mississippi State University, Mississippi State, MS.

Mihm, J. A. 1983. Efficient mass rearing and infestation techniques to screen for host plant resistance to fall armyworm, *Spodoptera frugiperda*. Internal report. International Maize and Wheat Improvement Center (CIMMYT), El Batan, Mexico.

Mihm, J. A., F. B. Peairs, and A. Ortego. 1978. New procedures for efficient mass production and artificial infestation with lepidopterous pests of maize. Internal report. International Maize and Wheat Improvement Center (CIMMYT), El Batan, Mexico.

Ng, S. S., F. M. Davis, and W. P. Williams. 1987. Coordinating supplies of southwestern corn borers (Lepidoptera: Pyralidae) and fall armyworms (Lepidoptera: Noctuidae) by manipulating development time using temperature. J. Econ. Entomol. 80: 1340–1344.

Ng, S. S., F. M. Davis, and J. C. Reese. 1993. Southwestern corn borer (Lepidoptera: Pyralidae) and fall armyworm (Lepidoptera: Noctuidae): Comparative developmental biology and food consumption and utilization. J. Econ. Entomol. 86: 394–400.

Parrott, W. L., and J. N. Jenkins. 1992. Equipment for mechanically harvesting eggs of *Heliothis virescens* (Lepidoptera: Noctuidae). J. Econ. Entomol. 85: 2496–2499.

Parrott, W. L., J. E. Mulrooney, J. N. Jenkins, and J. C. McCarty. 1986. Improved techniques for production of tobacco budworm (Lepidoptera: Noctuidae) eggs. J. Econ. Entomol. 79: 277–280.

Raina, A. K., and J. R. Adams. 1995. Gonad-specific virus of corn earworm. Nature 374: 770.

Riaz, M. N., E. W. Lusas, and G. Mohy-ud-Din. 1996. Applications and advantages of extrusion technology in the food industry. Pak. J. Food Sci. 6: 85–88.

Sikorowski, P. P., and A. M. Lawrence 1994. Microbial contamination and insect rearing. Amer. Entomol. 40: 240–253.

Sikorowski, P. P., and A. M. Lawrence. 1997. Major diseases of *Heliothis virescens* and *Helicoverpa zea* in Mississippi fields and insectaries. Mississippi Agric. & Forestry Exp. Stn. Tech. Bull. 218. Mississippi State University, Mississippi State, MS.

Soares, G. G. 1992. Problems with entomopathogens in insect rearing, pp. 289–324. *In*: T. E. Anderson and N. C. Leppla (eds.), Advances in insect rearing research & pest management. Westview Press, Boulder, CO.

Sparks, A. N., and E. A. Harrell. 1976. Corn earworm rearing mechanization. USDA-ARS Tech. Bull. 1554. US Government Printing Office, Washington, DC.

Tillman, P. G., G. McKibben, S. Malone, and D. Harsh. 1997. Form-fill-seal machine for mass rearing noctuid insects. Mississippi Agric. & Forestry Exp. Stn. Tech. Bull. 213. Mississippi State University, Mississippi State, MS.

Williams, W. P., and F. M. Davis. 1997. Maize germplasm with resistance to southwestern corn borer and fall armyworm, pp. 226–229. *In* J. A. Mihm (ed.), Proceedings of the International Symposium on Insect Resistant Maize: Recent Advances and Utilization, 27 November–3 December 1994, El Baton, Mexico. International Maize and Wheat Improvement Center (CIMMYT), Mexico City, Mexico.

Wiseman, B. R., F. M. Davis, and J. E. Campbell. 1980. Mechanical infestation device used in fall armyworm plant resistance programs. Florida Entomol. 63: 425–432.

Index

A

[1]Bold indicates a page that contains a definition of the term.

B

D

E

G

H

M

N

O

P

Q

R

S

X

Y

Z